Introduction to
ENGINEERING ECONOMY

PWS Series in Engineering

Anderson, *Thermodynamics*
Askeland, *Materials Science and Engineering Laboratory Manual*
Askeland, *The Science and Engineering of Materials*, Third Edition
Bolluyt, Stewart, and Oladipupo, *Modeling for Design Using SilverScreen*
Borse, *FORTRAN 77 and Numerical Methods for Engineers*, Second Edition
Clements, *6800 Family Assembly Language*
Clements, *Microprocessor Systems Design*, Second Edition
Clements, *Principles of Computer Hardware*, Second Edition
Das, *Principles of Foundation Engineering*, Second Edition
Das, *Principles of Geotechnical Engineering*, Third Edition
Das, *Principles of Soil Dynamics*
Fleischer, *Introduction to Engineering Economy*
Gere and Timoshenko, *Mechanics of Materials*, Third Edition
Glover and Sarma, *Power Systems Analysis and Design*, Second Edition
Janna, *Design of Fluid Thermal Systems*
Janna, *Introduction to Fluid Mechanics*, Third Edition
Kassimali, *Structural Analysis*
Keedy, *Introduction to CAD using CADKEY*, Third Edition
Knight, *F.E.m. Primer for Mechanical Design*
Knight, *The Finite Element Method in Mechanical Design*
Logan, *A First Course in the Finite Element Method*, Second Edition
McGill and King, *Engineering Mechanics: Statics*, Second Edition
McGill and King, *Engineering Mechanics: An Introduction to Dynamics*, Second Edition
McGill and King, *Engineering Mechanics: Statics and An Introduction to Dynamics*,
    Second Edition
Raines, *Software for Mechanics of Materials*
Reed-Hill and Abbaschian, *Physical Metallurgy Principles*, Third Edition
Reynolds, *Unit Operations and Processes in Environmental Engineering*
Sack, *Matrix Structural Analysis*
Schmidt and Wong, *Fundamentals of Surveying*, Third Edition
Segui, *Fundamentals of Structural Steel Design*
Segui, *LRFD Steel Design*
Shen and Kong, *Applied Electromagnetism*, Second Edition
Strum and Kirk, *Contemporary Linear Systems using MATLAB*
Sule, *Manufacturing Facilities*, Second Edition
Vardeman, *Statistics for Engineering Problem Solving*
Weinman, *VAX FORTRAN*, Second Edition
Weinman, *FORTRAN for Scientists and Engineers*

G	Arithmetic gradient: *amount* of cash flow increase/decrease from period to period.
g	Geometric gradient: *rate* of cash flow increase/decrease from period to period.
I	Interest paid on debt. Annual interest payment (uniform over each year).
i	Effective interest rate per interest period.
i_a	Effective interest rate per year (per *annum*).
i_m	Effective interest rate per subperiod.
i_s	Rate of interest "earned" by imaginary sinking fund.
$i*$	Internal rate of return.
i_e^*	External rate of return.
$\hat{i}*$	After tax (internal) rate of return.
IRR	(Internal) rate of return. Sometimes written as RoR for "rate of return."
k	Auxiliary interest rate used when computing the external rate of return. The minimum attractive rate of return.
$k*$	Cost of capital, reflecting inflation.
M	Number of compounding subperiods per period (each of which is assumed to be of equal length).
$MARR$	Minimum attractive rate of return.
N	Number of compounding periods (each of which is assumed to be of equal length): the length of the "planning horizon" (study period). Life of investment.
P	Initial investment. Equivalent present value of future cash flow(s). Loan principal.
P_j	Amount of loan principal unpaid at the start of period j.
Q	Amount of loan.
r	Nominal interest rate per period; usually, the nominal interest rate per *year*.
S	Net salvage value of capital investment.
SIR	Savings-investment ratio.
SYD	Sum of the years-digits.
t	Effective income tax rate.
$\varnothing$	The do-nothing alternative.

Introduction to
ENGINEERING ECONOMY

G. A. Fleischer
University of Southern California

PWS Publishing Company
Boston

PWS PUBLISHING COMPANY
20 Park Plaza, Boston, MA 02116-4324

PWS Publishing Company is a division of Wadsworth, Inc.

International Thomson Publishing
The trademark ITP is used under license.

 This book is printed on recycled, acid-free paper.

Fleischer, Gerald A.
    Introduction to engineering economy / G.A. Fleischer.
        p.   cm.
    Includes bibliographical references and index.
    ISBN 0-534-93108-1 (alk. paper)
    1. Engineering economy.   I. Title.
TA177.4.F58   1994
658.f1/5—dc20   93-11982
                                                        CIP

Sponsoring Editor: Jonathan Plant
Assistant Editor: Mary Thomas
Editorial Assistant: Cynthia Harris
Production Editor: Patricia Adams
Interior and Cover Designer: Patricia Adams
Interior Illustrator: Carl Brown
Cover Illustrator: Lyn Boyer Pennington .
Marketing Manager: Nathan Wilbur
Manufacturing Coordinator: Lisa Flanagan
Compositor: Doyle Graphics
Cover Printer: New England Book Components
Text Printer and Binder: R.R. Donnelley/Crawfordsville

Printed and bound in the United States of America.

94 95 96 97 98 99 — 10 9 8 7 6 5 4 3 2

To my children, L. Andrea and Adam S.,
with great love from their father,
always.

CONTENTS

3 MEASURES OF WORTH I: RANKING OF ALTERNATIVES BY INCREASING/DECREASING EQUIVALENT WORTH MEASURES 68

4

MEASURES OF WORTH II: RANKING OF ALTERNATIVES BY MARGINAL (INCREMENTAL) ANALYSIS 114

5

MEASURES OF WORTH III: SOME ADDITIONAL METHODS FOR DETERMINING ECONOMIC VALUE 167

6 DEPRECIATION AND CERTAIN OTHER NONCASH EXPENSES

10 ECONOMIC RISK ANALYSIS   351

11 INCORPORATING PRICE-LEVEL CHANGES (INFLATION) INTO THE ANALYSIS   **406**

PREFACE

The formal study of optimal investment decisions is of relatively recent origin. Although the general problem of allocating limited resources among a variety of competing alternatives must surely have existed for millenia, it is only within the last century that the evaluation process has been formalized and analytical procedures developed.

Engineering economy, the analysis of economic effects of engineering decisions, originated in Arthur M. Wellington's classic text, *The Economic Theory of Railway Location*, published in 1887.* Engineers subsequently expanded and refined the techniques of engineering economy, notably through the work of Professors J. C. L. Fish in the 1920s and E. L. Grant in the 1930s.† In parallel, certain (classical) economists and financial managers were developing scholarly material that both supplemented and complemented the work of the engineering economy community. The resulting literature can be classified under a variety of descriptive terms, usually depending on the professional discipline of the author and/or the target audience: engineering economy (or engineering economics), capital budgeting, economic analysis, life cycle costing, financial decision making, managerial economics, and capital allocation theory, among others. The common element shared by all terms is concern with the fundamental problem of allocating limited resources among competing investment alternatives.

This book has as its genesis *Capital Allocation Theory: The Study of Investment Decisions*, originally published in 1967. A substantially revised version, *Engineering Economy: Capital Allocation Theory*, was published in 1984. The initial title was selected to reflect the view that, although the notation and theoretical structure are most closely related

---

*A. M. Wellington, *The Economic Theory of Railway Location* (New York: John Wiley), 1887.
†E. L. Grant's first edition of *Principles of Engineering Economy* was published in 1930. He retired from Stanford University in 1962, and at this writing (July 1993) Grant is still active. The 8th edition of *Principles of Engineering Economy* was published in 1990.

to classical engineering economy, applications are by no means limited to investment alternatives arising from engineering and technological decisions. Capital allocation problems are concerned with alternative plans, programs, and projects, regardless of whether a technology component is present. Nevertheless, since the examples, problems, and exercises are most directly related to engineering decisions, the simpler title, *Introduction to Engineering Economy*, has evolved. The principal emphasis remains a consolidation of current and relevant views of engineering economists, financial managers, and others to present a unified theory of capital allocation appropriate to all levels within the business enterprise or government.

The theory presented here is as appropriate to individual (personal) investment decisions as to those of business entities, government agencies, or nonprofit organizations. These techniques are suitable whenever decisions must be made concerning selection from among alternative investment opportunities.

This book may be used as either a primary or a secondary source in a first course in engineering economy, financial management, capital budgeting, managerial economics, and the like. Most of these courses are given in schools of engineering and business. However, the text may also be used elsewhere when it is desired to explore concepts, principles, and procedures for examining the economic consequences of proposed plans, programs, and projects. Examples include a "systems analysis" course in a school of public administration or a short course in "economic analysis" offered by an industrial firm or government agency.

Introduction to Engineering Economy is intended for upper-division undergraduates or graduate-level students. There are no prerequisites, although prior exposure to accounting and microeconomics will be helpful. Knowledge of integral and differential calculus is not strictly necessary; with few exceptions, development of mathematical models uses only algebra. Some understanding of the elements of probability theory is useful in the discussion of risk and uncertainty (Chapter 10). For students who have not been exposed to this topic previously, an elementary presentation of probability and expectation is provided in Appendix 10-1. This material may also be omitted without loss of continuity.

This book begins with the role of engineering economy within the larger framework of systems analysis. Chapter 1 presents a qualitative discussion of certain principles from which a unified theory may be developed. The necessary mathematics of compound interest, including discrete and continuous assumptions for cash flows and discounting, are developed in Chapter 2. The principal methods of economic evaluation are presented in Chapters 3 and 4: Those measures of worth that permit direct preference ordering of alternatives (present worth, annual worth or cost, future worth) are summarized in Chapter 3; worth measures that must be based on incremental analysis (rate of return, benefit–cost ratio) are discussed in Chapter 4. Some additional measures for determining economic value (payback, return on investment, profitability indexes, approximations of rate of return) are presented in Chapter 5.

In Chapters 6 and 7 there is an extensive discussion of the effects of income taxes. Since cash flows for taxes are related to taxable income, and since taxable income is in part a function of depreciation and depletion expenses, procedures for determining these expenses are described in Chapter 6. In Chapter 7 we focus on procedures for determining the amounts and timing of cash flows for income taxes.

Many engineering textbooks include discussion of *personal* income taxes. I have elected to concentrate on the effects of taxes on the *corporate* business enterprise. The methodology for after-tax economy studies for corporations, however, is readily transferable to the noncorporate context. If the instructor wishes to include personal income taxes in his or her course, I suggest using the current edition of *Your Federal Income Tax*, IRS Publication 17, as a supplementary reference.

Chapter 8 deals with one of the most common engineering economy applications: retirement of assets from service with replacement by new assets. Problems of varying degrees of complexity are discussed from simple retirement (no replacement) to a generalized replacement model.

The revenue requirement method is presented in Chapter 9. Although this method is equivalent to the more widely used procedures given in Chapter 3—indeed, this is demonstrated in Chapter 9—the revenue requirement method is of special interest to the utility industry, and thus it is organized into a separate chapter. Some instructors may choose to omit Chapter 9 because of time constraints and/or their view that the revenue requirement method is limited in application. If so, this material may be omitted without loss of continuity.

The first nine chapters assume that all parameters are known with certainty. This artificiality is akin to assuming a frictionless plane in physics. So Chapter 10 explores a number of techniques for formally considering the noncertain future, including sensitivity analysis, risk analysis, and a variety of principles of choice from decision theory.

Chapter 11 is an extensive treatment of yet another real-world consideration that has become increasingly important in modern society: incorporating price-level changes (inflation) into the analysis.

Introduction to Engineering Economy incorporates a number of special features that, I believe, add significantly to the effectiveness of the text. These include:

1. *A partial summary of principal notation,* shown inside the front cover for ready reference. An expanded summary of symbols and key abbreviations is included in **Appendix A.** Moreover, the notation specific to the revenue requirement method (Chapter 9) is included in the appendix to that chapter.

2. *Definitions of key terms and principal notation.* These are generally consistent with those proposed by the American National

Standards Institute Committee on Industrial Engineering Terminology (ANSI Committee Z-94) as published in 1990.

3. *Extensive list of references.* An extensive list of references is provided to suggest additional reading opportunities across the general field of engineering economy. These are summarized in two groups: general textbooks and professional journals. In addition, references are provided for each of the chapters for those readers desiring more in-depth coverage of specific topics.

4. *A summary of principal mathematical models,* shown inside the back cover for ready reference. These models, accompanied by relevant cash flow diagrams, are keyed to the compound interest tables in Appendix B.

5. *Compound interest tables.* Tables are provided in Appendix B for 19 interest rates, ranging from 1% to 50%. Each table includes 11 factors: three single-payment factors, six uniform series factors, and two arithmetic gradient series factors. (Three of the 11 factors may be used when finding equivalent values of continuous cash flows under conditions of continuous compounding; these are shaded in the tables for ready identification.) The interest rates are *effective,* not nominal, in all cases. To assist the user in rapid selection of the appropriate table, the various interest rates are clearly marked within the margins.

6. *End of chapter problems.* There are 528 problems distributed over the 11 chapters. Some of these afford the student an opportunity to work through numerical exercises related directly to the text material, whereas others are extensions of the text in that they illustrate new applications. With the exception of Chapter 1, answers are provided for all odd-numbered problems. Complete solutions to all problems are provided in a separate **Instructor's Manual**.

7. *Applications illustrated by frequent numerical examples.* Following each step in the exposition of a theory or methodology, one or more numerical examples is included by way of illustration. These examples are numbered so that they may be readily cross-referenced.

8. *Illustrations of solutions using generic computer-based spreadsheets and special-purpose computer software.* Very early in the text (the Appendix to Chapter 2), the student is introduced to the notion that generic software may be used to generate appropriate spreadsheets for the solution of many problems arising in engineering economy analyses. Also, in the Appendix to Chapter 3 we introduce specific software, Cash Flow Analyzer (CFA), which may also be useful in this context. Indeed, for many applications, CFA is more efficient than general-purpose spreadsheets. (The CFA software is available to purchasers of this book; please see the card at the back of the book

for additional details.) Use of the software is further illustrated in the appendixes to Chapters 4 through 8.

9. *An extensive discussion of depreciation, depletion, amortization, and income taxes.* The discussion in Chapters 6 and 7 incorporates the principal features of the Tax Reform Act of 1986, including the Modified Accelerated Cost Recovery System (MACRS) required of most federal taxpayers. (Additional changes introduced by the Omnibus Budget Reconciliation Act of 1993 are also included. In addition, relevant changes introduced by subsequent federal legislation will be distributed to instructors adopting this book for their courses.)

10. *A discussion of the revenue requirement method in Chapter 9.* This material has been greatly simplified since the previous edition of *Engineering Economy.* To assist the reader, a separate glossary and summary of principal interest factors are included as appendixes to this chapter.

11. *Instructor's Manual.* As mentioned previously, complete solutions to all 528 problems are included in the *Instructor's Manual* that accompanies the instructor's copy of this book. Many of these solutions are displayed in generic spreadsheet format and/or using CFA, the special-purpose software. In addition, the *Instructor's Manual* includes:

- *Sample schedules* for an Engineering Economy course taught under different formats: two-unit, one semester; three-unit, one semester; three-unit, one quarter; and four-unit, one quarter.
- *Masters for overhead transparencies.*
- A *case study* that the instructor may introduce with Chapter 3 and then reference throughout the remaining chapters.

ACKNOWLEDGMENTS

As I look back at the four or five years this book has been in preparation, I am overwhelmed by an appreciation of the significant contributions of so many family, friends, students, and colleagues. I am especially grateful to Ms. Georgia Lum for her extraordinary typing skills and for her unfailing grace and good humor.

I would like to thank the following reviewers for their comments and suggestions:

James Alloway
Syracuse University

John R. Birge
The University of Michigan

John R. Canada
North Carolina State University

William F. Girouard
California Polytechnic State University

Dennis E. Kroll
Bradley University

Robert G. Lundquist
The Ohio State University

Geraldine M. Montag
Iowa State University

James S. Noble
*University of
Missouri-Columbia*

M. Wayne Parker
Mississippi State University

Richard Pike
Northeastern University

I am also indebted to my many colleagues, both past and present, without whose encouragement, inspiration, and critical judgment this text would not have been written.

G. A. Fleischer

1

An Introduction

PERSPECTIVE

In 1795 the English parliamentarian and man of letters, Edmund Burke, wrote a letter in defense of his own pension, proposed by the prime minister, the Younger Pitt:

> It may be new to his Grace, but I beg leave to tell him that mere parsimony is not economy. It is separable in theory from it; and in fact it may or it may not be a part of economy, according to circumstances. Expense, and great expense, may be an essential part in true economy. If parsimony were to be considered as one of the kinds of that virtue, there is, however, another and an higher economy. Economy is a distributive virtue, and consists, not in saving, but in selection. Parsimony requires no providence, no sagacity, no powers of combination, no comparison, no judgement. Mere instinct, and that not an instinct of the noblest kind, may produce this false economy in perfection. The other economy has larger views. It demands a discriminating judgement, and a firm, sagacious mind.[1]

Burke's commentary is a remarkably effective summary of the perspective underlying *Introduction to Engineering Economy*. This book rests on the proposition that refusing to expend scarce resources is rarely, if ever, the wisest course of action. Rather, the problem is choosing from among a variety of investment alternatives in order to best satisfy decision makers' immediate and longer-term objectives. The operative word is *economy*, and the essential ingredient in economy, according to Burke, is *selection*. This book is dedicated to the principles and procedures of the selection process, especially when the economic characteristics of alternatives are of principal or significant concern.

---

[1]Edmund Burke: *Selected Writings and Speeches*, ed. Peter J. Stanlis (New York: Doubleday Anchor, 1963).

1.2 THE ROLE OF ENGINEERING ECONOMY IN THE DECISION-MAKING PROCESS

In general, the process of making rational decisions of the types of interest to engineers consists of a number of well-defined steps:

1. *Problem definition.* This step stems from some recognition that our current system, the way we are doing things now, somehow fails to meet our goals and objectives. The problem may be defined for us by higher levels of management, or we may define the problem ourselves on the basis of our own observations. Examples are: "Shall we continue to manufacture a certain component or, alternatively, purchase the component from one or more outside suppliers?"; "Shall we purchase a fully automated materials handling system?"; "Shall we expand our plant now or delay expansion for another year or two?"

2. *Define alternatives.* Once the problem has been defined, it is now necessary to identify a set of alternative solutions. At this point we want to ensure that our candidate set is reasonably exhaustive. A possible solution that is *not* considered has *no* chance of being selected.

3. *Evaluate alternatives.* Appropriate evaluation model(s) must be selected that account for all significant consequences of the defined alternatives. *Constraints* must be established so as to determine which of the alternatives are *feasible*, and *criteria* must be identified so as to determine the optimal policy among the feasible alternatives. The criteria, of course, should reflect the fundamental goals and objectives that were originally employed at the problem definition stage. Relevant data are input to the evaluation model and the outcomes derived.

4. *Selection.* This is the point of decision. Based on the analytical results, the responsible individual(s) choose(s) from among the set of feasible alternatives. Note especially that the results of analysis alone may be insufficient to make a true choice. The decision makers' value system may play a role. That is, there may be certain esthetic, moral, philosophical, or other considerations that affect choice in a certain problem situation.

5. *Implementation and follow-up.* In this final step, the decision is implemented and the results monitored.

This process, greatly simplified, is summarized in **Figure 1.1**. Engineering economy,[2] it will be noted, occurs at step 3. The methodology of engineering economy is integral to the analysis phase because, properly

---

[2]Synonymous terms are *capital allocation theory, engineering economic analysis,* and, in a limited sense, *economic analysis.*

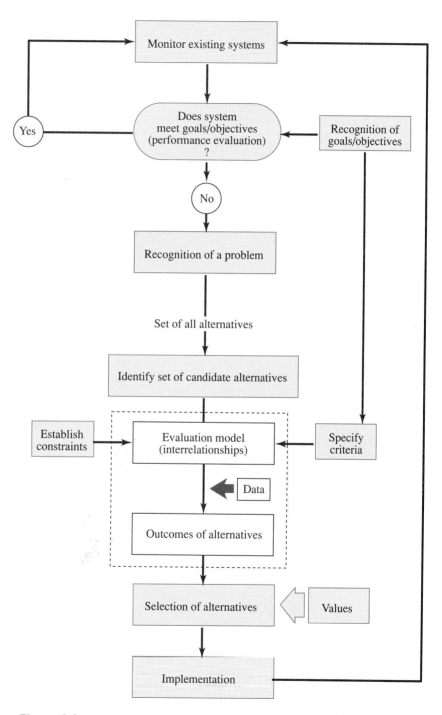

Figure 1.1   The Decision-Making Process

used, it evaluates the *economic* consequences of alternative plans, projects, or programs.

Capital allocation, the theme of this book, is *not* a decision-making process; it is a *decision-assisting* process. Decisions are rarely made on the basis of economic considerations alone. Nor should they be. Consider some common examples—which car to buy; which apartment to rent or which home to purchase; how to spend entertainment dollars; which university to attend, if any; and what portion of society's resources to spend on public education, health and welfare, defense, and public transportation. The choice of options should be based on consideration of all significant consequences that are likely to arise from each alternative.

Occasionally one hears the argument that, since there may be important noneconomic factors relevant to a given problem situation, the economic consequences should be ignored in favor of the "more important" factors. This viewpoint is especially prevalent with respect to public-sector investment decisions, although it is frequently encountered in the private sector as well. In any event, it is myopic and potentially wasteful of scarce resources.[3] Unless the economic consequences are clearly insignificant, decision makers should assess and consider them along with any and all other relevant data. Even though a more costly, or economically suboptimal, alternative may ultimately be selected, the analysis will yield the price being paid for the differential noneconomic consequences. This information will be valuable to decision makers as well as to stockholders (private enterprise) and the taxpayers (public sector) to whom decision makers are responsible.

1.3   THE CAPITAL ALLOCATION PROBLEM DEFINED

Many investment opportunities are normally available to individual investors, business firms, and government agencies. For private industry in particular, continued investment and reinvestment is a fact of economic life if a firm is to survive in a competitive economic environment. Supplies and materials must be purchased; new equipment must be acquired and obsolete equipment replaced; the physical plant must be maintained and, in many cases, expanded; new products must be researched, developed, and marketed; and so forth. Moreover, investment by one company in another in the form of stock ownership, purchase of government securities, and distribution of cash dividends to a firm's owners represent just a few of

---

[3]When considering capital allocation decisions within the context of a government agency, the taxpayers or citizens of the government unit in question may be viewed as the owners of the enterprise.

the investment possibilities that may be considered as alternatives of capital demand.[4]

Suppose that we lived in a world of unlimited capital resources. In such an ideal environment there would be no need for a book like this. All proposals that required the expenditure of funds would be acceptable, provided that a simple criterion were met: Total cash revenue must exceed total cash expenses. Indeed, even this simple test would be unnecessary in our make-believe world, because having unlimited resources destroys the incentive to increase existing wealth. A person who has plenty to eat—and who is assured of always having plenty to eat—does not devote his or her energies to acquiring more food.

Clearly, in our real world, resources are limited. In all but the most abnormal cases, the available capital supply is insufficient to meet all available demand. Although available capital may be increased by additional borrowing or, in some cases, by taking in additional partners, management is generally prevented from taking advantage of all available investment opportunities by limitations of capital. Thus, the cardinal problem of capital allocation is: *Which of the many available investment alternatives should be selected or rejected in order to maximize the long-term wealth of current owners of the enterprise?*[5] In other words, which alternatives should be budgeted, or funded, in order to produce the optimal capital budget? The following discussion addresses the development and application of a systematic and valid basis for answering this question.

Conceptually, the solution to the capital allocation problem may be viewed as a determination of the point where the marginal cost of capital supply is equal to the expected marginal revenue, or "payoff," of capital demand. Stated another way, alternatives should be accepted until the point where the last, or marginal, alternative accepted is expected to yield a return that is greater than the cost of the capital necessary to finance it. At this point, no additional capital funds should be obtained and no additional investment proposals should be accepted (see **Figure 1.2**).

Level Q_1 represents the total cost of all investments and reinvestments necessary to maintain the firm's existence. P_1 is the expected profitability from Q_1, stated as a rate of return on the invested capital, and C_1 is the cost of the capital funds necessary to support these investments. Now consider the increment of capital, $Q_2 - Q_1$, necessary to fund the

---

[4]Distribution of a firm's earnings in the form of cash dividends is frequently cited as an example of "disinvestment." The semantic distinction between investment and divestment is of little interest to us at this time; the difference will always be clear in context.

[5]Many authors have discussed the explicit objectives toward which capital management must be directed. Their consensus is that the relevant goal is to maximize the present worth of the current owners of the enterprise. As later chapters will show, *present worth* refers to the discounted present value of long-run net earnings.

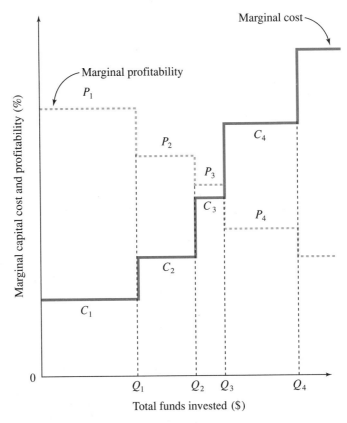

Figure 1.2 Marginal Profitability (Rate of Return) and Cost as a Function of Total Funds Invested

project, resulting in incremental return P_2. Q_2 represents the total number of investment dollars required for level Q_1 plus the next project under consideration. C_2 is the incremental cost of the incremental funds, $Q_2 - Q_1$. Since marginal return P_2 is greater than marginal cost C_2, this incremental project appears economically attractive or acceptable. By similar reasoning, investment of the next increment of funds, $Q_3 - Q_2$, is warranted, since marginal return P_3 exceeds marginal cost C_3. The next project, requiring incremental funds $Q_4 - Q_3$, should be rejected, because the cost of these funds, C_4, exceeds anticipated return P_4.

The downward-sloping curve—a step function in Figure 1.2—represents the marginal returns expected from the demand for capital, and the upward-sloping curve represents the cost of the supply of capital. In this simple illustration it is clear that not all projects are affordable; the firm must select certain projects and reject others. Thus, project selection (based on *economic* consequences) is the central feature of the capital allocation problem.

Having limited investment funds leads to another type of problem that will be detailed in later chapters: the selection decision when two or more alternatives are available to satisfy a given operational requirement and when each is otherwise affordable. To illustrate, suppose that a firm is considering acquiring production equipment to reduce manufacturing costs. Three alternatives are available: models X, Y, and Z. Each is affordable in that each is expected to produce an economic return exceeding the related investment capital. But because all three satisfy operational requirements, management must select the most desirable alternative.

Note then that there are two major types of problems that, together, are the focus of this book. The first type requires the selection of a subset, or budget, from apparently independent investment proposals. Some alternatives will be accepted and others rejected, because there are insufficient funds to support all the investment possibilities. Given limited funds, the alternatives are not truly independent because, at the margin, the acceptance of one or more alternatives precludes the remainder. The second type of problem requires selection from a set of mutually exclusive alternatives. Examples include which equipment to acquire, what size facility to construct, when to build the facility, whether to lease or purchase equipment, what degree of automation to establish, and where to locate the facility. Both types of problems, of course, are related to the universal condition of limited capital.

1.4   FUNDAMENTAL PRINCIPLES OF CAPITAL ALLOCATION

Before developing the mathematical models appropriate to evaluating capital proposals, it will be useful to identify the fundamental principles that give rise to the rationale of capital allocation. Moreover, some of these principles lead directly to the quantitative techniques developed in the later chapters.

Feasible alternatives only

1. *Only feasible alternatives should be considered.* For example, prospective investment in new equipment costing $100,000 is not a feasible alternative if the money is not available and cannot be obtained and if there are no other financing opportunities. Similarly, management need not consider distributing corporate profits as cash dividends if this action has been temporarily prohibited by lenders. The capital budgeting analysis begins with determination of all feasible alternatives, since courses of action that are not feasible, because of certain contractual or technological considerations, are properly excluded.

Common denominator for assessing alternatives

2. *Using a common unit of measurement (a common denominator) makes consequences commensurable.* Is a proposal to reduce direct labor by 3,000 labor hours superior to another that is expected to reduce raw-material requirements by 500 tons? Is an investment of $10,000 justified if it promises to shorten production time by 3 weeks? Clearly, these questions cannot be answered unless the prospective consequences of alternative

proposals have been described in common units. All decisions are made in a single dimension, and money units—dollars, francs, pesos, yen, and so forth—seem to be most generally suitable. Thus, after all consequences have been described for each of the feasible alternatives, they should be transcribed into a common denominator to the extent practicable. Of course, not all consequences may be evaluated in money terms. (See principle 9.)

Only differences relevant

3. *Only differences are relevant.* The prospective consequences that are common to all contending alternatives need not be considered in an analysis, because including them affects all alternatives equally. This is like adding or subtracting the same number to both sides of an equation; the equality remains undisturbed.

Ignore sunk costs

4. As a direct consequence of principle 3, an important axiom may be stated: *All sunk costs are irrelevant to an economic choice.* A **sunk cost** is an expense or a revenue that has occurred before the decision. When considering whether or not to replace manufacturing equipment, for example, the original cost of the equipment is "sunk" and hence of no direct interest.[6] By the same token, the cost of a major overhaul of equipment that occurred 3 weeks ago is also irrelevant. All events that take place before a decision are common to all the alternatives, so sunk costs are not differences among alternatives.

Common planning horizon

5. *All alternatives must be examined over a common planning horizon.* The **planning horizon** is the period of time over which the prospective consequences of various alternatives are assessed. (The planning horizon is often referred to as the *study period* or *period of analysis*.) If a certain alternative affects the flow of funds to and/or from an enterprise during a given period of time, the economic effects of all other alternatives during that same period should be estimated. The context of the decision—the planning horizon—must be uniform for all possible choices.

The time value of money

6. *Criteria for investment decisions should include the time value of money and related problems of capital rationing.* Which of the following is more attractive economically? Alternative A requires an expenditure of $100 today and promises a $200 return 5 years from now. Alternative B also requires an initial expenditure of $100, but a payoff of $500 can be expected in 10 years. There is no simple, direct answer to this problem, because the returns are expected at different points in time. B cannot be chosen simply because $500 is greater than $200. It should be

---

[6]As will be discussed in Chapter 6, the original cost of equipment may be of interest insofar as this information is needed to determine depreciation expenses for tax purposes.

evident that the appropriate solution must consider the opportunities available to the investor in the interval between year 5 and year 10. Because resources are limited in our finite world, money can be invested elsewhere, so it has a value, or worth, that increases over time. Therefore, the time value of money must be used as a criterion in selecting from among alternative investment opportunities.

Separable decisions considered separately

7. *Separable decisions should be made separately.* This principle requires the careful evaluation of all capital allocation problems to determine the number and type of decisions to be made. For example, an analyst may be called on to make recommendations concerning a technological alternative (the equipment to be acquired, for example) as well as a financial alternative (the source of funds to finance the acquisition). If separable decisions are not treated separately, optimal solutions may be obscured in the analysis.

   It is possible, of course, for an interrelationship to exist between certain investment decisions and related financing decisions. For example, the type of equipment selected may affect the range of financing alternatives available. When this occurs, the investment decisions are not entirely separable.

Consider noncertain future

8. *The relative degrees of uncertainty associated with various forecasts should be considered.* All capital budgeting decisions are based on a series of estimates concerning the future. Because estimates are only predictions of future events, it is probable that the actual outcomes will differ to a greater or lesser degree from the original estimates. The future is uncertain, and any analysis that fails to consider this uncertainty implies false omniscience. On the other hand, one cannot infer that specific solutions are invalid or useless simply because of the uncertainty associated with input data. Formal consideration of the type and degree of uncertainty ensures that the quality of the solution is evident to those responsible for capital allocation decisions.

Consider irreducibles

9. *Decisions should give weight to consequences that are not reducible to monetary units.* Selecting candidates for limited capital resources requires that prospective differences among alternatives be clearly specified. Whenever possible, of course, these differences should be reduced to a common unit of measurement, generally a monetary unit, to provide a basis for selection. But not all alternatives can be reduced to monetary units. One proposal may result in an increase in employee comfort and convenience; another may result in a production delay; still another may result in an increased share of the market. Although the analyst does not always have the time or resources to reduce such consequences to equivalent monetary units, decisions may be directly affected by their amount and timing. The irreducible as well as monetary consequences of

proposed alternatives should be clearly specified in order to give managers of capital all reasonable data on which to base their decisions.[7]

Procedures integrated throughout organization

10. *Efficacy of capital-budgeting procedures is a function of their implementation at various levels within the organization.* Responsibility for capital allocation decisions does not rest solely with corporate boards of directors or senior managers. Decisions affecting significant receipts and expenditures are made at many levels within the organization, especially among staff activities such as engineering design, marketing, and new product research and development. Moreover, the technical description of alternatives to be considered by top-level management generally undergoes preselection by lower-level personnel. Thus, capital allocation procedures must be clearly described and understood at all levels within the organization having responsibility, in whole or in part, for these decisions.

Note that a postdecision audit is inherently incomplete: If only one of several alternative courses of action is selected, we can never be sure what would have happened if another alternative had been chosen. "What might have been if..." is conjecture, and all postdecision audits should be made with this reservation in proper perspective.

SUMMARY

Capital allocation is the process of selecting investment or disinvestment proposals to optimize the long-term interest of current owners of an enterprise. Although capital allocation is usually thought of as the acquisition of physical assets (capital demand), the sources and amounts of investment funds (capital supply) are also relevant.

The principal problems of owner interest are of two general types. The first type relates to the fact that there is not enough capital to fund all prospective investment opportunities, even though the proposals are independent of one another. There simply is not enough money to do

---

[7]As the title of this book implies, we limit our concern to the economic implications of alternative investments. The simultaneous, or aggregate, consideration of monetary and other irreducible consequences is the proper focus of multiattribute analysis. There is substantial literature related to this problem. See, for example, Ralph L. Keeney and Howard Raiffa, *Decisions with Multiple Objectives* (New York: John Wiley & Sons, 1976); Detlof von Winterfeldt and Ward Edwards, *Decision Analysis and Behavioral Research* (New York: Cambridge University Press, 1986); and Robert T. Clemen, *Making Hard Decisions: An Introduction to Decision Analysis* (Boston: PWS-KENT, 1991).

everything that might otherwise be desirable. The second type of problem occurs when the decision maker must select from among two or more mutually exclusive alternatives. Both types of problems concern economy, Edmund Burke's "distributive virtue [that] consists, not in saving, but in selection."

Before going on to quantitative techniques and criteria for evaluating alternatives, let us review the principles that form the rational basis for capital allocation procedures:

1. Consider only feasible alternatives.
2. Use monetary units as a common denominator to make consequences commensurable.
3. Only differences are relevant to the decision-making process.
4. Sunk costs are irrelevant to decisions about the future.
5. All alternatives must be examined over a common planning horizon.
6. Criteria for investment decisions should recognize the time value of money and related problems of capital rationing.
7. Separable decisions should be made separately.
8. The relative degrees of uncertainty associated with various forecasts should be considered.
9. Decisions should give weight to consequences that are not reduced to monetary terms.
10. The efficacy of capital budgeting procedures is a function of their implementation at various levels within the organization.
11. Postdecision audits improve the quality of decisions.

| PROBLEMS (Discussion Questions)

Unlike the problems at the ends of the other chapters in this book, these are more properly discussion questions. There are no "right" or "wrong" answers. Rather, the questions that follow are intended to stimulate the student to think about issues germaine to the process of analyzing differences among alternatives, especially when there are economic consequences to be considered. These questions are meant to be provocative within the context of the principles described in the chapter.

1.1 Decisions are rarely, if ever, based on economic considerations alone. Can you think of any that are? List as many as you can, but remember to identify situations in which *only* economic considerations are relevant to the decision.

1.2 A high school senior is considering applying to several universities and colleges. List the various factors that might influence her decision.

1.3 The director of a university computer center is considering the purchase of 100 personal computers for use by undergraduate students. A number of options are available, including different manufacturers and different models. List the factors that might influence the decision. Discuss the significance of each factor.

1.4 F. R. Ward has been working for several years for a large manufacturing firm but has decided to go into business as an independent consultant providing engineering services. Ward is considering the purchase of a car that will be used for both business (80%) and personal (20%) purposes. List the factors that might influence the choice of car (make, model, new or used). Discuss the significance of each factor.

1.5 Both individuals and businesses have a variety of opportunities to invest their excess capital, the funds not needed for day-to-day expenditures. List 5 investment opportunities that might be available to investors in your community. (*Hint*: Consult local newspapers and magazines for advertisements of financial institutions. See especially the financial section of your newspaper.) For each opportunity, identify the rate of return available and discuss the potential risks.

1.6 Choose a city within 500 miles of the community where your school is located. Then prepare an exhibit that lists alternative means of travel from your community to the target city. List only alternatives that are feasible. Also list your constraints—the ways you have determined which alternatives are feasible. (*Example*: "Travel by roller skates is not feasible, because travel time would take 12 days and I must make the trip in 1 day.") Next, list your criteria for the trip. You might consider cost, time, comfort, and so forth. For each alternative, indicate its performance with respect to each criterion. Is one alternative clearly superior to the others? Explain.

1.7 Consequences that are truly irreducible to monetary units are rare. Although a marketplace may not exist to establish a monetary price for a particular good or service, it is often possible to infer monetary value through indirect measures. Consider "the value of human life." This issue frequently arises when evaluating the benefits of improving a public roadway, for example, as well as in many other contexts. How would you go about placing a monetary value on human life? (*Hint*: There is substantial literature on this subject. Consult your university library.)

1.8 How would you go about placing a monetary value on "the increased opportunity to use recreational resources"? Specifically, suppose that the Forest Service could improve an access road into a recreation area so that the number of visitors per day would increase from x to $x + \Delta x$. Can a monetary value be placed on Δx

persons per day? Refer to Problem 1.7. Explain your answer.

1.9 How would you go about placing a monetary value on "employee morale"? Refer to Problem 1.7.

1.10 How would you go about placing a monetary value on "environmental quality"? Refer to Problem 1.7.

1.11 Some years ago Mr. Anchor purchased 10 ounces of gold for $500 per ounce. His purchase was for investment purposes only; the gold has been stored in his safety deposit box until some indeterminate date when he could sell it for a "nice profit." Now, Anchor needs some money to buy a new car. He could sell the gold, but the current market value is only $350 per ounce. He is reluctant to sell because he will "take a big loss"—about $1,500. Instead, he will borrow money for the car from his credit union at an interest rate of 10% annually. What do you think of Anchor's reasoning? What should be the relevant considerations in his decision to keep or sell the gold?

1.12 Recently, the Southern California Rapid Transit District (SCRTD) completed construction of a surface rail line between Long Beach and downtown Los Angeles. Initial rider response suggested about 3.9 million round trips in the second year. The initial cost to build the rail line was $877 million. Operating costs were $38.6 million in the first year of operation and $44.0 million in the second year.

In order to find funds to build the new rail line, the SCRTD ended bus fare subsidies. Partially as a result, there was a decline of 96 million annual bus boardings in the first year. (Perhaps the decline in bus ridership was not entirely the result of ending the fare subsidies.) What do you think of the decision to build the rail line? What are the factors that might bear on this decision?

1.13 Mighty Tidy Car Wash offers a 10-wash plan for $59.50. That is, the motorist can buy a book of 10 tickets, each of which is good for a standard car wash any day of the week. Alternatively, a single car wash at Mighty Tidy costs $5.95 on Wednesdays and $6.95 on other days. (Mighty Tidy is open 7 days per week.) Is the 10-wash plan a good deal? What are the factors that might be considered?

1.14 Four years ago the University purchased a commercial hotel several blocks away from the campus to serve as a "residential college." Unfortunately, the acquisition has experienced financial difficulties, and there is now discussion to close it down and try to find a buyer for the property. The building was purchased for $12 million, and $3 million in improvements were necessary before the building could be occupied. The facility lost $7.3 million in its 4 years of operation. The building currently has a 17% vacancy rate during the school year, significantly higher than the University-wide average of 7%. A university official said that a sale is possible, but "we're not about to sell it at a discount. The University is losing money now, but the University will lose more money selling it at a discount." What are the factors to be considered that are relevant to the keep/sell decision? What action do you think is warranted at this time?

1.15 In recent years, significant increases in tuition costs have been of considerable concern to parents as well as academic administrators. Some universities are dealing with this problem by offering a form of "guaranteed tuition." There are a variety of schemes, one of which is a contractual arrangement wherein parents pay $X per year for Y years into a fund in order to lock in the tuition some years later. For example, soon after the birth of their daughter, Chris, the Fletcher's register their daughter at State U. and agree to pay $100 per month for 10 years. Then, assuming that Chris is admitted to State U., she will not be required to pay any tuition for 4 years of full-time attendance. (If Chris does not attend the university, all payments will be returned, with interest compounded at the rate of 3% per year.) The current tuition at State U. is $1,000 per semester. What do you think of this deal? What are the factors that might influence the decision?

APPENDIX: *Significant Figures*

Analysts frequently deal with cost or revenue estimates that are expressed, say, in thousands of dollars. When performing equivalence calculations of the sort discussed in the subsequent chapters, it is possible to generate solutions of the form $173,456.516. Even though your computer or pocket calculator may generate such apparent "precision," this kind of answer is silly. When dealing with engineering and financial data, you should always bear in mind the common sense rules of significance with respect to numerical values. This appendix is included to summarize appropriate guidelines.

Significant Figures

Significant figures (also called *significant digits*) are all accurate figures other than zeros needed to locate the decimal point.

- For *measurement*, all nonzero figures are significant.
- A zero between two other figures is always significant.
- Zeros to the left of all nonzero figures are not significant.
- Zeros that are both to the right of the decimal point and to the right of nonzero figures are significant.

EXAMPLES

0.3000	(4 significant figures)
0.04035	(5 significant figures)
748.520	(6 significant figures)

Addition or Subtraction

The results should contain no more figures to the right of the decimal point than the quantity that has the *least* number of figures to the right of the decimal point.

EXAMPLES

52.875	3 figures to the right
+ 64.13	2 figures to the right
+ 7.1234	4 figures to the right
124.1284	reported as 124.13 (2 figures to the right)

821	0 figures to the right
+ 3.7	1 figure to the right
+ 15.356	3 figures to the right
840.056	reported as 840 (0 figures to the right)

56.12365	5 figures to the right
− 7.375	3 figures to the right
48.74865	reported as 48.749 (3 figures to the right)

Multiplication, Division, and Roots of Numbers

Answers can have no more significant figures than the factor that has the *least* number of significant figures. The position of the decimal point is irrelevant.

EXAMPLES

20.6	3 significant figures
× 3.1416	5 significant figures
64.71696	reported as 64.7 (3 significant figures)

6.783	4 significant figures
÷ 4.0	2 significant figures
1.69575	reported as 1.7 (2 significant figures)

Defined Values

Some quantities are not measured but defined. For example, 1 metric ton = 1,000 kilograms. These values can be considered to have an infinite number of significant figures. These are exact values, and should be ignored when determining the number of significant figures in the answer to a problem. Use only the measured quantities.

Exact Counts

A count (e.g., 183 defective items) is treated as though it has an infinite number of significant figures. When a count is used with a measurement, the number of significant figures is the same as the number of significant figures in the measurement.

Exaggerated Precision

In some instances, a measurement may be reported as so much per standard unit, for example, "2.149 hours of direct labor time per 100 units produced." In such instances the required numbers of trailing decimals are as follows:

Standard Quantity	Required Trailing Decimals	Example	
1	.X	0.02	hr/unit
10	.XX	0.21	hr/10 units
100	.XXX	2.149	hr/100 units
1,000	.XXXX	21.4898	hr/1,000 units
10,000	.XXXXX	214.89783	hr/10,000 units

Equivalence and the Mathematics of Compound Interest

2.1 INTRODUCTION

The concept of **equivalence** is a critical element in the process of economic analysis. For example, would you prefer $100 or $200, everything else being equal? The obvious answer is $200. The question is deceptively simple, however. The *timing* of the two alternatives is not mentioned, but this information is certainly relevant to the decision.

Suppose, on the other hand, that you were given a choice between $100 today and $200 ten years from now. You might conclude that it is preferable to accept the $100 now so as to enjoy its use, either through consumption or investment, over the next ten years. Indeed, you could accept the $100 today, invest it immediately in a savings account at, say, 8% per year, and realize about $216 at the end of ten years. Thus the $100 received today is *equivalent* to $216 ten years from now if funds can be invested at 8% per year during the ten-year period. (This simple illustration ignores any tax consequences.)

The principle embedded in the above illustration is clear: In order to choose intelligently among economic alternatives, both the amounts and the timing of expected receipts and disbursements must be estimated, and appropriate adjustments must then be made to account for the differences in timing. These adjustments, effected through the mathematics of compound interest, result in equivalent values that can then be compared directly, as the remainder of this chapter shows.

2.2   CASH FLOWS AND INTEREST

Before going on, let us define and explain some key words and phrases that are commonly used in the calculation of compound interest. The first is **cash flow**, a term that refers to receipt or payment of an amount of money. Cash flow is actual receipts or actual expenditures, not merely a financial obligation or accrual.

It is undeniable that we live in a world of limited resources. Except for a few very special cases, companies (as well as individuals, family units, and governments) find it necessary to ration limited wealth. When assets are invested, it is reasonable to expect that a lender will charge a borrower a "rent" to reimburse the lender for the lost opportunity to invest elsewhere. **Interest**, the rent charged for the use of borrowed money, would have little importance to a lender who had unlimited capital.

Interest is stated in monetary units and should not be confused with the interest *rate*, which is a pure number. **Interest rate**, generally written as a percentage rather than a decimal, is the ratio of the interest charged during an **interest period** to the amount of money owed at the beginning of the interest period. The length of the period must be stated or understood. "Interest rate" is frequently abbreviated to "interest." For example, one may speak of 10% interest on a home mortgage, omitting "per year." (Although strictly improper, this should not be bothersome, because the meaning is usually clear in context.)

The relationships among interest, interest rate, and interest period are illustrated in **Figure 2.1**. The interest rate, i, is given by

$$i = \frac{(A_1 - A_0)}{A_0} = \left(\frac{A_1}{A_0}\right) - 1 \qquad (2.1)$$

where A_0 = amount owed at the start of the period

A_1 = amount owed at the end of the period

= A_0 + interest accrued during the period.

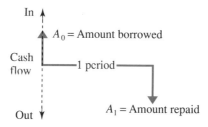

Figure 2.1

2.3 EQUIVALENT VALUES OF A SINGLE CASH FLOW

2.3.1 A Numerical Example Illustrating Equivalence

Consider an investment of $1,000, to be repaid at the end of 5 periods with interest computed at the rate of 10% per period. *Investment* can be interpreted in a broad sense. One may invest by depositing funds in a savings account, by making a business loan, by purchasing a piece of machinery, by adopting a certain management procedure, and so forth. All involve an original expenditure with possible future returns. **Table 2.1** calculates the amount owed at the end of 5 periods. One thousand dollars today is *equivalent* to $1,610.51 five periods from now if interest is accumulated at the rate of 10% per period. This notion of equivalence is critical to our understanding of the role that the mathematics of compound interest plays in the evaluation of alternative investments.

TABLE 2.1 Amount to Be Repaid on a $1,000 Loan with Interest Rate of 10% per Year

Period	Amount Owed at Beginning of Period	Interest During Period	Amount Owed at End of Period
1	$1,000.00	$100.00	$1,100.00
2	1,100.00	110.00	1,210.00
3	1,210.00	121.00	1,331.00
4	1,331.00	133.10	1,464.10
5	1,464.10	146.41	1,610.51

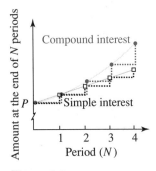

Figure 2.2 Comparison of Compound and Simple Interest

2.3.2 Compound and Simple Interest

Table 2.1 reflects **compound interest** calculations. That is, the interest for each period is based on the amount owed at the beginning of the period. But if the interest had been computed only on the amount originally borrowed, the **principal**, the cumulative debt would have grown more slowly—to only $1,500 after 5 periods, or $1,000 + ($1,000 × 10% × 5). This type of interest, known as **simple interest**, is of little or no practical concern, because it is rarely used in the real world. No rational investor would lend at simple interest. Compound interest is always greater than simple interest, because the former computes "interest on interest." See **Figure 2.2.**

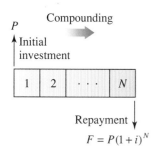

Figure 2.3

2.3.3　General Formulation

Let us turn now to the development of a general model that will enable us to compute equivalent values of a single cash flow. Let P represent the amount of the original investment, and let F represent the equivalent future amount at the end of N periods if interest is compounded (computed) at the rate i per period. (See **Figure 2.3.**) The calculations are shown, period by period, in **Table 2.2**.

TABLE 2.2　Initial Investment, P, to Be Repaid After N Periods with Interest Rate i

Period	Amount Owed at Beginning of Period	Interest During Period	Amount Owed at End of Period
1	P	iP	$P(1 + i)$
2	$P(1 + i)$	$iP(1 + i)$	$P(1 + i)^2$
3	$P(1 + i)^2$	$iP(1 + i)^2$	$P(1 + i)^3$
$\vdots$	$\vdots$	$\vdots$	$\vdots$
N	$P(1 + i)^{N-1}$	$iP(1 + i)^{N-1}$	$P(1 + i)^N = F$

Under *compound interest*, the relationship is clearly shown to be

$$F = P(1 + i)^N \tag{2.2}$$

Under *simple interest*,

$$F = P + NiP = P(1 + iN) \tag{2.3}$$

Figure 2.4

Equation (2.2), with a simple modification, can also be used to determine the equivalent present value, P, of a future cash flow, F, occurring at the end of the Nth period. (See **Figure 2.4.**) This process, known as **discounting**, results in

$$P = \frac{F}{(1 + i)^N} = F\left(\frac{1}{1 + i}\right)^N = F(1 + i)^{-N} \tag{2.4}$$

2.3.4　Some Numerical Examples

These fundamental relationships can now be used to solve a wide variety of practical problems.

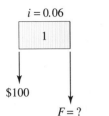

Figure 2.5   Example 2.1

EXAMPLE 2.1 *Future Value Unknown*

What is the equivalent future value 1 year from now of $100 invested at 6% per year? (See **Figure 2.5.**)

$$F = P(1 + i)^N$$

$$= \$100(1.06)^1$$

$$= \$106$$

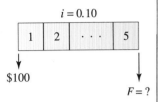

Figure 2.6   Example 2.2

EXAMPLE 2.2 *Future Value Unknown*

What is the equivalent value 5 periods from now of $100 invested at 10% per period? (See **Figure 2.6.**)

$$F = P(1 + i)^N$$

$$= \$100(1.10)^5$$

$$= \$161.05$$

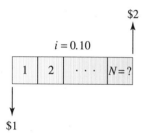

Figure 2.7   Example 2.3

EXAMPLE 2.3 *Number of Periods Unknown*

If interest is compounded at 10% per year, how long will it take to double an amount of money? (See **Figure 2.7.**)

$$F = P(1 + i)^N$$

$$\$2 = \$1(1.10)^N$$

$$N = 7.3 \text{ years}$$

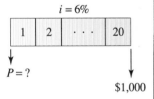

Figure 2.8   Example 2.4

EXAMPLE 2.4 *Present Value Unknown*

Using an interest (discount) rate of 6% per period, what is the equivalent present value of $1,000 flowing at the end of 20 periods? (See **Figure 2.8.**)

$$P = F\left(\frac{1}{1+i}\right)^N$$

$$= \$1,000\left(\frac{1}{1.06}\right)^{20}$$

$$P \simeq \$312$$

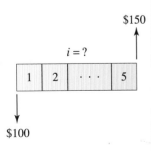

Figure 2.9   Example 2.5

EXAMPLE 2.5 *Interest Rate Unknown*

A current investment of $100 promises to yield $150 after 5 years. (See **Figure 2.9.**) To what compound interest rate does this investment correspond? (As will be shown in Chapter 4, this corresponds to the *internal rate* *of return* for the proposed investment.)

$$F = P(1 + i)^N$$

$$\$150 = \$100(1 + i)^5$$

$$i \simeq 0.084$$

In each of the above problems, four parameters are considered: a present cash flow, or equivalent present value, P; a future cash flow, or future equivalent value, F; the number of time periods, N; and the interest rate i, used for compounding or discounting. In each problem we are given three values and are asked to find the fourth:

Example	Given	Find
2.1	P, i, N	F
2.2	P, i, N	F
2.3	P, i, F	N
2.4	F, i, N	P
2.5	P, F, N	i

All problems, regardless of their complexity, are variations on this theme.

2.3.5 Compound Interest Tables

Numerical complexity can be demanding, especially without the use of an adequate calculator. In Example 2.4, it was necessary to compute $(1/1.06)^{20}$. To ease this burden, compound interest tables have been prepared in which values for various factors have been tabulated. A set of compound interest tables has been included here in **Appendix B**, with each table representing a specific interest rate. The rows within the tables are indexed on the number of periods, N. Each column represents a separate factor; the first three factors are related to the material discussed in this section, the single-payment models. (See **Figure 2.10**.)

	Single Payment		
	Compound Amount	Present Worth	
N	F/P	P/F	$P/\overline{F}$
1	1.010	0.9901	0.9950
2	1.020	0.9803	0.9852
3	1.030	0.9706	0.9754
4	1.041	0.9610	0.9658
5	1.051	0.9515	0.9562

Figure 2.10   Some Compound Interest Factors When $i = 0.01$

In particular, the first factor tabulates $(1 + i)^N$. In the literature of finance, this is known as the **compound amount factor**. It is used to find F given P, written F/P. Similarly, the reciprocal, $(1 + i)^{-N}$, is known as the **present worth factor**, or **discount factor**, because it is used to discount the future cash flow, F, in order to determine its equivalent present value, P. It is written P/F. (The third factor, $P/\overline{F}$, will be discussed in Section 2.9.)

If P, F, or N is unknown, the value of the factor can be selected directly from the table for a given value of i. When P, F, and N are given and i is unknown, as in Example 2.5, the solution can be obtained by searching the tables to find the particular i for which a value of the factor exists. In Example 2.5, for example, we determined that $(1 + i)^5 = \$150/\$100 = 1.500$. This is the compound amount factor, F/P, for $N = 5$. The tables in Appendix B show that

$$\text{For } i = 0.08 \qquad (1.08)^5 = 1.469$$

$$\text{For } i = 0.09 \qquad (1.09)^5 = 1.539$$

Using linear interpolation, $i \simeq 0.084$ when $F/P = 1.500$.

| 2.4   SOME USEFUL CONVENTIONS

The equivalence relationships for a single cash flow are relatively simple. However, as will be shown in the following sections, the mathematical relationships for a variety of other types of cash flows can be quite complex. Thus it will be useful to adopt certain conventions for the graphic display of these cash flows as well as a "shorthand" notation for the mathematical models.

2.4.1   Cash Flow Diagrams

In the literature of engineering economy, **cash flow diagrams** are frequently used to illustrate the *amount* and *timing* of cash flows. Generally, a bar or line is used to represent time, and arrows are used to represent positive or negative cash flows at the appropriate points in time. One such convention is illustrated in Figures 2.3 through 2.9. Throughout the remainder of this book we will use a time bar with upward arrows to represent positive cash flows (money received) and downward arrows to represent negative cash flows.

2.4.2   Functional Notation

As the algebraic form of the various equivalence factors can be complex, it is useful to adopt a standardized format that is easily learned and has a mnemonic connotation. The format which is in general use[1] is of the form:

$$(X/Y, i, N)$$

which is read as "to find the equivalent amount X given amount Y, the interest rate i, and the number of compounding/discounting periods N." For example, Eqs. (2.2) and (2.4) can be written in functional notation as

$$F = P(F/P, i, N)$$

and

$$P = F(P/F, i, N)$$

2.5 EQUIVALENT VALUES OF A UNIFORM SERIES OF CASH FLOWS

2.5.1 Present Value of a Uniform Series

Consider problems in which the equivalent present value of a series of end-of-period cash flows, $A_1, A_2, \ldots, A_N$, all of which are identical, must be found. (See **Figure 2.11**.) Letting $A = A_1 = A_2 = \cdots = A_N$, then

$$P = A_1(1 + i)^{-1} + A_2(1 + i)^{-2} + \cdots + A_N(1 + i)^{-N}$$

$$= A \sum_{j=1}^{N} (1 + i)^{-j}$$

$$= A \left[\frac{(1 + i)^N - 1}{i(1 + i)^N} \right] = A(P/A, i, N) \tag{2.5}$$

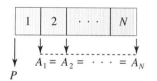

Figure 2.11

The term in brackets is known as the **series present worth factor**, because it is used to find the equivalent present value of a uniform series of cash flows.[2] Note that both the algebraic and the functional forms of the factor are shown in Eq. (2.5).

---

[1]This is the functional notation recommended in *Industrial Engineering Terminology*, Standard Z94.5 ("Engineering Economy"). This is an American National Standard approved by ANSI July 10, 1989. Copies are available from the Institute of Industrial Engineers, Norcross, Georgia 30092.

[2]In the derivation of Eq. (2.5), we use the relationship $\sum_{j=1}^{n} r^j = r(r^n - 1)/(r - 1)$, where $r < 1$.

In the literature of finance, an **annuity** is an amount of money paid to a beneficiary at regular intervals for a prescribed period of time out of a fund reserved for that purpose. Thus the terms *annuity* and *uniform series of cash flows* are often used interchangeably.

EXAMPLE 2.6

How much can we afford to spend now on a certain labor-saving device if the acquisition of this device is expected to result in savings of $2,000 per month for 30 months? Assume that all savings occur at end-of-month and our *opportunity cost*, the *discount rate*, is 2% per month.

Solution:

$$P = \$2,000(P/A, 2\%, 30)$$

$$= \$2,000(22.396) = \$44,792$$

Note that the (P/A) factor can be either calculated using Eq. (2.5) or taken directly from the 2% table in Appendix B.

2.5.2 Uniform Series (Annuity) from a Present Amount

It follows from Eq. (2.5) that

$$A = P\left[\frac{i(1+i)^N}{(1+i)^N - 1}\right] = P(A/P, i, N) \tag{2.6}$$

This is the **capital recovery factor**, used to convert an initial cash flow to a uniform series. The uniform series "recovers" the initial investment, P, plus an "opportunity cost" reflected by the interest rate, i. It is also known as the **annuity factor**, as it determines the amount of periodic annuity that may be paid out of a given fund.

EXAMPLE 2.7

A sum of $100,000 is to be spent on a certain cost-reduction program, the effects of which will be experienced over a 10-year period. If the firm's interest rate, the discount rate, is 15% per year, determine the minimum savings per year such that this investment would be warranted. Assume that all savings occur end-of-year. (See **Figure 2.12.**)

Solution:

$$A = \$100,000(A/P, 15\%, 10)$$

$$= \$100,000(0.1993) = \$19,930$$

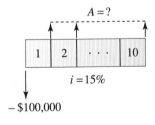

Figure 2.12   Example 2.7

As before, the (A/P) factor can be either calculated using Eq. (2.6) or taken directly from the 15% table in Appendix B.

EXAMPLE 2.8

A sum of $10,000 is deposited into a fund earning 3% per quarter (3 months). If $500 is withdrawn from the fund at the end of each quarter, how long will it take to exhaust the fund? (See **Figure 2.13**.)

Solution:

$$\$500 = \$10,000(A/P, 3\%, N)$$

$$(A/P, 3\%, N) = \frac{\$500}{\$10,000} = 0.0500$$

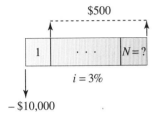

Figure 2.13   Example 2.8

Using the 3% table from Appendix B, $(A/P, 3\%, 31) = 0.0500$. Thus the fund will be exhausted after 31 quarters.

2.5.3 Future Value of a Uniform Series

For certain problems, it may be useful to compute the equivalent future value, F, given a uniform series of end-of-period cash flows, A. (See **Figure 2.14**.) This can be accomplished, of course, by first converting from A to P using Eq. (2.4). Then, by using the compound amount factor for a single payment, Eq. (2.2), we convert from P to F. This can be simplified to one step, however, by using

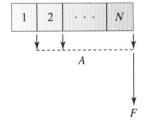

Figure 2.14

$$F = \{P\}(1 + i)^N$$

$$= \left\{A\left[\frac{(1 + i)^N - 1}{i(1 + i)^N}\right]\right\}(1 + i)^N$$

$$= A\left[\frac{(1 + i)^N - 1}{i}\right] = A(F/A, i, N) \qquad (2.7)$$

The factor in brackets is known as the **series compound amount factor**, because it finds the equivalent future value, or compound amount, of a uniform series of cash flows.

EXAMPLE 2.9

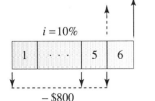

Figure 2.15   Example 2.9

A sum of $800 is deposited into a fund at the beginning of each year for 6 years. If the fund earns interest at the rate of 10% per year, what will be the value of the fund at the end of the sixth year? (See **Figure 2.15**.)

Solution:

$$F = \$800(F/A, \ 10\%, \ 6)(F/P, \ 10\%, \ 1)$$

$$= \$800(7.716)(1.10) = \$6,790$$

Note that two factors must be used in this case. The product [$800(F/A, 10%, 6)] results in an equivalent value at the start of the sixth year; this product must be carried forward one more year, i.e., to the end of the sixth year.

2.5.4 Annuity from a Future Sum

It follows from Eq. (2.7) that

$$A = F \left[\frac{i}{(1 + i)^N - 1} \right] = F(A/F, \ i, \ N) \ . \tag{2.8}$$

For reasons that will be explained later, the factor in brackets is known as the **sinking fund factor**. It is used to find the uniform series equivalent to an amount F at the end of the series.

EXAMPLE 2.10

A consulting engineer wants to buy certain computer equipment that will cost $4,000. How much should he deposit into a fund at the end of each month for 24 months if the fund is expected to earn 1% per month? (See **Figure 2.16**.)

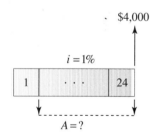

$$\$4,000$$
$$i = 1\%$$
$$A = ?$$

Figure 2.16   Example 2.10

Solution:

$$A = \$4,000(A/F, \ 1\%, \ 24)$$

$$= \$4,000(0.0371) = \$148$$

Again, the (A/F) factor can be evaluated using Eq. (2.8) or taken directly from the 1% Table in Appendix B.

2.5.5 Present Value of a Deferred Annuity

A **deferred annuity**, as shown in **Figure 2.17**, is a uniform series that begins at the end of period n and continues until the end of period N, $1 \le n \le N$. The equivalent present value, P, of the series of $N - n$ cash flows is given by

$$P = A(P/A, \ i, \ N - n + 1)(P/F, \ i, \ n - 1) \tag{2.9}$$

An alternative formulation is

$$P = A[(P/A, \ i, \ N) - (P/A, \ i, \ n - 1)] \tag{2.10}$$

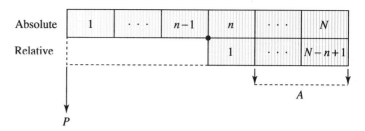

Figure 2.17   Deferred Annuity

EXAMPLE 2.11

If a fund generates interest at a rate of 12% per year, how much must be invested now in order to provide withdrawals of $10,000 per year at the ends of years 5 through 10? Here, $n = 5$ and $N = 10$. (See **Figure 2.18**.)

Solution:

$$P = \$10,000(P/A,\ 12\%,\ 10 - 5 + 1)(P/F,\ 12\%,\ 5 - 1)$$

$$= \$10,000(4.111)(0.6355)$$

$$= \$26,125 \quad \text{using Eq. (2.9)}$$

or

$$P = \$10,000[(P/A,\ 12\%,\ 10) - (P/A,\ 12\%,\ 5 - 1)]$$

$$= \$10,000(5.650 - 3.037)]$$

$$= \$26,130 \quad \text{using Eq. (2.10)}$$

The small difference in results is due to rounding of the factors in Appendix B.

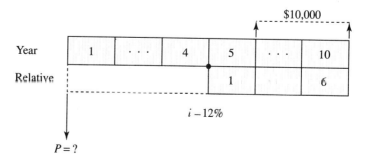

Figure 2.18   Example 2.11

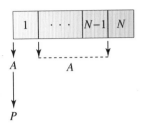

Figure 2.19

2.5.6 Annuity with Beginning-of-Period Cash Flows

When a uniform series of cash flows occur at the beginning, rather than the end, of each period, the previous formulations may be used but must be offset by one period. For example, consider the *present value* formulation as illustrated in **Figure 2.19**.

$$P = A[1 + (P/A, i, N - 1)] \qquad (2.11)$$

EXAMPLE 2.12

What is the equivalent present value of insurance payments of $750 each, occurring at the start of every 6-month period over 5 years, assuming a discount rate of 14% (nominal, per year) compounded semiannually? (See **Figure 2.20**.)

Solution:

$$P = \$750[1 + (P/A, 7\%, 9)]$$

$$= \$750[1 + (6.515)] = \$5,636$$

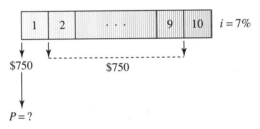

Figure 2.20 Example 2.12

2.6 EQUIVALENT VALUES OF A GRADIENT SERIES

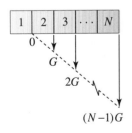

Figure 2.21 Arithmetic Gradient Series

2.6.1 Arithmetic Gradient

A relatively common condition is one in which cash flows increase by a uniform amount from period to period, resulting in an arithmetic progression. Compound interest factors for treating this case are useful. To derive these factors, let G denote the amount, or **arithmetic gradient**, by which the cash flows increase uniformly from one period to the next. (See **Figure 2.21**.) That is, $A_j = (j - 1)G$, for $j = 1, 2, \ldots, N$. This results in a sequence of cash flows, $0, G, 2G, \ldots, (N - 1)G$, at the end of periods 1, 2, 3, \ldots, N, respectively. As shown in **Figure 2.22**, this arithmetic gradient

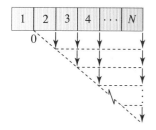

Figure 2.22

series can be described by $N-1$ uniform series, each of which has equivalent future value given by Eq. (2.7). Therefore, the equivalent value at the end of period N of the entire arithmetic gradient series is

$$F = G \sum_{j=1}^{N-1} \left[\frac{(1 + i)^j - 1}{i} \right]$$

$$= \frac{G}{i} \left[\sum_{j=0}^{N-1} (1 + i)^j \right] + \frac{G}{i}(-N)$$

Simplifying,[3]

$$F = G \left\{ \frac{1}{i} \left[\frac{(1 + i)^N - 1}{i} \right] - \frac{N}{i} \right\} \quad (2.12)$$

To convert to P, we multiply F by $(1 + i)^{-N}$, with the result

$$P = G \left[\frac{(1 + i)^N - iN - 1}{i^2(1 + i)^N} \right] = G(P/G, i, N) \quad (2.13)$$

The term in brackets is known as the **gradient present worth factor.**

It may also be shown that, to convert from an arithmetic gradient series to an equivalent uniform series, A, one can use the relationship

$$A = G \left[\frac{(1 + i)^N - iN - 1}{i(1 + i)^N - i} \right] = G(A/G, i, N) \quad (2.14)$$

The term in brackets is the **gradient uniform series factor**. Both factors, Eqs. (2.13) and (2.14), have been tabulated for various values of i and N and are included in the tables in Appendix B.

EXAMPLE 2.13 *Positive Arithmetic Gradient*

Consider a series of cash flows as shown in **Figure 2.23**. Assuming an interest rate of 10% per period, determine the equivalent present value.

The solution depends on recognizing that there are *two* series here: a uniform series of $100 per period and a gradient series where $G = \$10$.

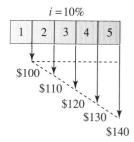

Figure 2.23 Example 2.13

End of Period	Cash Flow
1	$100 = \$100 + 0(\$10)$
2	$110 = \$100 + 1(\$10)$
3	$120 = \$100 + 2(\$10)$
4	$130 = \$100 + 3(\$10)$
5	$140 = \$100 + 4(\$10)$

---

[3]This derivation makes use of the relationship $\sum_{j=0}^{N-1} x^j = (x^N - 1)/(x - 1)$, $x > 1$.

Thus the solution is

$$P = \$100(P/A, \ 10\%, \ 5) + \$10(P/G, \ 10\%, \ 5)$$

$$= \$100(3.791) + \$10(6.862) = \$448$$

EXAMPLE 2.14 *Negative Arithmetic Gradient*

Consider a series of cash flows as shown in **Figure 2.24**. Assuming an interest rate of 10% per period, determine the equivalent uniform series.

As above, we note that there are two imbedded series: a uniform series of $700 per period and a gradient series where $G = -\$100$, a negative value.

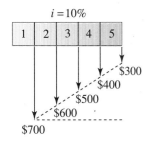

$i = 10\%$

$300
$400
$500
$600
$700

Figure 2.24   Example 2.14

End of Period	Cash Flow
1	$700 = \$700 + 0(-\$100)$
2	$600 = \$700 + 1(-\$100)$
3	$500 = \$700 + 2(-\$100)$
4	$400 = \$700 + 3(-\$100)$
5	$300 = \$700 + 4(-\$100)$

Thus the solution is

$$A = \$700 - \$100(A/G, \ 10\%, \ 5)$$

$$= \$700 - \$100(1.810) = \$519$$

2.6.2 Geometric Gradient

Another type of cash flow sequence that may be of interest is the **geometric gradient series**. (See **Figure 2.25**.) Here, the end-of-period cash flows increase at a constant rate, g, from period to period, such that, for a given value of A_1,

$$A_{j+1} = A_j(1 + g) \qquad \text{for } j = 1, 2, \ldots, N - 1 \qquad (2.15)$$

With cash flows discounted at rate i per period, it may be shown that the equivalent present value of the entire series of cash flows is given by

$$P = A_1 \left[\frac{1 - (1 + g)^N (1 + i)^{-N}}{i - g} \right] \qquad i \neq g \qquad (2.16a)$$

$$= A_1 N (1 + i)^{-1} \qquad i = g \qquad (2.16b)$$

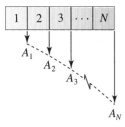

A_1
A_2
A_3
A_N

Figure 2.25   Geometric Gradient Series

As N tends to infinity, the expression in brackets in Eq. (2.16a) is *convergent* if $g < i$; the series is *divergent* if $g \geq i$. The proof, if desired, remains as an exercise for the reader.

The expression in brackets in Eq. (2.16a) is known as the **geometric series present worth factor**. The functional form is

$$(P/A_1, i, g, N)$$

Tables for this factor, based on selected values of i, g, and N, may be generated using Eq. (2.16). The factor may also be computed by using the more readily available compound amount (F/P) and present worth factors (P/F) by noting that, for $i \neq g$,

$$(P/A_1, i, g, N) = \left[\frac{1 - (F/P, g, N)(P/F, i, N)}{i - g}\right] \tag{2.17}$$

EXAMPLE 2.15

Manufacturing costs are expected to be \$100,000 in the first year, increasing by 5% each year over a 7-year period. Find the equivalent present value of these cash flows assuming a 10% discount rate and end-of-year cash flows. Here, $A_1 = \$100,000$, $i = 10\%$, $g = 5\%$, and $N = 7$. (See **Figure 2.26.**)

Evaluating the factor using Eq. (2.16a), the solution is

$$P = \$100,000(P/A_1, 10\%, 5\%, 7)$$

$$= \$100,000(5.5587) = \$555,870$$

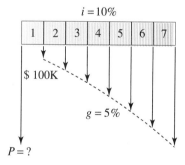

Figure 2.26   Example 2.15

2.7 EFFECTIVE AND NOMINAL INTEREST RATES

2.7.1 Periods and Subperiods

An interest rate is meaningful only if it is related to a particular period of time. Nevertheless, the "time tag" is frequently omitted in speech because it is usually understood in context. If someone tells you that he is earning 8%

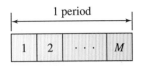

Figure 2.27

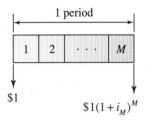

Figure 2.28

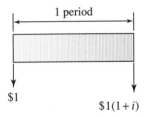

Figure 2.29

on his investments, for example, it is implied that the rate of return is 8% *per year*. However, in many cases the interest rate period is a week, a month, or some other interval of time, rather than the more usual year (*per annum*). At this point it would be useful to examine the process whereby interest rates and their respective "time tags" are made commensurate.

As before, let i represent the **effective interest rate** per period. As shown in **Figure 2.27**, let the period be divided into M subperiods of equal length. If interest is compounded at the end of each subperiod at rate i_M per subperiod, then \$1 flowing at the start of the first subperiod will have equivalent value of $\$1(1 + i_M)^M$ at the end of the Mth subperiod. (See **Figure 2.28**.) If our calculation is based on the period rather than the subperiod, then the end-of-period equivalent value of \$1 flowing at the start of the period, as shown in **Figure 2.29**, is simply $\$1(1 + i)$.

The relationship between the effective interest rates per period and per subperiod is easily determined. From the above:

$$\$1(1 + i_M)^M = \$1(1 + i)$$

or

$$i = (1 + i_M)^M - 1 \tag{2.18}$$

The **nominal interest rate** per period, r, is simply the effective interest rate per subperiod times the number of subperiods, or

$$r = Mi_M \tag{2.19}$$

Nominal and effective interest rates are most frequently used when the period is 1 year and M is the number of divisions within the year.

EXAMPLE 2.16   *Monthly Compounding*

Consider the case of consumer credit, say, a major oil company or bank charge card for which interest is compounded monthly at rate 1.5% on the unpaid balance. We would like to know the effective annual rate. Here, $i_M = 0.015$ and $M = 12$. The nominal rate per year is $12 \times 0.015 = 0.18$, and the effective rate per year, by Eq. (2.18), is

$$i = (1.015)^{12} - 1 = 0.1956$$

Except for the trivial case in which $M = 1$, the annual effective rate will always be larger than the annual nominal rate. Moreover, the difference between the nominal and effective rates increases as r and/or M increase, as illustrated in **Table 2.3**. Therefore, the appropriate interest rate to use in all equivalence calculations is the effective rate.

TABLE 2.3   Representative Values for Nominal and Associated Effective Interest Rates

Nominal Interest Rate (r)	Number of Compounding Subperiods (M)	Effective Interest Rate[a] (i)	Absolute Difference ($i - r$)	Relative Difference ($i - r)/(i$)
0%	Any	0%	0%	0.000
10	10	10.46	0.46	0.044
	20	10.49	0.49	0.047
20	10	21.90	1.90	0.087
	20	22.02	2.02	0.092
30	10	34.39	4.39	0.128
	20	34.69	4.69	0.135
40	10	48.02	8.02	0.167
	20	48.59	8.59	0.177

[a]Remember, $i = (1 + r/M)^M - 1$.

EXAMPLE 2.17   *Subperiod Cash Flows*

Labor costs for a certain operation are $8,000 per week, occurring at the end of each week over a 3-year period. If the effective discount rate is 15% per year, determine the equivalent present value of these costs. (See **Figure 2.30**.)

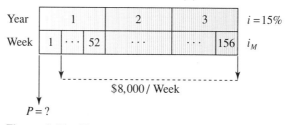

Figure 2.30   Example 2.17

The solution requires two steps. First, we must determine the equivalent subperiod (weekly) interest rate. From Eq. (2.18) with $i = 0.15$ and $M = 52$,

$$i_M = (1 + i)^{1/M} - 1 \tag{2.20}$$

$$= (1.15)^{1/52} - 1 = 0.00269$$

This value for the effective interest rate per week is now used to determine the equivalent present value of the uniform series of cash flows.

$$P = \$8,000(P/A, 0.269\%, 156)$$

$$= \$8,000(127.254) = \$1,018,000$$

[As there is no table for 0.269% in Appendix B, the value of the (P/A) factor is determined using Eq. (2.5).]

2.7.2 Periods and Superperiods

Example 2.17 illustrated the case where cash flows occur and are compounded more frequently than the defined period: The *period* is the year, but the *subperiod* is the week. Situations will arise in which the converse will be true, that is, cash flows will occur regularly but *less frequently* than the defined period. In such instances it will be helpful to think of *superperiod* cash flows, where the superperiod is some integer multiple of the defined period.

There are several approaches to problems of this type. Perhaps the easiest solution is based on converting the end-of-superperiod cash flow, A_S, to equivalent end-of-period cash flows, A, by using the sinking fund factor, Eq. (2.8):

$$A = A_s(A/F, i, N_s) \qquad (2.21)$$

where i is the interest rate per period and N_s is the number of periods in one superperiod. The remainder of the solution, then, is as described previously.

EXAMPLE 2.18 *Superperiod Cash Flows*

Major overhaul expenses of $20,000 each occur at the end of year 5 and continue, every 2 years, up to and including year 13. Assuming a 10% discount rate, determine the equivalent present value. The initial problem appears as in **Figure 2.31a**. The end-of-period cash flows, as shown in **Figure 2.31b**, are determined by

$$A = \$20,000(A/F, 10\%, 2)$$

$$= \$20,000(0.4762) = \$9,524$$

We now have a deferred annuity of $9,524 at the ends of periods 4 through 13. The equivalent present value is found from Eq. (2.9):

$$P = \$9,524[(P/A, 10\%, 13) - (P/A, 10\%, 3)]$$

$$= \$9,524(7.103 - 2.487) = \$43,963$$

2.8 CONTINUOUS COMPOUNDING

2.8.1 The Continuous Compounding Convention

Consider the case in which interest is compounded not once a period, not twice a period, not three times a period, but M times each period, where M is a very large number. Under these circumstances, it may be useful to

(a)   Cash Flows at the End of Each Superperiod

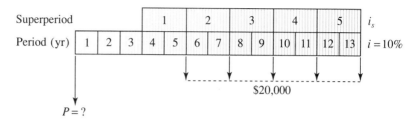

(b)   Equivalent Cash Flows at the End of Each Period

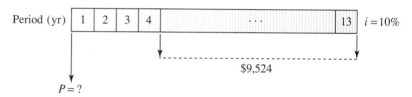

Figure 2.31   Example 2.18

define the relationship between the nominal interest rate, r, and the effective interest rate, i.

The derivation is straightforward. As before, divide the period into M equal parts. The relationship between i and i_M as given above is

$$i = (1 + i_M)^M - 1 \qquad \text{where } i_M = \frac{r}{M}$$

Rewriting,

$$i = \left[\left(1 + \frac{1}{M/r}\right)^{M/r}\right]^r - 1$$

Letting $k = M/r$, we have

$$i = \left[\left(1 + \frac{1}{k}\right)^k\right]^r - 1$$

For a given r, when M tends to infinity, k also tends to infinity. Thus,[4]

$$i = e^r - 1 \tag{2.22}$$

where e is the base of the Naperian logarithm system, the "exponential," approximately equal to 2.71828. Under the assumption that M is very

---

[4]You may recall that $e = \lim\limits_{k \to \infty} (1 + 1/k)^k$.

large, that is, $M \to \infty$, this is known as **continuous compounding (discounting)**.

EXAMPLE 2.19

If interest is compounded continuously, what effective rate per period, i, corresponds to a nominal rate, r, of 10.00%?

Solution:

$$i = e^{0.10} - 1 = 10.52\%$$

The **future worth** of a discrete cash flow under continuous compounding is given by

$$F = P(1 + i)^N = P[1 + (e^r - 1)]^N$$

$$= Pe^{rN} \tag{2.23}$$

where, from Eq. (2.22), $r = \log_e = \ln(1 + i)$.

EXAMPLE 2.20

A sum of $10,000 is invested in an account (e.g., a certificate of deposit at a thrift institution) earning interest at the nominal rate of 7.2% per year. Determine the value of this account at the end of 2 years if interest is compounded (a) quarterly, (b) monthly, and (c) continuously.

Solution:

(a) Quarterly ($M = 4$):

$$F = \$10,000\left[1 + \left(\frac{0.072}{4}\right)\right]^8 = \$11,534$$

(b) Monthly ($M = 12$):

$$F = \$10,000\left[1 + \left(\frac{0.072}{12}\right)\right]^{24} = \$11,544$$

(c) Continuously ($M \to \infty$):

$$F = \$10,000e^{0.072(2)} = \$11,549$$

Clearly, the terminal value increases as M increases. Moreover, the continuous compounding assumption yields a very good approximation to the "correct" values when M is a finite value of, say, 4 or more.

2.8.2 Rates and Yields

Thrift institutions (banks, savings and loans institutions) often publish two related rates paid on deposited funds: The **rate** corresponds to the nominal rate per year, r, as described above; and the **yield** corresponds to the effective rate per year, i. They are related as in Eq. (2.18) or (2.22).

EXAMPLE 2.21

A certain bank advertises that it will pay interest on a $10,000 certificate of deposit at a rate of 7.20%, with interest compounded "continuously." The corresponding yield is

$$i = e^{0.072} - 1 = 7.47\%$$

2.9 CONTINUOUS COMPOUNDING AND CONTINUOUS CASH FLOWS

2.9.1 The Funds Flow Conversion Factor

There are many instances in which an analyst will evaluate end-of-period cash flows as well as cash flows that occur frequently within a period. For example, suppose that a year is the basic period for an analysis, yet there are other cash flows, say, wages or material costs, that occur weekly or daily. One approach would be simply to accumulate the subperiod cash flows to compute a single value for the period. (See **Figure 2.32**.) So wages of $400 per week would be roughly equivalent to $20,800 per year, based on a 52-week year. This is only an approximation, however, because there may be compounding effects from one subperiod to the next.

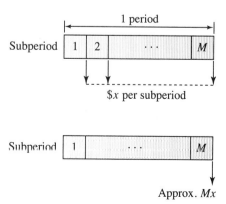

Figure 2.32 Approximate Equivalent

An alternative approach is based on the assumption that the number of subperiods, M, within the period is very large. (The question of how large is "very large" will be discussed in the following section.) Let $\bar{A}$ be an amount of money distributed uniformly over M subperiods within the period so that $\bar{A}/M$ is the cash flow occurring at the end of every subperiod. (See **Figure 2.33**.)

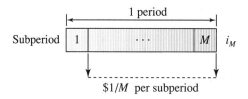

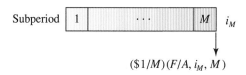

Figure 2.33 Exact Equivalent

From the series compound amount factor, Eq. (2.7), the equivalent cash flow, A, at the end of M subperiods is

$$A = \frac{\bar{A}}{M}\left[\frac{(1 + i_M)^M - 1}{i_M}\right] \tag{2.24}$$

where i_M represents the effective interest rate per subperiod. Of course, $i_M = r/M$, where r is the nominal interest rate per period.[5] Substituting in the above equation, rearranging terms, and taking the limit as M gets "very large," that is, as $M \rightarrow \infty$,

$$A = \bar{A}\left[\frac{e^r - 1}{r}\right] = \bar{A}\left[\frac{i}{\ln(1 + i)}\right] \tag{2.25}$$

where i is the effective interest rate per period. The factor in brackets in Eq. (2.25) is known as the **funds flow conversion factor**. It can be used to find the equivalent value, A_j, of an amount $\bar{A}_j$ flowing continuously and uniformly throughout period j.

EXAMPLE 2.22

It is expected that a total of $12,000 will be spent out of the firm's "petty cash" fund over the course of a year. Actual expenditures occur daily and are approximately the same from day to day. The continuous cash flow

---

[5]In some textbooks the term *nominal* rate is used only when the interest period is a year. But the nominal rate may always be used for an interest period of any duration, as long as compounding takes place over two or more subperiods within the period.

assumption would appear to be warranted here. If the firm's discount rate is 20% per year, determine the equivalent value at the end of the year.

Solution:

$$A = \$12,000\left(\frac{0.20}{\ln\ 1.2}\right) = \$13,164$$

2.9.2   Modified Discount Factors

The funds flow conversion factor is useful in modifying the end-of-period factors, previously discussed, to accommodate the "continuous" assumptions. To illustrate, consider the factor for determining the equivalent present value of a cash flow, $\bar{F}$, flowing continuously and uniformly during the Nth period hence (**Figure 2.34**). Combining Eqs. (2.4) and (2.25),

$$P = \bar{F}\left\{\left[\frac{i}{\ln(1+i)}\right]\left(\frac{1}{1+i}\right)^{N}\right\}$$

$$= \bar{F}(P/\bar{F},\ i,\ N) \tag{2.26}$$

Figure 2.34   Single Period Continuous Cash Flow

Similarly, consider the factor for determining the equivalent present value of a uniform series of continuous cash flows ($\bar{A}$) flowing during each and every period through N periods (**Figure 2.35**). Combining Eqs. (2.5) and (2.26),

$$P = \bar{A}\left\{\left[\frac{i}{\ln(1+i)}\right]\left[\frac{(1+i)^{N}-1}{i(1+i)^{N}}\right]\right\}$$

$$= \bar{A}(P/\bar{A},\ i,\ N) \tag{2.27}$$

The factors given by Eqs. (2.26) and (2.27) are included in Appendix B in the columns headed $P/\bar{F}$ and $P/\bar{A}$. (The factor $F/\bar{A}$ has also been incorporated into the tables.)

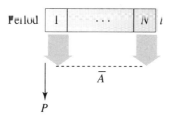

Figure 2.35   Uniform Series of Continuous Cash Flows

EXAMPLE 2.23 *Using the Continuous Single Payment Present Worth Factor*

A major overhaul is expected to cost $40,000 in the 30th month of ownership. If the discount rate is 2% per month, and assuming that these costs occur continuously and uniformly during the 30th month, determine the equivalent present value.

Solution:

$$P = \$40,000(P/\bar{F}, 2\%, 30)$$

$$= \$40,000(0.5576) = \$22,300$$

EXAMPLE 2.24 *Using the Continuous Uniform Series Present Worth Factor*

Repeat Example 2.17 using the continuous cash flow, continuous compounding approximation. Recall, cash flows are $8,000 each week over a 3-year period, and the annual discount rate is 15 percent. Assuming continuous cash flows, the present value is

$$P = (52 \times \$8,000)(P/\bar{A}, 15\%, 3)$$

$$= \$416,000(2.450) = \$1,019,200$$

This result is very close to the "true" value of $1,018,000.

2.10 THE MID-PERIOD CONVENTION

The mid-period convention assumes that all cash flows occurring at times within the period are aggregated and concentrated at the *midpoint* of the period. That is, we assume that the period is divided into two subperiods ($M = 2$) with effective rate i_s per subperiod. We note that i_s is related to the effective rate per period, i, by $i = (1 + i_s)^2 - 1$. It follows that $i_s = \sqrt{1 + i} - 1$. Let A_{jk} = cash flow at end of subperiod k in period j, $k = 1, 2, \ldots, M$. Assuming that all subperiod cash flows are concentrated at the midpoint of the period,

$$A_j = \text{equivalent cash flow at end of period } j$$

$$= \left(\sum_{k=1}^{M} A_{jk} \right) \sqrt{1 + i} \qquad (2.28)$$

This is illustrated in **Figure 2.36**. The mid-period convention provides a reasonably good approximation if the intraperiod cash flows are symmetrically distributed about the midpoint of the period, as is the case with uniform subperiod cash flows.

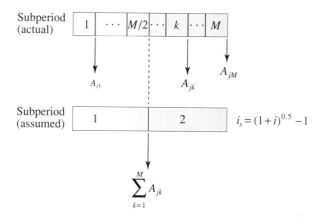

Figure 2.36 The Mid-Period Assumption

EXAMPLE 2.25

Cash flows of $1,000 per month are expected to occur at the end of each month throughout year j. If the interest rate is 20% per year, determine the equivalent end-of-year cash flow using the mid-period convention. (See **Figure 2.37**.)

Solution:

$$A_j = (12 \times \$1,000)\sqrt{1.20} = \$13,145$$

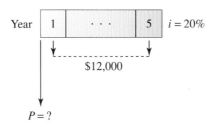

Figure 2.37 Example 2.25

EXAMPLE 2.26

Repeat Example 2.17 using the mid-period convention. (See **Figure 2.38c.**)

Solution:

$$P = (52 \times \$8,000)\sqrt{1.15}(P/A, 15\%, 3)$$

$$= \$446,110(2.283) = \$1,018,500$$

This result is very close to the "true" value ($1,018,000) as well as that obtained using the continuous cash flow convention ($1,019,200).

(a) Original Problem (Example 2.17)

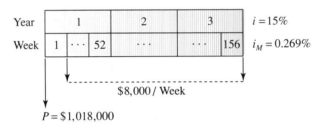

(b) Approximate Solution: Continuous Cash Flows (Example 2.24)

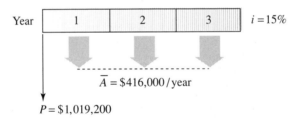

(c) Approximate Solution: Mid-Period Convention (Example 2.26)

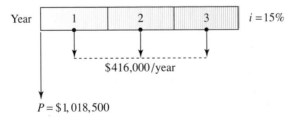

Figure 2.38 Examples 2.17, 2.24, 2.26

2.11 IMPORTANCE OF END-OF-PERIOD, MID-PERIOD, AND CONTINUOUS CASH FLOW ASSUMPTIONS

In theory, all economic analyses require prior specification of the interest period. In practice, this is normally taken to be the year or the quarter (3 months). Actual cash flows, however, occur at various times within the year, e.g., month, week, or day. Thus the analyst must address the issue of exactly how these intraperiod cash flows are to be treated.

Several conclusions are apparent from the preceding discussion. First, the *end-of-period assumption* always understates the true (exact) value, as the compounding effects of the nonzero interest rate are disregarded. Second, if the intraperiod cash flows are reasonably uniform throughout the period, both the *mid-period* and *continuous assumptions* will overstate the true value. Finally, in general, the relative errors introduced by using one of the approximation methods is a function of (1) the interest rate i, (2) the number of subperiods within the period M, and (3) the degree to which the cash flows are distributed uniformly within the period. (Of course, if absolute precision is required, then the exact method should be used as outlined above.) It is this author's view that, in those instances when approximation is acceptable, the mid-period convention usually offers the best balance between ease of use and precision.

2.12 ILLUSTRATING EQUIVALENCE

2.12.1 Loans

It may be useful at this point to examine one aspect of compound interest calculations of considerable interest to virtually everyone in our society: **loans**, the borrowing and lending of money. This examination will afford the opportunity to use several of the concepts discussed earlier in this chapter as well as to demonstrate the use of the functional format for the compound interest factors.

Consider an amount of money received by a borrower at some point in time. The sum advanced by the lender to the borrower is known as the **principal** of the loan and will be designated by the symbol P in the following discussion. In particular, let P_j represent the amount of principal remaining, or yet to be repaid, at the start of period j, for $j = 1, 2, \ldots, N$. Here, P_1 represents the amount of the original loan.

All loans have at least two critical pieces of information associated with them: (1) the amount and timing of cash flows between the borrower and the lender; and (2) the effective interest rate per period, which is a measure of the "rent" paid by the borrower for the temporary use of borrowed funds. In some instances, of course, only the cash flows and their timing are specified; the interest rate is only implied. In any case, the interest rate, whether explicit or implicit, represents the cost to the borrower and the return to the lender.

In general, the interest accumulated in period j, I_j is given by

$$I_j = iP_j \qquad (2.29)$$

where i is the effective interest rate per period.

Single Repayment

Consider the case in which the borrower and lender agree on a loan of amount P_1, with repayment of the principal and all accrued interest at the end of N periods. This is known as a "balloon payment." Interest is to be compounded at rate i per period. The amount of the terminal payment, then, is

$$F = P_1(F/P, i, N) \qquad (2.30)$$

EXAMPLE 2.27

Suppose that the original loan is $1,000, with payment due at the end of 24 months and with interest compounded monthly at the rate of 1% per month. Then

$$F = \$1,000(F/P, 1\%, 24)$$

$$= \$1,269.73$$

(Note the degree of precision here. This "penny accounting" is quite common in lending situations.)

Uniform Periodic Repayment

Another common repayment plan is one in which the borrower promises to repay the loan, P_1, with N equal payments. Some examples are home mortgages, car loans, and in many cases, student loans. If i is the interest rate per compounding period and if the loan payments, A, are to be made at the end of every compounding period, then

$$A = P_1(A/P, i, N) \qquad (2.31)$$

The portion of each payment that represents repayment of principal is simply the amount of the payment less the interest incurred in that period, or $A - I_j$. The cash flow diagram is shown in **Figure 2.39**. The principal remaining (i.e., the remaining debt) at the start of period j is the equivalent present value of the remaining payments. At the beginning of period j there are $N - j + 1$ payments remaining. Thus

$$P_j = A(P/A, i, N - j + 1) \qquad (2.32)$$

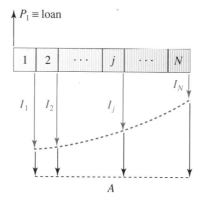

Figure 2.39   Loan with Uniform Periodic Repayment

EXAMPLE   2.28

A numerical example illustrating these calculations is summarized in **Table 2.4**. Here, a $1,000 loan is repaid over 5 periods with interest calculated at the rate of 10% per period. Consider the third payment. The total amount of the payment, from Eq. (2.31), is

$$A = \$1,000(A/P, 10\%, 5) = \$263.80$$

The amount owed at the start of period 3, from Eq. (2.32), is

$$P_3 = \$263.80(P/A, 10\%, 5 - 3 + 1)$$

$$= \$263.80(2.487) = \$656.07$$

TABLE 2.4   Repayment of $1,000 Loan Assuming 5 Uniform Payments at Interest Rate of 10% per Period

End of Period j	Amount of Payment,[a] A	Interest Portion,[b] I_j	Principal Portion,[c] $A - I_j$	Principal Remaining After Payment,[d] P_{j+1}
1	$263.80	$100.00	$163.80	$836.20
2	263.80	83.62	180.18	656.02
3	263.80	65.60	198.20	457.82
4	263.80	45.78	218.02	239.80
5	263.80	23.98	239.82	0

[a] $A = \$1,000(A/P, 10\%, 5) = \$263.80$
[b] $I_j = iP_{j-1}$, where $P_1 = \$1,000$
[c] $A - I_j$
[d] $P_{j+1} = P_j - (A - I_j)$

The amount of the uniform payment at the end of period 3 that is interest, from Eq. (2.29), is

$$I_3 = 0.10(\$656.07)$$
$$= \$65.61$$

2.12.2 Working Capital

Working capital is the additional funds required to start and support operating activities. Excluded are the funds required for assets such as land and physical plant. Working capital includes such elements as funds required for recruiting and training personnel, tools, spare parts, minimum levels of material inventories, petty cash, employee salaries, and the like. Broadly stated, working capital is the funds needed to meet the short-term cash requirements of the firm. Normally, outlays for working capital occur in the early stages of the project; funds are recovered, at least in part, at the end of the project life.

Cash flows for working capital can be significant, and they should be incorporated into the economic analysis wherever appropriate. These cash flows, once identified, are evaluated similarly to cash flows generated from other sources. (The effect of working capital on income taxes is generally somewhat different from that of other types of cash flows. This issue is addressed further in Chapter 7.)

EXAMPLE 2.29

The introduction of a new product requires changes in working capital over the product life cycle as follows:

End of Month	Cash Flow
0	− $50,000
3	− 30,000
6	− 10,000
48	75,000

If the firm's interest rate is 18% per year, determine the equivalent present value of these cash flows. (In **Figure 2.40**, the cash flow diagram for this problem, note that the beginning of month 1 is the end of month 0.)

Solution: First determine the equivalent interest rate per month.

$$i_M = (1.18)^{1/12} - 1 = 0.0139$$

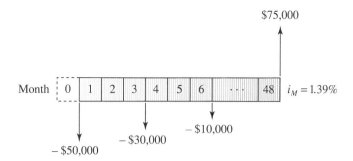

Figure 2.40   Example 2.29

Using this value for the discount rate,

$$P = -\$50{,}000 - \$30{,}000(1.0139)^{-3}$$

$$-\$10{,}000(1.0139)^{-6} + \$75{,}000(1.0139)^{-48}$$

$$-\$50{,}000 - \$28{,}784 - \$9{,}206 + \$38{,}684$$

$$= -\$49{,}300$$

| SUMMARY

Money amounts cannot be fairly compared unless they are evaluated at the same point in time. But since receipts and expenditures, positive and negative cash flows, resulting from alternative capital investments almost always yield dissimilar cash flow patterns, it is necessary to determine equivalent values through the use of compound interest calculations. Some of the factors most commonly employed for this purpose have been derived and are summarized in **Table 2.5**. (The various factors are also summarized inside the back cover of this book for ease of reference.) All of these factors, with the exception of the geometric series present worth factor, shown as the last equation in the table, are tabulated in Appendix B.

The distinction between nominal and effective interest rates should be clearly understood, because the relative difference between these two rates can be significant. In general, these differences will increase as the interest rate gets larger and as the number of compounding subperiods increases.

Compound interest factors have been developed under three separate sets of assumptions: (1) Receipts or disbursements occur only once during the interest period, at the end of the period, and the com-

TABLE 2.5 Compound Interest Factors and Their Functional Forms

Factor Name	Algebraic Form	Functional Format
Single-payment		
Compound amount	$(1 + i)^N$	$(F/P, i, N)$
Present worth	$\left(\dfrac{1}{1+i}\right)^N$	$(P/F, i, N)$
Continuous present worth	$\left\{\left[\dfrac{i}{\ln(1+i)}\right]\left[\dfrac{1}{1+i}\right]^N\right\}$	$(P/\bar{F}, i, N)$
Uniform series		
Compound amount	$\left[\dfrac{(1+i)^N - 1}{i}\right]$	$(F/A, i, N)$
Continuous compound amount	$\left\{\left[\dfrac{i}{\ln(1+i)}\right]\left[\dfrac{(1+i)^N - 1}{i}\right]\right\}$	$(F/\bar{A}, i, N)$
Present worth	$\left[\dfrac{(1+i)^N - 1}{i(1+i)^N}\right]$	$(P/A, i, N)$
Continuous present worth	$\left\{\left[\dfrac{i}{\ln(1+i)}\right]\left[\dfrac{(1+i)^N - 1}{i(1+i)^N}\right]\right\}$	$(P/\bar{A}, i, N)$
Sinking fund	$\left[\dfrac{i}{(1+i)^N - 1}\right]$	$(A/F, i, N)$
Capital recovery	$\left[\dfrac{i(1+i)^N}{(1+i)^N - 1}\right]$	$(A/P, i, N)$
(Arithmetic) gradient series		
Uniform series	$\left[\dfrac{(1+i)^N - iN - 1}{i(1+i)^N - i}\right]$	$(A/G, i, N)$
Present worth	$\left[\dfrac{(1+i)^N - iN - 1}{i^2(1+i)^N}\right]$	$(P/G, i, N)$
Geometric gradient series		
Present worth $i \neq g$	$\left[\dfrac{1 - (1+g)^N(1+i)^{-N}}{i - g}\right]$	$(P/A_1, g, i, N)$

pounding/discounting takes place at that time; (2) cash flow for the entire period is concentrated at the midpoint of the period, and interest is therefore compounded only during the remaining half-period; or (3) the cash flows continuously and uniformly throughout the interest period, and interest is compounded continuously. The appropriate assumptions for any given analysis should be those that best approximate the specific conditions for any given problem situation.

Principal and interest payments associated with loans provide an especially instructive opportunity to examine the mathematics of compound interest. Several features have been discussed in this chapter. The treatment of working capital provides another opportunity to illustrate compound interest calculations.

| PROBLEMS

Unless otherwise indicated, assume discrete cash flows, end-of-period compounding/discounting, and effective interest rates. When the interest period is unspecified, per year (*per annum*) is always implied.

The Power of Compound Interest

2.1 An amount of $100.00 is invested in an account earning interest at the rate of 13% per year. Determine the amount in the account at the end of 10 years.
[*Answer*: $339.46.]

2.2 If interest is compounded at the rate of 14% per period, determine how long it will take for an initial payment of $P to at least double in value.

2.3 [This problem was presented by Fibonacci in *Liber Abaci*, written in 1202 A.D. Also known as Leonardo of Pisa (1180–1250), Fibonacci traveled extensively and brought back to Italy a knowledge of Hindu numerals and the general learning of the Arabs.] A certain man puts one denarius at interest at such a rate that in 5 years he has 2 denarii, and in every 5 years thereafter the money doubles. How many denarii would he gain from this one denarius in 50 years?
[*Answer*: 1023 dinarii = 1024 − the original denarius.]

2.4 If interest is compounded at the rate of 1% per month, how long will it take an investment of $X to (approximately) double in value? (Give your answer to the nearest 0.1 month.)

2.5 How long will it take for a sum of money to at least double if it is invested in a fund earning
a. 8% per period?
b. 12% per period?
[*Answers*: a. 10 periods; b. 7 periods.]

2.6 Suppose that you have $1,000 to invest, but you need $3,000 to purchase a certain item. You can invest your funds ($1,000) at 3% per period, however, and wait until you have enough money in the fund to afford the investment. How many periods will you have to wait until you have at least $3,000 in the fund?

2.7 If interest is compounded at the rate of 10% per period, a present sum of money will triple after how many periods?
[*Answer*: 12 periods.]

2.8 A recent newspaper headline reads, "Los Angeles Woman Wins $1,000,000 in the State Lottery." On a closer reading, we learn that, under the rules of the lottery, the winner receives $100,000 immediately, with nine additional annual payments of $100,000 each. In other words, the winner receives $100,000 at the beginning of each year for 10 years. (Although income taxes are significant, we will ignore these in this problem.) If the winner's "opportunity cost" is 12%, what is the equivalent present value of the $1,000,000 lottery payoff?

2.9 The headline of an article appearing in the May 16, 1981, edition of the *Detroit Free Press* read, "Widow Gets $3.7 Million for Mate's Death on Job." The article described an out-of-court settlement in which the widow and children of a construction worker killed in an accident agreed to the following settlement:
a. Three lump-sum payments to the widow totaling $1,180,000 over the next 30 years
b. Another $1,620,000 payable to the widow in 540 consecutive monthly installments of $3,000 each
c. The sum of $150,000 payable to each of the six children after 20 years

Thus the totals are

Lump sum payments to widow	$1,180,000
Monthly payments to widow	1,620,000
Lump sum payments to children	900,000
	$3,700,000

If the family's "opportunity cost" averages, say, 8% per year, determine the equivalent present value of this future series of payments. In this calculation, assume that the three lump-sum payments to the widow will occur as follows:

Paid immediately	$ 180,000
Paid after 15 years	500,000
Paid after 30 years	500,000
	$1,180,000

[*Answer*: $1,032,000.]

2.10　At his death in 1790, Benjamin Franklin left 1,000 pounds each to the cities of Boston and Philadelphia on the condition that they would not touch the money for 100 years. Boston's bequest, which was equivalent to about $4,600, had ballooned to $332,000 by 1890.

a. Determine the equivalent annual rate of return between 1790 and 1890.

b. If the city of Boston had left this bequest in a fund earning 5% per year between 1890 and 1990, what would have been the value of the fund in 1990?

Evaluating Compound Interest Factors

2.11　The tables in Appendix B are truncated at $N = 100$, or when the single-payment present worth factor is equal to or less than 0.0001. Determine the limits of each of the 11 factors in the tables as the number of periods approaches infinity. That is, find:

a. $(F/P, i\%, \infty)$
b. $(P/F, i\%, \infty)$
c. $(P/\bar{F}, i\%, \infty)$
d. $(F/A, i\%, \infty)$
e. $(F/\bar{A}, i\%, \infty)$
f. $(P/A, i\%, \infty)$
g. $(P/\bar{A}, i\%, \infty)$
h. $(A/F, i\%, \infty)$
i. $(A/P, i\%, \infty)$
j. $(A/G, i\%, \infty)$
k. $(P/G, i\%, \infty)$

[*Answers*: a. ∞; b. 0; c. 0; d. ∞; e. ∞; f. $1/i$; g. $1/\ln(1+i)$; h. 0; i. i; j. $1/i$; k. $1/i^2$.]

2.12　The tables in Appendix B tabulate factors for a variety of typical interest rates. Suitable tables may not always be available, however. Prepare your own tables, if necessary, to compute:

a. $(P/A, 9.5\%, 2)$
b. $(P/\bar{A}, 9.5\%, 2)$
c. $(F/P, 9.5\%, 2)$
d. $(A/P, 9.5\%, 2)$
e. $(A/F, 9.5\%, 2)$

2.13　Assuming an effective interest rate, i, of 11% per period, compute the following to five significant digits:

a. $(F/P, 11\%, 10)$
b. $(P/F, 11\%, 10)$
c. $(P/\bar{F}, 11\%, 10)$
d. $(F/A, 11\%, 10)$
e. $(F/\bar{A}, 11\%, 10)$
f. $(P/A, 11\%, 10)$
g. $(P/\bar{A}, 11\%, 10)$
h. $(A/F, 11\%, 10)$
i. $(A/P, 11\%, 10)$
j. $(A/G, 11\%, 10)$
k. $(P/G, 11\%, 10)$

[*Answers*: a. 2.8394; b. 0.35218; c. 0.37122; d. 16.722; e. 17.626; f. 5.8892; g. 6.2075; h. 0.05980; i. 0.16980; j. 3.6544; k. 21.522.]

2.14　Repeat Problem 2.12 for $i = 13\%$.

2.15　Determine $(P/A, i, g, N)$ to five significant digits when:

a. $i = 10\%, g = 5\%, N = 5$
b. $i = 10\%, g = 10\%, N = 5$
c. $i = 10\%, g = 15\%, N = 5$

[*Answers*: a. 4.1506; b. 4.5455; c. 4.9779.]

2.16　Determine:

a. $(A/P, 0\%, 10)$
b. $(F/A, 0\%, 10)$
c. $(A/P, 0\%, 20)$
d. $(F/A, 0\%, 20)$

Equivalence: Discrete Cash Flows

2.17　What is the equivalent future value at the end of 5 years of $1,000 today, assuming interest compounded annually at the rate of 8% per year?

[*Answer*: $1,469.]

2.18 What sum of money now is equivalent to $1,000 four years hence if interest is compounded annually at 8% per year?

2.19 If $2,500 is invested in an account earning interest at the rate of 12% per year, determine the value of the account at the end of 8 years.
[*Answer*: $6,190.]

2.20 A certain investment is available that promises to return $5,000 five years from now. If the investor's "time value of money" is 8% per year, find the equivalent present value of this proposed investment.

2.21 Exactly $8,000 will be required 15 months from now in order to fund a certain activity. How much must be invested now in an account, earning at the rate of 1.5% per month, to meet this requirement?
[*Answer*: $6,398.81.]

2.22 A sum of $10,000 is deposited into an account earning at the rate of 15% per year. How much will be in the fund at the end of 7 years?

2.23 Find the equivalent present value of $500 six months hence assuming an interest rate of 2% per month.
[*Answer*: $444.]

2.24 Find the equivalent value at the end of 36 months of $1,000 invested now at an interest rate of 1% per month:
   a. Assuming compound interest
   b. Assuming simple interest

2.25 If interest is accumulated at the rate of 7.5% per period, what present sum is equivalent to $800 four periods hence?
[*Answer*: $599.]

2.26 For the following problems, assume an interest rate of 20% per period, with interest compounded at the end of each period.
   a. A sum of $1,000, invested now, will yield what amount at the end of 8 periods?
   b. What is the equivalent present value of a cash flow of $10,000 occurring 20 periods from now?

2.27 a. Find the sum of money now that will be equivalent to $1,000 four years from now if

interest is compounded annually at 8% per year.
   b. Find the sum of money now that will be equivalent to $1,000 four years from now if interest is compounded semiannually at 4% per 6-month period.
[*Answers*: a. $735.03; b. $730.69.]

2.28 Consider the cash flows associated with the following alternative projects:

End of Period	Alternative X	Alternative Y	Alternative Z
1	$1,000	$ 100	$ 200
2	0	0	200
3	0	0	200
4	0	0	200
5	0	900	200
Total	$1,000	$1,000	$1,000

   a. Assuming an interest rate of 5% per period, determine the equivalent future value (at the end of 5 periods) of each of the three projects.
   b. Find the interest rate, if there is one, at which the three projects are equivalent.

2.29 Here is a series of cash flows with an interest rate of 8% per period:

End of Period	Project X	Project Y
1–5	$1,000	$2,000
6–10	2,000	1,000

   a. Find the equivalent present values of the two projects.
   b. Find the equivalent values of the two projects at the end of 10 periods.
   c. Find the equivalent uniform series of the two projects.
[*Answers*: a. $9,428 and $10,703; b. $20,353 and $23,106; c. $1,405 and $1,595.]

2.30 Assuming an effective interest rate of 18% per year, determine the present worth of $1,000 flowing at the end of each month for 12 months.

2.31 Assuming an interest rate of 8% per period, determine the equivalent present value of each of the following three cash flow series:

End of Period	Series I	Series II	Series III
1	0	$100	$500
2	$100	200	400
3	200	300	300
4	300	400	200
5	400	500	100

[*Answers:* I. $737; II. $1,136; III. $1,259.]

2.32 Consider the following cash flow sequences:

End of Period	A	B	C	D	E
0	100	0	0	0	0
1	100	200	0	200	500
2	100	200	0	300	400
3	100	200	0	400	300
4	100	200	0	500	200
5	0	200	0	600	100
6	0	175	100	0	0
7	0	175	200	0	0
8	0	175	300	0	0
9	0	175	400	0	0
10	0	175	500	0	0

Assuming a discount rate of 10% per period, find the equivalent present value of each sequence.

2.33 Suppose the interest rate is 11% per period.
a. A present sum of $10,000 is equivalent to what value after 20 years?
b. $10,000 flowing at the end of each year for 20 years is equivalent to what equivalent present value?

[*Answers:* a. $80,623; b. $79,633.]

2.34 A sum of $1,000 is borrowed at an interest rate of 10% per year.
a. If repayment is made in a single lump sum at the end of 5 years, determine the amount of the repayment.
b. If repayment is made in 5 equal end-of-year payments, determine the amount of each payment.

2.35 Assuming an (effective) interest rate of 10% per annum:
a. How much must be invested today in order to provide an annuity of $20,000 per year for 4 years, with the first payment occurring exactly 10 years from now?
b. How much must be invested today in order to provide an annuity of $10,000 every 6 months for 4 years (8 payments) with the first payment occurring exactly 10 years from now?
c. A sum of $2,000 will be deposited into a savings account at the beginning of each year for 10 years. If the fund accrues interest at the rate of 10% per year, how much will be in the fund after 10 years?

[*Answers:* a. $26,888; b. $26,262; c. $35,061.]

2.36 Determine the following, assuming an interest rate of 10% per year:
a. The initial investment required to yield an income of $5,000 per year 10 years from now and continuing at the end of each year for years 10 through 15
b. The initial investment required to yield an income of $5,000 at the end of each year, in perpetuity ("In perpetuity" means "forever.")
c. The amount that must be deposited at the beginning of each year for 15 years in order to provide a fund of $10,000 at the end of 15 years
d. The equivalent effective interest rate per week

2.37 Assuming $i = 8\%$ per period, determine the equivalent present value of $100 flowing at the end of periods 1 through 10.

[*Answer:* $671.]

2.38 If interest is compounded at the rate of 12% per period, how much must be deposited at the beginning of each period for 20 periods in order to have $100,000 at the end of 20 periods?

2.39 Assuming $i = 15\%$ per period, determine the equivalent present value of $100 flowing at the end of periods 6 through 10.
[*Answer*: **$166.66.**]

2.40 If our minimum attractive rate of return (interest rate) is 15% per year, how much can we afford to invest now in overhauling a certain machine if that overhaul will have the effect of saving $20,000 at the end of each year for 5 years?

2.41 Assuming an interest rate of 12% per annum, determine the amount of an initial investment which would be justified by a savings of $10,000 per year, every year for 7 years.
[*Answer*: **$45,637.**]

2.42 Consider a cash flow of $100 at the start of each month for 20 months. With interest at the rate of 2% per month, find the equivalent value of these cash flows at the end of 24 months.

2.43 Find the present value of a deferred annuity of $500 per year for 12 years, deferred 6 years, if the current rate of interest is 10% annually. Note that the first cash flow occurs at the end of year 6.
[*Answer*: **$2,115.**]

2.44 (Infinite series; cash flows less frequently than compounding periods) Consider a series of cash flows in the amount of $1,000 each, flowing at the end of periods 6, 12, 18, and so forth. With an interest rate of 12% per period, find the equivalent present value of this infinite series.

2.45 With interest rate of 10% per period, determine the equivalent present value of an infinite series of $100 flowing at the beginning of every period.
[*Answer*: **$1,100.**]

2.46 How much money must be deposited today so as to yield an infinite series of cash flows (annuity) of $10,000 each beginning 5 years from now, assuming that funds earn interest at the rate of 10% per year?

2.47 If interest is compounded at the rate of 10% per year, how much must be deposited at the beginning of each year so as to have $10,000 in the fund at the end of 5 years? (These are uniform annual deposits.)
[*Answer*: **$1,489.**]

2.48 If the discount rate is 12% per year, determine the equivalent present value of maintenance costs of $20,000 per occurrence when such costs are incurred at the beginning of the 3rd, 5th, 7th, 9th, and 11th years.

2.49 What sum of money now is equivalent to $1,000 four years hence if interest is compounded semiannually at 4% per 6-month period?
[*Answer*: **$730.69.**]

2.50 Payments on the lease of data processing equipment are $100, payable at the start of each 2-month period.
a. If the effective interest rate is 10% per year, what is the equivalent end-of-year payment that corresponds to 6 bi-monthly payments?

2.51 Find the equivalent present value of $100 flowing at the start of every odd-numbered year, beginning in 1995, if the effective interest rate per year is 10%. Consider "the present" to be the start of 1995.
[*Answer*: **$576.19.**]

2.52 Determine the following, assuming an interest rate of 6% per year:
a. The initial investment required to yield an income of $1,000 per year 5 years from now and continuing at the end of each year for the subsequent 10 years.
b. The initial investment required to yield an income of $1,000 per year, at the end of each year, forever.
c. The amount that must be deposited at the beginning of each year for 20 years in order to provide a fund of $100,000 at the end of 20 years.

2.53 Determine the present worth of the following

cash flow sequence if interest is compounded at 8% per year:

End of Period	Positive Cash Flow (in millions of pesos)
1	300
2	400
3	500
4	600
5	600
6	600
7	600

[*Answer*: 2595 million pesos.]

2.54 Determine the equivalent present value of the following series of cash flows:

End of Period:	10	12	14	16	18	20
Cash Flow ($):	100	100	100	100	100	100

2.55 Consider the following series of cash flows:

End of Period	Cash Flow
1–5	$122,102
Total	$610,510

With $i = 0.08$, find the equivalent present value.
[*Answer*: $487,553.]

2.56 Consider the following series of cash flows:

J	A_j	
1	$ 20,000	
2	71,051	Note that
3	122,102	$G = \$51,051$.
4	173,153	
5	224,204	
Total	$610,510	

With $i = 0.08$, find the equivalent present value.

2.57 Consider the following sequences of cash flows:

End of Period	Sequence A	Sequence B	Sequence C
1	0	$ 300	$900
2	0	400	800
3	0	500	700
4	0	600	600
5	$400	700	500
6	400	800	400
7	400	900	300
8	400	1,000	200

In answering the following questions, use a discount rate of 8% per period.
a. Find the equivalent value of Sequence A at the start of Period 1.
b. Convert Sequence B to an equivalent uniform series over 8 periods.
c. Convert Sequence C to an equivalent uniform series over 8 periods.

[*Answers*: a. $974; b. $610; c. $590.]

2.58 Given the cash flows as indicated below, determine the equivalent present value. Assume an interest rate of 9% per period.

End of Period	Cash Flow
0–5	0
6	$800
7	775
8	750
9	725
10	700

2.59 Property taxes (*ad valorem* taxes) are to be paid at the beginning of each year for 10 years. The firm's discount rate is 12 percent.
a. If these taxes are uniform, $2,000 each year, determine the equivalent present value.
b. If these taxes are estimated to be $2,000 the first year, then decrease by $100 each year, determine the equivalent present value. *Note*: The negative cash flow is $2,000 at the start of year 1, $1,900 at the start of year 2, and so forth.

[*Answers*: a. $12,656; b. $10,388.]

2.60 With interest at 12% per period, find the equivalent present value of the following sequence of cash flows:

End of Period:	0	1	2	3	4	5	6	7
Cash Flow ($):	0	0	0	10	20	30	40	50

2.61 Assuming an effective interest rate of 12% per period, find the equivalent present value of the following series of cash flows:

End of Period:	8	9	10	11	12
Cash Flow ($)	100	120	140	160	180

[*Answer*: $220.94.]

2.62 Consider the following sequence of cash flows:

End of Period	Cash Flow
0–8	0
9	$125
10	115
11	105
12	95
13	85

Assuming an interest rate of 12% per period, find the equivalent present value.

2.63 Consider the following sequence of cash flows:

End of Period	Cash Flow
0–10	0
11	$80
12	75
13	70
14	65
15	60

Assuming an interest rate of 12% per period, find the equivalent present value.
[*Answer*: $82.57.]

2.64 Consider the following series of cash flows:

End of Period:	1	2	3	4	5
Cash Flow ($):	100	90	80	70	60

Assuming an interest rate of 5% per period, determine the value of these cash flows at the end of the fifth period.

2.65 Find the equivalent present value of the following sequence of cash flows using an interest rate of 20% per period:

End of Period:	0	1	2	3	4	5	6	7
Cash Flow ($):		100	90	80	70	60	50	40	30

[*Answer*: $341.9.]

2.66 Given the following cash flows, determine the equivalent value at the end of period 6, assuming that interest is accumulated at the rate of 8% per period.

End of Period:	1	2	3	4	5	6
Cash Flow ($):	100	90	80	70	60	50

2.67 With interest at 12% per period, find the equivalent present value of the following sequence of cash flows:

End of Period:	0	1	2	3	4	5	6	7
Cash Flow ($):	0	100	90	80	70	60	50	40

[*Answer*: $339.93.]

2.68 Using an interest rate of 7%, determine the

present value of the following cash flow sequence:

End of Period	Cash Flow (in thousands of francs)
0	− 100
1–5	+ 20
6	+ 18
7	+ 16
8	+ 14
9	+ 12
10	+ 10

2.69 With interest at 10% per period, convert the following sequence of cash flows to an equivalent uniform series. Use the series factors where appropriate, and write the simplest equation.

End of Period:	0	1	2	3	4	5	6	7	8
Cash Flow ($):		−10	20	18	16	14	12	10	8	20

[*Answer:* **$13.34.**]

2.70 Consider a series of cash flows increasing at the rate of 12% per period over 20 interest periods. The cash flow at the end of the first period is A_1. With interest compounded/discounted at an effective rate of 5% per period, find:
   a. The equivalent present value of these 20 cash flows
   b. The uniform annual series equivalent to this series of cash flows

2.71 A sum of money doubles in value after 60 months, indicating an approximate rate of return of what percent per month?
[*Answer:* **1.16%.**]

2.72 You are offered the opportunity to invest $2,500 with the promise of repayment of a total of $3,000 at the end of 6 months. What rate of return (per month) will you earn on this investment?

2.73 At what effective interest rate (per period) is $1,000 now equivalent to $10,000 twenty periods from now?
[*Answer:* **12.2%.**]

2.74 A section of land was purchased 25 years ago for $10,000 and it was just sold for $50,000. What rate of return (per year) was earned on this investment?

2.75 Wages in the first year of a 5-year project are expected to be $100,000. It is anticipated that wages will increase at the rate of 8% per year. If the appropriate discount rate is 10% per year, determine the equivalent present value of wage payments over 5 years. (The present worth factors for a geometric gradient series are not tabulated in Appendix B. These values can be obtained, however, if g, i, and N are given.)
[*Answer:* **$438,558.**]

2.76 The cost of capital (discount rate) for a certain manufacturer is 15%. Operating costs for a proposed manufacturing process are expected to be $200,000 the first year, then increase at the rate of 5% each year over the subsequent 4 years. That is, there are 5 years in the life cycle of this process.
   a. Determine the operating cost in the fifth year.
   b. Determine the equivalent present value of all operating costs over the 5-year life cycle.

Effective Versus Nominal Interest Rates; Continuous Compounding

2.77 Find the equivalent effective rate per year if the nominal rate is 10% per year and interest is compounded
   a. Weekly
   b. Continuously
[*Answers:* **a. 10.506%; b. 10.517%.**]

2.78 Suppose that $10,000 is placed into an account where the interest rate is 12% per year compounded continuously. (When stated in this way, the rate is assumed to be nominal.) What is the size of equal annual withdrawals over the next 4 years so that the account balance will be zero after the last withdrawal?

2.79 A sum of $100,000 is deposited into a fund that earns interest at the rate of 3% per quarter (3 months). Determine:
   a. The nominal rate per year
   b. The effective rate per year

c. How much must be withdrawn from the fund at the end of each quarter if the fund is to be exhausted (zero balance) after 12 quarters.

d. If $10,000 is withdrawn from the fund at the end of each quarter, how long it will take to exhaust the fund? (Note that the final withdrawal will be something less than $10,000.)

[*Answers*: a. 12.0% b. 12.6%; c. $10,050; d. 13 quarters.]

2.80   How many months will it take to pay off a $525 debt, with monthly payments of $15 at the end of each month, if interest is at the (nominal) annual rate of 18%, compounded monthly?

2.81   If interest is compounded continuously at a nominal rate of 12% per period, find the equivalent present value of $100 flowing 5 periods from now.

[*Answer*: $54.88.]

2.82   Assuming an (effective) interest rate of 12% per annum, determine the equivalent value at the end of 2 years of $1,000 flowing at the end of every month for 24 months. Assume that interest is compounded monthly.

2.83   You are offered the opportunity to invest $100 for 4 years with *simple* interest computed at the rate of 10% per year.

a. How much (principal plus accrued interest) will you receive at the end of 4 years?

b. What is your actual rate of return (per year) on this proposed investment?

c. If you invested the $100 elsewhere at a *nominal* rate of 10% per year compounded *continuously*, how much would you have at the end of 4 years?

d. If you invested the $100 elsewhere at an *effective* rate of 10% per year compounded at the end of each year, how much would you have at the end of 4 years?

[*Answers*: a. $140.00; b. 8.78%; c. $149.18; d. $146.41.]

2.84   What effective annual rate of interest corresponds to an annual nominal rate of 20%
a. Compounded quarterly?
b. Compounded continuously?

2.85   If interest is compounded monthly at the rate of 1.32% per month:

a. Find the nominal rate per year.
b. Find the annual effective rate.
[*Answers*: a. 15.84%; b. 17.04%.]

2.86   If interest is compounded at the (effective) rate of 20% per annum, compounded quarterly, determine:
a. The effective rate per quarter
b. The nominal rate per annum

2.87   If an interest rate of 2% per week is compounded weekly, determine:
a. The corresponding nominal rate per year
b. The corresponding effective rate per year
[*Answers*: a. 10.4%; b. 10.9%.]

2.88   Given that interest is compounded quarterly at an annual nominal rate of 12.25%, what is the annual effective rate?

2.89   A sum of $10,000 is invested in a fund earning *simple interest* at the rate of 1% per month.
a. Determine the amount in the fund at the end of 24 months.
b. Based on your answer to a. determine the rate of return (compound interest rate) per month.
c. Based on your answer to (b.) determine the nominal annual rate.
d. Based on your answer to (b.) determine the effective annual rate.
[*Answers*: a. $12,400; b. 0.90% c. 10.80%; d. 11.36%.]

2.90   If interest is compounded quarterly at the rate of 1.23% per quarter (3 months):
a. Find the nominal rate per year.
b. Find the annual effective rate.

2.91   Consider a series of cash flows in the amount of $100 per month, flowing at the end of each month for 18 months. If interest is compounded monthly at an effective interest rate of 12% per year, find the equivalent value at the end of 18 months.
[*Answer*: $1,953.]

2.92   A sum of $100 is invested at the end of every quarter (3-month period) for exactly 20 quarters. Invested funds accumulate interest at the rate of 1% per month.
a. Find the equivalent effective rate per quarter.
b. Find the equivalent nominal rate per year.
c. Find the equivalent effective rate per year.

d. What effective rate per year corresponds to a rate of 1% per month if interest is compounded continuously?

2.93 Find the effective interest rate per year that corresponds to a nominal rate of 12% per year:
a. Compounded semiannually
b. Compounded quarterly
c. Compounded monthly
d. Compounded continuously
[*Answers*: a. **12.36%**; b. **12.55%**; c. **12.68%**; d. **12.75%.**]

2.94 Find the effective interest rate per year that corresponds to a nominal rate of 10% per year:
a. Compounded semiannually
b. Compounded continuously

2.95 You are able to secure a $1,000 loan from the Happy Home Finance Company at an interest rate of 2% quarterly. Interest is compounded every 3 months. Find the nominal and effective annual rates.
[*Answers*: **8.00% and 8.24%**]

Continuous Compounding: Continuous Cash Flows

2.96 A firm's discount rate is 22% per year.
a. Find the equivalent effective rate per month.
b. Find the equivalent present value of $100 flowing during every year forever. These are continuous cash flows of $100 per year.
c. Find the equivalent present value of $100 flowing during every fifth year, beginning in year 5 and including the 50th year. The series is:

During Year (j):	5	10	15	...	50
Cash Flow ($\bar{A}_j$):	$100	$100	$100	...	$100

2.97 Maintenance expenses of M are expected to flow continuously and uniformly during every fourth year, beginning with year 4 and continuing indefinitely. Assuming a nominal interest rate, r, of 10% per year, find the equivalent present value of this infinite series.
[*Answer*: **2.138M$.**]

2.98 Consider a cash flow of $10,000 occurring continuously and uniformly during the fifth year of a 15-year planning horizon. Find the equivalent present value of this cash flow if the firm's nominal interest rate is 12% per year.

2.99 A sum of $1,000 flows continuously and uniformly during the 10th year of a 20-year planning horizon. Find the equivalent present value of this cash flow if the discount rate is 8% per year.
[*Answer*: **$481.48.**]

2.100 Determine:
a. $(P/F, 5\%, 5)$ e. $(P/F, 10\%, 5)$
b. $(P/\bar{F}, 5\%, 5)$ f. $(P/\bar{F}, 10\%, 5)$
c. $(P/F, 5\%, 10)$ g. $(P/F, 10\%, 10)$
d. $(P/\bar{F}, 5\%, 10)$ h. $(P/\bar{F}, 10\%, 10)$
By analyzing the differences among these values, what can be said about the importance of the choice of the timing assumption (end-of-period versus continuous) as the interest rate increases? As the number of periods increases?

2.101 Net receipts from a continuously producing oil well add up to $120,000 over one year. What is the equivalent present value of the well if it maintains steady output until it runs dry after 8 years, if $r = 10\%$?
[*Answer*: **$660,800.**]

2.102 It is expected that certain manufacturing equipment will require a major overhaul during the 18th month of its service life. The cost of this overhaul is expected to be $20,000. With a discount rate of 2% per month, and assuming that the $20,000 cost will be distributed continuously and uniformly over the 18th month, determine the equivalent present value of this cost.

2.103 For how many periods must an investment of $50,000 provide a continuous cash flow at the rate of $10,000 per period so that a (nominal) periodic interest rate of 12% compounded continuously is earned?
[*Answer*: **About 7.6 periods.**]

2.104 A certain project is expected to result in savings of $20,000 each year over a 5-year period. The firm's discount rate is 15% per year. Determine the equivalent present value of these cash flows (nearest dollar) if it is assumed that cash flows occur:

a. At the end of each year
b. At the midpoint of each year
c. Continuously and uniformly throughout each year

Loans

2.105 A loan of $5,000 requires monthly payments of $200 over a 3-year period. These payments include both principal and interest.

a. Determine the nominal rate of interest per year.
b. Determine the effective rate of interest per year.
c. Find the amount of the unpaid principal on the loan after 12 payments.
d. Find the total interest paid during the first 12 payments.

[*Answers*: a. 25.45%; b. 28.64%; c. $3,731; d. $1.111.]

2.106 In order to support his junior year in college, Ivan O. Universal obtains a $5,000 one-time loan from a federal agency. This is a 7% loan, which must be repaid in not more than 10 years from the date of graduation. I.O.U. intends to repay the loan in eight uniform annual payments, with the first payment due 3 years from the date of the loan.

a. Determine the amount of each uniform payment. Assume that the 7% cost of the loan begins at the time of graduation, which is 2 years hence, so Ivan will owe $5,000 in 2 years. If no payment is made, he will owe $5,000(1.07) = $5,350 in 3 years, and so on.
b. Determine the true cost of the loan, stated as a percentage, over the 10-year planning horizon. Note that the final (eighth) payment will be made 10 years after the date of the original loan. (*Hint*: The rate of return to the lender is the cost to the borrower.)

2.107 In order to finance the purchase of computer equipment, the S. C. Equipment Corporation borrows $100,000, to be repaid in 40 equal end-of-month payments at an interest rate of 1.5% per month.

a. Determine the amount of the monthly payments.

b. Determine the amount remaining on the loan after 2 years.
c. Determine the interest portion of the first payment.
d. Determine the principal portion of the 25th payment.

[*Answers*: a. $3,343; b. $47,241; c. $1,500; d. $2,634.]

2.108 A firm borrows $75,000 to be repaid in monthly installments of $720 at the end of every month for 30 years.

a. Determine the nominal interest rate per year. (*Hint*: Because there are $12 \times 30 = 360$ monthly payments, this uniform series may be approximated by an infinite series. That is, 360 is sufficiently large to use the perpetual payment models.)
b. Determine the effective interest rate per year.
c. At the end of the 5 years (60 payments), the firm wishes to make a lump-sum payment to settle its remaining debt. What should the amount of this payment be?

2.109 In order to meet tuition and related academic expenses, you are offered a $5,000 loan. The loan, plus $500 interest, is to be repaid at the end of 6 months. If you accept this loan:

a. Determine the equivalent nominal interest rate per annum.
b. Determine the equivalent effective interest rate per annum.

As an alternative to the above, another lender offers to lend you $5,000 with the loan, plus interest, to be repaid at the end of 1 year with interest compounded continuously at a nominal rate of 18% per annum. If you should accept this loan:

c. Determine the equivalent effective rate per annum.
d. Determine the total amount of the repayment after 1 year.
e. Determine the payment that would be owed after 6 months, even though repayment is not made at that time.

[*Answers*: a. 20.0%; b. 21.0%; c. 19.72%; d. $5,986; e. $5,471.]

2.110 Historically, 36-month car loans were "standard." In more recent years, payment periods of 48 months were not uncommon. Today, buyers of luxury cars can get car loans up to 84 months from certain lenders. Of course, the longer payment period has the effect of reducing monthly payments.

Consider a $20,000 car loan "at 12%." (When stated this way, the 12% is an annual nominal rate.) Assume end-of-month payments.

a. Determine the monthly payment for a 36-month loan.
b. Determine the monthly payment for a 72-month loan.
c. Assuming a 72-month loan, determine the amount of the original $20,000 principal remaining (unpaid) after 36 payments.

2.111 Assuming an effective interest rate of 10% per year, a "rule of thumb" for estimating the accumulated interest on any principal for any number of days is given by

$$\text{Interest} = \left(\frac{\text{principal}}{100}\right) \times \left(\frac{\text{no. of days}}{36}\right)$$

This is based on a 360-day year; daily compounding of interest is assumed.

a. Determine the estimated interest if the principal is $1,000 and interest is accumulated over 180 days.
b. Determine the effective interest rate per day corresponding to an effective rate of 10% per year. Assume a 360-day year. Show your answer as to the closest 0.01%.
c. If $1,000 is borrowed at an effective rate of 10% per year, interest compounded daily, how much interest will have accumulated after 180 days? Show your answer to the closest penny.

[*Answers*: a. $50.00; b. 0.026%; c. $48.81.]

2.112 Assuming an effective interest rate of 18% per year, a "rule of thumb" for estimating the accumulated interest on any principal for any number of days is given by

$$\text{Interest} = \left(\frac{\text{principal}}{100}\right) \times \left(\frac{\text{no. of days}}{20}\right)$$

This is based on a 360-day year; daily compounding of interest is assumed.

a. Determine the estimated interest if the principal is $1,000 and interest is accumulated over 180 days.
b. Determine the effective interest rate per day

corresponding to an effective rate of 18% per year. Assume a 360-day year. Show your answer to the closest 0.01%.
c. If $1,000 is borrowed at an effective rate of 18% per year, interest compounded daily, how much interest will have accumulated after 180 days?

2.113 In order to raise funds for plant expansion, a certain manufacturing firm is planning to issue $1,000 (face value) bonds, payable in 10 years, with bond interest payable semiannually at a (nominal annual) rate of 8%. Determine the amount that a "typical" purchaser would pay for this bond if the purchaser requires a 12% (effective annual) rate of return. (*Hint*: You will first have to determine the appropriate discount rate per 6-month period.)

[*Answer*: $787.16.]

2.114 Mr. Jones owes $24,730 on his home mortgage after his most recent installment payment. (Installments are due on the 15th day of each month.) Mr. Jones is currently paying 6% per annum, effective, with monthly payments of $246.00. After his most recent payment, there are exactly 141 payments remaining.

a. At an effective interest rate of 1% per month, determine the equivalent present value of 141 payments of $246 each.
b. If Mr. Jones will pay off his loan in full now, the lender will give him a *discount* equal to 15% of the remaining principal of the loan: $0.15 \times \$24,730 = \$3,710$. If Mr. Jones's opportunity cost is 1% per month, is this offer attractive? Why?

2.115 A sum of $1,000 is borrowed at an effective interest rate of 0.5% per week. At the end of 6 months (26 weeks), exactly $1,000 is repaid. There are no other intervening payments. How much debt remains at the end of 6 months? That is, how much does the borrower still owe on the debt?

[*Answer*: $138.46.]

2.116 a. Alice agrees to lend Bob $2,000 with full payment due at the end of 4 years, and with *simple interest* computed at the rate of 12% per year. What is the amount of the repayment (principal plus accrued interest)?

b. What is the "true" (effective) rate of return to Alice? Note that this is Bob's cost of the loan.

2.117 A local bank is advertising a 7% annual percentage rate, or APR, with interest compounded daily.
a. Determine the annual yield (effective annual rate) assuming a 360-day year.
b. Determine the approximate yield assuming that interest is compounded continuously.

[*Answers*: a. 7.25008%; b. 7.25082%.]

2.118 A loan of $10,000 is to be repaid at the end of 18 months.
a. If interest is compounded at the rate of 1% per month, what is the amount of the repayment?
b. If the accrued interest is to be determined on the basis of simple interest at the rate of 1%

per month, what is the amount of the repayment?
c. If the total amount of the repayment is $12,000, what is the effective interest rate per month paid by the borrower?

2.119 In order to finance the acquisition of certain manufacturing equipment, a firm is considering obtaining a $100,000 loan with interest computed at the rate of 4% per quarter. The loan will be repaid in 12 quarterly payments.
a. Determine the amount of the quarterly payments.
b. The firm intends to use this equipment for only 24 months (8 quarters). If the remainder of the debt will be paid at that time, how much debt will remain after the eighth quarterly payment has been made? That is, how much is owed at the beginning of the ninth quarter?

[*Answers*: a. $10,660; b. $38,677.]

APPENDIX: *Using Spreadsheets for Economic Analyses*

Many of the problems arising in economic analyses may be structured in tabular representations of data, or **spreadsheets**. As these types of problems occur in a variety of contexts, computer software has been developed to facilitate the algebraic manipulation of data. Numerous software packages are readily available, among the most popular of which are *SuperCalc*, *Lotus 1-2-3, Excel*, and *Quattro Pro*.[6]

This software is generic in the sense that it may be used to describe any tabular array of data. (Cells of the matrix may include a word or label, a number, or the result of a mathematical calculation or formula.) Special-purpose software—that is, software designed specifically to address certain problems of economic analysis—are discussed in the Appendix to Chapter 3.

The examples included in this appendix are intended to illustrate the use of spreadsheets to solve some of the problems discussed in the body of the chapter.

---

[6]*SuperCalc* is a registered trademark of Computer Associates International, Inc., *Lotus 1-2-3* is a registered trademark of the Lotus Development Corporation, *Excel* is a registered trademark of Microsoft, and *Quattro Pro* is a registered trademark of Borland International.

EXAMPLE 2.A.1 *Future Value of a Present Cash Flow*

Figure 2.A.1 is the printout of a spreadsheet used to determine the future value of a \$1,000 loan with interest rate of 10% per period and end-of-period compounding. (This is the example first presented in Section 2.3.1.) In keeping with the usual scheme for labeling rows and columns, rows are numbered $(1, 2, \ldots)$ and columns are lettered $(A, B, \ldots)$ starting in the upper left corner of the table. The key cell formulas in this particular example are:

$$B15 = \$F\$6$$

$$C15 = +B15*F\$8$$

$$D15 = +B15 + C15$$

$$B16 = +D15$$

EXAMPLE 2.A.2 *Present Value of an Irregular Series of Cash Flows*

Figure 2.A.2 is a printout of the present value of an irregular series of cash flows. The equivalent present value figures in column D are found from

$$A_j(1 + i)^{-j}$$

where $i = 0.10$.

Key cell formulas are

$$C15 = (1 + \$E\$7)^{\wedge}(-A15)$$

$$D15 = +B15*C15$$

$$E15 = +D15$$

$$E16 = +D16 + E15$$

EXAMPLE 2.A.3 *Discount Factors for the Present Value of the Geometric Gradient Series*

The geometric gradient series is discussed in Section 2.6.2. Recall Eq. (2.16):

$$P = A_1 \left[\frac{1 - (1 + g)^N (1 + i)^{-N}}{i - g} \right] \qquad i \neq g$$

$$= A_1 [N(1 + i)^{-1}] \qquad i = g$$

	A	B	C	D	E	F
1						
2	**EXAMPLE 2.A.1**					
3						
4	**COMPOUND INTEREST: FUTURE VALUE OF PRESENT CASH FLOW**					
5						
6	Given:	Present (initial) cash flow (P)				$1,000
7		Interest compounded at end of each period				
8		Compound interest rate (i)				0.10
9		Total number of periods (N)				5
10						
11		Amt. owed	Interest	Amt. owed		
12		at start	during	at end		
13	Period	of period	period	of period		
14	----------------	---------------------	----------------	---------------------		
15	1	$1,000.00	$100.00	$1,100.00		
16	2	1100.00	110.00	1210.00		
17	3	1210.00	121.00	1331.00		
18	4	1331.00	133.10	1464.10		
19	5	1464.10	146.41	1610.51		
20	----------------	---------------------	----------------	---------------------		
21						
22	Sample Cell Formulas:					
23	B15 = $F$6					
24	C15 = +B15*$F$8					
25	D15 = +B15+C15					
26	B16 = +D15					

Figure 2.A.1   Future Value of a Present Cash Flow

	A	B	C	D	E	F
1						
2	**EXAMPLE 2.A.2**					
3						
4	**PRESENT VALUE OF A SERIES OF CASH FLOWS**					
5						
6	Given:	Discounting at end of each period				
7		Discount rate (i)			0.10	
8		Total number of periods (N)			5	
9		End of period cash flows as shown (Aj)				
10						
11	End of	Cash	Discount	Equivalent	Cumulative	
12	Period	flows	factor	present	present	
13	j	Aj	$(1+i)^{\wedge}(-j)$	value	value	
14	----------------	---------------------	----------------	---------------------	-------------------	
15	1	-100.00	0.9091	-90.91	-90.91	
16	2	-20.00	0.8264	-16.53	-107.44	
17	3	20.00	0.7513	15.03	-92.41	
18	4	100.00	0.6830	68.30	-24.11	
19	5	50.00	0.6209	31.05	6.94	
20	----------------	---------------------	----------------	---------------------	-------------------	
21	Totals:	50.00		6.94		
22						
23						
24	Sample Cell Formulas:					
25	C15 = (1+$E$7)^(-A15)					
26	D15 = +B15*C15					
27	E15 = +D15					
28	E16 = +D16+E15					

Figure 2.A.2   Present Value of an Irregular Series of Cash Flows

The factor in brackets is evaluated using the spreadsheet in **Figure 2.A.3** for $i = 0.15$; $0.05 \leq g \leq 0.25$, and $1 \leq N \leq 10$. Key cell formulas are

$$B15 = ((1 - ((1 + B\$13)/(1 + \$E\$7))^{\wedge} \$A15)/(\$E\$7 - B\$13))$$

$$D15 = \$A15/(1 + \$E\$7)$$

	A	B	C	D	E	F
1						
2	EXAMPLE 2.A.3					
3						
4	DISCOUNT FACTORS TO FIND PRESENT VALUE OF					
5	GEOMETRIC GRADIENT SERIES (P/A1,i,g,N)					
6						
7	Given:	Discount Rate (i)			0.15	
8		Growth Rate (g)			various	
9		End of Period (N)			1-10	
10						
11						
12	Periods	g=	g=	g=	g=	g=
13	1-N	0.05	0.10	0.15	0.20	0.25
14	-------------	----------------	---------------	---------------	---------------	-------------
15	1	0.87	0.87	0.87	0.87	0.87
16	2	1.66	1.70	1.74	1.78	1.81
17	3	2.39	2.50	2.61	2.72	2.84
18	4	3.05	3.26	3.48	3.71	3.96
19	5	3.65	3.99	4.35	4.74	5.17
20	6	4.21	4.68	5.22	5.82	6.49
21	7	4.71	5.35	6.09	6.94	7.93
22	8	5.17	5.99	6.96	8.11	9.48
23	9	5.59	6.59	7.83	9.33	11.18
24	10	5.97	7.18	8.70	10.61	13.02
25	-------------	----------------	---------------	---------------	---------------	-------------
26						
27	Sample Cell Formulas:					
28	B15 = ((1-((1+B\$13)/(1+\$E\$7))^\$A15)/(\$E\$7-B\$13))					
29	D15 = \$A15/(1+\$E\$7)					

Figure 2.A.3 Discount Factors to Find the Present Value of Geometric Gradient Series

EXAMPLE 2.A.4 *Present Value of a Geometric Series*

Figure 2.A.4 shows a spreadsheet used to determine the equivalent present value of a geometric series of cash flows: $A_1 = \$100,000$, $g = 0.05$, $i = 0.10$, and $N = 7$. (This is Example 2.15 in the chapter.) Key cell formulas are

$$B15 = \$E\$6$$

$$B16 = +B15*(1 + \$E\$7)$$

$$C15 = +B15*((1/(1 + \$E\$8))^{\wedge}A15)$$

	A	B	C	D	E
1					
2	EXAMPLE 2.A.4				
3					
4	PRESENT VALUE OF A GEOMETRIC SERIES				
5					
6	Given:	Cash flow at end of 1st. period (A1)			$100,000
7		Growth rate (g)			0.05
8		Discount rate (i)			0.10
9		Total number of periods (N)			7
10					
11	End of	Cash	Equivalent		
12	period	flow	Present		
13	j	Aj	value		
14	-----------------	---------------------	----------------------		
15	1	$100,000	$90,909		
16	2	$105,000	$86,777		
17	3	$110,250	$82,832		
18	4	$115,763	$79,067		
19	5	$121,551	$75,473		
20	6	$127,628	$72,043		
21	7	$134,010	$68,768		
22	-----------------	---------------------	----------------------		
23	TOTAL:	$814,201	$555,870		
24					
25	Sample cell formulas:				
26	B15 = $E$6				
27	B16 = +B15*(1+$E$7)				
28	C15 = +B15*((1/(1+$E$8))^A15)				

Figure 2.A.4   Present Value of a Geometric Series

EXAMPLE   2.A.5   *Present Value of a Uniform Series of Subperiod Cash Flows*

Figure 2.A.5 is a spreadsheet for Example 2.17. Recall that labor costs are $8,000 per week, every week for 156 weeks (3 years). At a discount rate of 15% per year (0.269% per week), the present value is shown to be $1,018,030. The key cell formulas are:

$$D22 = ((1 + \$D\$8)^{\wedge}(1/\$D\$7)) - 1$$

$$B15 = D\$9*D\$8/D\$22$$

$$C15 = B15*(1/(1 + D\$8))^{\wedge}\$A15$$

	A	B	C	D	E
1					
2	EXAMPLE 2.A.5				
3					
4	PRESENT VALUE OF SUBPERIOD CASH FLOWS (UNIFORM SERIES)				
5					
6	Given:	Number of periods (N)		3	years
7		Number of subperiods per period (M)		52	weeks
8		Interest rate per period (i)		0.15	per year
9		Cash flow at end of each subperiod		$8,000	per week
10					
11					
12		Equivalent value	Present		
13	Year	at end of year	Value		
14	----------	----------------------------	----------------------------		
15	1	445,874	387,716		
16	2	445,874	337,145		
17	3	445,874	293,169		
18	----------	----------------------------	----------------------------		
19	TOTAL:	1,337,621	1,018,030		
20					
21					
22	Effective interest rate per subperiod (is) =			0.269%	per week
23					
24	Sample cell formulas:				
25	D22 = ((1+$D$8)^(1/$D$7))-1				
26	B15 = D$9*D$8/D$22				
27	C15 = B15*(1/(1+D$8))^$A15				

Figure 2.A.5 Present Value of a Uniform Series of Subperiod Cash Flows

EXAMPLE 2.A.6 *Uniform Period Repayment*
of a Loan

Figure 2.A.6 is a spreadsheet for the repayment of a $1,000 loan with 5 equal end-of-period payments. The effective interest rate is 10% per period. (This is Example 2.28 in the chapter.) In this tabular representation,

$$P_j = \text{principal remaining at the start of period } j$$

where

$$P_1 = \$1,000$$

and

$$P_j = P_{j-1} - (A - I_j), \qquad j = 2, 3, 4, 5$$

$$A = \text{uniform periodic payment}$$

$$= P_1(A/P, i, N)$$

$$= \$1,000*(i*((1 + i)^\wedge N))/(((1 + i)^\wedge N) - 1)$$

$$I_j = iP_j$$

$$= 0.10P_j \qquad j = 1, 2, \ldots, 5$$

	A	B	C	D	E	F
1	EXAMPLE 2.A.6					
2						
3	UNIFORM SERIES REPAYMENT OF A LOAN					
4						
5	Given:	Initial loan (P)			$1,000	
6		Interest rate (i)			0.1	per period
7		Number of Payments (N)			5	periods
8						
9	End of	Amount of	Interest	Principal	Principal Remaining	
10	Period	Payment	Portion	Portion	After Payment	
11	j	A	Ij	Aj	P(j + 1)	
12	----------------	------------------	----------------	----------------	-----------------------------------	
13	1	$263.80	$100.00	$163.80	$836.20	
14	2	$263.80	$83.62	$180.18	$656.03	
15	3	$263.80	$65.60	$198.19	$457.83	
16	4	$263.80	$45.78	$218.01	$239.82	
17	5	$263.80	$23.98	$239.82	$0.00	
18	----------------	------------------	----------------	----------------	-----------------------------------	
19	Sample cell formulas:					
20	B13 =	@PMT($E$5,$E$6,$E$7)				
21	C13 =	+$E$6*$E$5				
22	D13 =	+B13 - C13				
23	E13 =	+$E$5 - D13				
24	C14 =	+E13 * $E$6				
25	E14 =	+E13 - D14				

Figure 2.A.6   Uniform Series Repayment of a Loan

Of course, this spreadsheet gives the same results as those shown in Table 2.4.

In Figure 2.A.6, the loan payment amount, A, results from one of the special functions embedded in Release 2.2 of *Lotus 1-2-3*, namely

$$@\text{PMT} \ (\textit{principal, interest rate, term})$$

Here, the "principal" is $1,000, the "interest rate" is 0.10, and the "term" is 5 periods. When using other software for which a similar special function is not available, the loan payment amount may be determined using

$$A = P1*(i*((1 + i) \wedge N))/(((1 + i) \wedge N) - 1)$$

Measures of Worth I:
Selection or Ranking
of Alternatives
by Increasing/Decreasing
Equivalent Worth Measures

INTRODUCTION

The concept of equivalence was stressed throughout Chapter 2: A sum of money at one point in time can be equivalent to another amount at some other point in time, given (1) the number of intervening compounding interest periods and (2) an interest rate that represents alternative investment opportunities during that time interval. The interest rate is a measure of the minimum return that could be received if the available funds were invested elsewhere. This is the **minimum attractive rate of return** (MARR).

It is generally incorrect to compare two or more investment alternatives simply by summing their respective cash flows. The *timing* as well as the *amounts* of project cash flows must be considered. To illustrate this point, consider the cash flows associated with two alternative investment opportunities:

	Plan I	Plan II
Initial investment	−$100	−$100
Receipts at end of 1 year	50	100
Receipts at end of 2 years	100	50
Excess of receipts over disbursements	$ 50	$ 50

In this simple example it is clear that the decision maker would not be indifferent about the choice between the two alternatives merely because the cash flow totals are equal. Indeed, assuming no other significant differences between alternatives, plan II is preferable to plan I. Of course, real-world problems are substantially more complicated, and the relative desirability of alternative investments is not, in most cases, evident by inspection. The computation of equivalent values, using the mathematical techniques discussed in Chapter 2, must play a central role in any economic analysis. This chapter discusses three alternative methods for systematically comparing costs and benefits: the *present worth method*, the *annual worth method*, and the *future worth method*. For each of these methods, comparison of the measures of worth for the competing alternatives leads directly to their relative desirability based on an economic criterion.

Two additional methods are discussed in Chapter 4: the *rate of return method* and the *benefit–cost (ratio) method*. Proper allocation of these two methods is somewhat more complex than those of Chapter 3 because the measures of worth, rates of return and benefit–cost ratio, should not be compared directly when determining relative desirability. An *incremental* analysis is required.

All of these methods, properly used, lead to consistent results. That is, the rank ordering of competing investment alternatives is perfectly consistent, regardless of which method the analyst uses. The primary reason that all five are presented in this book is simply that they are widely used, in a variety of applications, by economic analysts. (The first four are most common to the private sector; the benefit–cost ratio method, because of its historical antecedents, is most widely used in the public sector, especially by government agencies.) All of these methods are commonly referenced in the professional literature, and no one method is used predominantly. Thus, it is useful for the analyst to understand thoroughly the proper application of all five procedures.

3.2   MUTUALLY EXCLUSIVE ALTERNATIVES

3.2.1   Classes of Investment Proposals

There are various ways to classify the types of investment proposals for which economic analysis may be appropriate. As discussed in Chapter 1, the methodology of engineering economy is applicable to any plans, projects, programs, and the like for which the amounts and timing of cash flows differ over a specified planning horizon and, further, these differences are expected to be significant. The specific type of investment (e.g., operating revenues and expenses, capital plant and equipment) is unimportant. What *is* relevant, however, is the logical relationships between alternatives. There are four such relation relationships, as outlined next.

Independent alternatives are those for which one or more of the set may be selected. For example, suppose that we are considering upgrading our computer workstation by acquiring a new (A) monitor, (B) printer, or (C) external disk drive. These options are independent in the sense that we can acquire one, two, or all three of these devices, finances permitting—or choose none at all (the "do nothing" alternative).

Mutually exclusive alternatives are those for which the acceptance of one precludes the acceptance of all others in the set. For example, suppose that we are considering the purchase of a monitor for our computer workstation: (A1) monochrome, (A2) VGA, or (A3) SVGA monitor. There is one, and only one, possible device. (The "do nothing" alternative is not a feasible option in this case because a monitor is a necessary component of the workstation.)

Financially mutually exclusive alternatives are really a subset of the above, but this condition results from limited available funds. Returning to our example, suppose that the costs of the three independent alternatives are as follows

A. Monitor (SVGA)	$ 350
B. Printer	650
C. Disk drive	200
Total	$1,200

If our budget is limited to $1,000, these are not truly independent alternatives. We can afford at most only two of the three devices.

Contingent alternatives occur when the selection of one depends on the selection of one or more others. For example, suppose that we are also considering the purchase of certain CAD software (D), but we will do so only if we also purchase an SVGA monitor (A3). Thus D is *contingent* on A3.

3.2.2 Establishing a Set of Mutually Exclusive Alternatives

The logical relationships among alternatives notwithstanding, it is always possible to redefine the alternatives in such a way as to identify an exhaustive, mutually exclusive set. To illustrate, consider the three independent alternatives, A, B, and C, along with the "do nothing" alternative, which we will label $\emptyset$. These four independent alternatives can be combined to yield eight mutually exclusive alternatives:

1. $\emptyset$ (do nothing)	5. A and B
2. A alone	6. A and C
3. B alone	7. B and C
4. C alone	8. A and B and C

In general, a set of x independent alternatives will yield 2^{x-1} mutually exclusive alternatives. Here, $x = 4$ and there are $2^3 = 8$ mutually exclusive alternatives.

This issue of mutual exclusivity is of great importance, as will be demonstrated. The evaluation methods discussed in Chapters 3 and 4 are designed to answer one or both of two questions: Is a proposed investment preferable to the "do nothing" alternative? And given two or more alternatives to the "do nothing" alternative, which one, or which combination, appears to be economically preferred? If the analyst can formulate a set of mutually exclusive alternatives, any of the evaluation methods discussed in this and the next chapter can be used to identify the preferred alternative(s).

3.3   PRESENT WORTH (PRESENT VALUE)

3.3.1   Present Worth Method

The **present worth method** is probably the most widely used economic evaluation technique. As illustrated in **Figure 3.1**, the essential feature of the present worth method is the discounting to present value of all cash flows expected to result from an investment decision. That is, in order to satisfy the basic requirement that alternatives be compared only if money consequences are measured at a common point in time, the "present date" is arbitrarily selected as the point of reference. (In practice, the "present date" is determined relative to the particular problem at hand. It is generally defined as the start of the planning horizon, the project's life cycle.) The net discounted value of all prospective cash flows is a direct

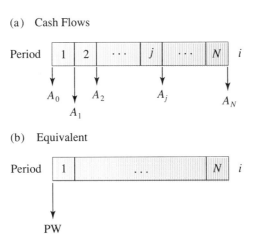

Figure 3.1   Present Worth—Discrete Cash Flows

measure of the relative economic attractiveness of the proposed investment.

There are two steps. First, the equivalent present worth, PW, of all expected cash flows is computed and aggregated for each alternative. For example, if all cash flows, A_j, are discrete, as shown in Figure 3.1, and the discounting takes place at effective rate i per period, then

$$PW = \sum_{j=0}^{N} A_j(1 + i)^{-j} \qquad (3.1)$$

If all cash flows are continuous, as indicated in **Figure 3.2**, and if discounting takes place continuously at rate i, then

$$PW = \sum_{j=1}^{N} \bar{A}_j \left[\frac{i(1 + i)^{-j}}{\ln(1 + i)} \right] \qquad (3.2)$$

where $\bar{A}_j$ is the cash flowing continuously and uniformly during period j. Various other models are used, depending on the assumptions related to the timing of cash flows and the type of discounting.

(a) Cash Flows

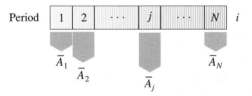

(b) Equivalent

Figure 3.2 Present Worth—Continuous Cash Flows

The second step in the present worth method is to rank-order alternatives on the basis of decreasing PWs. That is, assuming that there are no other differences between alternatives, the objective is *to maximize present worth*.

In the literature of finance, present worth is more generally known as **discounted cash flow** (DCF) or **net present value** (NPV). Indeed, most commercially available computer software uses NPV rather than PW because of the larger market in business applications. The terms will be used interchangeably throughout the remainder of this book.

3.3.2 Some Numerical Examples

EXAMPLE 3.1

The firm is considering the purchase of certain equipment that will require an initial investment of $12,000. It is expected to result in annual expenses of $6,000 occurring at the end of each year over 6 years and a residual (salvage) value of $3,000 at the end of 6 years. (See **Figure 3.3.**) In this example,

$$A_0 = -\$12,000$$

$$A_1 = A_2 = \cdots = A_5 = -\$6,000$$

$$A_6 = -\$6,000 + \$3,000$$

The discount rate is 12% per year.

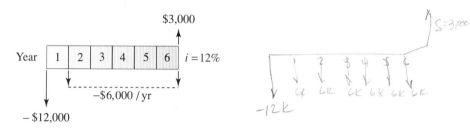

Figure 3.3 Example 3.1

Note that the six $6,000 annual expenses form a uniform series. Since we have access to a set of present worth factors for the uniform series, the computation can be simplified. Using the functional format, the solution equation can be written

$$PW = -\$12,000 - \$6,000(P/A, 12\%, 6) + \$3,000(P/F, 12\%, 6)$$

$$= -\$12,000 - \$6,000(4.111) + \$3,000(0.5066)$$

$$= -\$35,146$$

The negative net present worth indicates that the cost of this project, stated in equivalent current dollars, is $35,146.

At *zero interest rate*, that is, if money has *no* value over time, the present worth is

$$-\$12,000 - \$6,000(6) + \$3,000 = -\$45,000$$

This is a somewhat larger negative value because the present worth of future consequences is inversely related to the value of the discount rate. The net difference is $-\$9,854$.

EXAMPLE 3.2

This is the same as Example 3.1 with one important change: Assume that the annual expenses, $6,000 per year, flow continuously and uniformly throughout each year. (See **Figure 3.4.**) Then

$$\text{PW} = -\$12,000 - \$6,000(P/\bar{A}, 12\%, 6) + \$3,000(P/F, 12\%, 6)$$

$$= -\$12,000 - \$6,000(4.353) + \$3,000(0.5066)$$

$$= -\$36,598$$

The result is somewhat more negative than in Example 3.1 because, under the continuous cash flow assumption, the annual expenses occur earlier each year and thus are given greater weight when discounted.

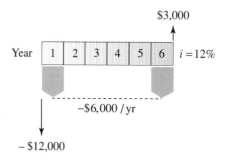

Figure 3.4   Example 3.2

3.3.3 Present Worth of the "Do Nothing" Alternative

If funds are not invested in a particular proposed project—if the investor chooses to "do nothing"—it is reasonable to assume that the funds will be invested elsewhere where they will yield a return of i per period. (Recall that i is a measure of the minimum return that could be received if the available funds were invested elsewhere. This is the "opportunity cost," stated as a percentage or decimal fraction.)

The net return of funds invested elsewhere, when discounted, is exactly zero. This may be easily demonstrated by observing that an amount P invested at rate i will be worth $P(1 + i)$ at the end of one period, $P(1 + i)^2$ at the end of two periods,..., and $P(1 + i)^N$ at the end of N periods. (See **Figure 3.5.**) Thus $A_0 = -P$ and $A_N = P(1 + i)^N$. From Eq. (3.1),

$$\text{PW} = A_0 + [A_N](1 + i)^{-N}$$

$$= -P + [P(1 + i)^N](1 + i)^{-N}$$

$$= 0$$

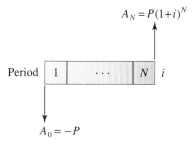

$$A_N = P(1+i)^N$$

$$A_0 = -P$$

Figure 3.5   The "Do Nothing" Alternative: Invest P Now and Recover A_N After N Periods

Thus, when comparing a single investment proposal with the "do nothing" alternative, the investment should be undertaken if its present worth is greater than zero. This indicates that the proposed investment is preferable to investing the same funds elsewhere at the minimum attractive rate of return, i.

3.3.4 Capitalized Cost

When the life of a proposed investment is perpetual and/or the planning horizon is extremely long, it may be necessary to determine the present value of an infinite series of cash flows. In the literature of engineering economy, this is known as the **capitalized cost** (CC). It represents the amount of money that must be invested today to yield a return of A at the end of each and every period forever, assuming rate of return i per period. Observe the limit of the uniform series present worth factor as N approaches infinity:

$$CC = A \left[\lim_{N \to \infty} (P/A, i, N) \right]$$

$$= A \left\{ \lim_{N \to \infty} \left[\frac{(1+i)^N - 1}{i(1+i)^N} \right] \right\}$$

$$= A \left(\frac{1}{i} \right) \tag{3.3}$$

EXAMPLE 3.3

To illustrate the use of Eq. (3.3), consider a proposed investment of $30,000 now that will require additional costs of $600 at the end of each and every year, perpetually. The discount rate is 12% per year. Therefore, the capitalized cost is

$$CC = \$600 \left(\frac{1}{0.12} \right) = \$5,000$$

and the total present worth of costs is

$$\$30,000 + \$5,000 = \$35,000$$

3.3.5 Ranking Multiple Alternatives

EXAMPLE 3.4

Consider four alternatives characterized by the following conditions: (1) They are *mutually exclusive*, in that accepting one necessarily precludes accepting any other; (2) they are *exhaustive*, in that there are no other candidate alternatives, except, of course, the do-nothing alternative; and (3) they are *feasible*, in that there are no considerations, other than economic, that preclude the acceptance of any one or more of them. The cash flows for the four investment opportunities are given in **Table 3.1**. All four have equal lives: 10 periods. Assume that the discount rate, the minimum attractive rate of return, i, is 20% per period.

TABLE 3.1   Example 3.4, Cash Flows for Four Mutually
             Exclusive Alternatives

End of Period	Alternative I	Alternative II	Alternative III	Alternative IV
0	−$1,000	−$1,000	−$1,100	−$2,000
1–10	0	300	320	550
10	4,000	0	0	0
Net cash flow	$3,000	$2,000	$2,100	$3,500

The preferred alternative, based on the economic considerations outlined in Table 3.1, can be determined by computing the equivalent present worths of the alternatives, using the appropriate discount rate. These calculations are summarized in **Table 3.2**. Ranking on the basis of decreasing present worth shows that:

- Alternative IV is preferred to alternative II because PW(IV) = $305 is greater than PW(II) = $258.
- Alternative II is preferred to alternative III because PW(II) = $258 is greater than PW(III) = $241.
- Alternative III is preferred to alternative ∅ (doing nothing) because PW(III) = $241 is greater than PW(∅) = $0.
- Alternative ∅ is preferred to Alternative I because PW(∅) = $0 is greater than PW(I) = −$354.

That is, $\text{IV} \succ \text{II} \succ \text{III} \succ \varnothing \succ \text{I}$. (The inequality $x \succ y$ indicates that Alternative x is *preferred* to Alternative y.)

TABLE 3.2 Present Worths of Cash Flows from Table 3.1

End of Period	Factor[a]	PWs for Alternatives			
		I	II	III	IV
0	1.000	− $1,000	− $1,000	− $1,100	− $2,000
1–10	4.192	0	1,258	1,341	2,305
10	0.1615	646	0	0	0
Net present worth		− $ 354	$ 258	$ 241	$ 305

[a]Factors: $(P/A, 20\%, 10) = 4.192$
$(P/F, 20\%, 10) = 0.1615$

Ranking on the basis of the sums of cash flows yields quite different results. From Table 3.1, $IV \succ I \succ III \succ II \succ \emptyset$. Ranking is changed by the influence of the timing of the cash flows and the opportunity costs as reflected in the discount rate.

The above example considered only five alternatives, including the do-nothing alternative. The procedure, however, can be extended to any number of alternatives. In each case, simply determine the PWs (NPVs) for all proposals and then rank the alternatives accordingly.

3.4 ANNUAL WORTH (ANNUAL COST)

3.4.1 The Annual Worth Method   *Do nothing to Annual $*

Annual Worth as a Figure of Merit

The essence of the **annual worth method** is the conversion of all cash flows associated with a project to a single figure, namely, the equivalent uniform annual worth (AW). Given any sequence of cash flows, A_j, an equivalent uniform series, A, is determined using the appropriate compound interest calculations. If all cash flows are end-of-period, then

$$AW = PW * (A/P, i, N)$$

$$= (A/P, i, N) \sum_{j=0}^{N} A_j(1 + i)^{-j} \tag{3.4}$$

where discounting takes place at effective interest rate i per period. There are various other models, of course, to accommodate other assumptions about the timing of cash flows and the type of discounting. For example, if all cash flows are continuous, and discounting takes place continuously at rate i, then

$$AW = (A/P, i, N) \sum_{j=1}^{N} \bar{A}_j \left[\frac{i(1 + i)^{-j}}{\ln(1 + i)} \right] \tag{3.5}$$

where $\bar{A}_j$ is the cash flowing continuously and uniformly during period j.

As used in this context, **annual worth** is the value of an investment opportunity expressed as an equivalent end-of-period series. The choice of a *uniform series* as a time reference has considerable appeal. Many costs (negative cash flows) and revenues (positive cash flows), such as rental costs of $500 per month, interest income of $200 per quarter, and property taxes of $2,000 per year, occur on a periodic basis. Moreover, even when cash flows do not, or will not, follow a uniform series, it is frequently convenient to refer to them as though they do or will. Budget planning and control, profit determination, and accounting for operations are examples of activities for which forecasts are generally stated in the form of uniform periodic results.

The annual worth method consists of two steps. First, compute the appropriate annual worth, AW, for each alternative. The same interest rate must be used to evaluate all alternatives. Second, rank the alternatives in decreasing order of AWs. Assuming no other differences among alternatives, our objective is to maximize AW.

The annual worth method is often cited in the literature of engineering economy as the **equivalent uniform annual cost** (EUAC or AC) method. Annual cost is the negative of AW, and thus the objective is to minimize this figure of merit.

Note that, although this is known as the *annual* worth or *annual* cost method, the time interval between equivalent values need not be in years. Days, weeks, months, and so on, do just as well. Thus the term *periodic worth* is more accurate. But we seem to be stuck with the somewhat misleading term *annual worth* because of historical precedent and common usage.

EXAMPLE 3.5

Recall Example 3.1: Initial investment of $12,000, expenses of $6,000 at the end of each year for 6 years, and salvage value of $3,000. The interest rate is 12%. Determine the annual worth.

Solution:

$$AW = -\$12,000(A/P, 12\%, 6) - \$6,000 + \$3,000(A/F, 12\%, 6)$$

$$= -\$12,000(0.24323) - \$6,000 + \$3,000(0.12323)$$

$$= -\$8,549$$

Of course, $AC = -AW = \$8,549$.

Significance of Annual Worth

In the Example 3.5, the original set of eight cash flows is equivalent to a uniform series of $8,549 occurring at the end of each year for 6 years. (See **Figure 3.6**.) Of course, this equivalence is valid only for an interest rate of 12% per year. Put somewhat differently, imagine a fund—a bank account,

(a)   Cash Flows

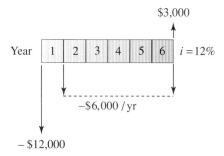

(b)   Equivalent Annual Worth

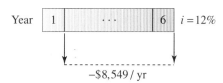

Figure 3.6   Example 3.5

for example—that earns 12% per year, using two alternative strategies. The first consists of an investment of $12,000 into the fund now, $6,000 into the fund at the end of each year for 6 years, and a withdrawal of $3,000 from the fund at the end of the sixth year. The second strategy consists of investments of exactly $8,549 into the fund at the end of each year for 6 years. Under both strategies, the amount remaining in the fund at the end of 6 years is identical.[1]

This method results in the substitution of a single figure, AC or AW, equivalent to a larger number of cash flows. Alternatives can now be readily compared using this figure. Suppose, for example, that the proposed equipment purchase presented previously is intended to replace a manual operation that now costs $8,000 per year. How would management know which alternative is economically preferable in the absence of an equivalent uniform annual cost computation? Note that the average annual cost per year for the proposed equipment is [$12,000 + 6($6,000) − $3,000]/6 = $7,500. Thus, the naive analysis, disregarding the effects of interest, would conclude that the new equipment is justified. But the true equivalent uniform annual cost, $8,549, is more expensive than the annual cost of the current procedure—$8,000. Thus the equivalent uniform annual cost may be thought of as a *weighted average*—weighted in the sense that the interest cost, the cost of opportunities

---

[1]You may verify that the "terminal value" of the fund is $69,374 in each case.

foregone, is included. This equivalent cost of interest is $8,549 − $7,500 = $1,049 annually.

3.4.2 Equivalence of the Annual Worth and Present Worth Methods

The annual worth and present worth methods are directly related. Given a discount rate and a sequence of cash flows occurring over N periods, ranking on the basis of decreasing AW (or increasing AC) is equivalent to ranking on the basis of decreasing PW. This is so because the alternatives' PWs are related to their AWs by the factor $(A/P, i, N)$, which is constant for given values of i and N. In a given problem situation, the interest rate and the planning horizon are common to all alternatives. Given any pair of alternatives (say, I and II), if

$$\text{PW(I)} > \text{PW(II)}, \quad \text{then} \quad \text{AW(I)} > \text{AW(II)}$$

because

$$\text{AW(I)} = \text{PW(I)} \times (A/P, i, N)$$

and

$$\text{AW(II)} = \text{PW(II)} \times (A/P, i, N)$$

Recall the four mutually exclusive alternatives summarized in Table 3.1. The annual worths of these alternatives are given in **Table 3.3**. The AW of the do-nothing alternative is also zero, because $\text{PW}(\emptyset) = 0$ and $\text{AW}(\emptyset) = \text{PW}(\emptyset) \times (P/A, i, N)$. Ranking the alternatives by decreasing AW, the results in Table 3.3 indicate

$$\text{IV} \succ \text{II} \succ \text{III} \succ \emptyset \succ \text{I}$$

This is the same result obtained by using the PW method, of course.

TABLE 3.3 Annual Worths of Cash Flows from Table 3.1

End of Period	Factor[a]	AWs for Alternatives: I	II	III	IV
0	0.2385	−$239	−$239	−$262	−$477
1–10	1.0000	0	300	320	550
10	0.0385	154	0	0	0
Annual worth		−$ 85	$ 61	$ 58	$ 73

[a]Factors: $(A/P, 20\%, 10) = 0.2385$
$(A/F, 20\%, 10) = 0.0385$.

The analysis reflected in Table 3.3 assumes that the present worths have not been computed. If, however, the PWs are known, then the AWs can be determined somewhat more simply:

Alternative	PW	(A/P, 20%, 10)	AW
$\emptyset$	0	0.2385	0
I	$-354	0.2385	$-84
II	258	0.2385	62
III	241	0.2385	57
IV	305	0.2385	73

(The above AW results differ slightly from those of Table 3.3 because of rounding.)

3.4.3 The Cost of Capital Recovery

The cost of capital recovery, generally referred to simply as **capital recovery** (CR), is defined as the uniform series equivalent of the original cost of an asset less its salvage value, if any, as indicated in **Figure 3.7**.

$$CR = P(A/P, i, N) - S(A/F, i, N) \qquad (3.6)$$

where P is the initial cost and S is the net salvage value after N periods.

(a)   Cash Flows

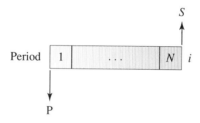

(b)   Equivalent

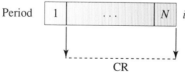

Figure 3.7   Capital Recovery

In our previous equipment purchase example, the cost of capital recovery is

$$CR = \$12,000(A/P, 12\%, 6) - \$3,000(A/F, 12\%, 6)$$

$$= \$12,000\,(0.24323) - \$3,000\,(0.12323)$$

$$= \$2,549$$

That is, the *weighted-average* cost per year (capital recovery) is $2,549. But this is not equal to the *average* cost per year, ($12,000 − $3,000)/6 = $1,500. The difference, $2,549 − $1,500 = $1,049 per year, represents the dollar cost of opportunities forgone. When money has value over time ($i > 0\%$), it is always true that the cost of capital recovery is greater than the average cost per period.

In Eq. (3.6), two factors are used to compute capital recovery, A/P and A/F. It is possible, however, to use a simplified form by observing that capital recovery consists of two elements: (1) one portion, $P − S$, is "consumed" and must therefore be recovered; and (2) the second portion, S, is costly only in the sense that interest on this portion, iS, is forgone during the N periods when S is invested in the asset.

$$CR = (\text{capital consumed}) + (\text{interest forgone})$$

$$= (P − S)(A/P, i, N) + iS \tag{3.7}$$

The term $(A/P, i, N)$ is known in the literature of engineering economy as the **capital recovery factor**. The reason, of course, is that it serves an important role in computing capital recovery. Because only one factor is used, A/P, Eq. (3.7) is somewhat easier to use than Eq. (3.6).

Still another equation used to determine capital recovery is

$$CR = (P − S)(A/F, i, N) + Pi \tag{3.8}$$

This alternative formula is occasionally employed in the public utility industry, although its use is otherwise extremely limited. When capital recovery is computed in this manner, it is generally described as **sinking fund depreciation plus interest on first cost**. The term $(A/F, i, N)$ is known as the **sinking fund factor**.

3.5 FUTURE WORTH

3.5.1 The Future Worth Method

The **future worth method** is based on the notion that the basic reference point is the *end* of the planning horizon. That is, the future worth (FW) is the equivalent value of all cash flows, A_j, at the end of N periods, with interest i per period. As illustrated in **Figure 3.8**,

$$FW = \sum_{j=0}^{N} A_j(1 + i)^{N − j} \tag{3.9}$$

When considering a set of mutually exclusive alternatives, that alternative with the maximum FW is preferred.

(a) Cash Flows

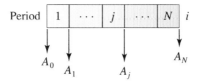

(b) Equivalent

Figure 3.8   Future Worth

3.5.2 Equivalence of the PW, AW, and FW Methods

The three figures of merit or measures of worth—PW, AW, and FW—are related as follows:

$$FW = (PW)(F/P, i, N)$$

$$= (AW)(F/A, i, N)$$

When considering a set of mutually exclusive alternatives, ranking of alternatives by FW must be consistent with ranking by PW and/or AW. This is so because i and N are constant, and the factors, F/P and A/P, are positive constants for all alternatives.

    To illustrate the above, return to Example 3.1; cash flows are summarized in Table 3.1. The future worths of the four projects are shown in **Table 3.4**. This tabular formulation reflects the equation

$$FW = P(F/P, 20\%, 10) + A(F/A, 20\%, 10) + S$$

where P is the initial cost, A is the annual cash flows, and S is the salvage value at the end of the planning horizon. The FW of the do-

TABLE 3.4   Future Worths of Cash Flows from Table 3.1

End of Period	Factor[a]	FWs for Alternatives:			
		I	II	III	IV
0	6.192	−$6,192	−$6,192	−$6,811	−$12,384
1 10	25.959	0	7,788	8,307	14,277
10	1.000	4,000	0	0	0
Future worth		−$2,192	$1,596	$1,496	$1,893

[a]Factors: $(F/P, 20\%, 10) = 6.192$
$(F/A, 20\%, 10) = 25.959$.

nothing alternative is also zero, because $PW(\emptyset) = 0$. Ranking the alternatives by decreasing FW, the results in Table 3.4 indicate that $IV \succ II \succ III \succ \emptyset \succ I$. The results are consistent with those obtained by the present worth and annual worth methods.

3.5.3 The Effect of Interest Rate on PW and FW

Present worth and future worth are functions of the interest rate, i, of the form:

$$PW(i) = A_0 + A_1(1 + i)^{-1} + A_2(1 + i)^{-2} + \cdots + A_N(1 + i)^{-N}$$

$$FW(i) = A_0(1 + i)^N + A_1(1 + i)^{N-1} + \cdots + A_N$$

Using differential calculus, it may be shown that the value of i that maximizes present worth is not in general the same as that which maximizes future worth. Depending on the amounts and timing of the cash flows, a certain change in the interest rate may result in a decrease in PW while concurrently resulting in an *increase* in FW.[2]

3.6 COMPARING WORTHS WHEN LIVES OF ALTERNATIVES ARE UNEQUAL

3.6.1 Importance of the Common Planning Horizon

One of the most troublesome complications arises when the service lives, or economic lives, of the various alternatives are unequal. It is essential that any specific capital allocation study be made within the context of a uniform planning horizon.

EXAMPLE 3.6

To illustrate this point, consider two alternative investment proposals, L and S, with the following expected cash flows:

End of Period	Alternative L	Alternative S
0	$-\$100$	$-\$120$
1	30	50
2	30	50
3	30	38
4	30	
Total cash flow	$\$ 20$	$\$ 18$

[2]G. A. Fleischer and L. C. Leung, "On Future Worth and Its Relation to Present Worth as an Investment Criterion," *The Engineering Economist*, Vol. 35, No. 4, Summer 1990, pp. 323–332.

Assume that the interest rate is zero. (This is an unrealistic assumption, of course, but it is useful in the context of this discussion. A nonzero interest rate merely complicates the arithmetic; the underlying lesson remains unaffected.)

Which alternative, L or S, is preferable? Since the present worth of L exceeds that of S—PW(L) = $20 and PW(S) = $18—it would appear that L is preferable to S. But the equivalent annual worth of L is *less* than that of S: AW(L) = $20/4 = $5 and AW(S) = $18/3 = $6. Thus it would appear that S is preferable to L. Which, then, is the correct choice?

The answer lies in the fact that the two alternatives have unequal lives. L generates a net of $20 over 4 periods, whereas S generates a net of $18 over only 3 periods. If we are to choose fairly between the alternatives, *the consequences of all alternatives must be evaluated over a common time interval, or planning horizon*—a fundamental principle of analysis. Thus the relevant question is: What would happen at the end of the third period if alternative S were selected? Would a replacement be purchased? If so, what would the cash flows associated with the replacement be? And are the total time periods associated with S and its replacement(s) now equal to the total time periods associated with L and its replacement(s)?

3.6.2   Alternative Assumptions Concerning the Common Planning Horizon

Identical Replication (Repeatability)

The **identical replication approach** rests on two assumptions: (1) At the end of their respective service lives, each alternative may be *repeated*, that is, a replacement will be available that will be identical in every respect to the project it replaces; and (2) there will be a *need* for each alternative that is at least as long as the least common multiple (LCM) of the lives of the competing alternatives.

EXAMPLE 3.6   *(Continued)*

To illustrate this approach, suppose that alternatives L and S may be identically replicated, and there is a need for either L or S, and their successors, over 12 periods. See **Figure 3.9**. The common planning horizon is the LCM, $3 \times 4 = 12$ periods. Our analysis now indicates that:

	Alternative I	Alternative S
Present worth	$3 \times \$20 = \$60$	$4 \times \$18 = \$72$
Annual worth	$\$60 \div 12 = \$5$	$\$72 \div 12 = \$6$

Thus both the PW and AW indicate that alternative S is preferred.

Alternative L

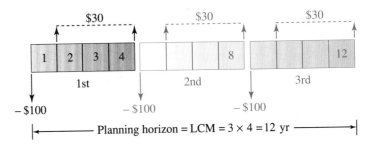

Alternative S

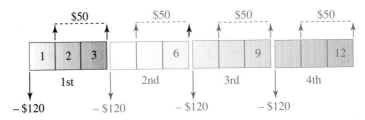

Figure 3.9   Example 3.6, Illustrating the Identical Replication Assumption

EXAMPLE   3.7

The firm's discount rate is 12%. Cash flows and service lives for two mutually exclusive alternatives, P and Q, are as follows:

	Alternative P	Alternative Q
Initial cost	$12,000	$20,000
Salvage value at end of life	$3,000	$2,000
Annual operating costs	$1,600	$ 900
Service life	6 yr	12 yr

These data are reflected in **Figure 3.10**.
   Using $N = 6$ for alternative P:

$$\text{PW}(P|6) = -\$12,000 - \$1,600(P/A, 12\%, 6) + \$3,000(P/F, 12\%, 6)$$

$$= -\$17,058$$

$$\text{AW}(P|6) = -\$12,000(A/P, 12\%, 6) - \$1,600 + \$3,000(A/F, 12\%, 6)$$

$$= -\$4,149$$

Alternative P

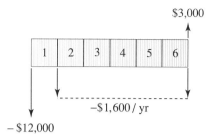

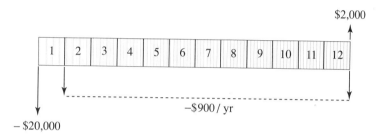

Alternative Q

Figure 3.10   Example 3.7

Using $N = 12$ for alternative Q:

$$\text{PW}(Q|12) = -\$20,000 - \$900(P/A, 12\%, 12) + \$2,000(P/F, 12\%, 12)$$

$$= -\$25,061$$

$$\text{AW}(Q|12) = -\$20,000(A/P, 12\%, 12) - \$900 + \$2,000(A/F, 12\%, 12)$$

$$= -\$4,046$$

Note the conflicting results: $\text{PW}(P|6) > \text{PW}(Q|12)$, but $\text{AW}(P|6) < \text{AW}(Q|12)$. This is so because we have failed to consider a common planning horizon.

Under the identical replication assumptions, the planning horizon, the LCM, is 12 years. The revised PW and AW values for alternative P are

$$\text{PW}(P|12) = (\text{PW of 1st 6 years}) + (\text{PW of 2nd 6 years})$$

$$= -\$17,058 - \$17,058(P/F, 12\%, 6)$$

$$= -\$17,058 - \$8,642 = -\$25,700$$

$$\text{AW}(P|12) = -\$25,700(A/P, 12\%, 12) = -\$4,149$$

The results are now consistent for this common planning horizon:

	Alternative P	Alternative Q
Present worth	− $25,700	− $25,061
Annual worth	− 4,149	− 4,046

Under the assumptions of identical replication, alternative Q is less costly than alternative P.

Study Period

The **study period approach** is much more general than the repeatability assumption described above. (This is sometimes known as the **cotermination method**.) Here we assume only that there is a finite time interval beyond which the economic consequences of the competing alternatives are no longer significant.

This planning horizon may be effected in a number of ways. For example, the investments in manufacturing equipment or machine tools may be associated with a certain product, and thus the planning horizon will be defined by the life of that product. Or improvements in physical plant are associated with a certain facility (e.g., a warehouse), and it is anticipated that the facility will be sold at a future date. In these instances, the context in which the analysis takes place will serve to define the study period of interest. The planning horizon will be the time interval appropriate to the decision under investigation.

EXAMPLE 3.7 *(Continued)*

The service lives of alternatives P and Q are 6 and 12 years, respectively. Assume that these two alternatives are manufacturing equipment associated with a certain product that is to be discontinued after 10 years. As indicated in **Figure 3.11**, the assumption of a 10-year planning horizon implies two new questions: If another P is purchased after 6 years, what will be the new P's salvage value after 4 years of ownership? If Q is purchased, what will be its salvage value after 10 years? Suppose that the required estimates are as follows:

$$S_4(P) = \text{salvage value for alternative P at end of 4 years}$$

$$= \$5,000$$

$$S_{10}(Q) = \text{salvage value for alternative Q at end of 10 years}$$

$$= \$4,500$$

The present worths of the two alternatives over this 10-year planning horizon are

$$PW(P) = -\$12,000 + (3,000 - \$12,000)(P/F, 12\%, 6)$$

$$-\$1,600(P/A, 12\%, 10) + \$5,000(P/F, 12\%, 10)$$

$$= -\$23,990$$

$$PW(Q) = -\$20,000 - \$900(P/A, 12\%, 10) + \$4,500(P/F, 12\%, 10)$$

$$= -\$23,664$$

The equivalent uniform annual costs for P and Q can further be shown to be \$4,246 and \$4,183, respectively.

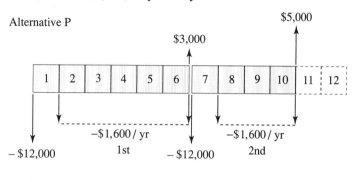

Alternative P

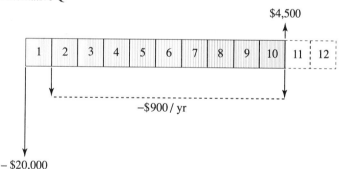

Alternative Q

Figure 3.11   Example 3.7, Illustrating the Cotermination (Study Period) Approach

3.6.3   Conditions Under Which Annual Worths May Be Compared Directly

You may have noted that the ranking of alternatives in Examples 3.6 and 3.7 under the identical replication assumption, on the basis of PWs or AWs, is exactly the same as the ranking that results from the AW method for the original problem. This is so because of the special character of the replacement assumption. Given a sequence of cash flows over N periods, the equivalent annual worth remains constant if and only if the identical set of cash flows is replicated over $2N$ periods, $3N$ periods, ..., and so on. This replication is shown in **Figure 3.12**.

Annual Worth or (Equivalent Uniform) Annual Cost

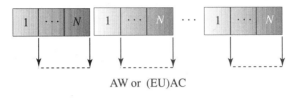

Cash Flows

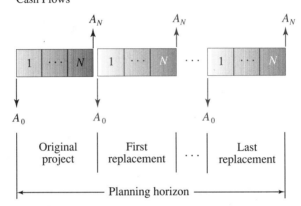

Figure 3.12 Converting Cash Flows to Equivalent Annual Worth (or Cost) Assuming Identical Replacements

It is tempting to conclude that the problem of unequal project lives is easily solved if the analyst uses the annual worth method for all alternatives. But this is true only if the identical replication assumption holds. Otherwise, a complete evaluation should be made over the common planning horizon.

Although these assumptions allow us to use the original solutions for annual worth without modification, one should resist the temptation to use them, explicitly or implicitly, without carefully considering likely future consequences. If annual costs are compared without considering differences in project lives, the solutions imply identical repetition and a planning horizon equal to a common multiple of project lives. These assumptions should always be questioned in real situations.

3.7 THE PROBLEM OF PRELIMINARY SELECTION

Consider the selection of one project from each of four units. (In general, a "unit" is an operating department, a section, a division, a plan, or a group. It is a budgeting unit wherein a single project must be selected from a set of mutually exclusive alternatives. Proposals are independent *between* units but are mutually exclusive *within* units.) The relevant data and analysis are summarized in **Table 3.5**. All dollar amounts are equivalent present values; costs, in this case, are assumed to be initial costs.

Clearly, based on maximum PW, alternatives A2, B2, C2, and D2 are preferable to A1, B1, C1, and D1, respectively. The total cost of these four "preferred" alternatives is $1,850. Now suppose that only $1,400 is available for the total capital budget. Considering only these four alternatives, the optimal feasible combination (budget "package") as shown in **Table 3.6** is A2 + B2 + C2. The PW is $1,120 at a total cost of

TABLE 3.5   Two Mutually Exclusive Alternatives in Each
of Four Independent Units

Unit	Alternative	Initial Cost	Present Worth
A	A1	$300	$300
	A2	400	320 max
B	B1	150	150
	B2	200	160 max
C	C1	600	630
	C2	800	640 max
D	D1	350	350
	D2	450	370 max

TABLE 3.6   Present Worth Analysis Given a $1,400 Budget
Constraint (Considering Independent Alternatives
A2, B2, C2, and D2)

Package	Alternatives	Costs	PW
1	A2	$ 400	$ 320
2	B2	200	160
3	C2	800	640
4	D2	450	370
5	A2 + B2	600	480
6	A2 + C2	1,200	960
7	A2 + D2	850	690
8	B2 + C2	1,000	800
9	B2 + D2	650	530
10	C2 + D2	1,250	1,010
11	A2 + B2 + C2	1,400	1,120 max
12	A2 + B2 + D2	1,050	850
13	A2 + C2 + D2	1,650	Not feasible
14	B2 + C2 + D2	1,450	Not feasible
15	A2 + B2 + C2 + D2	1,850	Not feasible

$1,400. However, reviewing the original data, the same expenditure for A1 + B1 + C1 + D1 produces a PW of $1,430, or an increase of $310.

Alternative	Costs	PW
A1	$ 300	$ 300
B1	150	150
C1	600	630
D1	350	350
Total	$1,400	$1,430

This is an illustration of the *preselection error*, proving that preselection can lead to a suboptimal allocation of resources.

Analysts must be alert to the possibility that selection of an apparent "best alternative" at some preliminary design or budget level may preclude some future combination that, if examined, may yield a higher present worth. When financially mutually exclusive alternatives are being considered, correct analysis may be based on the definition of an exhaustive set of alternative capital budget packages in which all possible combinations are considered, and each of which is mutually exclusive.[3] Once all the packages have been completely described, the optimum package can be selected by any of the several methods that take into account the amounts and timing of cash flows and the time value of money.

3.8 ILLUSTRATING WORTH

The present worth, annual worth, and future worth methods can be used to solve a wide variety of problems. To illustrate these techniques, three classes of problems of particular interest are discussed in this section.

3.8.1 Property Valuation

Owners of assets may have an opportunity to dispose of their property— or, conversely, potential buyers may have the opportunity to purchase property, either new or used. In either case, sellers/buyers are interested in determining the value of that property at the time of the prospective sale/purchase. (Of course, the value to the seller is not necessarily equal to the value to the buyer in any given instance.) This activity is known as **property valuation**.

From the point of view of the decision maker, the value of the property is the equivalent present value, at the time of the sale/purchase, of

---

[3]The "budget package" approach described here is relatively inefficient because it is an exhaustive search procedure. Other mathematical programming procedures for this capital budgeting problem are more efficient but are beyond the scope of this book.

all estimated future cash flows, discounted by an interest rate that represents the cost of opportunities foregone. The interest rate, the *minimum attractive rate of return* desired by the decision maker, represents the "cost" of the foregone opportunity to invest one's wealth elsewhere rather than to tie it up in the particular property under consideration.

The valuation process is a straightforward application of the present worth method. The property value is given by

$$PW = \sum_{j=1}^{N} A_j(1 + i)^{-j} \qquad (3.10)$$

where A_j is the estimated cash flow at the end of period j and i is the discount rate.

EXAMPLE 3.8

A certain component is currently being purchased from a local supplier (vendor) at a cost of $30,000 per month. The firm now has an opportunity to acquire the vendor with the result that net costs can be reduced to $10,000 per month. If acquired, this activity will be retained for 60 months, then sold for an estimated $300,000. If the firm's minimum attractive rate of return is 2% per month, what is the current value of this property to the firm?

Solution:

$$PV = (\$30,000 - \$10,000)(P/A, 2\%, 60) + \$300,000(P/F, 2\%, 60)$$

$$= \$20,000(34.761) + \$300,000(0.3048)$$

$$= \$786,700$$

From the firm's point of view, the investment is attractive if the cost to acquire the vendor does not exceed $786,700.

3.8.2   Bond Valuation

A **bond** is a financial instrument issued by private corporations, governmental agencies, and certain other organizations to obtain funds through borrowing. The bond is an "IOU": It indicates that the borrower will pay periodic interest as well as a terminal value at the bond's specified *maturity date*. The lender, the bond holder, generally has the opportunity to sell the bond in the bond market at any time prior to the maturity date.

In addition to the maturity date, each bond bears a *par value* (also known as *face value* or *stated value*), which is the amount due at the maturity date. Typically, the par value is $1,000. The price of a bond in the market place is almost never equal to the par value. Each bond also indicates two pieces of information necessary to determine the amount and timing of the periodic interest payments to the bond holder: the *nominal interest rate* and the *frequency of payments* each year, e.g., quarterly or monthly.

The cash flows associated with a bond are reflected in **Figure 3.13**. Here,

$$F = \text{par (face) value of the bond}$$

$$r_b = \text{bond's interest rate (nominal, per year)}$$

$$M = \text{number of payments per year}$$

$$N = \text{total number of payments to maturity}$$

$$i_M = \text{investor's effective interest rate per interest period}$$

The interest paid per year $= r_b F$, and the interest paid per interest period $= r_b F / M$. Thus the present value of the bond is given by

$$\text{PW} = \left(\frac{r_b F}{M}\right)(P/A, i_M, N) + F(P/F, i_M, N) \qquad (3.11)$$

(Other characteristics of bonds, namely, coupon rate, current yield, and rate to maturity, are discussed in Chapter 4.)

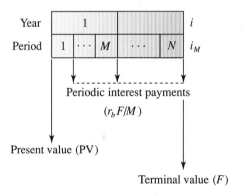

Figure 3.13 Cash Flows for a Bond

EXAMPLE 3.9

An investor has an opportunity to purchase a $1,000, 6% bond with quarterly payments. The bond's maturity date is such that there are exactly 40 quarters remaining to maturity. Here:

$$F = \$1,000 \qquad r_b = 6\% \text{ per year}$$

$$N = 40 \text{ quarters} \qquad M = 4 \text{ payments/year}$$

If the investor's required discount rate is 10% per year, determine the present value of this bond to this investor.

Solution:

$$i_M = (1.10)^{1/4} - 1 = 0.0241$$

$$r_b F/M = \frac{0.06(\$1,000)}{4} = \$15$$

$$PW = \$15(P/A, i_M, 40) + \$1,000(P/F, i_M, 40)$$

$$= \$15(25.48) + \$1,000(0.3855)$$

$$= \$768$$

From the investor's viewpoint, this bond would be an attractive investment if it could be purchased for less than $768.

3.8.3 Lease Versus Purchase

A **lease** can be described generally as a rental agreement for a period of more than 1 month. Most students are familiar with apartment or home leases. In the industrial world, leases are also quite common. The leasing of office buildings, aircraft, cars and trucks, furniture, and warehouse space are but a few of the examples where leasing is widespread.

One of the principal advantages of leasing is that it frees up capital that would otherwise be required. In some instances the investor has no choice in the matter. He or she may not have the funds currently available to purchase certain required productive assets. Nevertheless, when such options are feasible, rational decision making requires careful analysis of the economic consequences of the different policies.

EXAMPLE 3.10

A temporary warehouse is required in Kansas City to meet the firm's storage and distribution needs over the next 5 years. A present worth analysis has been completed for the purchase of the warehouse, resulting in a PW of costs of $1.86 million. A similar warehouse can be leased with semiannual payments of $300,000, payable at the start of each 6-month period for 5 years (10 periods). There are no other differences between the alternatives. If the firm's minimum attractive rate of return is 9% per period (18.81% per year), determine the present worth of costs of the lease.

Solution:

From **Figure 3.14** at the top of page 96

$$PW \text{ of costs} = \$300,000[1 + (P/A, 9\%, 9)]$$

$$= \$2,098,600$$

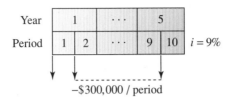

Figure 3.14   Example 3.10, Semiannual Lease Payments

Thus it would appear that leasing is somewhat less expensive than ownership.

Note: This example is naive in that it ignores the effects of income taxes on both the ownership and lease alternatives. This observation applies to all our examples through Chapter 5, but cash flows for income taxes are especially important with respect to lease-versus-buy decisions. As will be discussed in Chapters 6 and 7, the tax consequences can be very significant indeed.

SUMMARY

In the present worth method, all cash flows are discounted to date zero; that is, an equivalent present value is obtained. It is essential that planning horizons for each alternative be of equal length. Of particular importance to understanding this method is the fact that funds invested elsewhere at rate i result in zero present worth.

The annual worth method, also known as the equivalent uniform annual cost method, requires the translation of all cash flows into an equivalent uniform series by using appropriate compound interest factors. Alternatives may be compared directly if annual worth (net annual benefits) or annual costs have been distributed uniformly over a common time period. This method results in a weighted-average worth, or cost, per unit of time.

The future worth method utilizes the end of the planning horizon as the principal time of reference. Properly applied, the PW, AW, and FW methods lead to the same rank-ordering of a set of mutually exclusive alternatives.

It is essential that alternatives be compared over a common planning horizon. If the identical replication assumption is valid, as discussed in Section 3.6, then the AWs of the competing alternatives may be evaluated and compared directly. However, this assumption is rarely justified other than in textbook examples and problems.

Analysts must be alert to the error of preliminary selection. The selection of an apparent "best alternative" at some preliminary design or budget level may preclude some future combination that, if examined, may yield a higher present worth.

| PROBLEMS

Unless otherwise indicated, assume discrete cash flows, end-of-period compounding/discounting, and effective interest rate.

Digital computers or programmed calculators may be used to determine the solutions to most of the following problems. (See the Appendix to this chapter and the Appendix to Chapter 2.) At least some of the problems should be solved manually, however, in order to improve understanding of the underlying procedures.

Present Worth

3.1 Consider the following series of cash flows:

End of Period	Series A	Series B	Series C
1	$ 50	$ 60	0
2	52	58	$ 66
3	54	56	66
4	56	54	66
5	58	52	66
6	60	50	66
Totals	$330	$330	$330

Assuming a 10% interest rate, find the equivalent worth of the three series.
[*Answers:* $237, $242, $227.]

3.2 Consider the following cash flow sequences:

End of Period	Series D	Series E	Series F
0	0	$100	− $100
1	$ 20	80	0
2	40	60	0
3	60	40	100
4	80	20	100
5	100	0	100
6	0	− 100	100
Total	$300	$200	$300

Assuming an effective interest rate of 10% per period, find the equivalent present worth of each sequence.

3.3 Consider two mutually exclusive alternatives, G and H, with the following expected cash flows:

End of Period	Alternative G	Alternative H
0	− $100	− $120
1	20	50
2	30	50
3	40	50
4	50	50
5	60	80
Totals	$100	$160

Assuming that $i = 0.10$, find the present worth for each alternative.
[*Answers:* PW(G) = $44; PW(H) = $88.]

3.4 With a 12% discount rate, determine the present worths of each of the following four series of end-of-period cash flows:

Period	Series I	Series J	Series K	Series L
1	$180	0	$325	0
2	175	0	325	0
3	170	0	325	0
4	165	$260	325	0
5	160	260	0	0
6	155	260	0	0
7	150	260	0	0
8	145	260	0	0
9	0	0	0	0
10	0	0	0	1,300
Totals	$1,300	$1,300	$1,300	$1,300

3.5 A firm is considering two competing investment projects, M and N. The after-tax cash flows are given next. The firm's after-tax MARR is 6%.

End of Period j:	0	1	2	3	4	5
Alternative M ($):	−205	50	50	50	50	50
Alternative N ($):	−350	100	90	80	70	60

Using the present worth method, determine the relative desirability of the alternatives.

[*Answer*: $M \succ \varnothing \succ N$.]

3.6 Assuming an effective interest rate of 12% per year, find the equivalent present value of the following cash flows:

a. $1,000 flowing at the start of each year for 10 years

b. $250 flowing at the end of each quarter (3 months) for 40 quarters

c. $2,000 flowing at the end of each 2-year period: years 2, 4, 6, 8, and 10

3.7 Consider a prospective investment in a warehouse having a first cost of $300,000, operating and maintenance costs of $35,000 per year, and an estimated net disposal value of $50,000 at the end of 60 years. Assume an 8% interest rate.

a. What is the present value of costs of this investment if the planning horizon is 60 years?

b. If replacement structures will have the same first cost, life, salvage value, and operating and maintenance costs as the original building, what is the capitalized cost of perpetual service? Note the relatively small difference between a 60-year life and an infinite life.

[*Answers*: a. $732,700; b. $740,000.]

3.8 The Bigditch Construction Company has a contract to build a major hydroelectric project over a 3-year period. Bigditch has most of the requisite equipment, but it will need a mobile van to serve as a field office. If it is purchased, the van will be sold at the end of 3 years. Relevant data are:

Initial cost	$54,000
Physical life	6 years
Salvage value at end of 3 years	$20,000
Salvage value at end of physical life	$10,000

Property taxes, payable at the *end* of each year	3% of first cost
Insurance, payable at the *start* of each year	2% of first cost
Operating costs occurring *during* each year	$1,000

The firm's MARR is 10% per year. Find the net present worth of costs of this proposed purchase.

3.9 The payroll for an accounting process is currently $2,000 per week. It is now possible to purchase an electronic data processing system at an initial cost of $100,000 that will have the effect of reducing payroll costs to only $1,200 per week. This system, if purchased, will be maintained by the vendor at an additional cost of $3,000 paid at the beginning of each year. At the end of 5 years, the equipment may be returned to the vendor for a trade-in value equal to 10% of the original cost. Assuming a 10% discount rate per year, determine the present worth of the proposed equipment as compared to the do-nothing alternative. Inasmuch as the payroll costs occur weekly, use the continuous cash flow assumption during each 52-week year. Assume a 5-year planning horizon.

[*Answer*: $59,156.]

3.10 A firm spends $10,000 per year on materials, with the cost spread continuously and uniformly over each year. Annual rental payments for the building and equipment total $30,000 per year, with the payments being made at the beginning of each year. Using an interest rate of 10% per year, find the equivalent present worth of 10 years of activity.

3.11 A specialized welding machine may be purchased for $20,000. The expected excess of receipts over disbursements is $4,000 a year for the first 4 years and $3,000 in the fifth year. It is expected that the machine will be sold at the end of the fifth year for $12,000. Treat the annual receipts and disbursements using the uniform flow, continuous compound convention. If the effective interest rate is 10% per year before taxes, what is the net present worth of this series of cash flows?

[*Answer*: $2,709.]

3.12 A forklift truck may be purchased for $10,000. If

it is purchased, it is expected that the equipment will be retained for 10 years and then sold for an expected $1,000. Costs of operation, such as costs for gas, oil, and normal maintenance costs, are expected to be $1,200 per year. Labor costs are expected to be $6,000 per year in the first year, with an increase of $500 per year, each year, over the life of the equipment. The company's minimum attractive rate of return is 15%. Determine the equivalent present worth of costs of this proposed investment.

3.13 A firm's opportunity cost (effective interest rate) is 24% per year. Find the capitalized cost of an infinite series of $1,000 payments made at the end of every 3-month period. That is, $1,000 payments are made at the end of months 3, 6, 9,..., forever.
[*Answer*: **$18,099.**]

3.14 Consider a project having a first cost of $1,000, a 10-year life, and no salvage value. Labor expenses during each year of operation are estimated to be $100. A major overhaul will be required 5 years after the project starts, will last for 1 year, and will cost a total of $500. (Of course, the project will not be in operation during the overhaul period.) Using the continuous compounding, continuous cash flow convention for labor and overhaul expense, what is the total cost of the project in present dollars if the interest rate is 6% per year?

3.15 Two years ago, the White Trucking Company leased a terminal site from the Eastern Pacific Railroad and prepaid the rent for a period of 5 years. The total prepaid rent was $60,000. The terms of the lease permit White Trucking to continue to rent the site for a period of 5 additional years by paying $10,000 at the beginning of each year of the second 5-year period.

 The railroad has now decided that it needs funds to finance legal action in connection with a possible merger, and it has proposed that White Trucking prepay the rent that was to have been paid year by year in the second 5-year period. Assuming an interest rate of 8% per year, what is a fair payment to be made now in lieu of the 5 annual payments?
[*Answer*: **$34,231.**]

3.16 A manufacturing firm is considering the purchase of a heavy-duty punch press that will cost $32,000 and that may be sold at the end of 8 years for an expected $16,000. Labor costs are expected to be $20,000 per year to operate the press. Insurance premiums for the press, payable at the start of each year, are expected to be $600 the first year and to decline by $50 each subsequent year. Find the equivalent present value of this proposed investment assuming an interest rate of 15% per year. Use the continuous cash flow convention for the labor costs.

3.17 It is proposed that a facility be constructed to store the firm's landscaping equipment. The site may be purchased for $40,000. Construction will take 1 year and will cost $100,000, distributed uniformly over the year. Over the following 10 years, that is, over years 2 through 11, costs to operate and maintain the facility are estimated to be about $1,150 per week, or $60,000 per year. Property taxes are expected to be $1,000 per year, payable at the end of each year. Insurance will cost $2,000 per year, payable at the start of each year, that is, at the beginning of years 1 through 11. It is expected that the land and the facility will be sold at the end of 11 years for $200,000. Find the present worth of cost for this proposed project. Assume an interest rate of 10% per year.
[*Answer*: **$444,200.**]

3.18 A manufacturing plant and its equipment are insured for $700,000. The present annual insurance premium, payable at the beginning of each insured year, is $0.86 per $100 of coverage. A sprinkler system with an estimated life of 20 years and no salvage value at the end of that time can be installed for $18,000. Annual operation and maintenance costs are estimated to be $360. Property taxes, payable at the end of each year, are 1.0% of the initial cost of the plant and equipment. If the system is installed and maintained, the premium rate will be reduced to $0.38 per $100 of coverage. If the firm's MARR is 10% before income taxes, should the sprinkler system be installed? Find the present worth of costs of both (a) continuing with the present system, that is, paying annual premiums of (0.86/$100) × ($700,000) = $6,020, and (b) purchasing the sprinkler system and taking advantage of the lower insurance rate.

3.19 The firm is considering investment in certain manufacturing equipment that will cost $75,000, have a useful life of 6 years, and will then be sold for an estimated $15,000. Operating expenses are expected to be $30,000 each year. Assuming an annual effective discount rate of 12%, and also assuming that cash flows for expenses are end-of-year, determine the present worth of costs of this proposal.
[*Answer*: $190,743.]

3.20 a. The firm's (effective) discount rate (per annum) is 14%. Find the equivalent present value of labor costs of $10,000 per month over 3 years. Here, labor costs are $10,000 at the end of each and every month for 36 months.
  b. Suppose that the labor costs in (a) are assumed to flow continuously and uniformly during each year over the 3-year period. That is, costs would be $120,000 per year for 3 years. Find the equivalent present value of these costs.

3.21 The firm is considering the acquisition of a certain materials handling device that is expected to save $50,000 each year over its 7-year life. (Savings are assumed to flow continuously and uniformly throughout each year.) The initial cost is $90,000. If it is sold at the end of 7 years, the expected salvage value is $10,000. The firm's pretax MARR is 12%.
  a. Determine the present worth of this proposed investment.
  b. Determine the imputed salvage value if the device is kept in service only 5 years. The "imputed salvage value" is defined as the equivalent value at the time of removal from service of the cash flows foregone as the result of the premature disposal.
  c. Given the imputed salvage from (b) above, determine the present worth if the device is kept in service only 5 years.
[*Answers*: a. $156,144; b. $97,449; c. $156,144.]

3.22 A large manufacturing firm is considering the acquisition of a robot for use in its painting department. The robot will take about 1 year to set up and test before it is fully functional. Then it will be used in connection with the manu-facture of a certain part that will be produced over the subsequent 4 years. The planning horizon is 5 years: 1 year for setup and 4 years for operation. The original cost of the painting robot is $300,000. At the end of 5 years its estimated value will be $100,000. The firm's MARR is 15%.
  a. Property taxes, $6,000 per year, will be paid at the end of each year. Determine the present worth of these costs.
  b. Insurance costs are payable at the start of each year over the 5-year planning horizon. The cost will be $10,000 the first year, then increasing by $500 in each of the subsequent 4 years. Determine the PW of these costs.
  c. Material costs will occur at the start of each 6-month period, beginning at the start of the second year and ending at the midpoint of the fifth year. Each payment for materials will be $50,000. There will be 8 of these payments. Determine the PW of the costs.
  d. Labor costs will be $10,000 the first month, increasing at the rate of 0.5% per month over the 60-month planning horizon. That is, labor costs will be $10,000 the first month, $10,050 the second month, $10,100.25 the third month, and so on. Assuming that these are end-of-month payments, determine the PW of the costs.
  e. Determine the total present worth of costs for this project.

3.23 Assuming a discount rate of 20%, find the present worth of each of the following:
  a. Development costs of $50,000 per month, assumed to be incurred at the end of each month over the first 2 years of the project (24 months)
  b. Manufacturing costs of $200,000 per year, beginning in year 3 and continuing through year 10
  c. Revenue of $600,000 in year 3, declining by $30,000 each year from years 4 through 10. That is, revenue would be $600,000 in year 3, $570,000 in year 4, $540,000 in year 5,..., $390,000 in year 10
  d. The total PW for the project, assuming the above consequences, over the 10-year planning horizon
[*Answers*: a. −$997,900; b. −$532,900; c. $1,392,900; d. −$137,900.]

3.24 Determine the following assuming an effective discount rate of 15% per year and a planning horizon of 10 years:

a. The equivalent present value of insurance costs of $2,000 per year paid at the beginning of each year for 10 years

b. The equivalent present value of cost of materials, assuming that these payments are $50,000 payable at the beginning of each 6-month period (i.e., at the end of months 0, 6, 12,..., 108, 114)

c. The equivalent present value of property taxes, assuming that these taxes are expected to be $5,000 at the end of the first year and will increase at the rate of 3% per year

d. The equivalent present value of labor costs of $100,000 per year, assuming that these costs are distributed uniformly and continuously throughout each year

e. The equivalent present value of revenue, assuming that these end-of-year receipts are expected to be $300,000 at the end of the first year and increasing by $20,000 each year (so that receipts will be $480,000 at the end of the 10th year).

3.25 The University is considering the acquistion of a van to be used by the Physical Plant Department for moving work crews between jobs. The initial cost is $15,000; it will probably be kept in service for 5 years, then sold for an estimated $5,000. The University's cost of capital (opportunity cost) is 7% per annum. If acquired, the van will result in net saving in operating costs of $4,000 per year. Assuming these savings to occur at the end of each year, determine the (total) present worth of the proposed acquisition.

[*Answer*: **$4,966.**]

3.26 A certain project is under consideration that requires an initial investment of $25,000 with an expected salvage value of $3,000 at the end of 10 years. Annual savings in operating costs, assumed to flow continuously and uniformly throughout each year, are estimated to be $13,500. If the firm's MARR is 10%, determine the present worth of this project.

3.27 A certain investment in manufacturing equipment is expected to result in a savings of $10,000 per month, each month, over the next 8 years.

The investor's discount rate (MARR) is 12% per annum.

a. Assuming that these cash flows are concentrated at the *end* of each *year*—$120,000 per year—determine the equivalent present value.

b. Assuming that these cash flows occur uniformly and continuously *during* each *year*, determine the equivalent present value.

c. Assuming that these cash flows occur at the *end* of each *month*—$10,000 per month—determine the equivalent present value.

[*Answers*: **a. $596,160; b. $631,200; c. $628,232.**]

3.28 From an article in the *Los Angeles Times* (1988), "A Costa Mesa man won $2.2 million from the California Lottery, officials announced Thursday,... [the winner] will pay 20% of his winnings to the Internal Revenue Service but will not pay any state tax, the spokeswoman said.... The winnings will be paid out to [the winner] in annual installments of $88,000 over the next 20 years...." That is,

$$\text{Annual payment} = (1.00 - 0.20)\left(\frac{\$2,200,000}{20}\right)$$

$$= \$88,000$$

Assume that the payments will be made at the start of each year.

a. Find the equivalent present value of these after-tax payments assuming that the winner's minimum attractive rate of return is 12% per year.

b. Suppose that the California Lottery were to offer the winner a one-time, lump-sum payoff, now, of $1.0 million before tax withholding. This would be $800,000 after the 20% payment to the IRS for federal income taxes. From the winner's point of view, is this an attractive offer? Explain.

3.29 A manufacturing firm is considering the purchase of a production facility that it will keep for 15 years. Insurance premiums of $40,000 per year are to be paid at the start of every 3-year interval. That is, $120,000 will be paid at the start of years 1, 4, 7, 10, and 13. If the firm's MARR is 12% per year, find the equivalent present value of these insurance premiums.

[*Answer*: **$340,280.**]

3.30 Investment in an industrial crane is expected to produce net cash flows from rentals, as shown below, over the next 6 years.

End of Year:	1	2	3	4	5	6
Net cash flow ($):	15,000	12,500	10,000	7,500	5,000	2,500

If the firm's discount rate is 12%:

a. Determine the equivalent present worth of this sequence of cash flows.

b. The crane may be purchased for $30,000. It will be sold at the end of 6 years for an expected $10,000. Determine the total present worth of this proposed investment, that is, rental receipts less capital costs.

3.31 The XYZ Company produces a certain component that is subsequently used in the manufacture of the firm's final product. The equipment used in the production of this component was purchased 20 years ago at a cost of $200,000. The equipment is now quite old, however, and currently has no market value. The cost of labor last year was $40,000; it is expected to be $50,000 next year and to increase at the rate of $10,000 per year thereafter. The firm's pretax minimum attractive rate of return is 20% per annum. Because wages are paid weekly, the continuous cash flow assumption is appropriate; for example, the labor cost next year will be $50,000 flowing continuously and uniformly during the year. It is expected that there will be a continuing need for this component over the next 5 years. Determine the present worth of operating costs over 5 years.
[*Answer*: $217,848.]

3.32 The United Services Corporation (USC) currently purchases certain components from a subcontractor at a cost of $200,000 per year. These components will be needed over the next 10 years. As an alternative, USC could manufacture the parts "in-house." To do so would require the acquisition of a heavy-duty milling machine at a cost of $100,000; the expected salvage value at the end of 10 years is $25,000. Property taxes and insurance are estimated to be $5,000 the first year, declining by $200 per year

each year over the 10-year period. Annual costs of labor and materials are expected to be $150,000 the first year, increasing by $5,000 each year over the 10-year period. The firm's minimum attractive rate of return is 20% per year. Determine the present worth of the proposed investment.

3.33 The General Manufacturing Corporation (GMC) is currently spending $5,000 at the end of every 6 months for the periodic maintenance of its materials handling system. As an alternative, GMC can contract with the WeDoIt Company to provide this maintenance at a cost of $700 per month, with monthly payments at the start of each month. The materials handling system will be in service for 8 years. Assume that there will be maintenance at the end of the final 6-month period. This will be necessary to condition the equipment prior to resale. GMC's discount rate is 10% per year. Find the present worth of costs of the two alternatives.
[*Answers*: PWOC (in-house) = $54,652 and PWOC (contracted) = $47,206.]

3.34 a. The firm is considering the purchase of the component parts from an outside supplier (vendor). Under the terms of the contact agreement, parts may be purchased at a cost of $70,000 at the start of each and every year over the 5-year planning horizon. Determine the present worth of costs of this purchase option over 5 years.

b. Another alternative is the purchase of a semiautomatic machine to produce these parts. The initial cost is $300,000, and the expected salvage value after 5 years is expected to be $100,000. Costs of operation are expected to be $20,000 per year flowing continuously and uniformly each year throughout the 5-year planning horizon. Determine the present worth of costs of this alternative.

3.35 Marlene Corp. is a medium-sized producer of industrial fasteners. The company is considering the acquisition of certain manufacturing equipment with an initial cost of $80,000, an expected life of 6 years, and expected salvage value of $10,000 at the end of 6 years. Operating costs associated with the equipment are estimated to be $24,000 per year. Maintenance costs in year j are expected to be

$$\$2,500 + \$500j, \qquad j = 1, 2, 3, 4, 5$$

and zero otherwise. The firm's pretax MARR is 20%.

a. Assuming that operating costs flow continuously and uniformly throughout the year, determine the present worth of these operating costs.

b. Assuming that maintenance costs flow at the end of year j, determine the present worth of these maintenance costs.

c. Determine the total PW of costs of the proposed investment.

d. If the operating costs are assumed to occur at the end of each month, that is, $2,000 per month, determine the present worth of these operating costs.

[*Answers*: a. $87,551; b. $11,425; c. $175,627; d. $86,888.]

3.36 The ABC Company plans to acquire a certain unit of production equipment that will require periodic maintenance at a cost of $1,000 every 3 months after acquisition. The firm's discount rate is 2% per month.

a. The firm plans to use this equipment for 5 years, and maintenance costs will be experienced every quarter except the last one. That is, maintenance costs of $1,000 will be experienced at end of months 3, 6, 9,..., 54, 57. Find the equivalent present value of these periodic maintenance costs.

b. The cost of operation of this equipment is expected to be $100,000 per year over the 5 years of ownership. Assuming that these operating costs flow continuously and uniformly throughout each year, find the equivalent present value.

3.37 A company is considering three investment proposals, S, T, and U. They must implement one or more of the proposals to resolve the situation under consideration. Proposals S and U are mutually exclusive, and proposal T is contingent on S. The cash flow data for the investments over a 10-year planning horizon are given. The company has a budget limit of $850,000 for investments of the type being considered.

	Proposal S	Proposal T	Proposal U
Initial investment	$300,000	$450,000	$600,000
Life	10 years	10 years	10 years
Salvage value	$ 50,000	$ 50,000	$100,000
Annual receipts	$350,000	$350,000	$450,000
Annual disbursements	$200,000	$100,000	$200,000

Assume that $i = 12\%$. Determine the present worths of the feasible combinations of proposals.
[*Answer*: PW($\varnothing$) = $0, PW(S) = $563,600, PW(U) = $844,752, and PW(S + T) = $1,542,285.]

3.38 A company is considering engineering proposals P, Q, and R, each with a service life of 8 years, and with cash flows estimated to be as follows:

	Proposal P	Proposal Q	Proposal R
Investment	$800,000	$600,000	$400,000
Annual revenue	$450,000	$400,000	$300,000
Annual costs	$200,000	$180,000	$150,000
Salvage value	$100,000	$ 80,000	$ 60,000

Proposals P and Q are mutually exclusive, and Proposal R is contingent on Proposal Q. The investment limit is $1,000,000. If the firm's MARR is 12%, determine the present worth of each feasible mutually exclusive investment "package."

3.39 A firm has available $30,000 for investment. The after-tax cost of this capital is 6%. The firm is considering four investments as summarized below:

Project	Initial Investment	Net Annual Benefits
X	$10,000	$1,628
Y1	10,000	1,558
Y2	20,000	2,981
Z	10,000	1,457

Projects Y1 and Y2 are mutually exclusive. Otherwise, X, Y (either Y1 or Y2), and Z are independent investments. The expected life of each project is 10 years; the expected salvage value of each proposal is zero.

a. If the manager of Department Y must choose between Y1 and Y2, which should be chosen?
b. Determine which projects should be in the firm's optimum capital budget. What is the present worth of this optimum set of projects?
[*Answers*: a. Y2; b. PW(X + Y1 + Z) = $4,119.]

Annual Worth or (Equivalent Uniform) Annual Cost

3.40 Refer to Problem 3.1. Find the annual worth of series A, series B, and series C.

3.41 Refer to Problem 3.2. Find the annual worth of series D, series E, and series F.
[*Answers*: $48.92, $48.12, and $37.19.]

3.42 Refer to Problem 3.3. Find the annual worth of alternative G and alternative H.

3.43 Refer to Problem 3.4. Find the annual worth of series I, J, K, and L.
[*Answers*: $145, $118, $175, and $74.]

3.44 Find the (equivalent uniform) annual cost of the Bigditch construction project described in Problem 3.8.

3.45 Find the (equivalent uniform) annual cost of the electronic data processing system described in Problem 3.9.
[*Answer*: $15,600.]

3.46 Find the (equivalent uniform) annual cost of the welding machine described in Problem 3.11.

3.47 Find the (equivalent uniform) annual cost of the forklift truck described in Problem 3.12.
[*Answer*: $10,835.]

3.48 Find the (equivalent uniform) annual cost of the project described in Problem 3.14.

3.49 Find the equivalent uniform annual cost of the facility described in Problem 3.17.
[*Answer*: $68,400.]

3.50 Find the (equivalent uniform) annual costs of (a) continuing with the present system and (b) purchasing the sprinkler system as described in Problem 3.18.

3.51 Find the (equivalent uniform) annual cost of the manufacturing equipment described in Problem 3.19.
[*Answer*: $46,394.]

3.52 Determine the (equivalent uniform) annual cost of the painting robot described in Problem 3.22.

3.53 Determine the annual cost of the project described in Problem 3.23.
[*Answer*: $32,892.]

3.54 Determine the total annual worth of the project described in Problem 3.24.

3.55 Determine the annual worth of the project described in Problem 3.26.
[*Answer*: $10,285.]

3.56 Determine the annual worth of the industrial crane described in Problem 3.30.

3.57 Determine the EUAC of the operating costs described in Problem 3.31.
[*Answer*: $72,844.]

3.58 Determine the EUAC of the heavy-duty milling machine described in Problem 3.32.

3.59 Determine the EUAC of the Marlene Corp. manufacturing equipment described in Problem 3.35.
[*Answer*: $52,813.]

3.60 A certain data processing system may be purchased for $100,000. Annual operating costs are expected to be $20,000 the first year, increasing by $2,000 each year. (Assume that operating costs occur uniformly and continuously throughout each year.) Preventive maintenance costs of $5,000 are expected at the end of every 6-month period throughout the life of the system. It is anticipated that the system may be sold for $25,000 at the end of its 5-year service life. The interest rate (opportunity cost) is 10% per year.
a. Find the capital recovery cost for the system.
b. Find the (equivalent uniform) annual cost of the preventive maintenance.
c. Find the total EUAC for this system.

3.61 An international funding agency is considering the construction of a road in a certain country. The climate is extremely arid, and road use is expected to remain low, so an infinite life is assumed. ("Infinite life" is often assumed when $N \geqslant 50$ years.) If the road is constructed, it is expected that net costs to roadway users (fuel costs and users' time) will be $1,000,000 per year. Maintenance costs of $400,000 will be experienced every 5 years. Assuming the during-year convention for user costs and the end-of-year convention for maintenance costs, find the equivalent uniform annual cost of this project if the discount rate is 10% per year.

[*Answer*: **$1,114,700.**]

3.62 The Black Company is considering two alternative plans for erecting a fence around its new central city plant. Galvanized steel "chicken-wire" fencing requires a first cost of $35,000 and estimated annual upkeep costs of $300. The expected life is 25 years. A concrete block wall requires a first cost of only $20,000, but it will need minor repairs every 5 years at a cost of $1,000 and major repairs every 10 years at a cost of $5,000. Assuming an interest rate of 10% before taxes and a perpetual (continuing) need, determine the (equivalent uniform) annual costs for the two plans. Specify all other assumptions.

3.63 The National Park Service is developing a certain recreation facility that will require periodic major maintenance at a cost of $10,000 every 18 months. This federal agency uses an effective interest rate of 10% per year. Assume an infinite planning horizon.
a. Find the (equivalent uniform) cost per month.
b. Find the equivalent uniform annual cost.

[*Answers*: **a. $519; b. $6,506.**]

3.64 Consider a proposed investment with an expected life of 8 years, an initial cost of $20,000, and an expected salvage value of $5,000 at the end of 8 years. If acquired, this proposal will result in savings of $3,000 the first year, increasing by $400 per year; that is, the savings will be $3,000, $3,400, $3,800,..., $5,800. Determine the annual worth of this proposal if the interest rate is 15% per year.

3.65 VKC Manufacturing Company is considering an investment in new materials handling equipment that is expected to generate substantial savings. Estimates are as follows:

Initial cost	$200,000
Service life	7 years
Net salvage value after 7 years	$40,000
Major maintenance cost at the start of the fifth year	$25,000
Net annual savings: $30,000 the 1st year, increasing by $4,000 per year, to a maximum savings of $54,000 in the seventh year (assume "end-of-year" cash flows)	

The firm's minimum attractive rate of return is 12% per year.
a. Determine the average net cash flow per year.
b. Determine the annual worth (AW).

[*Answers*: **a. $15,571; b. −$3,133.**]

3.66 The Uncle Charlie Manufacturing Company is considering the acquisition of certain metal-forming equipment. The initial cost of the equipment is $80,000. If it is acquired, it will be used for 6 years, then sold for an estimated $20,000. Property taxes and insurance expenses will be $800 semiannually, payable at the start of each 6-month period. Periodic maintenance will be required at a cost of $6,000 at the end of years 1, 2, 3, 4, and 5. The firm's minimum attractive rate of return is 12%.
a. Determine the average annual cost (negative cash flow) per year over the 6-year study period.
b. Determine the (equivalent uniform) annual cost.

3.67 GAF Contractors is a medium-sized general construction company located in central California. The firm is considering the acquisition of certain earth-moving equipment to be used on various projects over the next 4 years. Initial cost of the equipment is $90,000; expected salvage value at the end of 4 years is $15,000. The firm's pretax MARR is 20%. Insurance for the equipment, payable at the start of each year, is $1,000 per year. Determine the pretax (equivalent uniform) annual cost.

[*Answer*: **$33,173.**]

3.68 The University is considering the purchase of a commercial-grade dishwasher for the cafeteria. The cost is $8,000. If it is purchased, the dishwasher will be kept in service for 7 years and then sold for an expected $2,000. The University's "cost of capital," as estimated by the rate of return on its endowment funds, is 8% per year. The cost of operating the dishwasher is estimated to be $600 per month. Find the (equivalent uniform) annual cost of this proposal.

3.69 Certain manufacturing equipment will be purchased for $100,000, kept in service for 4 years, and then sold for an expected $25,000 salvage value. Taxes and insurance will be $2,000 per year, paid at the start of each year. Operating costs will be $1,000 per week, or $50,000 per year based on a 50-week year. The firm's minimum attractive rate of return is 12% per year. Determine the (equivalent uniform) annual cost for this investment.

[*Answer*: $82,876.]

3.70 The Brown Manufacturing Company is considering the purchase of a numerically controlled (NC) machine. The initial cost is $60,000. If it is purchased, it will be kept in service for 6 years, then sold for an expected $10,000. The firm's minimum attractive rate of return is 20% before income taxes. Other expected consequences of this proposed investment are as follows:

- Periodic maintenance—$3,000 per year, payable at the beginning of years 2 through 6
- Property taxes and insurance—$4,000 per quarter, payable at the beginning of quarters 1 through 24
- Operating costs—$40,000 per year, assumed to flow continuously and uniformly during each year

Find the EUAC for this proposed investment.

3.71 A welding robot may be purchased at a cost of $50,000. It will be used for 8 years, at the end of which time it will be sold for $10,000. Operating costs, assumed to flow continuously and uniformly during each year, are expected to be $20,000 per year. Property taxes and insurance, payable at the start of each year, are expected to be $1,000 the first year and increase by $200 per year. The firm's minimum attractive rate of return is 20% per year. Determine the EUAC for this proposed investment.

[*Answer*: $36,179.]

3.72 The Trojan Corporation is a mid-sized manufacturer of materials handling equipment. The firm's pretax minimum attractive rate of return is 20% (per annum). Trojan's engineering design department has developed a new product that will require an initial investment of $950,000 for manufacturing equipment. This equipment will be used for 5 years and then sold for an expected $50,000. The new product will require expenditures for materials of $200,000 at the start of each quarter over the 5-year life of the project (20 quarters). Costs of labor are expected to be $50,000 at the end of each month over the 5 years (60 months).
   a. Determine the total EUAC for the proposed project.
   b. Suppose that the cash flows for material and labor are assumed to flow at the midpoints of each year. That is, materials costs of $800,000 ($= 4 \times \$200,000$) and labor costs of $600,000 ($= 12 \times \$50,000$) are assumed to flow at the midpoints of years 1 through 5. Determine the total EUAC under this assumption.

3.73 The Plant Engineering Department is considering the modification of a certain machine that will cost $3,000 now. Alternatively, if the modification is not made, operating expenses will increase by $300 per year over the next 5 years. The machine will be used for exactly 5 years and will have no salvage value under either alternative at the end of that time. Assume that operating expenses will occur at the end of each year. The firm's minimum attractive rate of return is 15%.
   a. Determine the EUAC if the modification is not made.
   b. Determine the EUAC if the modification is made.

[*Answers*: **a.** $895; **b.** $817.]

3.74 A certain methods improvement in an assembly operation will result in labor savings of $10,000 per month for 60 months. The initial cost of installing these improvements will be $500,000; there is no residual value after 60 months. The firm's pretax MARR is 20% per year.
   a. Determine the equivalent uniform annual worth of this proposal.
   b. Determine the minimum monthly labor savings such that the methods improvement would be warranted.

3.75 A firm is considering the installation of a sprinkler system in its warehouse. The initial cost of the sprinkler system is $4,000, the expected life is 15 years, and the expected salvage value is zero after 15 years. The firm's pretax MARR is 15%. Preliminary engineering analysis yields the following estimates:

	Without Sprinklers	With Sprinklers
Cost of fire insurance to provide $100,000 insured coverage	$700	$400
Probability of loss (fire) in one year	0.006	0.003
Probability of no loss (no fire) in one year	0.994	0.997
Expected uninsured loss if fire occurs	$50,000	$20,000

a. Determine the EUAC without sprinklers.
b. Determine the EUAC if the sprinkler system is acquired.
c. Determine the maximum initial cost of the sprinkler system such that its purchase would be warranted.
[*Answers*: a. **$1,000**; b. **$1,144**; c. **$3,158.**]

3.76 The Public Works Department of Central City is responsible for the selection of air-conditioning units to be installed at City Hall. (There is, at present, no central air conditioning in the building.) Twenty offices will be equipped: one air-conditioning unit per office. Two units are under consideration:

	Type A	Type B
Initial cost per unit	$600	$800
Installation cost per unit	$50	$50
Output (BTUs)	35,000	40,500
Efficiency rating (BTUs output per watt of electrical consumption)	7.0	7.5
Expected life (years)	8	8
Salvage value after 8 years	None	None

The units will be used an average of 1,000 hours annually. (Assume that usage is distributed evenly throughout the year.) Both units are equally likely to require maintenance and repairs. The City's opportunity cost is 10% per annum. The cost of energy is $0.05 per kilowatt-hour. Determine the (equivalent uniform) annual cost for the Type A and Type B units.

3.77 Two possible routes for a power transmission line, one direct and one indirect, are under consideration. Relevant data are:

	Indirect Route	Direct Route
Length	15 miles	5 miles
Initial cost	$5,000/mile	$25,000/mile
Annual maintenance	$200/mile/year	$400/mile/year
Useful life	15 years	15 years
Salvage value	$3,000/mile	$5,000/mile
Annual cost of power loss	$500/mile	$500/mile
Annual property taxes	2% of first cost	2% of first cost

Assume that all cash flows are end-of-period. Using an 8% interest rate, determine the annual costs of each of the two routes.
[*Answers*: **AC(direct) = $19,105**; **AC(indirect) = $20,683.**]

Future Worth

3.78 Refer to Problem 3.1. Find the future worth of series A, series B, and series C.

3.79 Refer to Problem 3.2. Find the future worth of series D, series E, and series F.
[*Answers*: **FW(D) = $377**; **FW(E) = $371**; **FW(F) = $287.**]

3.80 Refer to Problem 3.3. Find the future worth of alternative G and alternative H.

3.81 Refer to Problem 3.4. Find the future worths of series I, J, K, and L.
[*Answers*: **FW(I) = $2,552**; **FW(J) = $2,072**; **FW(K) = $3,066**; **FW(L) = $1,300.**]

3.82 Refer to Problem 3.5. Find the future worths of investment projects M and N.

3.83 Refer to Problem 3.9. Find the future worth of the electronic data processing system.
[*Answer*: **$95,271.**]

Unequal Lives

3.84 A certain operation can be performed satisfactorily by both machine X and machine Y. Pertinent data for the two alternatives are:

	Machine X	Machine Y
First cost	$6,000	$14,000
Salvage value	$0	$2,800
Service life	12 years	18 years
Annual disbursements	$4,000	$2,400

Compare the equivalent total annual cost, assuming a minimum attractive rate of return of 12% before taxes. Specify all assumptions.

3.85 Two gas-powered electric generators are being considered for purchase. Relevant data are:

	Economy Generator	Deluxe Generator
Initial cost	$12,000	$20,000
Annual operating expense (during year)	$1,600	$ 900
Salvage value	$3,000	$2,000
Operating life	6 years	12 years

Assuming a 10% interest rate, determine which alternative is more economical. State all necessary assumptions.
[*Answers*: AC(Economy) = $4,045 and AC(Deluxe) = $3,786, assuming 12-year planning horizon and identical replication.]

3.86 Two types of conveyor systems, System I and System II, are being considered by the Blue Corporation:

	System I	System II
Initial investment	$5,000	$12,000
Net salvage value	$1,000	0
Annual disbursements	$900	$600
Estimated life	10 years	15 years

Assuming a pretax interest rate of 20 percent, which system appears to be the better investment, everything else being equal? State all necessary assumptions.

3.87 A firm is considering two investment alternatives, Alpha and Beta. Alpha has an initial cost of $1,000, a 3-year life, zero salvage value after 3 years, and it returns $500 at the end of each year over the 3-year life. At the end of 3 years, an identical Alpha will be available for purchase.

Beta has an initial cost of $1,000, a 6-year life, and a $3,700 salvage value after 6 years. There will be an additional cost of $1,000 occurring continuously and uniformly during the third year of ownership.

The firm's pretax minimum attractive rate of return is 10%. Use the present worth method to determine which alternative is preferable.
[*Answer*: PW(Alpha) = $426; PW(Beta) = $300.]

3.88 Ms. Tami Trojan is considering the purchase of an office machine for her consulting firm to replace a manual operation that currently costs $900 per year. She is considering two alternatives.

Alternative A has an initial cost of $2,000, operating costs of $400 per year, service life of 6 years, and $300 salvage value after 6 years. (Note that this results in a savings of $500 per year over the current manual method.)

Alternative B1 has an initial cost of $1,000, operating costs of $500 per year, service life of 3 years, and zero salvage value after 3 years. At the end of 3 years, a successor machine will be available—call this B2—that will cost only $600, have operating costs of $500 per year, a service life of 3 years, and a $300 salvage value at the end of its service life.

The business activity requiring these machines will last over the next 6 years. Ms. Trojan's minimum attractive rate of return is 10% before income taxes.

The cash flows associated with the alternatives are as follows:

End of Year	A	B1	B2	
0	−$2,000	−$1,000		Note that positive
1–3	500	400		cash flows in
3	—	0	−$600	years 1–6
4–6	500		400	represent savings
6	300		300	over the current manual operation.

Determine the present worth of the alternative policies over the 6-year planning horizon.

3.89 Consider the following end-of-period cash flows for two mutually exclusive alternatives:

Period (j):	0	1	2	3	4	Total
Project I ($):	−100	40	40	40	40	60
Project II ($):	−110	55	55	55		55

Note that the life of Project I is 4 years; the life of Project II is 3 years. The investor's minimum attractive rate of return is 5% per period.
a. Determine the PW of Project I ($N = 4$ years) and the PW of Project II ($N = 3$ years).
b. With the identical replication assumption, and assuming a planning horizon of exactly 12 periods, determine the annual worths of Project I and Project II.
c. Assume that there will be no future replacement(s). That is, if Project I is selected, there will be no replacement after 4 years; if Project II is selected, there will be no replacement after 3 years. If this is so, which alternative is preferred? Explain.
[*Answers*: **a. $41.84, $39.78; b. $11.80, $14.61; c. PW(I) > PW(II).**]

Bonds

3.90 Ms. Jones has an opportunity to purchase a bond from the Los Angeles Public Utility Company on February 1, 19xx. The terms of the bond are as follows:

- $1,000 face value due and payable exactly 3 years later
- 8% coupon rate
- Interest payable quarterly each January 31, April 30, July 31, October 31, up to and including the day the bond becomes due

(That is, there are 12 future interest payments.)

If Ms. Jones' effective opportunity cost (discount rate) is 3% per quarter, determine the equivalent present value of this bond to her.

3.91 Mr. Master is considering the purchase of a new bond that is soon to be issued by the Comma Corporation. The bond will have a $1,000 par value and will be due and payable in exactly 10 years. Bond interest payments will be semi-annual; the bond rate is 8%. Mr. Master's pretax opportunity cost is 12% nominal per annum. The Comma Corporation's pretax opportunity cost is 14% nominal per annum.
a. From Mr. Master's point of view, what is the minimum cost of the bond that would make it an attractive investment?
b. From the Comma Corporation's point of view, what is the equivalent present value of the bond payments?
[*Answers*: **a. $771; b. $682.**]

Leases

3.92 The L.A. Leasing Company is planning the purchase of a vehicle that is intended for subsequent leasing to the firm's clients. The initial cost of this vehicle is $10,000. It will be kept for 3 years, at the end of which time it will be sold for an expected $4,000. Insurance for the vehicle is $400 per year, with insurance premiums payable at the start of each year. The firm's minimum attractive rate of return is 10% per annum.
a. Determine the (equivalent uniform) annual cost to the firm.
b. If the firm desires to earn a rate of return of 10% per annum, determine the monthly lease payments it should charge. Assume that lease payments will be made at the end of each month.
[*Answers*: **a. $3,253; b. $259.40.**]

A P P E N D I X: *Cash Flow Analyzer (CFA)*

In the Appendix to Chapter 2 we indicated that a variety of general-purpose software may be used effectively for engineering economic analysis. In particular, when cash flow data are displayed in tabular format, spreadsheet programs are available that permit rapid and accurate calculations with little or no additional programming. *SuperCalc*, *Lotus 1-2-3*, and *Excel*,[4] among others, are widely used for this purpose.

In addition to generic spreadsheet software, a wide variety of special-purpose software is commercially available to assist the analyst in evaluation of proposed investments. To mention just a few:

- *Economic Analysis*, available from the Institute of Industrial Engineers, 25 Technology Park/Atlanta, Norcross, GA 30092.

- *Business Wits*, available from DECISUS Inc. (A Xerox subsidiary), 9938 Via Pasar, Suite A, San Diego, CA 92126.

- *TL-5310 Business Manager*, available from Texas Instruments, Customer Response Center, M.S. 57, P.O. Box 655012, Dallas, TX 75265. (Not computer software, strictly speaking, the TL-5310 is representative of a variety of pocket calculators that have been "hard-wired" to perform certain investment calculations.)

There are literally hundreds of these special-purpose programs and devices, far too many to review here. An excellent summary may be found in *The Individual Investor's Guide to Computerized Investing*, published by The American Association of Individual Investors, 625 North Michigan Avenue, Department CI, Chicago, IL 60611. The 1993 edition of this book listed 513 software products and 162 information services.

A specific special-purpose software that you may find particularly useful is *Cash Flow Analyzer* (CFA), developed especially for use by students of engineering economics.[5] This software, suitable for use with IBM or compatible equipment, has the following principal features:

- CFA provides the user with the ability to calculate several figures of merit—present worth (PW, NPV), annual worth (AW, EUAC), future worth (FW), and internal rate of return (IRR)—using a variety of standard cash flow models.

- Once data have been entered, CFA sums the present, annual, and future worths of individual sequences of cash flows and thus enables the analyst to find internal rates of return, to plot present worth versus a specified range of discount rates, to print reports, and to perform sensitivity analyses on discount rate, planning horizon, and individual cash flow parameters.

---

[4]*SuperCalc* is a registered trademark of Computer Associates International, Inc., *Lotus 1-2-3* is a registered trademark of the Lotus Development Corporation, and *Excel* is a registered trademark of Borland International.

[5]CFA is available from the author, G. A. Fleischer and Associates, P.O. Box 5611, Los Alamitos, CA 90721-5611. The price, $19.95, includes California sales tax and shipping and handling. When ordering, please specify whether a $5\frac{1}{4}''$ or $3\frac{1}{2}''$ floppy disk is desired.

- Pop-up menus display all user options, prompt for all necessary data for the selected calculations, and provide error messages, when appropriate, to advise the user as to the source of the error.
- The user may abort any entry or selection and may edit or delete cash flow entries at any time.
- An on-line help facility gives the user a brief tutorial.
- CFA has been written to support both monochrome and color monitors, with or without graphics, and both IBM and EPSON printers and their compatibles.
- CFA handles a wide variety of cash flow patterns, including uniform (annuity) series, arithmetic and geometric gradient series, and cash flows that occur more frequently or less frequently than the compounding period. It also handles both discrete and continuous cash flows.
- CFA permits planning horizons up to 99 periods and interest rates up to 50%. Cash flows may be entered with up to 8 significant digits. Internal rates of return may be calculated up to three-place decimal accuracy, that is, to 0.001%.
- CFA can detect cases in which multiple rates of return may be present and provides a procedure for identifying all real positive rates. This is an important issue, as will be discussed in Chapter 4.

To illustrate CFA, consider Example 3.1 presented in the body of the chapter. Recall that the initial investment in certain equipment is $12,000, salvage value at the end of 6 years is $3,000, annual operating expenses are $6,000 (end-of-years 1–6), and the discount rate is 12% per year. CFA's cash flow table for this problem is shown as **Figure 3.A.1**; the printout of the solution is shown as **Figure 3.A.2**.

MARR= 12.00%   Example 3.1   Horizon= 6

T	1st	End	A(1st)	Suppl.Data	PW	Descrip
S	0		-12,000.0		-12,000.0	initial cost
U	1	6	-6,000.0		-24,668.4	operating costs
S	6		3,000.0		1,519.8	salvage value
ΣA(j)=			-45,000.0	Σ PW (j)=	-35,148.6	IRR=

Figure 3.A.1 CFA Cash Flow Table for Example 3.1

*** EXAMPLE 3.1 ***

MARR (Discount Rate) : 12.00%

Planning Horizon :   6 periods

Cash Flow Information

#	T	1st	End	A(1st)	Suppl. Data	Description
1	S	0	—	-12,000.0		initial cost
2	U	1	6	-6,000.0		operating costs
3	S	6		3,000.0		salvage value

Cash Flow Analysis

#	PW	AW	FW
1	-12,000.0	-2,918.7	-23,685.8
2	-24,668.4	-6,000.0	-48,691.1
3	1,519.8	369.6	3,000.0

Sum of Present Worths : -35,148.6

Sum of Annual Worths : -8,549.1

Sum of Future Worths : -69,376.9

Sum of All Cash Flows Under 0.0% MARR : -45,000.0

Internal Rate of Return : Not Calculated

Figure 3.A.2   Solutions to Example 3.1 Using CFA

Of course, the same problem can be analyzed using a generic spreadsheet program. **Figure 3.A.3** is an analysis of Problem 3.1 using *Lotus 1-2-3*. It should be noted that the present worth of the cash flows in this illustration (column B) is obtained by first computing the equivalent present value of each A_j (column C) and then summing the values. The result may also be obtained more directly in *Lotus 1-2-3* by using the specialized NPV command:

@ NPV (interest rate, cash flows)

Of course, both CFA and *Lotus 1-2-3* yield the correct solution, − $35,149.

	A	B	C	D	E	F
1						
2	EXAMPLE 3.A.1					
3						
4	CASH FLOW TABLES FOR PRESENT, ANNUAL AND FUTURE WORTHS					
5						
6	Given:	Discount rate (i)				0.12
7		Study period, planning horizon (N)				6
8		End of period cash flows (Aj)				See table
9						
10			Equivalent	Equivalent	Equivalent	
11	End of	Cash	Present	Annual	Future	
12	Period	Flow	Value	Value	Value	
13	j	Aj	Pj		Fj	
14						
15	0	-12,000	-12,000		-23,686	
16	1	-6,000	-5,357		-10,574	
17	2	-6,000	-4,783		-9,441	
18	3	-6,000	-4,271		-8,430	
19	4	-6,000	-3,813		-7,526	
20	5	-6,000	-3,405		-6,720	
21	6	-3,000	-1,520		-3,000	
22						
23	TOTALS:	-45,000	-35,149	-8,549	-69,377	
24						
25	Sample Cell Formulas:					
26	C15 = +B15*(1+$F$6)^(-A15)					
27	E15 = +B15*((1+$F$6)^($F$7-A15))					
28	B23 = @SUM(B15..B21)					
29	C23 = @SUM(C15..C21)					
30	D23 = +C23*($F$6/(1-(1+$F$6)^(-$F$7)))					
31	E23 = @SUM(E15..E21) also = +C23*((1+$F$6)^($F$7))					

Figure 3.A.3   Analysis of Example 3.1 Using *Lotus 1-2-3*

4

Measures of Worth II: Ranking of Alternatives by Marginal (Incremental) Analysis

4.1 INTRODUCTION

The methods discussed in Chapter 3 permit direct rank ordering of alternatives on the basis of their respective measures of worth: PW, AW (or AC), and FW. This chapter now turns to two other methods, rate of return and benefit–cost ratio, for which such an approach is not appropriate. It will be shown that mutually exclusive alternatives may not, in general, be rank-ordered on the basis of rate of return or benefit–cost ratio. To do so is a major theoretical error and can lead to misallocation of resources. To emphasize this point, the methods of Chapters 3 and 4 have been separated in the organization of this book.

4.2 RATE OF RETURN

The **rate of return method** is widely used in the private sector. Surveys of industrial practice indicate that this method, along with the present worth (net present value) method, is most common among the various discounted cash flow techniques.

4.2.1 (Internal) Rate of Return Defined

The **(internal) rate of return** for a given investment proposal is that interest rate, i^*, for which the net present value of all expected cash flows is precisely zero. When all cash flows are discounted at rate i^*, the equivalent present worth of all benefits exactly equals the equivalent present worth of project costs. One mathematical definition of the internal rate of return is the rate, i^*, that satisfies the equation

$$PW = \sum_{j=0}^{N} A_j(1 + i^*)^{-j} = 0 \qquad (4.1)$$

This formula assumes discrete cash flows A_j and end-of-period discounting in periods $j = 1, 2, \ldots, N$.

An alternative definition of rate of return is that it is the particular interest rate, i^*, for which the present worth of all future cash flows is exactly equal to the initial investment, P. With the assumptions underlying Eq. (4.1), for example, with $P = -A_0$, we have

$$PW = A_0 + \sum_{j=1}^{N} A_j(1 + i^*)^{-j} = 0$$

or

$$P = \sum_{j=1}^{N} A_j(1 + i^*)^{-j} \qquad (4.2)$$

Still another definition of rate of return is that interest rate, i^*, for which the present worth of all receipts (positive cash flows) is exactly equal to the present worth of all disbursements (negative cash flows). Note here that $A_j = R_j - C_j$, where R_j and C_j are the cash receipts and disbursements, respectively, incurred in period j. Again, using the assumptions underlying Eq. (4.1), we have

$$PW = \sum_{j=0}^{N} (R_j - C_j)(1 + i^*)^{-j} = 0$$

or

$$\sum_{j=0}^{N} R_j(1 + i^*)^{-j} = \sum_{j=0}^{N} C_j(1 + i^*)^{-j} \qquad (4.3)$$

When rate of return is computed in this manner, the amount as well as the timing of expected costs and revenues are taken into consideration. It is, in a sense, a measure of the profitability of the project. Thus, it is an *internal* rate of return. It arises solely from the amounts and timing of the cash flows associated with the investment; there is no relationship to any external factors. It is *not* a rate of return on the initial investment.[1] The terms *internal rate of return* (IRR) and *rate of return* (RoR), are used

---

[1] For further discussion of the properties of the (internal) rate of return, see Lynn E. Bussey, *The Economic Analysis of Industrial Projects* (Englewood Cliffs, NJ: Prentice-Hall, 1978), pp. 213–217. Given a cash flow sequence in which the initial cash flow is negative and all others are positive, Bussey demonstrates that the internal rate of return is the rate of interest earned on the time-varying, unrecovered balances of an investment such that the final balance is zero at the end of the project life.

interchangeably in the literature of finance and engineering economy. Both terms are used in this book.

4.2.2 The Rate of Return Method

The first step in the rate of return method is to determine the internal rate of return for the proposed investment. Next, in order for the investment to be acceptable, the rate of return must be compared to some minimum attractive rate that could be earned if the proposed project were rejected and the funds invested elsewhere.

A variety of names have been given to the method of calculation described above as the rate of return method. The most prominent are the **interest rate of return** and the **investor's method**. These methods are equivalent and, when properly applied, produce precisely the same solutions as those that result from the annual worth method and the present worth method. The rate of return method is frequently misused, however. Indeed, surveys of real-world applications suggest that misuse is more often the rule than the exception. Examples of abuse of the concept of rate of return are discussed in some detail in Chapter 5.

EXAMPLE 4.1

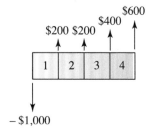

Figure 4.1 Example 4.1

Consider the cash flows given in **Table 4.1** for a proposed investment, Project P. The cash flow diagram is shown in **Figure 4.1**. Assuming end-of-period discounting, the internal rate of return is the value of i^* that satisfies the equation

$$\sum_{j=0}^{4} A_j (1 + i^*)^{-j} = 0$$

$$= -\$1,000 + \$200(1 + i^*)^{-1} + \$200(1 + i^*)^{-2}$$
$$+ \$400(1 + i^*)^{-3} + \$600(1 + i^*)^{-4}$$

There is no simple algebraic solution to this equation, so a trial-and-error approach is necessary. Using the 12% and 15% tables from Appendix B, the present worth of the cash flows is $4 when $i = 12\%$; it is −$69 when an

TABLE 4.1   Rate of Return Solution for Project P

End of Period	Cash Flow	Assuming $i = 12\%$		Assuming $i = 15\%$	
j	A_j	$(P/F, 12\%, j)$	PW@12%	$(P/F, 15\%, j)$	PW@15%
0	−$1,000	1.0000	−$1,000	1.0000	−$1,000
1	200	0.8929	179	0.8696	174
2	200	0.7972	159	0.7561	151
3	400	0.7119	285	0.6575	263
4	600	0.6355	381	0.5718	343
Totals	$400		$4		−$69

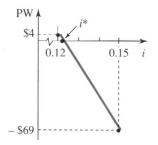

Figure 4.2 Determining IRR for Example 4.1

interest rate of 15% is used. Since the solving interest rate should result in zero present worth, it is clear that we have bracketed the solution with the tables available to us in the Appendix. The P/F factor is a monotone decreasing function of the interest rate, i. It is approximately linear over a small range of the interest rate. Thus, using straight-line interpolation, as indicated in **Figure 4.2**,

$$i^* = 12\% + (15\% - 12\%)\left[\frac{4}{4 - (-69)}\right]$$

$$= 12\% + 3\%\left(\frac{4}{73}\right) = 12.2\%$$

The internal rate of return for this proposed investment should now be compared to the minimum attractive rate of return, that is, the rate of return that could be earned by the investor if the funds were invested elsewhere. This is the value i used in the annual worth and present worth methods. Here, the proposed investment of $1,000 is justified if the minimum attractive rate of return is less than 12.2%. Put somewhat differently, AW > 0 and PW > 0 for this sequence of cash flows with $i < 12.2\%$.

In this example we have used available tables of discount factors for our trial-and-error calculations. There are, however, a variety of computer software available which, when given a set of cash flows, will generate the solving rate of return quickly and accurately. See, for example, *Lotus 1-2-3* or *Cash Flow Analyzer (CFA)*, described in the Appendix to this chapter. In addition, several hand-held pocket calculators will generate similar results for relatively simple cash flows. However, if appropriate computer software or a programmed pocket calculator is not used, the analyst will find the trial-and-error approach to be laborious. The next two sections, therefore, are intended to give some guidance as to how the (internal) rate of return, i^*, may be determined with minimal effort.

4.2.3   Shortcut Solutions Using One or Two Factors

In certain instances the pattern of cash flows may be such that solutions can be determined directly. For example, suppose that you want to determine the prospective rate of return from an initial investment of $100,000, which results in net revenues of $40,000 per year for 4 years. The equation that gives the prospective rate of return is

$$\sum_{j=0}^{4} A_j (1 + i^*)^{-j} = 0$$

$$= -\$100,000 + \$40,000 \sum_{j=1}^{4} (1 + i^*)^{-j}$$

$$= -\$100,000 + \$40,000(P/A, i^*, 4)$$

Thus

$$(P/A, i^*, 4) = \frac{\$100,000}{\$40,000}$$

$$= 2.5$$

From the compound interest tables in Appendix B, $(P/A, 20\%, 4) = 2.589$ and $(P/A, 25\%, 4) = 2.362$. Using linear interpolation, the proposed rate of return is found to be approximately 21.9%.

The factors for the uniform series and arithmetic gradient series should be used whenever possible to simplify calculations. Consider, for example, the following pattern of cash flows:

End of Period (j)	Cash Flow (A_j)
0	−$ 850
1–20	48
20	1,000

The *solution equation*, the equation that yields the prospective rate of return, is

$$\sum_{j=0}^{20} A_j(1 + i^*)^{-j} = 0$$

$$= -\$850 + \$48(P/A, i^*, 20) + \$1,000(P/F, i^*, 20)$$

Unlike in the previous example, a trial-and-error approach is required here because there is more than one factor in the solution equation. However, using the uniform-series present worth factor does simplify the problem; the solution equation is reduced from 20 terms to only two:

$$PW @ 6\% = -\$850 + \$48(11.470) + \$1,000(0.3118)$$

$$= +\$12$$

$$PW @ 7\% = -\$850 + \$48(10.594) + \$1,000(0.2584)$$

$$= -\$84$$

Interpolating,

$$i^* = 6.125\%$$

4.2.4 Estimating the Rate of Return: First Approximations

When trial-and-error calculations are required, it is obviously helpful if initial trials can result from an "educated guess" rather than from a random choice. There are several ways to improve the first approximation. First, *examine the cash flow tables to find out whether the solving rate of return is relatively low or high*, or, for that matter, whether there is a meaningful solution. Consider the following three examples:

End of Period	Project X	Project Y	Project Z
0	−$100	−$100	−$100
1	20	30	10
2	30	60	30
3	40	40	30
4	20	70	20
Totals	$ 10	$100	−$ 10

With an original investment of $100, Project X results in a net gain of $10 after 4 years, Project Y yields a net gain of $100, and Project Z results in a net loss. It follows that the rate of return will likely be low for X and relatively high for Y. (While there *is* a solution to the present worth equation for Project Z, what does it mean? If the cash flows shown represent those of the lender, then the solution rate of return is that obtained by the borrower. The lender's return is negative in this case.)

Second, *an irregular series of cash flows may be approximated by a uniform series.* Returning to Project P:

End of Period	Actual Cash Flow	Approximate Cash Flow
0	−$1,000	−$1,000
1	200	350
2	200	350
3	400	350
4	600	350
Totals	$ 400	$ 400

The approximate rate of return may now be found.

$$\sum_{j=1}^{4} A_j(1 + i^*)^{-j} = 0$$

$$= -\$1,000 + \$350(P/A, i^*, 4)$$

From which

$$(P/A, i^*, 4) - 2.857$$

From the compound interest tables, $(P/A, 15\%, 4) = 2.855$, so i^* is approximately 15%.

Third, *in conjunction with the result of the uniform-series approximation, observe the location of the largest cash flows.* The largest cash flows associated with Project P, for example, occur near the end of the project life, during the third and fourth periods. Thus, the actual rate of return should be something less than 15%, and our first trial might be 12%. (As determined earlier, the solution is 12.2%.)

4.2.5 Comparing Alternatives Using the Internal Rate of Return

Single Alternative to "Do Nothing"

The discount rate used in present worth calculations is the opportunity cost—a measure of the return that could be earned on capital if it were invested elsewhere. Thus a given proposed project should be economically attractive if and only if its internal rate of return (IRR) exceeds the cost of opportunities forgone as measured by the firm's discount rate, the minimum attractive rate of return (MARR). The proposed investment is preferred to "do nothing" if IRR > MARR.

In Example 4.1 we found that the IRR of the investment is 12.2%. If the MARR is, say, 10%, the proposed investment would be justified on an economic basis.

Two Alternatives to "Do Nothing"

The rationale underlying the rate of return method is that any incremental expenditure of funds must earn a rate of return at least as great as that which could be earned "elsewhere." Thus, when comparing two investment alternatives, it is necessary to determine the rate of return on the differences in cash flows between the alternatives. This is known as **incremental** (or **marginal**) **analysis**.

EXAMPLE 4.2

Suppose that there is an alternative to Project P presented in Example 4.1; this is Project Q, with cash flows as given in **Table 4.2**. The differences in cash flows, as shown in the table, yield an incremental rate of return as follows:

$$PW = 0 = -\$900 + \$400(1 + i^*)^{-1} + \$400(1 + i^*)^{-2} + \$200(1 + i^*)^{-3}$$

Solving,

$$i^* = 6.1\%$$

We conclude, therefore, that Project Q is preferred to Project P if the minimum attractive rate of return is less than 6.1%. The additional $900 investment required for Q yields a rate of return of 6.1%. This incremental investment is justified only if the MARR < 6.1%. (Of course, we would be interested in comparing Q with P only if P were preferred to the do-nothing alternative or, in certain cases, if the do-nothing alternative is not feasible.)

TABLE 4.2   Cash Flows for Two Mutually Exclusive
Alternatives

End of Period j	End-of-Period Cash Flows		
	Project P $A_j(P)$	Project Q $A_j(Q)$	Difference $\Delta A_j = A_j(Q) - A_j(P)$
0	−$1,000	−$1,900	−$900
1	200	600	400
2	200	600	400
3	400	600	200
4	600	600	0
Totals	$ 400	$ 500	$100
IRR	12.2%	10.0%	6.1%

It may be shown in this example that the IRR for Project Q is approximately 10.0%. Although $i^*(Q) < i^*(P)$, there are conditions for which Project Q is in fact the preferred alternative, namely, MARR $< 6.1\%$. That is, ranking alternatives by their respective rates of return may lead to incorrect solutions. This "ranking error" is discussed further in the next section.

Multiple Alternatives

Proper use of the rate of return method when considering two or more alternatives is a straightforward application of the incremental approach discussed above. The differences in cash flows for any pair of alternatives yield an **incremental rate of return**, $i^*(\Delta)$. The incremental investment in the more costly alternative is justified if $i^*(\Delta) > $ MARR. Rank ordering of a set of mutually exclusive alternatives results from successive application of this principle. (Note that the alternatives are listed in order of increasing costs. This is not a strict requirement, but otherwise, the analyst must exercise caution in interpreting the solving rate of return.)

EXAMPLE 4.3

Consider the four alternatives discussed previously in Example 3.4. Recall that MARR $= 20\%$ is assumed. The cash flows and present worths at $i = 20\%$ are summarized in **Table 4.3**.
    The IRR for alternative I is computed as:

$$PW(I) = -\$1,000 + \$4,000(P/F, i^*, 10) = 0$$

$$(P/F, i^*, 10) = 0.25$$

$$i^*(I) \simeq 14.9\%$$

TABLE 4.3 Example 4.3: Multiple Alternatives

End of Period j	Cash Flows			
	$A_j(I)$	$A_j(II)$	$A_j(III)$	$A_j(IV)$
0	−$1,000	−$1,000	−$1,100	−$2,000
1–10	0	300	320	550
10	4,000	0	0	0
Totals	$3,000	$2,000	$2,100	$3,500
PW @ 20%	−$ 354	$ 258	$ 241	$ 305
IRR(i^*)	14.9%	27.3%	26.3%	24.4%

Since the opportunity cost in this example is 20%, it is clear that Alternative I is *not* attractive; the funds should be invested elsewhere. The IRR for Alternative II is computed as:

$$PW(II) = -\$1,000 + \$300(P/A, i^*, 10) = 0$$

$$(P/A, i^*, 10) = 3.333$$

$$i^*(II) \simeq 27.3\%$$

Similarly, the IRRs for Alternatives III and IV may be shown to be 26.3% and 24.4%, respectively. Both III and IV are also preferable to doing nothing inasmuch as their respective IRRs exceed 20%.

At this point, it is tempting to rank-order Alternatives II, III, and IV according to their IRRs. But that would be a serious methodological error, because the incremental effects between pairs of projects have not been considered.

We know that II $\succ \varnothing$ and III $\succ \varnothing$. But is III preferable to II? To resolve this question, note that an incremental cost of $100 yields an incremental positive cash flow of $20 per period, each and every period, at the end of periods 1 through 10.

	Cash Flows		Difference
End of Period	Alternative II	Alternative III	III − II
0	−$1,000	−$1,100	−$100
1–10	300	320	20

The rate of return on this incremental investment is

$$PW(III - II) = -\$100 + \$20(P/A, i^*, 10) = 0$$

$$(P/A, i^*, 10) = 5.0$$

$$i^*(III - II) \simeq 15.1\%$$

Since this incremental rate of return is less than the minimum required

(20%), the additional investment in Alternative III is not justified. Alternative III is *not* preferable to Alternative II.

Now we must determine whether IV is preferable to II. (The question of whether IV is preferable to III is irrelevant at this point.) The incremental rate of return is determined by

	Cash Flows		Difference
End of Period	Alternative II	Alternative IV	IV − II
0	− $1,000	− $2,000	− $1,000
1–10	300	550	250

$$PW(IV - II) = -\$1,000 + \$250(P/A, i^*, 10) = 0$$

$$(P/A, i^*, 10) = 4.0$$

$$i^*(IV - II) \simeq 21.4\%$$

Alternative IV is preferable to alternative II since its incremental rate of return (21.4%) exceeds the opportunity cost (20%).

The analytical procedure outlined above is summarized in **Table 4.4**. The arrows shown in the first column $(x \to y)$ indicate that we are evaluating the rate of return on the incremental investment of y over x, that is, the incremental rate of return resulting from the cash flows

$$\Delta A_j = A_j(y) - A_j(x) \tag{4.4}$$

$$PW = \sum_{j=0}^{N} (\Delta A_j)(1 + i^*)^{-j} = 0 \tag{4.5}$$

Observe that this incremental procedure yields the conclusions $IV > II > III > \varnothing > I$. This is the identical ranking that resulted from applying the present worth method. Ranking will always be identical if the analyst ensures that only *incremental* rates of return are considered.

TABLE 4.4 (Internal) Rate of Return Analysis of
Alternatives from Table 4.3

Step	Comparison of Alternatives	Incremental Rate of Return	Conclusion (Assuming MARR = 20%)
1	$\varnothing \to I$	14.9%	$I < \varnothing$
2	$\varnothing \to II$	27.3	$II > \varnothing$
3	$\varnothing \to III$	26.3	$III > \varnothing$
4	$\varnothing \to IV$	24.4	$IV > \varnothing$
5	$II \to III$	15.1	$III < II$
6	$II \to IV$	21.4	$IV > II$

Again, it is incorrect to rank-order alternatives solely on the basis of their respective internal rates of return, thereby concluding that alternatives having the higher rate are economically superior. This is known as the *ranking error*. IRRs are not direct measures of relative worth of an investment! As the example problem shows, ranking by IRRs yields incorrect results.

Alternative	IRR (i^*)	Rank by IRR	Correct Ranking
Ø	(MARR = 20%)	4	4
I	14.9%	5	5
II	27.3	1	2
III	26.3	2	3
IV	24.4	3	1

Correctly applying the (internal) rate of return method requires determination of the prospective return on the incremental investment, but irrelevant increments need not be evaluated. Note that steps 3 and 4 in Table 4.4 are unnecessary; if II $\succ$ Ø, then we need not also compare III and IV with the do-nothing alternative. After step 2 we learn that II $\succ$ Ø $\succ$ I. All that remains is to compare III and IV with II. In step 5 we learn that III $\prec$ II. Since it is now unnecessary to compare IV with III, we compare IV with II in step 6, concluding that IV $\succ$ II. Now we learn that IV $\succ$ II $\succ$ III and II $\succ$ Ø $\succ$ I, so we conclude that IV is the preferred alternative. To complete the rank ordering, if desired, we must compare III with Ø, as shown in step 3 of Table 4.4. Since III $\succ$ Ø, the completed rank order is IV $\succ$ II $\succ$ III $\succ$ Ø $\succ$ I.

4.2.6 Multiple Interest Rates and the External Rate of Return

When applied properly, the internal rate of return method yields the same solutions as those obtained from the annual worth and present worth methods, yet the algebraic structure is such that one may easily be led to incorrect solutions. This does not mean that the rate of return method is inherently incorrect—rather, the analyst should be aware of errors that may result from certain situations, such as the ranking error discussed above. Our discussion at this point centers on the condition known as the *multiple interest rate problem*.

Consider the end-of-period model described by Eq. (4.1):

$$\sum_{j=0}^{N} A_j (1 + i^*)^{-j} = 0$$

This expression may also be written as

$$A_0 + A_1 x + A_2 x^2 + \cdots + A_N x^N = 0 \qquad (4.6)$$

where $x = (1 + i*)^{-1}$. Solving Eq. (4.6) for x leads to $i*$, so we want to find the roots, x, of this Nth-order polynomial expression. Only the real, positive roots are meaningful, of course.

Descartes' rule of signs states that there may be as many real, positive roots, or solutions, as there are changes in sign of the coefficients, the A_j's. Thus, in theory, there may be many possible solutions for x and, by extension, for the rate of return, $i*$.

Another sufficient (but not necessary) condition for a *single* positive rate of return is **Norstrom's criterion**, based on the number of sign changes of the cumulative cash flows.[2] Let

$$F_j = \text{accumulated cash flows up to and including period } j$$

Norstrom's criterion states that there exists a unique solution if there is only one sign change in the sequence $F_0, F_1, F_2, \ldots, F_j, \ldots, F_N$.

In most problems there is only one sign change in the cash flow sequence. That is, the initial investment (a negative cash flow) generally results in a sequence of net revenues, or cost savings (positive cash flows). Such a situation leads to a unique solution. On the other hand, consider the following sequence of cash flows:[3]

End of Period (j)	Cash Flow (A_j)	Cumulative (F_j)
0	$ 1,600	$1,600
1	− 10,000	−8,400
2	10,000	1,600

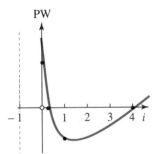

PW

Figure 4.3 Present Worth as a Function of the Discount Rate Showing Two Solutions for the IRR

The internal rates of return are those values of $i*$ that satisfy the equation

$$\$1,600 - \$10,000(1 + i*)^{-1} + \$10,000(1 + i*)^{-2} = 0$$

The present worth as a function of the discount rate, i, is graphed in **Figure 4.3**.

You may readily verify that there are two solutions to this equation: $i* = 0.25$ and $i* = 4.00$. But what is the significance of these results? If the minimum attractive rate of return (MARR) is, say, 0.30, should the proposed investment be accepted or rejected?

One possible solution to this problem requires consideration of an **auxiliary interest rate**, the return available from reinvested capital. Assume that the positive cash flow at date zero, $1,600, may be reinvested at the minimum attractive rate of return. Thus, its value at the end of the second

---

[2] For a more extensive discussion of the Norstrom criterion and related issues, see Richard H. Bernhard, "Unrecovered Investment, Uniqueness of the Internal Rate of Return and the Question of Project Acceptability," *Journal of Finance and Quantitative Analysis*, Vol. 12, No. 1, March 1977, pp. 33–38.

[3] This cash flow sequence is based on an example provided by Ezra Solomon, "The Arithmetic of Capital Budgeting Decisions," *Journal of Business*, vol. 29, 1956, p. 124.

Original Problem

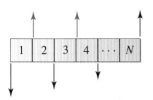

Revised Problem

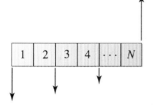

Figure 4.4 Cash Flow Sequence Revised by Ensuring that the Equivalent Values of All Positive Cash Flows Are Shifted to the End of the Study Period

period is $1,600(1.30)^2 = \$2,704$, and the net cash flow at the end of the second period is $12,704\ (=\$2,704 + \$10,000)$. This manipulation is illustrated in **Table 4.5**. The revised problem has only one variation in sign:

$$-\$10,000(1 + i^*)^{-1} + \$12,704(1 + i^*)^{-2} = 0$$

which has for its solution $i^* = 27.04\%$. This result is known as the **external rate of return** (ERR). It is "external" in the sense that it is the result of the amounts and timing of cash flows of the original investment as well as of the influence of the auxiliary interest rate. Assuming that the auxiliary interest rate is identical to the minimum attractive rate of return, the investment should be rejected, because ERR < MARR ($27.04\% < 30\%$).

The preceding example is relatively simple—only three cash flows with two sign changes $(+, -, +)$. The solution procedure can be generalized to any number of sign changes simply by ensuring that all positive cash flows occur at the end of the last period, as illustrated in **Figure 4.4**. As before, let A_j represent the cash flow at the end of period j, for $j = 0, 1, 2, \ldots, N$. Further, let

$$A_j = \begin{cases} R_j & \text{if } A_j > 0 \\ -C_j & \text{if } A_j < 0 \end{cases}$$

The R_j's are the positive cash flows and the C_j's are the negative cash flows. Let k represent the auxiliary interest rate, the rate of return on positive cash flows R_j over the remaining $N - j$ periods. Thus it is reasonable to assume that k is the MARR. The equivalent value of all positive cash flows at the end of period N is given by

$$\sum_{j=0}^{N} R_j(1 + k)^{N-j}$$

and the external rate of return, i_e^*, is the solution to the equation

$$\sum_{j=0}^{N} C_j(1 + i_e^*)^{N-j} = \sum_{j=0}^{N} R_j(1 + k)^{N-j} \qquad (4.7)$$

In words, the external rate of return is that interest rate for which the equivalent value of all *negative* cash flows at the end of period N is exactly equal to the equivalent value of all *positive* cash flows at the end of period

TABLE 4.5   Using the Auxiliary Interest Rate to Create a Sequence of Cash Flows with Only One Sign Change

End of Period	Original Problem (Two Sign Changes)	Revised Problem (One Sign Change)
0	$ 1,600	—
1	− 10,000	−$10,000
2	10,000	$10,000 + 1.600(1.30)^2$

N, where the equivalent value of the latter is determined by the auxiliary interest rate, k. The investment is justified if $i_e^* > k$.

EXAMPLE 4.4

To demonstrate this procedure, consider the following sequence of cash flows:

End of Period (j)	Cash Flow (A_j)	Cumulative (F_j)	Positive Cash Flow (R_j)	Negative Cash Flow (C_j)
0	$-$100	$-$100	$ —	$100
1	$+50$	-50	50	—
2	-20	-70	—	20
3	$+100$	$+30$	100	—
4	-80	-50	—	80
5	$+100$	$+50$	100	—

There are five sign changes in the cash flow sequence, indicating that there may be more than one real positive IRR. Moreover, there is more than one sign change in the cumulative sequence (F_j), so there may exist more than one solution. Both sign tests indicate that multiple solutions for IRR are possible.

   Let $k = 0.20$. To find the external rate of return from Eq. (4.7),

$$\$100(1 + i_e^*)^5 + \$20(1 + i_e^*)^3 + \$80(1 + i_e^*)$$
$$= \$50(1.2)^4 + \$100(1.2)^2 + \$100$$

There is a unique solution to this revised formulation: $i_e^* = 17.3\%$. Since this is less than the minimum attractive rate of return, it follows that this investment is not economically justified. (You may find it interesting to determine the PW at $i = 20\%$ for this sample problem. The present worth, of course, will be negative.)

4.2.7 Project Balances and Rates of Return

In the previous section we used Descartes' Rule of Signs to determine if there may be more than one internal rate of return (IRR). That is, there may be as many IRRs as there are sign changes in the sequence of cash flows (A_0, A_1, A_2, ..., A_N). A second, more discriminating test, is Norstrom's criterion, based on the number of sign changes in the sequence of cumulative cash flows (F_1, F_2, ..., F_j, ..., F_N).

$$F_j = \sum_{k=0}^{j} A_j$$

Assuming that the sequence starts negatively, $A_0 < 0$, there exists a unique positive value for IRR if there is only one change in sign in the sequence.

Summarizing, given the sequence of cash flows (A_j) with n sign changes:

- If $n = 1$, then the IRR is unique.
- If $n > 1$, then go to Norstrom's criterion.

Given the sequence of cumulative cash flows (F_j) with n sign changes:

- If $n = 1$, then the IRR is unique.
- If $n > 1$, then there may be $0, 1, \ldots, n$ IRRs.

We turn now to the question of what to do if the project's cash flows are such that the sign tests indicate the possible presence of multiple IRR solutions. If we can determine that there is indeed only a single IRR solution, the external rate of return (ERR) approach, as outlined in the previous section, is unnecessary. We now describe a test that indicates whether there is a unique IRR.

Consider the investor's net investment in the project at any point in time. This is known as the **project balance** (PB). Specifically,

$\text{PB}(i)_j$ = project balance at end of period j, assuming that all

funds are earning interest at rate i

$$= [\text{PB}(i)_{j-1} \times (1 + i)] + A_j \qquad \text{for } j = 1, 2, \ldots, N$$

and $\text{PB}(i)_0 = A_0$.

The **net investment test** begins with any interest rate (i^*) that satisfies the equation $\text{PW}(i^*) = 0$, given a set of cash flows for the project. We are about to determine if this solution, i^*, is in fact the only solution for the IRR. To do so, we compute the project balances over the life of the project. With $A_0 < 0$,

$$\text{if } \text{PB}(i^*)_j \leqslant 0, \qquad j = 1, 2, \ldots, N$$

this project is a pure investment in the sense that the investor has an investment in the project at all times. At no point has the investor taken out of the project any funds in excess of those to be returned assuming a rate of return of i^*. In such instances, it is unnecessary to compute the external rate of return.

EXAMPLE 4.5

Recall the sequence of cash flows given in the previous example:

End of Period:	0	1	2	3	4	5
Cash flows ($):	-100	$+50$	-20	$+100$	-80	$+100$
Cumulative ($):	-100	-50	-70	$+30$	-50	$+50$

Both the cash flow and cumulative cash flow sign tests indicate that there may be multiple rates of return. It may be shown that one solution for i^* is at 14.216%. (See the Appendix to this chapter.)

To determine whether this is the only solution, we examine the project balances, as shown in **Table 4.6**. Since the project balances are ≤ 0 throughout the project life, we conclude that $i^* = 14.216\%$ is a unique IRR for this project. It is to be compared to the investor's MARR as discussed previously.

TABLE 4.6   Project Balances for Example 4.5

End of Period	Cash Flow	Project Balance at $i = 14.216\%$
0	$-\$100$	-100.000
1	50	$-100.000(1.14216) + 50 = -64.216$
2	-20	$-64.216(1.14216) - 20 = -93.345$
3	100	$-93.345(1.14216) + 100 = -6.615$
4	-80	$-6.615(1.14216) - 80 = -87.555$
5	100	$-87.555(1.14216) + 100 = -0.000^a$

[a]Any difference is due to rounding.

If there is one or more project balances that are positive, the investor is acting as a borrower from the project. That is, if $PB(i^*)_j > 0$ for any j, the project is providing net funds to the investor. This type of project is known as a **mixed investment**, and in such cases the external rate of return approach should be used to produce a unique value of i_e^* to be compared to the MARR.

EXAMPLE 4.6

The two-period example presented in Section 4.2.6 can be shown to represent a mixed investment. Our sign tests indicate that as many as two solutions are possible for the IRR.

End of Period:	0	1	2	Sign Changes
Cash flows ($):	1,600	$-10,000$	10,000	2
Cumulative ($):	1,600	$-8,400$	1,600	2

One IRR is located at $i^* = 25\%$. To determine if this solution is unique, we examine the sequence of project balances using this value. The procedure is summarized in **Table 4.7**. There are two sign changes in the project

TABLE 4.7   Project Balances for Example 4.6

End of Period	Cash Flow	Project Balance at $i = 25\%$
0	$\$1,600$	1,600
1	$-10,000$	$1,600(1.25) - 10,000 = -8,000$
2	10,000	$-8,000(1.25) + 10,000 = 0$

balances. The investor is a *borrower* from the project at the start of the first period and a *lender* to the project at the start of the second period. Thus this is a mixed project. There is a second IRR; it is found at $i* = 400\%$.

4.2.8 Use of the PW(i) Graph

Some projects have patterns of cash flows that require the most careful interpretation of their internal rate(s) of return. For example, as in the previous section, solution of the present worth equation may lead to multiple rates of return. There are other aberrations as well: There may be "reverse solutions" resulting from positive cash flows followed by a sequence of negative cash flows; or the cash flows may be such that there are no solutions (no real roots) to the PW equation.

If there is any question about the interpretation of IRR results, the analyst should prepare a graph of PW as a function of the interest rate. Graphing provides meaningful insight because the relationship between PW and the rate(s) of return, if any, will be clearly indicated.

The (PW(i) graph is illustrated in **Figure 4.5** using the data from Example 4.1, Project P. It is clear from the graph that any discount rate less than $i* \simeq 0.12$ will yield a positive present worth. That is, PW > 0 when MARR < IRR.

Another illustration of the PW(i) graph was given in Figure 4.3. In that example, the two internal rates of return, 25% and 400%, are readily apparent. Moreover, it is clear from the graph that a positive present worth results only when $i < 0.25$ or $i > 4.00$. If, say, MARR = 0.30, the proposed investment should be rejected since the PW at that rate is less than zero.

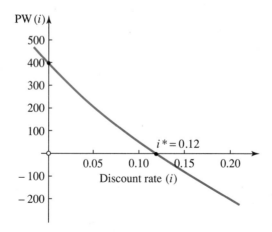

Figure 4.5 Present Worth as a Function of the Discount Rate, Project P from Example 4.1

4.2.9   Determining Bond Yields

The *bond*, a financial instrument used to raise debt capital, was discussed in Chapter 3. In Section 3.10.5 we demonstrated how the current value of a bond can be determined, given the minimum attractive rate of return of the investor. We will now show an alternate analysis procedure. Specifically, given a set of cash flows associated with the purchase of the bond, the internal rate of return, $i*$, may be calculated. When $i*$ is the effective annual interest rate, it is known as the *yield to maturity*, or simply the *bond yield*. Specifically,

$$i* = \text{yield to maturity}$$
$$= (1 + i_M^*)^M - 1 \tag{4.8}$$

where i_M^* satisfies the equation

$$-P + \left(\frac{r_b F}{M}\right)(P/A, i_M^*, N) + F(P/F, i_M^*, N) = 0 \tag{4.9}$$

given

$$F = \text{par (face) value of the bond}$$
$$r_b = \text{bond interest rate (nominal, per year)}$$
$$M = \text{number of interest payments per year}$$
$$N = \text{total number of payments to maturity}$$
$$P = \text{current cost of the bond}$$

The bond is an attractive investment if $i*$ is greater than the investor's minimum attractive rate of return.

EXAMPLE 4.7

As originally presented as Example 3.9, an investor has an opportunity to purchase a $1,000 bond at a current cost of $920. Interest is payable quarterly at a 6% rate. The bond's maturity date is such that there are 40 quarters remaining to maturity. Here

$$P = \$920 \qquad M = 4 \text{ quarters/year} \qquad N = 40 \text{ quarters}$$
$$F = \$1,000 \qquad r_b = 6\% \text{ per year}$$

From Eq. (4.9),

$$-\$920 + \left[\frac{0.06(\$1,000)}{4}\right](P/A, i_M^*, 40) + \$1,000(P/F, i_M^*, 40) = 0$$

Solving, $i_M^* = 1.78\%$. Using this value in Eq. (4.8):

$$i* = (1.0178)^4 - 1 = 7.3\%$$

Thus this bond is an attractive investment if the investor's minimum attractive rate of return is less than 7.3%.

4.3 THE BENEFIT–COST RATIO METHOD

4.3.1 Public-Sector Decision Making

The need for economic analysis in the public sector is not unlike that in the private sector. Funds available to government agencies are ultimately derived from the citizens, and government decision makers have a responsibility for allocating these scarce resources in a manner that is consistent with the goals and aspirations of the society. In the United States, elected representatives are analogous to the board of directors in a private corporation; government officials are analogous to "management" in the private sector. Our society has a right to expect government agencies to husband resources intelligently and responsibly.

Of course, decisions in the public sector should not be influenced solely by economic considerations. Issues of social equality, quality of life, security, health, education, and other considerations are frequently of paramount importance. (Private-sector decisions are also influenced by noneconomic considerations, at least in the near term, such as desire to increase market share and to enhance the general reputation of the firm.) Nevertheless, careful assessment of the economic consequences of alternative policies should play at least some part in public-sector decisions.

Costs, Benefits, and Disbenefits

Just as *income* and *expenses* are used in the private sector, so are *costs*, *benefits*, and *disbenefits* used in the public sector to signify the economic consequences of economic decisions. Of course, what are important are the differences between alternatives, not the labels that are used to classify these differences. Nevertheless, these three terms are used quite frequently in the literature of public-sector decision making.

Costs are generally regarded as the economic consequences to the government agency, the sponsoring organization, resulting from the initial investment, annual operating costs, and including any receipts (e.g., user fees). **Benefits** are the positive effects on the users as well as other members of society who may be impacted by the programs or projects under consideration. For example, a highway improvement project will affect the motorists who use that highway as well as businesses and home owners located nearby. **Disbenefits** are the negative consequences. Construction of an expressway, for example, may adversely affect the small businesses located along the old street or highway from which the new expressway will generate its traffic.

Benefits and disbenefits include economic as well as noneconomic consequences. However, in the remainder of this section, we will assume that *benefits* describes only those effects that can be measured in monetary terms.

The Appropriate Discount Rate

In the private sector, as mentioned previously, the discount rate used in economic analyses is the decision maker's minimum attractive rate of return, the cost of opportunities forgone as the result of investing in the proposed project. Although there is some debate about the appropriate numerical value of the rate to use in the public sector, there appears to be general consensus that the discount rate should reflect the opportunity costs to those members of the society served by the governmental organization. (Alternative viewpoints include the interest rate paid on capital borrowed by government to fund the project and the opportunity cost of capital to the governmental agency.) Stated somewhat differently, the minimum rate to be "earned" by public-sector investments should, at the margin, be at least as great as the return available on those same funds if they were to remain in the hands of the citizens.[4]

4.3.2 The Acceptance Criterion

The essential element of the **benefit–cost ratio method** is almost trivial, but it can be misleading in its simplicity. An investment is justified only if the incremental benefits, B, resulting from it exceed the resulting incremental costs, C. Of course, all benefits and costs must be stated in equivalent terms, that is, with measurement at the same point(s) in time. Normally, both benefits and costs are stated as "present values" or are "annualized" by using compound interest factors as appropriate.

Clearly, if benefits must exceed costs, then the ratio of benefits to costs must exceed unity. That is, if $B > C$, then $B:C > 1.0$. This statement of the **acceptance criterion** is true only if the incremental costs, C, are positive. It is possible, when evaluating certain alternatives, for the incremental costs to be negative, that is, for the proposed project to result in a reduction of costs. Negative benefits arise when the incremental effect is a reduction in benefits. These possibilities require a somewhat more complex statement of the acceptance criterion, as given in **Table 4.8**. In summary,

For $C > 0$, if $B:C > 1.0$, accept; otherwise reject.

For $C < 0$, if $B:C > 1.0$, reject; otherwise accept.

In both cases, if the ratio $B:C = 1.0$, the implication is that benefits equal

---

[4]In 1972 the federal government specified that a 10% rate should be used when evaluating certain investment projects [Office of Management and Budget, "Discount Rates to be Used in Evaluating Time-Distributed Costs and Benefits," Circular No. A-94 (rev.), March 27, 1972].

TABLE 4.8   The Acceptance Criterion for the Benefit–Cost Ratio

Numerator (B)	Denominator (C)	Ratio (B : C)	Decision
Positive	Positive	> 1.0	Accept
"	"	< 1.0	Reject
Positive	Negative	Any	Accept
Negative	Positive	Any	Reject
Negative	Negative	> 1.0	Reject
"	"	< 1.0	Accept

costs, so we would be indifferent about accepting or rejecting the proposed project.

4.3.3 Examples Illustrating the Importance of Incremental Comparisons

EXAMPLE 4.8   *Comparing Two Alternatives*

Consider two alternatives, T and U:

Alternative	Benefits	Costs	B : C
T	$ 7	$2	3.5
U	12	6	2.0

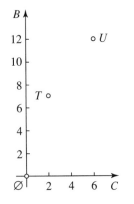

Figure 4.6 Benefits versus Costs for Example 4.8

The data are plotted in **Figure 4.6**. Note that the origin represents the do-nothing alternative, represented by $\emptyset$: Invest in neither T nor U but invest the funds elsewhere. The benefits and costs for the two alternatives are measured against the option of doing nothing.

The benefit–cost ratios, 3.5 and 2.0 for T and U, respectively, do not imply that T is preferable to U. Because each ratio exceeds unity, we conclude only that both T and U are preferable to the third alternative, do nothing (and invest the funds elsewhere). We cannot conclude from these values and calculations alone that T is preferable to U.

To choose between T and U, we must note the consequences if U is selected rather than T. If so, the incremental effects are:

Alternative	Incremental Benefits	Incremental Costs	Incremental B : C
U rather than T	$12 − $7 = $5	$6 − $2 = $4	$5/$4 = 1.25

Because the incremental ratio exceeds unity, we conclude that U is in fact preferable to T.

This example is graphed in **Figure 4.7**. Since the scales for benefits (ordinate) and costs (abscissa) are identical, the line drawn at 45 from the

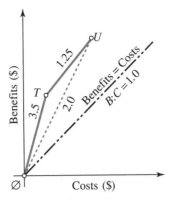

Figure 4.7   Benefit–Cost Analysis for Example 4.8

origin represents the locus of all points at which benefits are exactly equal to costs. Notice that the slope of this line, the ratio of the ordinate to the abscissa, is precisely unity. All alternatives lying above this line are acceptable because their slopes are greater than unity; points below the 45° line are unacceptable because their slopes (benefit–cost ratios) are less than unity. The point at the origin represents the do-nothing alternative. Note too that the slopes that of interest are represented by the lines $\varnothing$–T and T–U. In the former instance, T is preferable to doing nothing ($\varnothing$) and U is preferable to T. The slope of the line $\varnothing$–U is irrelevant in this case.

EXAMPLE 4.9   *Comparing Multiple Alternatives*

Recall the four mutually exclusive alternatives, plus the do-nothing alternative, originally presented in Example 3.4 (present worth method) and continued as Example 4.3 (rate of return method). We will now use the benefit–cost ratio method to rank-order these alternatives. The solution is outlined in **Table 4.9**. The benefits shown in the Table are the equivalent

TABLE 4.9   The Benefit–Cost Ratio Method in Evaluating Alternatives from Table 4.3

Step	Comparison of Alternatives	Incremental Benefits	Incremental Costs	Incremental $B:C$	Conclusion
1	$\varnothing \rightarrow$ I	$ 646	$1,000	0.65	I $\prec \varnothing$
2	$\varnothing \rightarrow$ II	1,258	1,000	1.29	II $\succ \varnothing$
3	$\varnothing \rightarrow$ III	1,341	1,100	1.22	III $\succ \varnothing$
4	$\varnothing \rightarrow$ IV	2,305	2,000	1.15	IV $\succ \varnothing$
5	II $\rightarrow$ III	83	100	0.83	III $\prec$ II
6	II $\rightarrow$ IV	1,047	1.000	1.05	IV $\succ$ II

Summary: IV $\succ$ II $\succ$ III $\succ \varnothing \succ 1$

present values of the positive cash flows given in Table 4.3; the costs are the equivalent present values of the negative cash flows. The benefit–cost analysis summarized in the Table indicates that $IV \succ II \succ III \succ \emptyset \succ I$, the identical ranking obtained by the present worth and rate of return methods.

Note that steps 3 and 4 in Table 4.9 are irrelevant. In step 2 we determined that $II \succ \emptyset$; thus we need only compare the next proposal, alternative III, with alternative II. If $III \succ II$, then it must also be true that $III \succ \emptyset$. The same argument holds for alternative IV.

4.3.4   A Proof that Incremental Analysis Always Produces Results Consistent with the Present Worth Method

Incremental analysis always results in maximum present worth, as illustrated by **Figure 4.8**. Here, the curved line $\emptyset MN$ represents the set of all alternatives under consideration. (Of course, in real-world applications, this line is not continuous because there is a finite number of alternatives and their incremental benefit–cost ratios are irregular. Nevertheless, this simplification illustrates the point and in no way violates the conclusion.) Benefits and costs are measured by their equivalent present worths.

Assuming that line $\emptyset MN$ is a continuous, differentiable function, we can show that the maximum excess of benefits over costs occurs at the point where the rate of change of benefits relative to costs, the incremental benefit–cost ratio, is exactly equal to unity. That is, we maximize $P = B - C$ by taking the first derivative with respect to costs and setting it equal to zero:

$$\text{Max } P = B - C$$

$$\frac{dP}{dC} = \frac{dB}{dC} - \frac{dC}{dC} = 0$$

Thus

$$\frac{dB}{dC} = 1.0$$

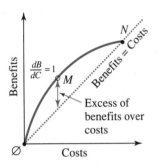

Figure 4.8 The Point (M) of Maximum Excess of Benefits over Costs

Here, dB/dC is the marginal benefit–cost ratio. Present worth, the excess of benefits over costs, is maximized at the point where dB/dC equals unity. Or, stated somewhat differently, maximum present worth results as long as increments of costs continue to be justified by their respective increments of benefits.

4.3.5   The Numerator–Denominator Issue in Calculating Benefit–Cost Ratios

Some authors are critical of the benefit–cost ratio method on the grounds that the magnitude of the ratio depends on whether a particular economic consequence is considered in the numerator as a benefit or in the denominator as a "negative cost." (Similarly, one may choose between including an economic consequence in the denominator as a cost or in the numerator as a "negative benefit.")

This question occurs frequently with respect to facilities designed to result in user savings (benefits) but that require annual costs, such as the cost of maintenance. In particular, consider three major consequences of a proposed highway improvement: (1) capital costs of construction, (2) benefits accruing to users because of an improved level of service, and (3) costs of facility operation and maintenance. The issue here is whether the operation and maintenance expenses should be deducted from user benefits (the numerator), or conversely, added to capital costs (the denominator), because each approach results in a different benefit–cost ratio.

As will be demonstrated in the following numerical example, the choice of numerator or denominator is not significant. The only character- istic of the benefit–cost ratio that is relevant to the decision is whether the ratio is greater than unity. Otherise, the value of the ratio is irrelevant. To show that this is true in general, let

B = economic consequences that definitely appear in the numerator, e.g., user benefits

C = economic consequences that definitely appear in the denominator, e.g., capital costs of construction

K = economic consequences to be included either in the numerator or denominator, e.g., costs of facility operation

If it is true that

$$\frac{B - K}{C} > 1 \tag{4.10}$$

it is also true that

$$\frac{B}{C + K} > 1 \tag{4.11}$$

This result arises from the fact that the direction of the inequality cannot be reversed merely by adding or subtracting a constant from both sides.[5]

---

[5]The argument given here holds only if both $C > 0$ and $C + K > 0$. Otherwise, it may be shown that the conditions for acceptance are

$$\frac{B - K}{C} > 1 \quad \text{or} \quad \frac{B}{C + K} < 1, \quad \text{if} \quad C > 0 \quad \text{and} \quad C + K < 0$$

$$\frac{B - K}{C} < 1 \quad \text{or} \quad \frac{B}{C + K} > 1, \quad \text{if} \quad C < 0 \quad \text{and} \quad C + K > 0$$

$$\frac{B - K}{C} < 1 \quad \text{or} \quad \frac{B}{C + K} < 1, \quad \text{if} \quad C < 0 \quad \text{and} \quad C + K < 0$$

Thus an "accept' conclusion using inequality (4.10) will also signal "accept" using inequality (4.11). Of course, the converse is true: If

$$\frac{B - K}{C} < 1$$

then

$$\frac{B}{C + K} < 1$$

EXAMPLE 4.10

Consider two mutually exclusive alternatives with expected economic consequences as follows:

	Project R	Project S
B = PW of user benefits	$150,000	$240,000
C = PW of capital costs of the facility	80,000	200,000
K = PW of operation and maintenance expenses (O & M)	50,000	0

In the event that operation and maintenance costs are treated as a disbenefit, i.e., subtracted from the numerator,

$$B : C(R) = \frac{B - K}{C} = \frac{\$150,000 - \$50,000}{\$80,000} = 1.25$$

$$B : C(S) = \frac{B - K}{C} = \frac{\$240,000 - \$0}{\$200,000} = 1.20$$

Alternatively, if O&M expenses are treated as costs in the denominator,

$$B : C(R) = \frac{B}{C + K} = \frac{\$150,000}{\$80,000 + \$50,000} = 1.15$$

$$B : C(S) = \frac{B}{C + K} = \frac{\$240,000}{\$200,000 + \$0} = 1.20$$

If one were to compare benefit–cost ratios, it would appear that $R \succ S$ in the first instance but $R \prec S$ in the second. This is the ranking error, however, and should be avoided in either case. All that we have determined so far in the analysis is that $R \succ \varnothing$ and $S \succ \varnothing$, because all benefit–cost ratios exceed unity.

Correct analysis requires consideration of the differences between alternatives. Here, if project S were to be selected instead of project R:

	Incremental (R → S)
ΔB	$240,000 − $150,000 = $ 90,000
ΔC	200,000 − 80,000 = 120,000
ΔK	0 − 50,000 = −50,000

Now determine the incremental benefit–cost ratio using either formulation:

$$\Delta B : C = \frac{\Delta B - \Delta K}{\Delta C} = \frac{\$90,000 - (-\$50,000)}{\$120,000} = 1.17$$

or

$$\Delta B : C = \frac{\Delta B}{\Delta C + \Delta K} = \frac{\$90,000}{\$120,000 - \$50,000} = 1.29$$

Because the ratios exceed unity (with positive denominators), we conclude in both cases that project S is preferred. In summary, either formulation leads to the conclusion that $S \succ R \succ \varnothing$.

Also note that our results are consistent with that obtained using the present worth method:

$$PW(R) = \$150,000 - \$80,000 - \$50,000 = \$20,000$$

$$PW(S) = \$240,000 - \$200,000 - \$0 = \$40,000$$

4.3.6 Variations

Several variations on the basic benefit–cost ratio have been introduced from time to time. As will be demonstrated, when they are applied properly, all these approaches lead to correct, consistent conclusions. Ratios must not be compared directly; *incremental analysis is required.*

The Net Benefit–Cost Ratio

The **net benefit–cost ratio**, $NB : C$, is the ratio of discounted net benefits to the initial cost of the project. Specifically,

$$NB : C = \frac{B - C}{C} = \frac{B}{C} - 1 \qquad (4.12)$$

where B is the present value of all cash flows over the planning horizon ($j = 1, 2, \ldots, N$) and C is the initial cost of the project.

Note here that our criterion for acceptance is that the net benefit–cost ratio exceed zero. Requiring that $B : C > 1$ is the same as requiring that $NB : C$ be greater than zero.

EXAMPLE 4.11

Consider two mutually exclusive alternatives, X and Y:

Alternative	B	C	$B - C$	$NB : C$
X	$250	$100	$150	1.50
Y	320	150	170	1.13

Although $NB:C(X) > NB:C(Y)$, this result does not imply that X is preferable to Y. Indeed, since $PW = B - C$, $PW(X) < PW(Y)$, so it follows that Y is preferable to X.

To use net benefit–cost ratios properly, we should examine the differences between Y and X. The incremental benefits and costs are

$$\Delta B = \$320 - \$250 = \$70$$

$$\Delta C = \$150 - \$100 = \$50$$

Thus $\Delta NB:C = (\$70 - \$50)/\$50 = 0.40$. Since the resulting $\Delta NB:C$ is greater than zero, the incremental investment is justified and Y is preferable to X.

The Savings–Investment Ratio

Although rarely found in the private sector, the **savings–investment ratio** (SIR) has enjoyed some prominence in the public sector in recent years, especially in the U.S. Department of Defense.[6] The SIR is yet another variation of the more familiar benefit–cost ratio ($B:C$). When it is used solely as a profitability index, the SIR can lead to incorrect solutions. If it is used properly, incremental SIRs may be used to find valid results.

The savings–investment ratio is defined as

$$SIR = \frac{P(E)}{P(I) - P(S)} \tag{4.13}$$

where $P(E)$ is the discounted present value of all earnings over the life of the project, $P(I)$ is the present value of the initial investment, and $P(S)$ is the discounted value of the project's terminal salvage value, if any.

Parenthetically, the term *earnings* appears in Eq. (4.13), whereas in other formulations, particularly in those used by the Department of Defense, the term *savings* is used instead. The difference in terminology results from the fact that government investment projects are not expected to produce income the way it is produced in the private sector. However, our concern is with the effect of the project on cash flows. Whether positive cash flows are generated by increasing revenues or by reducing costs is irrelevant in this context. Therefore, *earnings* means net cash flows produced by the project, other than the initial investment and the terminal salvage value.

Note that the present worth (PW) of the proposed investment is given by

$$PW = P(E) - P(I) + P(S) \tag{4.14}$$

---

[6]An early reference is *Economic Analysis Handbook* (Alexandria, VA: Naval Facilities Engineering Command, 1975), pp. 23–28.

If the present worth is greater than zero, it follows that

$$P(E) > P(I) - P(S)$$

and, if the right-hand side of the inequality is positive, then

$$\frac{P(E)}{P(I) - P(S)} > 1.0$$

In other words, if $P(I) - P(S) > 0$, then the investment is warranted if the SIR > 1.0.

EXAMPLE 4.12

Consider a certain process that currently costs $40,000 per year and is expected to continue at this level over the next 15 years. An improved process, System Alpha, which costs $60,000, has been proposed. System Alpha will reduce costs to $30,000 per year but is not expected to have any terminal salvage value at the end of 15 years. The initial investment, $60,000, will occur entirely at the start of the first year, and savings will occur continuously and uniformly during each year over the 15-year planning horizon. The discount rate is 10% per year. Based on Eq. (4.13),

$$P(E) = (\$40,000 - \$30,000)(P/\bar{A}, 10\%, 15)$$

$$= \$10,000(7.980)$$

$$= \$79,800$$

$$P(I) = \$60,000$$

$$P(S) = \$0$$

Therefore,

$$SIR(Alpha) = \frac{\$79,800}{\$60,000} = 1.33$$

Since the SIR exceeds unity, it appears that investment in System Alpha is warranted. Alpha is preferable to the do-nothing alternative, that is, preferable to continuing with the current process.

Consider a second alternative, System Beta, with $P(E) = \$125,000$, $P(I) = \$110,000$, and $P(S) = \$10,000$. Here

$$SIR(Beta) = \frac{\$125,000}{\$110,000 \quad \$10,000} = 1.25$$

We should not conclude that Alpha is preferable to Beta simply because SIR(Alpha) > SIR(Beta). Indeed, the correct incremental analysis shows the opposite to be true:

	System Alpha	System Beta	Incremental Alpha → Beta
P(E)	$79,800	$125,000	$45,200
P(I)	60,000	110,000	50,000
P(S)	0	10,000	10,000

If we were to adopt Beta rather than Alpha, the incremental SIR would be

$$\Delta \text{SIR} = \frac{\$45,200}{\$50,000 - \$10,000} = 1.13$$

Since $\Delta \text{SIR} > 1.0$, the incremental investment in Beta is warranted.

SUMMARY

Central to the problem of economic analysis is the concept of mutually exclusive alternatives, which are investment proposals such that the selection of one necessarily precludes all others. As discussed in Section 3.2, mutually exclusive alternatives arise from two fundamental situations. First, there are those that are inherently mutually exclusive because of their form or function; these are technologically mutually exclusive. Second, some alternatives are mutually exclusive at the margin because limited funds do not permit all otherwise worthy proposals to be accepted; these are financially mutually exclusive. Independent alternatives that are financially mutually exclusive may be structured into an exhaustive set of mutually exclusive budget packages from which the most worthy package(s) can be identified.

Regardless of whether alternatives are mutually exclusive technologically or financially, valid procedures must be employed for selecting the preferred alternative. Three of these procedures are summarized in **Table 4.10**. The present worth method permits the rank ordering of alternatives on the basis of their respective present worths. Also, as discussed in Chapter 3, the annual worth and future worth methods produce rank ordering consistent with present worth analysis. However, in general, competing alternatives may not be similarly rank-ordered on the basis of their respective IRRs and $B:C$s.

The (internal) rate of return and benefit–cost ratio methods both require careful consideration of the incremental consequences between pairs of alternatives. The IRR and $B:C$ for a given project reflect only the differences between that alternative and the do-nothing case. It is necessary to examine the differences between pairs of alternatives to ensure that the incremental IRR exceeds the opportunity cost or, alternatively, to ensure that $B:C$ exceeds unity. The tests are equivalent. Successive applications of this test ensure that the net present worth is maximized.

TABLE 4.10   Summary of Investment Criteria

Assumptions: Discrete cash flows
                 End-of-period discounting/compounding
Notation: i = effective interest rate per period (opportunity cost);
               may be written as k (cost of capital)
         i^* = internal rate of return
         B_j = cash flow at end of period j for consequences considered
             to be "benefits" $(j = 0, 1, 2, \ldots, N)$
         C_j = cash flow at end of period j for consequences considered
             to be "costs" $(j = 0, 1, 2, \ldots, N)$

Present Worth Method

Given: B_j, C_j, N, and i

Figure of merit: $PW = \sum\limits_{j=0}^{N} (B_j - C_j)(1 + i)^{-j}$

Test: Accept if $PW > 0$

Remarks: Projects can be rank-ordered by decreasing PWs.
           Annual worths (AWs) are directly related:

$$AW = PW(A/P, i, N)$$

      As in all methods, N must be constant for all alternatives.

(Internal) Rate of Return Method

Given: B_j, C_j, N, and k

Figure of merit: i^*, where $= \sum\limits_{j=0}^{N} (B_j - C_j)(1 + i^*)^{-j} \equiv 0$

Test: Accept if $i^* > k$

Remarks: Projects cannot be rank-ordered by decreasing i^*'s.

Benefit–Cost Ratio Method

Given: B_j, C_j, N, and i

Figure of merit: $B:C$, where $B = \sum\limits_{j=0}^{N} B_j(1 + i)^{-j}$

$$C = \sum\limits_{j=0}^{N} C_j(1 + i)^{-j}$$

Test: Accept if $B:C > 1.0$

Remarks: Projects cannot be rank-ordered by decreasing $B:C$s

| PROBLEMS

Note: Unless otherwise indicated, assume discrete cash flows, end-of-period compounding/discounting, and effective interest rates.

(Internal) Rate of Return

4.1 A firm is considering investment in one of two lathes. The relevant data are:

	Lathe A	Lathe B
Initial cost	$10,000	$15,000
Service life	5 years	10 years
Salvage value	$2,000	$3,000
Annual receipts	$5,000	$7,000
Annual disbursements	$2,200	$4,300

Determine the rate of return on the incremental investment in lathe B. (That is, since lathe B is $5,000 more expensive than lathe A, what is the rate of return on the additional $5,000?) State any necessary assumptions.
[*Answer*: **8.95%, assuming identical replication of lathe A after 5 years and a 10-year planning horizon.**]

4.2 A certain data processing operation is now being done manually at a cost of $100,000 annually. The lease of automatic equipment that will reduce labor costs to $30,000 annually has been proposed. Assume that labor costs are distributed continuously and uniformly over the year. Lease payments are $60,000 per year, payable at the start of each year. It is expected that there will be a need for this operation over the next 5 years.
a. Determine the rate of return on the proposed lease.
b. If the firm's MARR is 15%, should the 5-year lease be undertaken?

4.3 The Green Corporation is considering the installation of a semiautomatic freight conveyor requiring an immediate investment of $300,000. It is estimated that if the conveyor is purchased, it will be kept for 10 years and then sold for $12,000. It is also believed that the purchase will result in the following cost savings:

Year	Net Savings	Year	Net Savings
1	$60,000	6	$55,000
2	60,000	7	50,000
3	60,000	8	45,000
4	60,000	9	40,000
5	60,000	10	35,000

Determine the prospective rate of return of this investment before taxes.
[*Answer*: **0.132.**]

4.4 A firm is considering an investment in data processing equipment that has an initial cost of $10,000 and an expected salvage value of zero if it is retired after 10 years. The equipment will result in end-of-year savings of $1,500 each year throughout its 10-year life.
a. Determine the internal rate of return for this proposed investment.
b. If the firm's minimum attractive rate of return is 12%, is this proposal acceptable?

4.5 Consider the following end-of-period cash flows:

i	A_i
0	− $10,000
1–20	1,000
20	5,000

Determine the internal rate of return.
[*Answer*: **9.03%.**]

4.6 The Universal Manufacturing Company is considering the purchase of an industrial robot for a certain welding operation. The cost of the robot is $50,000; it has an expected service life of 8 years, with a salvage value of $10,000 at the end of its service life. Major maintenance will be required at a cost of $5,000 at the end of the 2nd, 4th, and 6th years. If the robot is acquired, the company can expect to reduce its costs for this operation by $25,000 per year over 8 years. (*Note*: Operating cost savings can be assumed to occur continuously and uniformly over each year.) The Company's pretax minimum attractive rate of return is 20% per year. Determine the (internal) rate of return for this proposed investment.

4.7 Our manufacturing firm is considering a proposal to increase plant capacity from 100,000 to 120,000 units annually. The initial cost will be $800,000 for additional space and equipment. The net economic benefits to be derived from this plant expansion are expected to be $300,000 the first year, decreasing by $50,000 each year. (The net benefits will be $50,000 in year 6 and zero in year 7, it will be noted.) The additional capacity will be used in connection with a product to be produced over the next 7 years, at the end of which time the additional space and equipment is expected to have a salvage value of $80,000. The firm's MARR is 15%. Assume all cash flows are end-of-year. Determine the (internal) rate of return for this proposal.

[*Answer*: 13.2%.]

4.8 The Southwest Manufacturing, Inc., plant engineering department is considering the purchase of certain material handling equipment that, if acquired, will result in labor savings of $10,000 per month, assumed to flow at end of each month. Preventive maintenance for this equipment will take place at the end of each year, except the final year, and will cost $10,000 per occurrence. The initial cost of the equipment is $400,000. The equipment will be kept in service for 6 years, then sold for an estimated $20,000 salvage value. Determine the (internal) rate of return of this equipment.

4.9 The current costs associated with a certain operation are $100,000 per year, assumed to be flowing continuously and uniformly during each year, and will continue over the next 5 years. Alternatively, equipment may be purchased at a cost of $400,000 and which will have a salvage value of $100,000 at the end of 5 years. If it is purchased, this equipment will have the effect of reducing the operating costs to $20,000 during each year of the 5-year period. Determine the (internal) rate of return for this proposed investment.

[*Answer*: 8.0%.]

4.10 As an engineer in the production engineering department, you are asked to evaluate a pro-

posed design for a new "machining cell." The firm's MARR is 10%.

The initial cost of the equipment is $300,000. It will be sold at the end of 5 years for an estimated $50,000. Operating costs are expected to be $100,000 the first year, increasing by $10,000 per year over each of the next 4 years. This new system, if implemented, will have the effect of saving $180,000 each year over the next 5 years. Determine the rate of return for this project.

4.11 The manufacturing engineering department of a medium-sized electronics firm has been asked to consider the purchase of certain automated tooling for use in an electronics assembly operation. The tooling has an initial cost of $40,000; it will be used for 3 years (36 months), then sold for an estimated $4,000. The cost of materials and supplies associated with this tooling is expected to be about $5,000 per month. The firm's pretax cost of capital (MARR) is 2% per month. If it is acquired, the tooling is expected to generate savings in labor costs of $7,000 per month. Determine the rate of return (per month) for the proposed investment.)

[*Answer*: 3.79%.]

4.12 The Environmental Protection Agency (EPA) is examining a municipal proposal for a new sewer system. The system has an initial cost of $4,000,000, a life of 30 years, and zero expected salvage value. Expected annual costs of operations are $200,000; annual benefits to users are $780,000. The EPA has established a minimal acceptable rate of return of 8%. Determine the (internal) rate of return of this proposed investment.

4.13 The firm is considering the purchase of an automatic controller that has an initial cost of $20,000, an estimated service life of 7 years, and an expected salvage value of $3,000 at the time of disposal. The firm's MARR is 15% (before income taxes). If it is purchased, this device will replace an operator whose cost is $25,000 per year. Assume that all cash flows are end-of-year, with the initial cost at the start of the first year.
a. Determine the (internal) rate of return of the proposed investment.

b. Is the proposed investment economically justified? Why?

[*Answers*: a. 11.4%; b. no, since IRR < MARR.]

4.14 A firm is considering an investment in certain data processing equipment that has an initial cost of $10,000 and expected salvage value of zero if it is retired after 10 years. It is expected that if it is purchased, the equipment will result in end-of-year savings of $1,500 each year throughout the 10-year life of the equipment.
a. Determine the (internal) rate of return for this project.
b. If the firm's minimum attractive rate of return is 12%, should this project be purchased? Explain.

4.15 In order to provide financial support for his junior year in college, Ivan O. Universal obtains a $5,000, one-time loan from a federal agency. This is a "7% loan," which must be repaid within 10 years from the date of graduation.

It is IOU's intention to repay the loan in eight uniform annual payments, with the first payment 3 years from the date of the loan.
a. Determine the amount of each uniform payment. Assume here that the "7% cost" of the loan begins at the time of graduation, which is 2 years hence. That is, he will owe $5,000 in 2 years; if no payment is made, he will owe $5,000(1.07) = $5,350 in 3 years; and so on.
b. Determine the true cost of the loan over the planning horizon, expressed as a percentage. Note that the final (eighth) payment will be made 10 years after the date of the original loan.

[*Answers*: a. $837.50; b. 4.70%.]

4.16 Consider the following data for two mutually exclusive alternatives:

	Alternative A	Alternative B
First cost	$10,000	$12,000
Expected life	6 years	6 years
Net salvage value	0	$3,000
Annual operating costs	$1,500	$1,600

Suppose that either A or B must be selected.
a. What is the expected rate of return on the additional (incremental) $2,000 investment in alternative B?
b. If the MARR for the company is 12% before taxes, is the additional investment justified?
c. Which alternative should be selected? Is this the same solution as that obtained from the annual worth and present worth methods?
d. At what interest rate(s), if any, would B be less costly than A?

4.17 Consider two mutually exclusive alternatives, C and D, with the following end-of-period cash flows:

End of Period	Alternative C	Alternative D
0	−$100	−$150
1–5	30	43

a. Find the (internal) rates of return for the alternatives.
b. If the firm's MARR is 8%, use the rate of return method to determine which alternative should be selected.
c. For what range of values of MARR is alternative D preferable to alternative C?

[*Answers*: a. 15.2%, 13.3%; b. Alt. D; c. MARR < 9.4%.]

4.18 Consider the following cash flows:

End of Year	Alternative E	Alternative F
0	−$10	−$20
1	15	28

Over what range of values of the minimum attractive rate of return (MARR) is alternative F preferable to alternative E?

4.19 Refer to Problem 3.2.
a. For each cash flow sequence, plot present worth as a function of the interest rate, i. Display the results over the range $-1 < i < 0.50$.
b. What is the rate of return for each cash flow sequence?
c. For what values of i is D preferable to E?
d. For what values of i is F preferable to E?
e. For what values of i is F preferable to D?

Note: Precise answers are not required. You may note the answers by inspecting the graph developed in (a).
[*Answers*: b. none, -21.8%, 38.0%; c. $i < 10.5\%$; d. $i < 6.0\%$; e. $i < 0\%$.]

4.20 The Jupitur Manufacturing Company is required by the Occupational Safety and Health Administration (OSHA) to provide certain safety equipment in its plant operations. The prospective cash flows for the four plans under consideration are given below. The firm's pretax minimum attractive rate of return is 20% per year.

End of Year	Plan A	Plan B	Plan C	Plan D
0	$-\$10,000$	$-\$13,000$	$-\$8,000$	$-\$12,000$
1–5	$-2,000$	$-1,000$	$-3,000$	$-1,000$
5	4,000	5,000	0	0

a. Choose any method—present worth, annual worth, or rate of return—to rank-order the four alternatives.
b. Find the interest rate at which the firm would be indifferent between Plan C and Plan D.

4.21 A firm is considering three mutually exclusive alternatives as part of a production improvement program:

	Plan A	Plan B	Plan C
Initial cost	$10,000	$15,000	$20,000
Uniform annual benefit	$1,625	$1,625	$1,890
Useful life	10 years	20 years	20 years

The salvage value at the end of the useful lives of all three alternatives is zero. At the end of 10 years, alternative A could be replaced with an identical replacement. The firm's minimum attractive rate of return is 6%. The planning horizon is 20 years. Use the rate of return method to determine which alternative should be selected.
[*Answer*: $B > A > C$.]

4.22 A firm is considering two investment alternatives, X and Y. The after-tax cash flows are shown next. The firm's after-tax MARR is 10%.

	Cash Flows	
End of Period	Alt. X	Alt. Y
0	$-\$200$	$-\$320$
1	50	100
2	50	90
3	50	80
4	50	70
5	50	60

a. Determine the (internal) rates of return for alternatives X and Y.
b. If the do-nothing alternative ($\emptyset$) is not feasible, determine the preferred alternative.

4.23 State law mandates that the United Services Corporation take certain measures to abate pollution at one of its plants. Two alternatives are under consideration. The expected cash flows (in $1,000s) are as follows:

End of Year	Alternative I	Alternative II
0	$-1,000$	$-1,500$
1	-100	-50
2	-200	-100
3	-300	-150
4	-400	-200
5	-500	-250

USC's pretax minimum attractive rate of return is 20%. Use the rate of return method to determine which of the two alternatives is preferable.
[*Answer*: I is preferred to II.]

4.24 Consider the cash flows associated with the following three mutually exclusive projects:

j	$A_j(X)$	$A_j(Y)$	$A_j(Z)$
0	$-\$20,000$	$-\$12,000$	$-\$15,000$
1–5	5,000	0	0
5	0	24,000	28,000

The firm's MARR is 10%. Using the rate of return method, properly applied, determine the correct rank ordering of alternatives.

4.25 Consider two alternative investments. Alternative Uno has an initial cost of $15,000 and is expected to return $4,000 at the end of each year for 7 years. Alternative Dos has an initial cost of $20,000 and is expected to return $5,000 at the end of each year for 7 years.

a. Determine the (internal) rate of return for Uno.

b. For what values of the MARR(k) is Dos preferred to Uno?

[*Answers*: a. 18.6%; b. $k < 9.2\%$.]

4.26 A certain manufacturing firm is considering two alternative processes for parts assembly. The economic consequences (cash flows) are summarized below. The negative cash flows at end of year 0 are due to initial costs. The positive cash flows in years 1–5 represent savings over the current assembly process. The firm's discount rate is 12%.

	Cash Flows	
End of Year	Alt. I	Alt. II
0	−$700,000	−$650,000
1	200,000	160,000
2	200,000	180,000
3	200,000	200,000
4	200,000	220,000
5	200,000	240,000

a. Determine the (internal) rate of return for alternatives I and II.

b. Which alternative is preferred? Explain.

4.27 Mr. Green is considering the investment of $10,000 in a project that would generate before-tax cash flow of $15,000 at the end of 2 years. Ms. Blue already operates a similar project and, to avoid the competition, is willing to pay Mr. Green $1,000 if he will not invest in this project. In summary, the pretax cash flows from Green's point of view are:

End of Year	Alternative I (Green Invests in Project)	Alternative Ø (Green Does Not Invest)
0	−$10,000	$1,000
2	15,000	0

It may be shown that the IRR for alternative I is approximately 22.47%; the IRR for alternative Ø is infinitely large. Determine Mr. Green's maximum before-tax MARR such that he would accept Ms. Blue's offer. That is, for what values of his MARR would Mr. Green consider Ms. Blue's offer attractive?

[*Answer*: MARR > 16.8%.]

4.28 Four mutually exclusive alternatives are under consideration: "do nothing" (Ø) and alternatives X, Y, and Z. Analysis of these alternatives yields the following results:

	X	Y	Z
Initial cost	$25,000	$30,000	$50,000
Sum of cash flows	15,000	24,000	20,000
PW @ 12%	3,000	−2,000	2,000
IRR(i^*)	17%	11%	14%

The minimum attractive rate of return (k) is 12%.

a. Using the rate of return method, determine the preferred alternative.

b. For what values of k would the do-nothing alternative be preferred?

c. For what values of k would alternative X be preferred to alternative Z?

4.29 Four projects are under consideration for a 1-year study period. The initial costs and the net revenues generated by the projects are as follows:

Project	Initial Cost	Revenues at End of Year
A	$1,000	$1,120
B	1,200	1,380
C	1,500	1,695
D	2,000	2,250

The minimum attractive rate of return is 10%.

a. Compute the rate of return for each project.

b. Assume that the projects are mutually exclusive and that the firm has unlimited funds. Which project should be chosen?

c. If the projects are independent and funds are unlimited, which one(s) should be selected?

[*Answers:* a. 12.0%, 15.0%, 13.0%, 12.5%; b. B; c. all.]

4.30 [This problem is adapted from J. L. Riggs, *Engineering Economics*, 2nd ed. (New York: McGraw-Hill, 1982).] Six mutually exclusive alternatives are being considered. They are listed in order of increasing costs in the following table. All the alternatives have been evaluated over a common planning horizon.

Alt.	Initial Cost	Net Annual Revenues	(Internal) Rate of Return
P_1	$ 60,000	$ 8,540	7%
P_2	80,000	14,160	12
P_3	100,000	16,980	11
P_4	180,000	29,290	10
T_1	100,000	15,580	9
F_1	20,000	3,990	15

a. Which of the four mutually exclusive projects should be preferred by the Phoenix terminal manager?

b. If $300,000 is available to fund these projects, which should be funded for the Phoenix, Tucson, and Flagstaff terminals?

c. If the firm has available only $200,000 for capital investment, which projects should be funded?

d. Now let's take another look at alternatives P_2 and P_3. Note that P_3 is somewhat more expensive than P_2 ($100,000 - $80,000 = $20,000) and yields a somewhat lower rate of return. Determine the largest value of the MARR such that P_3 is preferable to P_2.

[*Answers:* a. P_4; b. P_4, T_1, F_1; c. P_2, T_1, F_1; d. 6.8%.]

Alternative	IRR on Overall Investment	IRR on Increments of Investment Compared with Alternative:				
		I	II	III	IV	V
I	1%	—	—	—	—	—
II	8	21%	—	—	—	—
III	11	15	12%	—	—	—
IV	15	22	19	17%	—	—
V	13	19	16	15	9%	—
VI	14	21	18	16	14	21%

Let k = minimum attractive rate of return.

a. For what values of k is II preferred to I?

b. If the do-nothing alternative ($\emptyset$) is also feasible, and if alternatives IV, V, and VI are not feasible, which feasible alternative is preferred when $k = 15\%$?

c. If the do-nothing alternative is not feasible, for what values of k is alternative IV preferred?

d. If $k = 9\%$, which alternative is preferred?

4.31 Arizona Transport, Inc., is a small trucking company with terminals located in Phoenix, Tucson, and Flagstaff. The firm's minimum attractive rate of return is 8%.

The terminal managers at Phoenix (P), Tucson (T), and Flagstaff (F) have been asked to forward requests for capital funding during the firm's next fiscal quarter. The Phoenix manager submits four mutually exclusive plans for enlarging the loading dock at his terminal: P_1, P_2, P_3, P_4. The Tucson and Flagstaff managers submit one proposal each: T_1 and F_1.

The economic life (planning horizon) for each of the proposals is 10 years. All of the economic data, including internal rates of return, are summarized next.

4.32 Consider an investment proposal requiring a $220 investment at the end of the first period and resulting in net revenues of $80 at the beginning of the first period and $150 at the end of the second period:

End of Period:	0	1	2
Cash Flow ($):	80	−220	150

a. Find the (internal) rate(s) of return for this investment. *Hint:* You may use the quadratic formula:

$$x = \frac{-b \pm \sqrt{b^2 - 4ac}}{2a}$$

for $ax^2 + bx + c = 0$

b. If the firm's MARR is 0.30, find the external rate(s) of return for this project.

c. If the MARR = 0.30, should the project be accepted? Explain your answer.

4.33 An investor is considering two mutually exclusive alternatives with the following economic data:

	Project G	Project H
Initial capital investment	$58,500	$ 48,500
Net uniform annual benefits	6,648	0
Salvage value 10 years from now	30,000	138,000
Internal rate of return	8%	11%

Both projects are based on a 10-year study period.
a. Compute the incremental interest rate(s) from investing in project G rather than in project H. (*Hint*: There are two solving rates of return. One is about 19%. Find the other.)
b. Over what range of minimum attractive rate of return (MARR) is G preferable to H?
[*Answers*: a. 60%; b. 19% < MARR < 60%.]

4.34 (A relatively common form of investment is the so-called "interest only with balloon payment." For such investments the investor assumes a debt whose principal is not repaid until some future date. In the interim, periodic interest payments are made on the outstanding debt.)

Mr. Venture has an opportunity to acquire certain commercial property that is currently available at a price of $100,000. The seller will accept a note payable in full at the end of 4 years, with interest due and payable each year at the rate of 10% per year, or $10,000 annually for 4 years. The income to Mr. Venture, before interest payments, is estimated to be $15,000 annually. (For simplicity, assume that these receipts will occur at the end of each year.) Mr. Venture expects to keep the property for 5 years, then sell it for an estimated $100,000. The net cash flows for this proposed investment are:

End of Year:	1	2	3	4	5
Cash Flow ($1,000):	5	5	5	-95	115

a. Find the internal rate(s) of return.
b. Assuming a reinvestment rate of 30% per year, find the external rate of return.

4.35 The cash flows for a certain project are:

End of Period:	0	1	2	3
Cash Flow ($):	-200	300	100	-100

The firm's discount rate (MARR) is 10% per period.
a. Determine the external rate of return.
b. Is the project acceptable? Why?
[*Answers*: a. 23.1%; b. yes, because ERR > MARR.]

4.36 A proposed investment requires a disbursement of $3,000 now and $5,000 two years hence. It will yield an income of $2,200 at the end of each year for 6 years. The income will be reinvested at 10%.
a. Find the external rate of return of this investment.
b. Is this investment attractive? Why?

4.37 Consider two investment opportunities, X and Y. The expected cash flows are as follows:

End of Year	X	Y
0	-$ 5,500	-$ 8,000
1	-10,000	-4,250
2	-10,000	-13,300

The firm's MARR(k) is 15%.
a. Use the rate of return method to determine the preferred alternative.
b. For what values of the MARR is Y preferred to X?
c. Determine the external rate of return of X versus Y.
[*Answers*: a. X < Y; b. 0.10 < k < 0.20; c. 14.9%.]

4.38 An investment opportunity is available that promises to return $1,000 at the beginning of years 1 through 4 in return for an investment of $3,000 at the end of year 1. That is,

End of Year j	Cash Flow A_j
0	$1,000
1	1,000 - 3,000 = -2,000
2	1,000
3	1,000

a. Determine the (internal) rate(s) of return for this investment.

b. Assuming a reinvestment rate of 20% per year, find the external rate of return.

c. Given your answer in (b), is the proposed investment attractive? Why?

4.39 [Based on Tung Au and Thomas P. Au, *Engineering Economics for Capital Investment Analysis* (Boston: Allyn & Bacon, 1983), p. 158.]

A certain investment project has a total negative cash flow of $292,600. Of this amount, the cost will be $76,600 at the start of the project life, with the remaining $216,000 to be paid at the end of the 5-year project life. The project, if acquired, will result in positive cash flows of $56,000 at the end of each year over the 5-year project life. Note that the net cash flow at the end of the fifth year will be −$160,000 (=$56,000 − $216,000). This results in the following cash flows:

End of Year:	0	1	2	3	4	5
Cash Flow ($1,000):	−76.6	56	56	56	56	−160

It can be shown that there are two internal rates of return for this project: 8.0% and 33.6%.

a. If the firm's MARR is 12.0%, find the external rate of return for this project.

b. For what values of the MARR is the project desirable?

[*Answers*: a. 12.8%; b. 8.0<MARR<33.6%.]

The Benefit–Cost Ratio Method

4.40 A 400-m tunnel must be constructed as part of a new aqueduct system for a major city. Two alternatives are being considered:

● Alternative A: Build a full-capacity tunnel now for $500,000.

● Alternative B: Build a half-capacity tunnel now for $300,000 and then build a second half-capacity tunnel 20 years from now for $400,000.

The cost of repairing the tunnel lining at the end of every 10 years is estimated to be $20,000

for the full-capacity tunnel and $16,000 for each half-capacity tunnel.

Determine whether alternative A or alternative B should be constructed now. Use the benefit–cost ratio method with a discount rate of 5% and a 50-year planning horizon. (Of course, there will be no tunnel-lining repair at the end of 50 years.) Here, assume that "benefits" are defined as the reduction in tunnel-lining repair costs.

4.41 A municipal agency is considering a proposal from the Jack Janitorial Services to provide service for a number of the city's public facilities. The cost of this service is $100,000 per year, but the contract would save the city $180,000 annually above current costs.

A somewhat more ambitious proposal has been received from the Queen Cleaning Company. This contract would cost the city $150,000 annually, but benefits in the amount of $250,000 per year could be expected.

In summary, the benefits and costs of these proposals are:

Alternative	Benefits	Costs
Jack Janitorial	$180,000	$100,000
Queen Cleaning	250,000	150,000

Use the benefit–cost ratio method to determine which alternative is preferable.

[*Answer*: Queen.]

4.42 Consider two mutually exclusive alternatives, R and S. The economic consequences resulting from these two investments are:

	Alternative R	Alternative S
PW of annual benefits to user (B)	$85,000	$110,000
PW of annual operating expenses (K)	0	10,000
PW of initial investment (C)	45,000	50,000

As discussed in Section 4.3.5, there are two ways of computing benefit–cost ratios:

1. Subtract K in the numerator, or

$$\frac{B - K}{C}$$

2. Add K in the denominator,

$$\frac{B}{C + K}$$

Use both definitions of the benefit–cost ratio method to determine which investment is preferable.

4.43 Terry Trojan is an operations analyst for a local municipal agency in which the benefit–cost ratio method is commonly employed for project justification. She has been asked to evaluate a forms processing operation that currently costs the agency $50,000 annually. She is considering two alternatives:

1. System X requires an initial investment of $10,000 for equipment that has a service life of 5 years. Operating costs will be reduced to $45,000 annually. The equipment will have no salvage value at the end of 5 years.

2. System Y requires an initial investment of $15,000 for equipment that has a service life of 5 years. Operating costs will be reduced to $41,000 annually. Annual maintenance costs for the equipment will be $1,000 annually over the 5-year service life. There is no salvage value at the end of 5 years. The agency uses a 10% interest rate as its cost of capital.

a. One formulation of the benefit–cost ratio is to define benefits as the savings in operating costs minus the maintenance costs. That is, the maintenance costs are in the numerator of the $B:C$ ratio. With this formulation, find the $B:C$ ratios for Systems X and Y.

b. An alternative formulation of the $B:C$ ratio is to define benefits solely as the savings in operating costs. That is, maintenance costs are in the denominator of the $B:C$ ratio. With this formulation, find the $B:C$ ratios for Systems X and Y.

c. Consider the three alternatives: do nothing, choose X, or choose Y. Using the benefit–cost ratio method, determine which alternative is preferable. Explain your answer.

d. Suppose that a new alternative, System Z, is available. System Z is a fully automated system that costs $160,000, has a useful life of 5 years, and has zero salvage value at the end of 5 years. If it is acquired, it will entirely eliminate the $50,000 annual operating costs currently being experienced. Is Z preferable to X and Y? Explain your answer.

e. Mr. Bruin, administrative director for the municipal agency, is concerned about the inherent uncertainties associated with the estimates in this analysis. He proposes that raising the cutoff $B:C$ ratio from 1.00 to 1.25 would be prudent, because it would provide a 25% "safety margin" for the agency. If Mr. Bruin's proposal is adopted, would System Z be an acceptable alternative? Is Mr. Bruin's proposal reasonable? Explain your answer.

[*Answers*: a. 1.89, 2.02; b. 1.89, 1.82; c. Y≻X≻∅; d. Z≻Y≻X≻∅; e. yes, because $\Delta B:C = 1.26$, but the proposal is not reasonable.]

4.44 Consider three mutually exclusive alternatives with the following cash flows:

End of Year	Alternative A	Alternative B	Alternative C
0	−$100	−$150	−$110
1–4	47	66	49

The firm's minimum attractive rate of return is 10%. Use the benefit–cost ratio method to determine the preferred alternative.

4.45 Consider four mutually exclusive alternatives:

	D	E	F	G
Initial cost	$75	$50	$15	$90
Uniform annual benefit	18.8	13.9	4.5	23.8

Each alternative has a 5-year useful life and no salvage value. The minimum attractive rate of return is 10%. Use the benefit–cost ratio method to determine which alternative should be selected.

[*Answer*: Alternative E.]

4.46 The Municipal Department of Parks and Recreation is considering four proposals for new landscaping equipment. Analyzing the relevant costs and benefits of these mutually exclusive alternatives yields:

Alternative	Equivalent Annual Benefits	Equivalent Annual Costs
H	$182,000	$91,500
I	167,000	79,500
J	115,000	88,500
K	95,000	50,000

Benefits and costs have been annualized using a 10-year assumed life and an 8% discount rate. Using the benefit–cost ratio method, determine which alternative is preferable.

4.47 A public agency considering five alternatives for the design of a certain flood-control culvert. The discounted present values of expected benefits and costs are:

Alternatives	Benefits	Costs
L	$ 80,000	$100,000
M	148,000	120,000
N	130,000	110,000
P	180,000	150,000
Q	185,000	160,000

Using the benefit–cost ratio method, determine which alternative, if any, is preferable.
[Answer: Alternative P.]

4.48 Five mutually exclusive alternatives, each with a 20-year useful life and zero salvage value, are summarized below. Assuming a 6% discount

Alternatives:	I	II	III	IV	V
PW of Costs ($):	4,000	2,000	6,000	1,000	9,000
Uniform Annual Benefits ($):	639	410	761	117	785

rate, use the benefit–cost ratio method to determine which alternative should be selected.

4.49 The Public Works Department for the State of Chaos has prepared an economic analysis for seven different configurations (designs) for a proposed project. These alternatives are mutually exclusive. The design engineers have determined the economic consequences for each of these alternatives as measured over a 25-year planning horizon, discounted the expected cash flows using a 10% discount rate, and summarized the results in the following table. (All figures in the table are equivalent present values in millions of dollars.)

Consequences	Alternatives						
	T	U	V	W	X	Y	Z
Initial investment	0.8	3.0	0	2.2	1.0	1.0	3.0
Cost of operations and maintenance	0.2	0	0.5	0.3	1.0	0.4	1.0
Benefits to users	0.5	2.7	2.0	4.0	3.0	3.5	5.2

a. Determine the alternative(s) that is (are) not preferable to the do-nothing alternative.
b. Find the preferred alternative.
c. Depending on how one defines the benefit–cost ratio, there are two possible ratios for the preferred alternative. What are they?
[Answers: a. Alternatives T and U; b. Alternative Y; c. 2.5, 3.1.]

4.50 The Public Works Department of Small City is considering the construction of a bridge over a flood-control channel. Four alternative designs are being considered. There are three categories of economic consequences: (1) construction costs, (2) expenses associated with periodic maintenance, and (3) benefits to the users of the bridge. All future cash flows over a 30-year planning horizon have been discounted at a 10% discount rate, and the resulting equivalent present worths are summarized next. If the benefit–cost ratio method is used properly, determine the rank ordering of the five alternatives (including "do nothing").

	Present Values			Benefit – Cost Ratios	
Design	Construction Costs (C)	Periodic Maintenance (M)	User Benefits (B)	$\dfrac{B-M}{C}$	$\dfrac{B}{C+M}$
I	$100	$50	$200	1.50	1.67
II	300	0	240	0.80	0.80
III	400	20	480	1.15	1.14
IV	300	60	540	1.60	1.50

4.51 Consider a set of mutually exclusive alternatives for which pertinent data are:

Alternative	Initial Cost	Annual Benefits
∅	0	0
S	$100,000	$30,000
T	120,000	29,000
U	150,000	38,000
V	150,000	35,000
W	200,000	40,000
X	180,000	40,000

Benefits are measured over a 20-year planning horizon. The appropriate discount rate is 10% before taxes. Use the benefit–cost ratio method to determine which of the alternatives is preferable.
[*Answer*: **Alternative U.**]

4.52 A public agency is considering seven mutually exclusive alternatives. The discounted present values of benefits and costs, in thousands of dollars, are:

Alternative	Benefits	Costs
P	40	10
Q	50	30
R	30	20
S	30	40
T	75	40
U	90	60
V	95	70

Using the benefit–cost ratio method in the correct manner, determine the preferred alternative.

4.53 Consider a set of five mutually exclusive alternatives. The equivalent present values of earnings, $P(E)$, initial investments, $P(I)$, and salvage values, $P(S)$, are

Alternative	$P(E)$	$P(I)$	$P(S)$
V	$17,000	$ 6,000	$2,000
W	14,000	10,000	3,000
X	9,000	10,000	8,000
Y	24,000	15,000	5,000
Z	26,000	15,000	0

Using the *savings–investment ratio method*, find the preferable alternative.
[*Answer*: **Alternative Y.**]

4.54 Refer to Problem 4.52.
  a. Find the net benefit–cost ratios for the seven alternatives.
  b. Using the net benefit–cost ratio method, determine the preferred alternative.

4.55 The Public Works Department of Central City has been asked by the city council to develop plans for a waste-disposal plant. Seven mutually exclusive plans have been developed, with benefits and costs as summarized next. All figures are discounted present worths over a 25-year planning horizon using an 8% discount rate.

Plan	Benefit	Costs
P	$4,000,000	$ 500,000
Q	4,000,000	2,000,000
R	7,000,000	2,000,000
S	6,000,000	5,000,000
T	9,000,000	6,000,000
U	2,000,000	4,000,000
V	7,000,000	8,000,000

a. Which alternative has the largest benefit–cost ratio?
b. Which is the most expensive alternative with $B:C > 1.0$?
c. Which is the least expensive alternative with $B:C > 1.0\%$?
d. Using the benefit–cost ratio method properly, find the most worthy alternative.
[*Answers:* a. P; b. T; c. P; d. R.]

4.56 The Department of Airports is considering four plans for processing arriving passengers through the International Terminal. "Benefits" are defined as the savings over the current (do-nothing) process; "costs" are the capital and operating costs of each of the plans, beyond those incurred under the do-nothing case. Analysis of the alternatives yields the following present worths in $100,000s:

Alternative Plan	Benefits	Costs
W	300	150
X	80	100
Y	400	240
Z	500	400

The benefit–cost ratio method is used by the Department of Airports.
a. Which alternative has the largest benefit–cost ratio?
b. Which alternative(s) is (are) not preferable to the do-nothing alternative?
c. Which alternative is the most expensive alternative with $B:C$ greater than unity?
d. Which alternative is the preferred alternative?

4.57 The Nation of Upncoming has applied to the World Bank for a loan to fund an irrigation project in the Rojo River Valley. Eight different construction plans have been submitted, only one of which may be funded. Using a 50-year planning horizon and a 6% interest rate, the discounted present value of all expected benefits and costs, in millions of equivalent U.S. dollars, are as follows:

Plan	Benefits	Costs
P	$10	$ 2
Q	16	5
R	14	7
S	6	8
T	17	9
U	18	11
V	22	14
W	19	20

a. Which plan has the largest benefit–cost ratio?
b. Which plan(s) have/has $B:C$ less than unity?
c. Which is the most costly alternative with a $B:C$ greater than unity?
d. If the benefit–cost ratio method is used properly, which plan is the preferred alternative?
[*Answers:* a. P; b. S and W; c. V; d. Q.]

4.58 Consider the cash flows associated with the following mutually exclusive projects:

j	$A_j(X)$	$A_j(Y)$	$A_j(Z)$
0	−$20,000	−$12,000	−$15,000
1–5	5,000	0	0
5	0	24,000	28,000

The firm's MARR is 10%. If the benefit–cost ratio method is applied properly, determine the appropriate ranking of the alternatives.

4.59 The Public Works Department of Peach County has been asked to evaluate four mutually exclusive programs for the planting and maintenance of trees in parkways within the County. As summarized in the following table, there are three principal classes of consequence associated with these plans. All dollar values in the table are in thousands; they are equivalent present values assuming a 50-year planning horizon and a 10% discount rate.

	Plan A	Plan B	Plan C	Plan D
Benefits to community	$6,000	$2,200	$7,000	$7,000
Operations and maintenance	1,500	300	0	2,200
Initial costs	3,000	2,000	5,000	3,000

Use the benefit–cost ratio method to determine the preferred alternative.

[*Answer*: $C > D > A > \emptyset > B$.]

4.60 Three mutually exclusive alternatives are under consideration by the Public Works Department of Gotham City. The equivalent present values of associated costs and benefits (in $1,000s) are as follows:

Alternative	Benefits	Costs
X	450	150
Y	200	50
Z	570	300

a. Use the benefit–cost ratio method (properly) to determine which of these three alternatives is preferred.

b. The Public Works Department is also considering two plans for improving traffic flow over a bridge in the city. There are three consequences: initial cost, annual maintenance, and annual savings to users. The equivalent present values (in $1,000s) for the two alternatives are as follows:

Alternative	Initial Cost	Annual Maintenance	User Savings
P	50	30	100
Q	100	0	130

Assemblyman Black claims that alternative P should be preferred, since

$$B:C(P) = \frac{100 - 30}{50} = 1.4$$

and

$$B:C(Q) = \frac{130}{100} = 1.3$$

Mayor White claims that alternative Q should be preferred, since

$$B:C(P) = \frac{100}{50 + 30} = 1.25$$

and

$$B:C(Q) = \frac{130}{100} = 1.3$$

Which is the preferred alternative? Explain.

4.61 The Roads Department of Orange City is considering four (mutually exclusive) design alternatives for a highway interchange. Estimates of the economic consequences yield the following results:

Alternative	PW of Benefits to Motorists	PW of Maintenance and Repair Expenses	PW of Construction Costs
I	$1,500,000	$200,000	$ 500,000
II	700,000	100,000	400,000
III	2,500,000	600,000	1,200,000
IV	1,000,000	500,000	1,000,000

All of the above present worth equivalences were determined using a 9% discount rate and a 50-year study period.

a. Using the benefit–cost ratio method, determine the preferred alternative.

b. It is the practice of the Roads Department to consider maintenance and repair expenses as negative benefits; that is, only construction costs are considered as "costs" to appear in the denominator of the benefit–cost ratio. With this definition, is III preferred to II? Why?

[*Answers*: a. Alternative I; b. yes, because $B:C(II \rightarrow III) = 1.63$.]

4.62 The Capricorn Manufacturing Company is required by the Occupational Safety and Health Administration to provide certain safety equipment in its plant operations. The prospective cash flow for the four plans under consideration are given next. The firm's pretax minimum attractive rate of return is 20% per annum.

End of Year	Cash Flows ($)			
	Plan A	Plan B	Plan C	Plan D
0	−$10,000	−$13,000	−$8,000	−$12,000
1–5	− 2,000	− 1,000	− 3,000	− 1,000
5	4,000	5,000	0	0

Use the rate of return method to rank-order the four alternatives.

4.63 Semiautomated manufacturing equipment is available at a cost of $250,000, will last for 5 years, and is expected to have a salvage value of $50,000 at the end of 5 years. Costs for direct labor, material, and incremental overhead would be $150,000 in the first year and then decrease by $20,000 each year thereafter. The firm's MARR is 12% per annum. Determine the benefit–cost ratio if the annual operating savings, the benefits, are $200,000 each year for 5 years. Is the proposal economically justified? Explain.

[*Answer:* Yes, because $B:C = 1.14 > 1.00$.]

A P P E N D I X: *Computer Software Applications*

The use of generic spreadsheet software (e.g., *Lotus 1-2-3* and *Excel*) for the analysis of engineering economy problems was introduced in the Appendix to Chapter 2. Certain special-purpose software, *Cash Flow Analyzer (CFA)*, was discussed initially in the Appendix to Chapter 3. Both of these approaches are now illustrated with respect to the types of problems discussed in Chapter 4.

EXAMPLE 4.A.1 *Determining the Internal Rate of Return*

Recall the set of cash flows first introduced as Example 4.1:

End of Period:	0	1	2	3	4
Cash Flow:	−$1,000	$200	$200	$400	$600

The spreadsheet approach to the solution of the (internal) rate of return is shown in **Figure 4.A.1** on page 158. In this case we find the present value at a variety of trial interest rates and then use linear interpolation when the solution brackets zero. Specifically,

$$\left. \begin{array}{l} \text{PV @ } 12\% = 4 \\ \text{PV @ } 13\% = -22 \end{array} \right\} \quad \text{Interpolating, IRR} = 12.16\%$$

The CFA solution to this same problem is illustrated in **Figure 4.A.2** on page 159. With maximum precision for CFA, the solution is shown to be 12.157%.

	A	B	C	D	E	F
1						
2	EXAMPLE 4.A.1					
3						
4	DETERMINING THE INTERNAL RATE OF RETURN					
5						
6	Assumes no "IRR" command					
7						
8						
9						
10	End of	Cash	P.V. @	P.V. @	P.V. @	P.V. @
11	period	Flows				
12	j	Aj	0.10	0.11	0.12	0.13
13	0	-1,000	-1000	-1000	-1,000	-1,000
14	1	200	182	180	179	177
15	2	200	165	162	159	157
16	3	400	301	292	285	277
17	4	600	410	395	381	367
18						
19	Totals	400	57	30	4	-22
20	IRR = 0.122					
21						
22	Sample Cell Formulas:					
23	C13 = B13*((1+C12)^(-A13))					
24	B19 = @SUM(B13..B17)					
25	A20 = E12+((F12-E12)(E19/(E19-F19))					

Figure 4.A.1 Spreadsheet Solution for the IRR of Example 4.A.1

EXAMPLE 4.A.2 *Using the Rate of Return Method to Choose Between Alternatives*

Recall Example 4.2 in the text. There are two alternatives, P and Q, with cash flows as follows:

	End of Period				
	0	1	2	3	4
Cash flow (alt. P)	$1,000	$200	$200	$400	$600
Cash flow (alt. Q)	− 1,900	600	600	600	600
Cash flow (difference)	−$ 900	$400	$400	$200	0

The spreadsheet solution to this problem is shown in **Figure 4.A.3** on page 160. Note here that, unlike the previous example in which the trial-and-error approach was illustrated, we now take advantage of a special command in the software. Specifically, we are using Release 2.2 of *Lotus 1-2-3*, in which there is a special function for finding the internal rate of return generated by a set of cash flows:

$$@IRR(guess, cash flow)$$

*** Example 4.A.1 ***

MARR (Discount Rate) : 10.00%

Planning Horizon :   4 periods

Cash Flow Information

#	T	1st	End	A(1st)	Suppl. Data	Description
1	S	0	—	-1,000.0		
2	S	1		200.0		
3	S	2		200.0		
4	S	3		400.0		
5	S	4		600.0		

Cash Flow Analysis

#	PW	AW	FW
1	-1,000.0	-315.4	-1,464.1
2	181.8	57.3	266.2
3	165.2	52.1	242.0
4	300.5	94.8	440.0
5	409.8	129.2	599.9

Sum of Present Worths : 57.3

Sum of Annual Worths : 18.0

Sum of Future Worths : 84.0

Sum of All Cash Flows Under 0.0% MARR : 400.0

Internal Rate of Return : 12.157%

Figure 4.A.2   CFA Solution for the IRR of Example 4.A.1

The "guess" argument is typically a percent, between zero and one. The "cash flows" are the cell labels for the range of relevant cash flows. The first cash flow in the sequence must be a negative amount. In this example, the rate of return for the differences in cash flows between P and Q are:

Cell Label:	D15	D16	D17	D18	D19
Cell Value ($):	−900	400	400	200	0

	A	B	C	D	E
1					
2	**EXAMPLE 4.A.2**				
3					
4	**RATE OF RETURN ANALYSIS: CHOOSING BETWEEN ALTERNATIVES**				
5					
6	Assume "IRR" command available in software				
7					
8	Guessed value of IRR =			0.30	
9					
10					
11		End of	Alt-P	Alt-Q	Difference
12		Period			
13		j	Aj(P)	Aj(Q)	Aj(Q-P)
14	-------------------	---------------------	--------------------	-----------------------	
15		0	-1000	-1900	-900
16		1	200	600	400
17		2	200	600	400
18		3	400	600	200
19		4	600	600	0
20	-------------------	---------------------	--------------------	-----------------------	
21	TOTALS:		400	500	100
22					
23	IRR =		0.122	0.100	0.061
24					
25	Sample Cell Formulas:				
26					
27		B21=	@SUM(B15..B19)		
28		C21=	@SUM(C15..C19)		
29		D21=	@SUM(D15..D19)		
30		B23=	@IRR($D$8,B15..B19)		
31		C23=	@IRR($D$8,C15..C19)		
32		D23=	@IRR($D$8,D15..D19)		
33					

Figure 4.A.3 Spreadsheet Solution for Example 4.A.2

The "guess" value is arbitrarily chosen as 0.30 and placed in cell D8. Thus the command in cell D23 is

$$@IRR(\$D\$8, D15..D19)$$

The solution is shown to be 6.1%.

The CFA solution is given in **Figure 4.A.4**. The IRR for the difference in cash flows is 6.085%. As this is less than the MARR (10.00%), the incremental investment in alternative Q is not justified.

EXAMPLE 4.A.3 *External Rate of Return*

Recall Example 4.4:

End of Period:	0	1	2	3	4	5
Cash Flow ($):	−100	50	−20	100	−80	100

As there are multiple sign changes for the cash flows, there may be as many internal rates of return as there are changes in direction of sign.

The external rate of return (ERR), assuming an auxiliary interest rate of 20%, is determined using the spreadsheet shown in **Figure 4.A.5**. The solution, 17.25%, indicates that the project is not acceptable, because the ERR is less than the MARR, 20%.

*** Example 4.A.2 ***

MARR (Discount Rate) : 10.00%

Planning Horizon :   4 periods

Cash Flow Information

#	T	1st	End	A(1st)	Suppl. Data	Description
1	S	0		-900.0		
2	S	1		400.0		
3	S	2		400.0		
4	S	3		200.0		

Cash Flow Analysis

#	PW	AW	FW
1	-900.0	-283.9	-1,317.6
2	363.6	114.7	532.4
3	330.5	104.2	484.0
4	150.2	47.4	220.0

Sum of Present Worths : -55.7

Sum of Annual Worths : -17.6

Sum of Future Worths : -81.2

Sum of All Cash Flows Under 0.0% MARR : 100.0

Internal Rate of Return : 6.085%

Figure 4.A.4   CFA Solution for Example 4.A.2

	A	B	C	D	E	F
1						
2	**EXAMPLE 4.A.3**					
3						
4	**EXTERNAL RATE OF RETURN**					
5						
6	Given:	Auxiliary interest rate (k)			0.20	
7		Planning horizon (N)			5	
8		Guessed value of ERR =			0.10	
9						
10					FW of	
11	End of	Cash	Negative	Positive	Positive	Modified
12	Period	Flows	Cash	Cash	Cash	Cash
13	j	Aj	Flows	Flows	Flows	Flows
14	--------------	----------------	----------------	----------------	----------------	----------------
15	0	-100	-100		0.00	-100
16	1	50		50	103.68	0
17	2	-20	-20		0.00	-20
18	3	100		100	144.00	0
19	4	-80	-80		0.00	-80
20	5	100		100	100.00	347.68
21	--------------	----------------	----------------	----------------	----------------	----------------
22	TOTAL:	50.00	-200.00	250.00	347.68	147.68
23						
24	External rate of return (ERR)					0.1725
25						
26	Sample Cell Formulas:					
27	E15 = +D15*(1+$E$6)^($E$7-$A15)					
28	F20 = $E$22					
29	B22 = @SUM(B15..B20)					
30	C22 = @SUM(C15..C20)					
31	D22 = @SUM(D15..D20)					
32	E22 = @SUM(E15..E20)					
33	F22 = @SUM(F15..F20)					
34	F24 = @IRR($E$8,F15..F20)					

Figure 4.A.5 Spreadsheet Solution for Example 4.A.3—External Rate of Return

It is of interest to note that, although there are five sign changes for the cash flows in this problem, there is only one internal rate of return! This is somewhat surprising, because the Norstrom criterion indicates three sign changes of the cumulative cash flows:

End of Period:	0	1	2	3	4	5
Cumulative Cash Flow ($):	−100	−50	−70	+30	−50	+50

This illustrates that multiple sign changes of the cumulative cash flows do not preclude the possibility that there may be fewer real positive values for the internal rate of return.

The comparable CFA solution for this problem is summarized in **Figure 4.A.6**. The ERR is shown to be 17.256%.

Figure 4.A.7 on page 164, generated by CFA, indicates that there is only a single zero root to the PW equation over the range $0 < i < 100\%$. **Figure 4.A.8** on page 165, the CFA solution summary, indicates that the IRR is 14.216%.

*** Example 4.A.3 ***

MARR (Discount Rate) : 20.00%

Planning Horizon :   5 periods

Cash Flow Information

#	T	1st	End	A(1st)	Suppl. Data	Description
1	S	0		-100.0		
2	S	1		0.0		revised
3	S	2		-20.0		
4	S	3		0.0		revised
5	S	4		-80.0		
6	S	5		347.7		revised

Cash Flow Analysis

#	PW	AW	FW
1	-100.0	-33.4	-248.8
2	0.0	0.0	0.0
3	-13.8	-4.6	-34.5
4	0.0	0.0	0.0
5	-38.5	-12.9	-96.0
6	139.7	46.7	347.7

Sum of Present Worths : -12.6

Sum of Annual Worths : -4.2

Sum of Future Worths : -31.6

Sum of All Cash Flows Under 0.0% MARR : 147.7

External Rate of Return : 17.256%

Figure 4.A.6   CFA Solution for Example 4.A.3—External Rate of Return

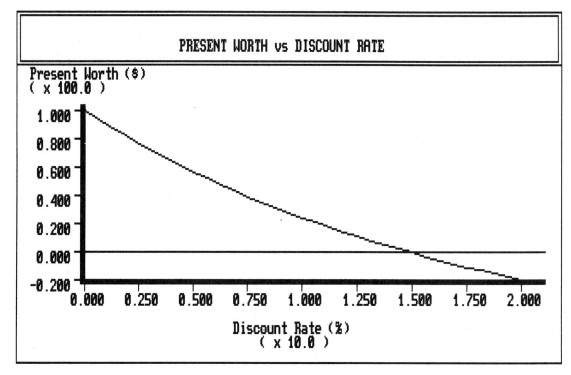

PRESENT WORTH vs DISCOUNT RATE

Figure 4.A.7   CFA-Generated Plot of PW(i) for Example 4.A.3

EXAMPLE 4.A.4 *Benefit–Cost Analysis*

Recall the four mutually exclusive alternatives discussed in Example 4.7:

Alternative	Benefits	Costs
I	$ 646	$1,000
II	1,258	1,000
III	1,341	1,100
IV	2,305	2,000

The appropriate benefit–cost analysis, using the spreadsheet approach, is illustrated in **Figure 4.A.9** on page 166. The optimal policy, it will be noted, is

$$IV \succ II \succ III \succ \varnothing \succ I$$

*** Example 4.A.3 ***

MARR (Discount Rate) : 20.00%

Planning Horizon :   5 periods

Cash Flow Information

#	T	1st	End	A(1st)	Suppl. Data	Description
1	S	0		-100.0		
2	S	1		50.0		
3	S	2		-20.0		
4	S	3		100.0		
5	S	4		-80.0		
6	S	5		100.0		

Cash Flow Analysis

#	PW	AW	FW
1	-100.0	-33.4	-248.8
2	41.6	13.9	103.6
3	-13.8	-4.6	-34.5
4	57.8	19.3	144.0
5	-38.5	-12.9	-96.0
6	40.1	13.4	100.0

Sum of Present Worths : -12.8

Sum of Annual Worths : -4.3

Sum of Future Worths : -31.7

Sum of All Cash Flows Under 0.0% MARR : 50.0

Internal Rate of Return : 14.216%

Figure 4.A.8   CFA Solution for Example 4.A.3—Internal Rate of Return

	A	B	C	D	E	F
1						
2	**EXAMPLE 4.A.4**					
3						
4	**BENEFIT COST ANALYSIS: MULTIPLE ALTERNATIVES**					
5						
6	Notes:	Alt-O = "do nothing" alternative				
7		Alternatives ranked in order of increasing cost				
8						
9	Input Data:					
10						
11		Alternative	Benefits ($)	Costs ($)		
12		------------------------	-------------------	----------------		
13		I	646	1000		
14		II	1258	1000		
15		III	1341	1100		
16		IV	2305	2000		
17		------------------------	-------------------	----------------		
18	Analysis:	B/C resulting from going from i to j				
19						
20		O	I	II	III	IV
21	---------------	------------------------	-------------------	----------------	----------------	----------------
22	O		0.65	1.26	1.22	1.15
23	I			#DIV/0!	6.95	1.66
24	II				0.83	1.05
25	III					1.07
26	IV					
27	---------------	------------------------	-------------------	----------------	----------------	----------------
28	Results:		I<O	II>O	III>O	IV>O
29				#DIV/0!	III>I	IV>I
30					III<II	IV>II
31						IV>III
32	---------------	------------------------	-------------------	----------------	----------------	----------------
33	**Sample Cell Formulas:**					
34	C22 = +C13/D13	E24 = (C15-C14)/(D15-D14)				
35	D22 = +C14/D14	F24 = (C16-C14)/(D16-D14)				
36	E22 = +C15/D15	F25 = (C16-C15)/(D16-D15)				
37	F22 = +C16/D16	C28 = @IF(C22=1,"I=O",@IF(C22>1,"I>O","I<O"))				
38	D23 = (C14-C13)/(D14-D13)					
39	E23 = (C15-C13)/(D15-D13)					
40	F23 = (C16-C13)/(D16-D13)					

Figure 4.A.9 Spreadsheet Solution for Example 4.A.4—Benefit–Cost Analysis

5

Measures of Worth III:
Some Additional Methods
for Determining Economic Value

INTRODUCTION

Although the capital allocation techniques discussed in Chapters 3 and 4 are demonstrably valid, they are not universally adopted by the business community. Significant evidence exists to indicate that, although sound methodology is being used more frequently, faulty techniques remain widespread in government and industry. (The extent of current usage in industry is discussed in Section 5.4.)

There are a number of reasons for this unhappy state of affairs. First, introduction of these techniques into college curricula is a relatively recent development. (In contrast, academic institutions have long taught the principles of double-entry accounting.) Second, in many cases business people call on their "golden nugget of experience" to make decisions pertaining to the allocation of limited capital. Differences among alternatives may appear to be sufficiently large that the appropriate course of action either is immediately obvious or follows from a few preliminary calculations. In any event, the decision maker's intuition, based on accumulated experience, frequently leads to the choice of investment alternatives. Finally, certain incorrect and/or approximation methods are often used to select from alternatives in the mistaken view that such procedures are valid. Unfortunately, these approaches have been institutionalized because of their apparent simplicity. The purpose of the following discussion is to present several of the most common of these methods in order to establish their advantages and disadvantages when employed in the capital allocation process.

5.2 THE PAYBACK METHOD

The payback method is widely used in American industry to determine the relative attractiveness of investment proposals. The essence of this technique is determination of the *number of periods required to recover an initial investment*. Once this has been done for all alternatives under consideration, a comparison is made on the basis of respective payback periods.

As illustrated in **Figure 5.1**, **payback**, or **payout** as it is sometimes known, is the number of periods required for cumulative benefits to exactly equal cumulative costs. Costs and benefits are usually expressed as cash flows, although discounted present values of cash flows may be used. In either case, the payback method is based on the assumption that the relative merit of a proposed investment is measured by this statistic. The smaller the payback (period), the better the proposal.

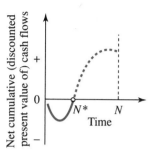

Figure 5.1 Payback (Payout)

5.2.1 Conventional (Undiscounted) Payback

The payback method is most often used only in connection with cash flows. In this case, the number of payback periods, or payback, is that value of N^* such that

$$P = \sum_{j=1}^{N^*} A_j \tag{5.1}$$

where P is the initial investment and A_j is the cash flow in period j. This conventional approach uses undiscounted cash flows.

EXAMPLE 5.1 *Uniform Cash Flows*

Consider a proposed project that has a first cost of $10,000 and net revenues (excess of receipts over disbursements) of $2,000 per year each year over a 10-year period. In this case the number of periods required to "pay back" the original investment is $10,000/$2,000 per year = 5 years.

Now consider another project that has a first cost of $9,000 and net revenues of $1,500 per year for 7 years. The payback for this proposal is $9,000/$1,500 = 6 years. The conclusion, according to the payback method, is that the first alternative is superior to the second because its initial cost will be recovered more quickly.

EXAMPLE 5.2 *Irregular Cash Flows*

Another example of payback appears in **Table 5.1**. Here, cash flows are not uniform from period to period, and further, N^* is not an integer value. The column of cumulative cash flows indicates that $6 < N^* < 7$ periods. Note that a solution for $N^* = 6.25$ periods is not strictly correct, as the cash

TABLE 5.1   Determining Payback for an
Initial Investment of $10,000

Period	Net Cash Flow	Cumulative Cash Flow
0	− $10,000	− $10,000
1	− 2,000	− 12,000
2	− 1,000	− 13,000
3	0	− 13,000
4	2,000	− 11,000
5	4,000	− 7,000
6	6,000	− 1,000
7	4,000	3,000
8	2,000	5,000

flows are end-of-period. However, in practice, decimal fractions are commonly used.

5.2.2 Discounted Payback

An obvious shortcoming of the undiscounted payback method described above is that the time value of money, as reflected in the discount rate, is ignored. We can adjust for this problem by discounting all cash flows before determining payback. Assuming that all end-of-period cash flows A_j are discounted at effective rate i per period, N^* is defined by the relationship

$$P = \sum_{j=1}^{N^*} A_j(1 + i)^{-j} \tag{5.2a}$$

or, since $P = -A_0$,

$$\sum_{j=0}^{N^*} A_j(1 + i)^{-j} = 0 \tag{5.2b}$$

EXAMPLE 5.3

Consider the three mutually exclusive alternatives, Q, R, and S, described in **Table 5.2**. Based on cash flows alone, we see that

$$N^*(Q) = 4 \quad N^*(R) = 2 \quad N^*(S) = 3$$

So it would appear that R is the preferred alternative. Discounted payback, however, presents a very different set of results. Alternative Q results in no definable payback. With a discount rate of 15% per period, the initial investment is not entirely recovered over the 5-period planning horizon. Furthermore, S now has a shorter discounted payback period than R:

$$N^*(R) = 5 \quad \text{and} \quad N^*(S) = 4$$

TABLE 5.2   Determining Discounted Payback for Three Mutually
Exclusive Alternatives

End of Period	Cash Flows			Present Value @ $i=15\%$			Cumulative Present Value		
	Q	R	S	Q	R	S	Q	R	S
0	−$1,000	−$1,000	−$1,000	−$1,000	−$1,000	−$1,000	−$1,000	−$1,000	−$1,000
1	100	100	900	87	87	783	− 913	− 913	− 217
2	200	900	50	151	680	38	− 762	− 233	− 179
3	300	100	50	197	66	33	− 565	− 167	− 146
4	400	100	255	229	57	146	− 336	− 110	0
5	500	221	100	249	110	50	− 87	0	50
Totals	$ 500	$ 421	$ 355	−$ 87	$ 0	$ 50			
N^*	4	2	3				None	5	4

The explanation for this shift in the rank order is that the major positive cash flow is received earlier by S than by R.

5.2.3 Theoretical Errors

There are several objections to the payback method. Two are relatively minor; the third is critically important.

The first objection concerns the fact that *the conventional definition of payback considers cash flows only.* But as we saw, this problem can be overcome by using discounted cash flows to determine the payback period.

The second objection concerns the possibility that *the payback period, N*, may appear to be greater than the project life, N.* This can occur when Eqs. (5.2) are used without regard to N. Consider, for example, a project with initial cost of $10,000 that is expected to yield net returns of $1,000 per year each and every year for 6 years. The conventional (undiscounted) payback is

$$N^* = \frac{\$10,000}{\$1,000 \text{ per year}} = 10 \text{ years}$$

Assuming a 9% discount rate, the discounted payback is found from

$$(P/A, 9\%, N^*) = \frac{\$10,000}{\$1,000} = 10.0$$

from which $N^* \simeq 28$ years. In either case, the results are meaningless because the expected life of the project is less than the total number of periods required for payback. In general, N^* is meaningful only if $0 < N^* \leq N$, where N is the project life (planning horizon).

The third objection is that *the payback method ignores consequences that are beyond the end of the payback periods.* That is, this approach disregards all A_j's for $N^* < j \le N$. It limits information to the decision maker that may be vital to the decision.

EXAMPLE 5.4

Consider two alternatives, Projects T and U, with cash flows as summarized in **Table 5.3**. Assuming a 10% discount rate, the present values of these cash flows are also shown. The payback periods, N^*s, are:

Approach	Project T	Project U
Conventional	2	3
Discounted	3	4

(These values of N^* are meaningful, since $N = 5$.) Using either approach, it would appear that T is preferred to U because $N^*(T) < N^*(U)$; that is, T has the shorter payback. However, using the present worth method, Project U is the preferred alternative, because

$$PW(T) = \$13 \quad \text{and} \quad PW(U) = \$621$$

Why the inconsistency? Because the payback method ignores the consequences beyond the end of the payback periods. In this example, the cash flow $A_5(U) = \$1,000$ proved to be decisive.

TABLE 5.3 Cash Flows and Present Values at $i = 10\%$ for Example 5.4

End of Period j	Cash Flows (A_j)		Cumulative		Discounted Cash Flows		Cumulative	
	T	U	T	U	T	U	T	U
0	−$1,000	−$1,000	−$1,000	−$1,000	−$1,000	−$1,000	−$1,000	−$1,000
1	500	400	−500	−600	455	364	−545	−636
2	500	400	0	−200	413	331	−132	−305
3	176	200	176	0	132	150	0	−155
4	10	227	186	227	7	155	7	0
5	10	1,000	196	1,227	6	621	13	621
Totals	$196	$1,227			$13	$621		
N^*			2	3			3	4

5.2.4 Reciprocal of Payback as a Measure of Rate of Return

Under certain conditions, the reciprocal of the payback period is an approximate measure of the project's rate of return. That is, $i \simeq 1/N^*$. There are two necessary conditions: (1) the periodic cash flows must be uniform; and (2) the number of periods must be very large. Recall that, given uniform cash flows, A, and $N \to \infty$,

$$\text{PW} = A_0 + A(P/A, i, N \to \infty)$$

$$= A_0 + A(1/i)$$

The internal rate of return, i^*, is determined when $\text{PW} = 0$. Thus

$$i^* = \left(\frac{-A_0}{A}\right)^{-1} = \left(\frac{P}{A}\right)^{-1} = \frac{1}{N^*} \tag{5.3}$$

Of course, the validity of Eq. (5.3) depends on the extent to which the two conditions are met.

EXAMPLE 5.1 *(Continued)*

Recall that the project has an initial cost of $10,000 and uniform net revenues of $2,000 per year for 10 years. Here, $N^* = 5$. Using Eq. (5.3),

$$i^* \simeq \frac{1}{5} = 0.20$$

Note that the project life is fairly short, 5 years. It may be shown that the true IRR is only 0.15.

EXAMPLE 5.3 *(Continued)*

Recall alternative Q with irregular cash flows over 5 periods as shown in Table 5.2. The conventional payback is $N^* = 4$, and

$$i^* \simeq \frac{1}{4} = 0.25$$

It may be shown, however, that the true IRR is only 0.12.

5.2.5 Payback and the Internal Rate of Return

The payback method may be useful under the special conditions that (1) the lives of the alternatives are equal and, (2) other than the initial investment, the periodic cash flows are uniform. However, even under these conditions, results may be spurious.

Given an initial investment, P, a uniform series of cash flows, A, and a project life, N, the internal rate of return, i^*, is the solution to the equation

$$P = A(P/A, i^*, N)$$

Also note that, under these conditions, the conventional payback is found from

$$N^* = \frac{P}{A}$$

Thus payback and rate of return are directly related:

$$N^* = (P/A, i^*, N) \tag{5.4}$$

Since $(P/A, i^*, N)$ decreases as i^* increases, larger rates of return yield shorter payback periods. Put somewhat differently, given two alternatives, I and II, each with uniform cash flows over a common planning horizon, N, then if

$$N^*(\text{I}) < N^*(\text{II})$$

it will also be true that

$$\text{IRR(I)} > \text{IRR(II)}$$

EXAMPLE 5.5

Consider alternative V, which has an initial investment of $100 and net annual receipts of $20 each year for 10 years. The initial cost of alternative W is $80, and it is expected to produce net receipts of $10 each year for 10 years. The conventional payback analysis yields:

$$N^*(\text{V}) = \frac{\$100}{\$20/\text{year}} = 5 \text{ years}$$

$$N^*(\text{W}) = \frac{\$80}{\$10/\text{year}} = 8 \text{ years}$$

The rates of return for the two alternatives are determined as follows.

For V: $(P/A, i^*, 10) = 5.0$, so $i^* = 15.1\%$.
For W: $(P/A, i^*, 10) = 8.0$, so $i^* = 4.3\%$.

Thus the alternative with the shorter payback is also the one with the higher internal rate of return.

It is particularly important to emphasize, however, that one alternative is not preferable to another simply because it has a higher internal rate of return. This is the ranking error discussed in Chapter 4. Returning to our example, it may be shown that

MARR(k)	Correct Ranking
$k < 4\%$	$\text{V} \succ \text{W} \succ \varnothing$
$4\% < k < 15\%$	$\text{V} \succ \varnothing \succ \text{W}$
$15\% < k < 49\%$	$\varnothing \succ \text{V} \succ \text{W}$
$49\% < k$	$\varnothing \succ \text{W} \succ \text{V}$

Thus rank ordering by payback is a function of the discount rate appropriate to the problem.

5.2.6   Possible Usefulness

Since the payback method is a measure of how fast original investments are repaid, it yields information that may be of considerable interest. For example, when future consequences of a prospective investment are highly uncertain, managers may want to mitigate the associated risk by recovering the initial cost as quickly as possible. Under these conditions, payback period computations provide an index of one aspect of the proposed investment.

The payback method may also be useful when a firm's current capital is extremely limited and cash flow estimates indicate that similar limitations will persist for the next few budgeting periods. In addition, if the firm expects unusually profitable investment opportunities to arise in the near future, then it will want to recover capital as quickly as possible in order to take advantage of these potential reinvestments. Information provided by the payback method should be useful under these conditions, but, in any case, this method should be used only to provide information *in addition* to that given by the valid methods discussed in Chapters 3 and 4.

5.3   ACCOUNTING METHODS FOR COMPUTING RETURN ON INVESTMENT

Chapter 4 presented the appropriate procedure for determining the (internal) rate of return, based on the analysis of cash flows. A number of other approaches use accounting data (income and expenses) rather than cash flows to determine "rate of return," where income and expense are reflected in the firm's accounting statements. Because the accounting approach to project evaluation is so widely used in industry, this section examines several of the more common procedures.

Although there is no universally accepted terminology for "rate of return" using accounting data, the term **return on investment (ROI)** is commonly used in such instances. As will be shown, IRR and ROI are superficially similar ideas, but their differences are significant. IRR is based on cash flows; ROI is based on accounting notions of income and expenses. "Income" (or "revenue") is the accountant's term to describe a net increase in assets due to the sale of goods or services. "Expense" is the accountant's term for the cost incurred during the period related to the production of income. Income is not necessarily a positive cash flow; expense is not necessarily a negative cash flow.

EXAMPLE   5.6

To illustrate the differences between IRR and ROI, consider a relatively simple numerical example. There are two mutually exclusive alternatives, X and Y, with cash flows as summarized in **Table 5.4**. The (internal) rate of return for alternative X is found from

$$-\$50,000 + \$5,000(P/A, i_X^*, 5) + \$5,000(P/G, i_X^*, 5) = 0$$

from which $i_X^* = 12.0\%$. For alternative Y,

$$-\$50,000 + \$14,000(P/A, i_Y^*, 5) = 0$$

from which $i_Y^* = 12.4\%$.

TABLE 5.4   Cash Flows Before Income Taxes for
Two Mutually Exclusive Alternatives

End of Year	Alternative X	Alternative Y
0	−$50,000	−$50,000
1	5,000	14,000
2	10,000	14,000
3	15,000	14,000
4	20,000	14,000
5	25,000	14,000
Total cash flow	$25,000	$20,000

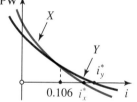

Figure 5.2   PW As a Function of the Discount Rate, Example 5.6

Once again, Y is not preferable to X simply because $i_Y^* > i_X^*$. Project preference depends on the minimum attractive rate of return, k. Indeed, it may be shown that the present worth of the differences between alternatives $(Y - X)$ is

$$PW = -\$9,000(P/A, i, 5) + \$5,000(P/G, i, 5)$$

The PW is equal to zero when $i = 10.6\%$. Therefore, as shown in **Figure 5.2**, X is preferable to Y when $k < 10.6\%$; otherwise Y is preferable to X.

5.3.1   Original Book Method

Using the **original book method**, the project's return on investment is found by dividing the average annual accounting profit by the original book value of the asset. That is,

$$\text{ROI} - \frac{\overline{AP}}{B_0} \tag{5.5}$$

where $\overline{AP}$ = average annual accounting profit and B_0 = original book value of the asset.

Profit, an accounting term, is the excess of income over expenses during the period. Both *income* and *expenses* may include cash as well as noncash items. An example of the latter is *depreciation*, an expense that reflects the distribution of the initial cost of a depreciable asset, less expected salvage value, if any, over its estimated useful life. (Depreciation is discussed in detail in Chapter 6.) Accounting profit is a combination of both cash and noncash elements.

Book value, another accounting concept, is the original cost basis less accumulated depreciation. For our purposes, assume that the original book value, or cost basis, is simply the initial cost of the asset, P.

EXAMPLE 5.7

Assume that alternatives X and Y described in Example 5.6 are equipment that will be depreciated using the *straight line depreciation method*. As will be explained in Chapter 6, the annual depreciation expenses, D, are given by

$$D = \frac{P - S_d}{N_d} \qquad (5.6)$$

where

P = initial cost

S_d = salvage value after N periods for depreciation purposes

N_d = asset life for depreciation purposes

Using data from the example,

$$D = \frac{\$50,000}{5} = \$10,000$$

for both alternative X and alternative Y.

The contributions to the firm's average annual profits for the two projects are determined in **Table 5.5**: $\overline{AP}(X) = \$5,000$ and $\overline{AP}(Y) = \$4,000$. Therefore the returns on investment are

$$ROI(X) = \frac{\$5,000}{\$50,000} = 0.10$$

$$ROI(Y) = \frac{\$4,000}{\$50,000} = 0.08$$

Compared to the correct values for X and Y (12.0% and 12.4%, respectively), this method substantially understates the true rates of return and, incidentally, reverses their relative attractiveness.

TABLE 5.5   Average Annual Accounting Profits for Alternatives X
and Y in Table 5.4

	Alternative X			Alternative Y		
Year	Cash Flow	Depreciation	Profit	Cash Flow	Depreciation	Profit
1	$ 5,000	$10,000	− $ 5,000	$14,000	$10,000	$ 4,000
2	10,000	10,000	0	14,000	10,000	4,000
3	15,000	10,000	5,000	14,000	10,000	4,000
4	20,000	10,000	10,000	14,000	10,000	4,000
5	25,000	10,000	15,000	14,000	10,000	4,000
Totals	$75,000	$50,000	$25,000	$70,000	$50,000	$20,000
Annual average ($\overline{AP}$)			$ 5,000			$ 4,000

5.3.2   Average Book Method

Proponents of the **average book method** argue that the value of the
investment is reduced from year to year, so the rate of return should be
based on the *average* book value. In the case of straight line depreciation,
the average book value is simply half the sum of the first cost and the
expected salvage value, or

$$\text{ROI} = \frac{\overline{AP}}{0.5(P + S_d)} \tag{5.7}$$

So, for alternatives X and Y,

$$\text{ROI(X)} = \frac{\$5,000}{0.5(\$50,000)} = 0.20$$

$$\text{ROI(Y)} = \frac{\$4,000}{0.5(\$50,000)} = 0.16$$

By comparing these results to those found by the original book method,
we see that the differences arise entirely from the values in the denominators. Moreover, these figures are considerably higher than the true rates
of return obtained by using proper compound interest methods.

5.3.3   Year-by-Year Book Method

Still another method, the **year-by-year book method**, is based on year-by-
year ratios of profit after depreciation to book value at the start of the
year. Using alternative X, for example:

Year	Profit After Depreciation	Book Value at Start of Year	ROI
1	− $ 5,000	$50,000	− 10%
2	0	40,000	0
3	5,000	30,000	17%
4	10,000	20,000	50%
5	15,000	10,000	150%

A major problem here lies in applying the results in any useful way. What do these values mean? Perhaps the analyst could average the year-by-year rates to compare alternatives, but using the average return would mask the yearly variation.

5.4 INDUSTRY USE OF DISCOUNTED CASH FLOW, PAYBACK, AND ACCOUNTING METHODS

The evaluation methods described in Chapters 3 and 4 are generally known, collectively, as *discounted cash flow techniques*. As was shown, when they are used properly, they lead to valid, consistent results. The two methodologies presented so far in this chapter, payback and accounting-based return on investment, suffer from significant theoretical difficulties. An interesting question to address at this point is the extent to which these methods are actually used in industry.

There have been a number of surveys of industry practice over the past 25 years. The results from two of the more recent surveys are summarized in **Table 5.6**. One of the surveys was conducted in Great Britain, the other in the United States. In both instances, payback is reported as the evaluation method most commonly used; rate of return is somewhat more widely used than present worth (net present value); and accounting rate of return (return on investment) is also frequently encountered.

5.5 PROFITABILITY INDEXES

The term **profitability index** is found in the literature of economic analysis in a variety of contexts, but it has no generally accepted definition. To one analyst, a profitability index is simply the internal rate of return on

TABLE 5.6   Current Usage of Economic Analysis Methods[a]

Method	100 Large Firms in the United Kingdom[b]	117 Large Firms in the United States[c]
Payback	92%	75%
Internal Rate of Return (IRR)	75%	73%
Net Present Value (NPV)	68%	59%
Accounting Rate of Return (ROI)	56%	34%
Benefit–cost ratio	Not Reported	38%

[a]Table figures show percentage of sample reporting usage.
[b]Richard Pike, "Do Sophisticated Capital Budgeting Approaches Improve Investment Decision-Making Effectiveness?" *The Engineering Economist*, Vol. 34, No. 2, Winter 1989, pp. 149–161. This survey of 100 large U.K. firms was conducted in 1986.
[c]Thomas J. Cook and Ronald H. Rizzuto, "Capital Budgeting Practices for R&D: A Survey and Analysis of Business Week's R&D Scoreboard," *The Engineering Economist*, Vol. 34, No. 4, Summer 1989, pp. 291–304. This survey, conducted in 1985, includes a sample of 117 usable responses from firms listed in *Business Week*'s annual scoreboard of major R&D firms in the United States. The data shown here are reported for the non-R&D projects of these firms.

investments; to another it represents the present worth of a set of cash flows; and to a third it may bear a different meaning. In this section we will examine a specific definition of the profitability index (PI) that is often encountered in one form or another: *the ratio of the net present worth of all cash flows to the initial investment*, known as the **premium worth percentage**,

$$PI = \frac{PW}{P} \tag{5.8}$$

where

$$PW = \sum_{j=0}^{N} A_j(1+j)^{-j}$$

$$P = \text{initial investment} = -A_0$$

Proponents of this technique assert that the PI is useful in those instances in which the present worths of competing proposals are produced by investments of differing sizes.

EXAMPLE 5.8

Consider two mutually exclusive alternatives, I and II. For alternative I:

Present worth of future cash flows	$242,000
Initial investment (P)	200,000
Net present worth (PW)	$ 42,000

For alternative II,

Present worth of future cash flows	$130,000
Initial investment (P)	100,000
Net present worth (PW)	$ 30,000

With these results, it follows from Eq. (5.8) that

$$PI(I) = \frac{\$42,000}{\$200,000} = 0.21$$

and

$$PI(II) = \frac{\$30,000}{\$100,000} = 0.30$$

It does *not* follow that alternative II is preferable to alternative I simply because $PI(II) > PI(I)$.

As discussed in Section 3.3, the present worth of funds invested elsewhere at rate i, when discounted at rate i, is exactly zero. This result makes it possible to compare directly present worths without adjusting for differences in initial costs. In the present example, we are considering two choices: *either* (1) invest $200,000 in alternative I, yielding a present worth of $42,000; *or* (2) invest $100,000 in alternative II, yielding a present worth of $30,000, and invest $100,000 elsewhere at rate of return i, yielding zero present worth.

The profitability index and its variants are irrelevant at best. They add nothing to the analysis and can be misleading to decision makers. They are frequently encountered in industry, however, and you would do well to understand the nature of the deficiencies.

5.6 APPROXIMATIONS OF RATE OF RETURN

As discussed in Chapter 4, the internal rate of return (IRR) is the solution to the equation $PW(i^*) = 0$. This can be a burdensome calculation, especially when cash flows do not follow a regular pattern. Various "short-cut" or simplified procedures have been proposed to provide reasonable approximations to the true IRR. Two of these formulations are described below.

5.6.1 The "Truth in Lending" Formula for Rate of Return

At the urging of various consumer groups, the federal government and many states have enacted "truth in lending" legislation requiring that interest rates charged by lenders be clearly and uniformly stated. Since these rates represent returns to the lenders as well as costs to the borrowers, they are, in a sense, measures of rate of return. One commonly used definition of the **truth-in-lending formula** is presented here. We use the notation:

P = amount of original debt (the original cost, or price, less any down payment)

M = number of payments in one year

N = total number of payments necessary to repay the debt

A = dollar amount of each payment, assuming a uniform series of payments

r = nominal interest rate per year

$\dfrac{r}{M}$ = interest rate per compounding period

I_j = portion of the jth payment that is interest

D = total interest paid

Assuming uniform payments each period,

I_1 = portion of the first payment that is interest

$$= P\left(\frac{r}{M}\right)$$

I_N = portion of the last payment that is interest

$$= \left(\frac{P}{N}\right)\left(\frac{r}{M}\right)$$

Now, let $\bar{I}$ represent the average interest paid over N payments, and assume it to be the average of the first and last payments.

$$\bar{I} = 0.5(I_1 + I_N)$$

The total interest paid over N periods, D, is

$$D = N\bar{I} = 0.5N(I_1 + I_N)$$

$$= 0.5N\left[P\left(\frac{r}{M}\right) + P\left(\frac{r}{MN}\right)\right]$$

Solving for r, we have

$$r = \frac{2MD}{P(N + 1)} \tag{5.9}$$

EXAMPLE 5.9

Assume that a certain item is purchased for which the buyer agrees to repay a debt of $1,000 with 12 equal monthly payments. The buyer is told by the seller that the "interest rate" is 10% per year and that finance charges will be added to the purchase price. (We will assume here that the seller will also provide the financing, so the seller is also the lender.) Total interest charges are 0.10($1,000) = $100. This, added to the $1,000 original debt, requires the buyer (borrower) to make monthly payments of (1/12)($1,000 + $100) = $91.67. The cash flow diagram for this exchange is shown in **Figure 5.3**. In this example,

$$M = 12 \qquad P = \$1{,}000$$

$$D = \$100 \qquad N = 12$$

Therefore, from Eq. (5.11),

$$r = \frac{2(12)(\$100)}{\$1{,}000(12 + 1)} = 0.185$$

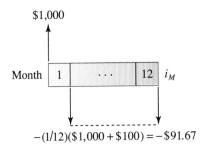

Figure 5.3   Cash Flow Diagram for Example 5.8

In this case, the true annual interest rate per month (i_M) follows from the relationship

$$\$1{,}000 - \$91.67(P/A, i_M, 12) = 0$$

or

$$(P/A, i_M, 12) = 10.9087$$

Solving, $i_M = 0.015$, so the annual rate is

$$(1.015)^{12} - 1 = 0.186$$

This is very close to the approximate rate given by the truth-in-lending formula. (Note that both the approximate and true rates are considerably higher than the original 10 percent rate told to the borrower by the lender!)

5.6.2   The Rule of 72

How long does it take for a sum of money, P, to double if it is invested in a fund earning at rate i per period? The answer can be determined precisely by using the compound amount factor from Chapter 2 and noting that, here, $F/P = 2$.

$$F = P(1 + i)^N \qquad (5.10)$$

from which

$$N = \frac{\ln(F/P)}{\ln(1 + i)} = \frac{\ln 2}{\ln(1 + i)} \qquad (5.11)$$

Conversely, if a sum of money doubles in N periods, what is the rate of return on this investment? Again, solving Eq. (5.10) for i;

$$i = \left(\frac{F}{P}\right)^{1/N} - 1 \qquad (5.12)$$

Although Eqs. (5.11) and (5.12) lead to accurate solutions, they do require the ability to manipulate natural logarithms and/or to solve for the Nth root. The rule of 72, although it provides only approximate solutions, is much easier to use.

The **rule of 72** specifies that the approximate number of periods for a sum of money to double, given that it is invested at rate i, is

$$N \simeq \frac{72}{100i} \qquad (5.13)$$

This is based on the assumptions that $\ln 2 \simeq 0.72$ (actually, $\ln 2 = 0.693$) and, second, that $\ln(1 + i) \simeq i$. This latter assumption is true, of course, when $i = 0\%$, but the assumption is less valid as i increases. For example, when $i = 10\%$, $\ln 1.10 = 0.095$; when $i = 20\%$, $\ln 1.20 = 0.182$.

EXAMPLE 5.10

A sum of money is invested at 8% per year. How long will it take to double? Using Eq. (5.12), the correct solution is

$$N = \frac{\ln 2}{\ln(1.08)} = \frac{0.693}{0.077} = 9.01 \text{ years}$$

Using Eq. (5.13), the rule of 72 yields

$$N \simeq \frac{72}{100(0.08)} = 9 \text{ years}$$

In like manner, the rule of 72 can be used to determine the approximate rate of return for an investment that doubles in N periods:

$$i \simeq \frac{72}{100N} \qquad (5.14)$$

EXAMPLE 5.11

A sum of money doubles in 5 years. What is the approximate rate of return? Using Eq. (5.14), the correct solution is

$$i = (2)^{1/5} - 1 = 14.9\%$$

Using Eq. (5.14), the rule of 72 yields

$$i \simeq \frac{72}{100(5)} = 14.4\%$$

Parenthetically, since $\ln 2 = 0.693$, why don't we use the "rule of 69," as it would provide more accurate approximations? The answer is that the number 72, an integer multiple of months (12) and quarters (4), is somewhat more convenient. As a "shortcut" technique, and bearing in mind that it is only an approximation, the rule of 72 can be quite useful.

SUMMARY

The evaluation methods presented in Chapters 2 and 3 (PW, AW, FW, IRR, and $B:C$), known collectively as discounted cash flow methods, consider the amount and timing of cash flows as well as the effects of the cost of opportunities foregone reflected by the interest rate used in these analyses. These methods yield consistent results. They provide correct guidance when selecting from among alternative investment alternatives when the maximization of wealth (worth or value) is the criterion.

Certain other measures of worth are also used in industry. Principal among these is the *payback* (or *payout*) method, which measures the number of periods necessary to recover an initial investment. There are two formulations of the payback method; one uses cash flows, the other uses discounted values of the cash flows. In either case, the payback method does provide some valuable information under certain circumstances, but it should never be used as the sole criterion. Its principal shortcoming is that consequences beyond the end of payback are ignored.

A number of methods based on accounting data have been used to determine project profitability. In such cases, the "rate of return" is generally known as *return on investment (ROI)*. These methods are fundamentally incorrect in that the ROI is a function of an accounting device for allocating first cost (less salvage value, if any) over the life of the asset. Thus it is possible to alter apparent profitability by changing the depreciation schedule for the asset in question. (Depreciation schedules *can* affect profitability, but only because of the effect on cash flows for taxes. This matter is discussed in Chapter 7.)

So-called *profitability indexes* are often employed in industry to measure the relative desirability of investment proposals. Perhaps the most common of these is the ratio of present worth to initial cost, an attempt to adjust PWs for differences in project size. This is unnecessary and misleading, however, because the present worths of the differences in initial cost are zero.

The *truth-in-lending* formula presented in this chapter is one of several attempts to standardize computation of the cost of consumer loans, but the result is approximate and generally understates the true cost.

The *rule of 72* is a convenient shortcut for determining the number of periods over which a sum of money will double in value, given the interest rate; or, conversely, for determining the approximate rate of return on an investment that has doubled in value over N periods.

| PROBLEMS

Payback

5.1 Proposal I has a first cost of $40,000, expected excess of receipts over disbursements of $10,000 each year for 7 years, and zero salvage value at the end of that time. Proposal II has a first cost of $30,000, net revenues of $6,000 per year for 7 years, and is expected to be sold for 100% of the first cost at the end of the seventh year.
   a. Assuming an interest rate of 5%, use the present worth method to determine the most economically desirable alternative.
   b. Solve the same problem using the payback method.
   c. Comment on the results of (a) and (b).
[*Answers*: a. PW(I) = $17,860, PW(II) = $26,039; b. N^*(I) = 4 years, N^*(II) = 5 years.]

5.2 Mister Miser is considering investing $10,000 in one of two alternatives. (Assume that he has other investment opportunities elsewhere that, with similar risk, will return at least 10%.) His first alternative requires an initial cost of $10,000 and additional annual expenditures of $3,000. However, the investment is expected to yield revenues of $6,000 per year for 5 years, at the end of which time it will be sold for a net salvage value of $100.

The second alternative requires an original investment of $10,000 also, but there are no other cash flow consequences until the 20th year, at which time it is expected that the asset will be sold for $60,000.

a. Use either the present worth or rate of return method to determine which of the alternatives, if any, is most desirable. Assume that replacements for the first alternative will be made with identical assets.
b. Which alternative is indicated by the payback method?
c. Comment on the results of (a) and (b). In particular, discuss the effects of the replacement assumption in (a).

5.3 Consider the problem presented by the following cash flows for alternatives Alpha and Beta:

End of Period	Alpha	Beta
0	−$100	−$220
1	50	100
2	50	100
3	50	100

a. Determine the best alternative using the payback method.
b. Assuming a minimum attractive rate of return of 15%, determine the best alternative by either the present worth or the annual cost method.
c. The payback method is independent of the interest rate. However, what must the value of the minimum attractive rate of return be in order to reverse the decision in (b)?
[*Answers*: a. N^*(Alpha) = 2.0 years, N^*(Beta) = 2.2 years; b. PW(Alpha) = $14.16, PW(Beta) = $8.32; c. Approximately 12%.]

5.4 The anticipated cash flows for two mutually exclusive alternatives are:

End of Period	Alternative I	Alternative II
0	$-$300	$-$300
1	100	200
2	100	100
3	100	100
4	300	100

a. If the appropriate discount rate is 5% per year, determine the present worth of both alternatives.

b. Determine the undiscounted payback (periods) for both alternatives.

c. Determine the payback (periods) for both alternatives after adjusting for discounting of future benefits.

5.5 The payback method may be modified by subtracting annual straight line depreciation from the annual cash flows before income taxes. For alternatives Alpha and Beta given in Problem 5.3, determine the undiscounted payback period after depreciation. Assume $N_d = 3$ and $S_d = 0$. What may be said about the double counting of depreciation?

[*Answer*: $N^*(\text{Alpha}) = 6.0$ years, $N^*(\text{Beta}) = 8.25$ years.]

5.6 A firm is considering two projects, X and Y, each of which will result in a savings of $400 during each year over a 5-year study period. Project X will cost $1,000 and will have no salvage value. Project Y will cost $1,200 initially and will have a 100% salvage value ($1,200) at the end of 5 years. The firm's pretax MARR is 20% per year.

a. Find the undiscounted payback for projects X and Y.

b. Find the discounted payback for the two projects.

c. Using the present worth method, which project is preferred?

5.7 Consider three mutually exclusive alternatives with the following cash flows:

End of Year	Alternative A	Alternative B	Alternative C
0	$-$100	$-$150	$-$110
1–4	47	66	49

The firm's minimum attractive rate of return is 10%.

a. Apply the (undiscounted) payback period method to determine the preferred alternative.

b. Use the benefit–cost ratio method to determine the preferred alternative.

[*Answers*: a. $B>C>A$; b. $B>A>C$.]

5.8 Three mutually exclusive alternatives (R, S, and T) are under consideration. The anticipated cash flows are as follows:

End of Year	$A_i(R)$	$A_i(S)$	$A_i(T)$
0	$-$50,000	$-$ 70,000	$-$70,000
1	$-$ 20,000		30,000
2	10,000		25,000
3	20,000		20,000
4	30,000		15,000
5	40,000		10,000
6	50,000		5,000
7	60,000	200,000	0

a. Determine the (undiscounted) payback for each of the three alternatives.

b. Assuming a 10% minimum attractive rate of return, determine the present worth of each of the three alternatives.

5.9 Three alternative plans for warehousing raw materials are under consideration, X, Y, and Z. Preliminary analysis indicates that the estimated cash flows, in $1,000s, are as follows:

End of Year j	$A_i(X)$	$A_i(Y)$	$A_i(Z)$
0	$-$400	$-$1,200	$-$300
1	$-$200	400	$-$300
2	0	400	$-$300
3	200	400	300
4	400	400	300
5	600	400	300
6	800	400	300
7	1,000	400	2,000

a. Determine the (undiscounted) payback periods for the three plans.

b. Assuming a 12% discount rate, determine the present worth of the three plans.

[*Answers*: a. $N^*(X) = 4$, $N^*(Y) = 3$, $N^*(Z) = 5$; b. PW(X) = $1,016, PW(Y) = $626, PW(Z) = $824.]

5.10 In order to reduce operating costs in a certain manufacturing operation, the Collage Company is considering the acquisition of used equipment costing $50,000. It will be kept in service for 5 years, at the end of which time it will have zero salvage value. Alternatively, the Collage Company can buy new equipment at a cost of $60,000 and that is expected to have a $20,000 salvage value after 5 years. Either the used or the new equipment should result in net annual savings of $2,000 per month for 60 months. In summary, the expected cash flows are as follows:

End of Month	(Alternative I) Used Equipment	(Alternative II) New Equipment
0	−$50,000	−$60,000
1–60	2,000	2,000
60	0	20,000

a. Find the (undiscounted) payback for the two alternatives.
b. Determine the value of the MARR such that alternative II is preferred to alternative I.

5.11 The Bruin Company is considering two mutually exclusive projects that are expected to produce savings in material handling costs over the next 10 years. The first, Plan LO, has a somewhat lower initial cost. The second, Plan HI, has a higher initial cost, but the annual savings and the terminal salvage value are also somewhat higher. The firm's discount rate is 12%. Cash flow forecasts for the two plans are as follows:

End of Year	Plan LO	Plan HI
0	−$100,000	−$300,000
1–10	20,000	50,000
10	10,000	200,000

a. Determine the (undiscounted) payback for the two plans.
b. Determine the discounted payback for the two plans.
c. For what values of the MARR is HI preferred to LO?

[*Answers*: a. 5 years and 6 years for LO and HI; b. 8–9 years for LO, no payback up to 10 years for HI; c. MARR < 14.8%.]

5.12 Three mutually exclusive alternatives are under consideration with annual cash flows (in $1,000s) as follows:

	Alt. X	Alt. Y	Alt. Z
Initial cash flow (A_0)	−600	−800	−800
Uniform, continuous ($\bar{A}_j$ for $j = 1, \ldots, 5$)	200	400	400
Salvage value, end of year 5 (A_5)	1000	200	0

a. Determine the (undiscounted) payback for each of the three alternatives.
b. If the effective interest rate is 20% per year with interest compounded continuously, determine the present worths of the three alternatives.

5.13 Consider an initial investment of $10,000 yielding returns of $2,500 at the end of each year for 10 years. The MARR is 23% per year.
a. Determine the (undiscounted) payback.
b. Determine the (internal) rate of return.
c. The **payback rate of return** is defined as the reciprocal of the (undiscounted) payback. Determine the payback rate of return in this case.
d. At what interest rate (i) is the *discounted payback* (N^*) exactly equal to 10 years?
e. Is the proposed investment preferred to the do-nothing alternative? Why?

[*Answers*: a. 4 years; b. 21.5%; c. 25.0%; d. 21.5%; e. no, because IRR < MARR.]

Return on Investment

5.14 [This problem was adapted from E. L. Grant and W. G. Ireson, *Principles of Engineering Economy*, 5th ed. (New York: The Ronald Press, 1970), p. 181.] Two investment proposals, a vacuum still and a product terminal, are competing for limited funds in an oil company. Each requires an immediate disbursement of $110,000, all of which will be capitalized on the books of account. The after-tax cash flows and the after-tax profits for the two alternatives are estimated to be those shown in the following table.

End of Year	After-Tax Cash Flows			After-Tax (Accounting) Profits	
	Vacuum Still	Product Terminal		Vacuum Still	Product Terminal
1	$ 38,000	$ 5,000		$27,000	−$ 6,000
2	34,000	9,000		23,000	− 2,000
3	30,000	13,000		19,000	2,000
4	26,000	17,000		15,000	6,000
5	22,000	21,000		11,000	10,000
6	18,000	25,000		7,000	14,000
7	14,000	29,000		3,000	18,000
8	10,000	33,000		− 1,000	22,000
9	6,000	37,000		− 5,000	26,000
10	2,000	41,000		− 9,000	30,000
	$200,000	$230,000	Total profit	$90,000	$120,000
			Average annual profit	$9,000	$12,000

a. If the after-tax minimum attractive rate of return is 15% per year, determine which alternative is preferable.

b. Find the returns on investment (ROIs) for both alternatives, using the *original book* method.

c. Find the ROIs for both alternatives, using the *average book* method.

d. What is the true after-tax internal rate of return for the vacuum still?

5.15 A company is considering investment in depreciable property. Estimates for the two alternatives under consideration are:

	Location M	Location N
First cost of property	$1,000,000	$1,400,000
Salvage value	$ 200,000	$ 200,000
Annual cash revenues (net)	$ 120,000	$ 160,000
Useful life	25 years	25 years

a. Determine the true rates of return for both proposals.

b. If the minimum attractive rate of return for the company is 9% before income taxes, are either or both of the alternatives economically acceptable?

c. Using a minimum attractive rate of return of 9%, determine whether the extra investment in location N ($400,000 increment) is justified. That is, find the rate of return on the incremental investment and compare it to the minimum attractive rate of return.

d. Determine the rates of return for locations M and N using the *original book* method. Compare your results with a 9% minimum attractive rate of return.

[*Answers*: a. 11.3% and 10.6% for M and N; b. both are acceptable; c. $i^*\Delta = 8.8\%$; d. ROI(M) = 8.8%, ROI(N) = 8.0%.]

Solve Problem 5.15 using the *average book* method.

5.17 Solve Problem 5.15 by using the *average of the year-by-year rates of return* as measured by the ratio of profit after depreciation to book value at starts of year. Assume straight line depreciation. [*Answer*: ROI(M) = 17.02%, ROI(N) = 17.24%.]

5.18 In our discussion of accounting methods for computing rate of return, we discussed alternative X with cash flows as given in Table 5.4. The IRR may be shown to be about 12.0% for this project.) The original cost of the asset is $50,000; it has a life of 5 years, with zero salvage value. In Example 5.7 we assumed straight line depreciation, $10,000 per year. Now assume that the depreciation expenses are as given below. Use the *average book method* to determine the average return on investment.

Year	Cash Flow	Depreciation	Profit
1	$ 5,000	$ 7,500	− $ 2,500
2	10,000	11,000	− 1,000
3	15,000	10,500	4,500
4	20,000	10,500	9,500
5	25,000	10,500	14,500
Totals		$50,000	$25,000

Hint: You will want to find the average book value each year ($\bar{B}_j$), then take the average of the averages. For example,

$$\bar{B}_1 = 0.5[\$50,000 + (\$50,000 - \$7,500)]$$
$$= \$46,250$$

Profitability Indexes

5.19 Mr. Bearish, the owner of a small business, has an opportunity to purchase a Model A press for $10,000. It is anticipated that the press will result in a savings of $2,500 per year over its 10-year life. There is no salvage value; the press will be worthless after 10 years. Mr. Bearish's minimum attractive rate of return is 20% per year.
   a. Find the present worth of Model A.
   b. Find the profitability index for Model A.
   c. Find the true rate of return for this proposed investment.
[*Answers:* a. $481; b. 0.048; c. 21.4%.]

5.20 Refer to Problem 5.19. Mr. Bearish is considering an alternative press, Model B, which has an initial cost of $25,000 and will result in annual savings of $6,200 over its 10-year life. The expected salvage value after 10 years is negligible.
   a. Find the present worth of Model B.
   b. Find the profitability index for Model B.
   c. Is Model B preferable to the do-nothing alternative? Why?

5.21 Refer to Problems 5.19 and 5.20. Which press, A or B, if any, is preferable? Explain your answer. [*Answer:* **B ≻ A ≻ ∅.**]

5.22 Two plant layouts are under consideration. The initial costs are $200,000 and $300,000 for plans I and II, respectively. An engineering economic analysis indicates that the present worth of plan I is $408,000 and the PW of plan II is $456,000.
   a. Determine the premium worth percentages for plans I and II.
   b. Which plan is preferable? Why?

5.23 Refer to the two mutually exclusive projects of the Bruin Company described in Problem 5.11: plan LO and plan HI.
   a. Determine the premium worth percentage for plan LO and for plan HI.
   b. Determine the present worths for plans LO and HI.
[*Answers:* **a. 0.162 for LO and 0.156 for HI;**
b. PW(LO) = $16,200 and PW(HI) = $46,900.]

5.24 Partial analysis of three mutually exclusive alternatives (U, V, S) is as follows:

Alternative	U	V	S
Initial Cost ($)	200,000	327,000	530,000
Net Annual Receipts ($)	80,000	100,000	150,000
PW of Annual Receipts @ 10% ($)	389,000	487,000	730,000
Net PW @ $i=$10% ($)	189,000	160,000	200,000

   a. Determine the net benefit–cost ratio for each of the three alternatives.
   b. Determine the premium worth percentage for each of the three alternatives.
   c. Determine the appropriate ranking of these three alternatives, considered together with the do-nothing alternative.

5.25 Consider the cash flows associated with the following mutually exclusive alternatives:

End of Period	Alt. X	Alt. Y	Alt. Z
0	− $24,000	− $21,000	− $50,000
1–8	4,000	3,000	0
8	4,800	21,000	135,000

a. Determine the undiscounted payback period (N^*) for each of the three alternatives.
b. Assuming a 12% discount rate, determine the premium worth percentage (PWP) for each of the three alternatives.
c. Given the four alternatives—the do-nothing alternative ($\varnothing$) is the fourth alternative—what is the proper ranking?

[*Answers*: a. 6, 7, and 8 for X, Y, and Z; b. -0.09, 0.11, and 0.09; c. $Z > Y > \varnothing > X$.]

5.26 Consider three mutually exclusive alternatives (X, Y, Z) as follows:

Alternative:	X	Y	Z
Initial investment:	$1,200	$1,500	$2,000

Discounted *present value* of cash flows occurring at end of period j:

j	X	Y	Z
1	$ 200	$500	$400
2	1,000	500	500
3	100	500	600
4	100	300	500
5	100	300	700

a. Determine the (discounted) payback for each of the three alternatives.
b. Determine the Savings Investment Ratio for each of the three alternatives. (The SIR is discussed in Chapter 4.)

"Truth in Lending" and the Rule of 72

5.27 A $1,500 item is purchased with a $500 down payment. The balance owed, $1,000, is to be paid in 12 equal end-of-month payments. The seller advertises a "1% per month plan," but in practice the monthly payments are determined by adding the total finance charge to the original debt and then dividing the total by the number of payment periods.

a. Determine the amount of the monthly payments.
b. Determine the annual cost of this debt using the truth-in-lending formula.
c. Determine the true annual cost of this debt.

[*Answers*: a. $93.33; b. 22.2%; c. 23.7%.]

5.28 A manufacturing firm is considering the purchase of a certain quality inspection device at an initial cost of $10,000. The firm will make a down payment of $2,500, then pay off the remaining $7,500 in 36 equal end-of-month payments. The vendor will charge 10% per year "simple interest," or a total of $2,250 in interest over the 3 years: 10%/year × 3 years × $7,500 = $2,250. The interest payments will also be spread uniformly over the 36 months. Thus each end-of-month payment will be ($7,500 + $2,250)/36 = $270.83.

a. Determine the effective cost per month of this "loan."
b. Determine the effective cost per year.
c. Find the rate of return (cost per year) using the truth-in-lending formula.

5.29 A computer workstation is purchased for an initial cost of $10,000. The buyer agrees to pay a down payment of $2,000 and to pay the remaining $8,000 debt, with interest, over the next 8 quarters. That is, payments will be made at the end of months 3, 6,..., 24. The vendor charges "12% per year" on the debt: total interest charges will be 0.12 × $8,000 × 2 years, or $1,920 spread evenly over the 8 quarters. Thus each quarterly payment will be

$$\left(\frac{1}{8}\right)(\$8,000 + \$1,920) = \$1,240$$

a. Determine the true cost of this loan per quarter.
b. Determine the true cost of this loan per year.
c. Determine the cost of this loan using the truth-in-lending formula.

[*Answers*: a. 5.04%; b. 21.8%; c. 21.3%.]

5.30 Certain laboratory equipment may be purchased at a cost of $18,000. A 20% down payment ($3,600) is required, with the remaining 80%

($14,400) to be paid in 48 equal end-of-month installments. The vendor of this equipment says that monthly payments are based on a 12% annual rate, so the total interest is $6,912 ($=0.12 \times \$14,400 \times 4$). Thus the amount of the monthly payments will be:

($14,400 + $6,912)/48 months = $444 per month

a. Using the truth-in-lending formula, determine the approximate annual cost, expressed as an interest rate, of this "borrowed" money. (*Hint:* It's a lot more than 12%!)
b. Determine the true effective cost per month.
c. Determine the true effective cost per year.

5.31 A $1,500 item is purchased with a $500 down payment, with the balance ($1,000) to be repaid in 24 equal monthly installments. The seller advertises a "1% per month" plan, but in practice the monthly payments are determined by adding the total finance charges to the original debt and then dividing the total by the number of repay-

ment periods. Here, the total financial charges are 0.01 × $1,000 × 24 months = $240.
a. What is the cost of this debt as determined by the truth-in-lending formula?
b. What is the true cost of this debt per month?
c. What is the true cost of this debt per year?
[*Answers:* **a. 23.04%; b. 1.798%; c. 23.84%.**]

5.32 If interest is compounded at the rate of 6% per period:
a. How long will it take for a sum of money to double (non-integer solution ok)?
b. Approximately how long will it take for a sum of money to double according to the "rule of 72"?
c. How long will it take for a sum of money to quadruple (non-integer solution ok)?
d. Approximately how long will it take for a sum of money to quadruple according to the "rule of 72"?
e-f. Rework parts (a) and (b) assuming an effective compounding rate of 24% per period.

APPENDIX: *Computer Software Applications*

EXAMPLE 5.A.1 *Payback and Discounted Payback*

Recall Example 5.4. Cash flows are given for two mutually exclusive alternative investments:

End of Period:	0	1	2	3	4	5
Alt. T cash flows ($):	−1,000	500	500	176	10	10
Alt. U cash flows ($):	−1,000	400	400	200	227	1,000

Payback is 2 and 3 periods for T and U, respectively. At a 10% discount rate, discounted payback is 3 and 4 periods for T and U, respectively.

The analysis of this problem, in spreadsheet form, is shown in **Figure 5.A.1**. Note that the appearance is virtually identical to Table 5.3 in the text.

	A	B	C	D	E	F	G	H	I
1									
2	EXAMPLE 5.A.1								
3									
4	PAYBACK AND DISCOUNTED PAYBACK								
5									
6	Cash Flows and Present Value at i =						0.10		
7									
8									
9									
10		Cash Flows (Aj)				Discounted Cash Flows			
11	End of								
12	Period			Cumulative				Cumulative	
13	j	Alt T	Alt U	Alt T	Alt U	Alt T	Alt U	Alt T	Alt U
14									
15	0	-1000	-1000	-1000	-1000	-1000	-1000	-1000	-1000
16	1	500	400	-500	-600	455	364	-545	-636
17	2	500	400	0	-200	413	331	-132	-306
18	3	176	200	176	0	132	150	0	-156
19	4	10	227	186	227	7	155	7	0
20	5	10	1000	196	1227	6	621	13	620
21									
22	TOTAL:	196	1227			13	620		
23									
24	PAYBACK (N*) =		The first value in column A for which value						
25			in column D >= 0 (positive number)						
26			Similarly for columns E,H and I.						
27									
28									
29	Sample Cell Formulas:								
30	D15 = +B15					I15 = +G15			
31	D16 = +B16+D15					I16 = +G16+I15			
32	E15 = +C15					B22 = @SUM(B15..B20)			
33	E16 = +C16+E15					C22 = @SUM(C15..C20)			
34	F15 = +B15*(1+$G$6)^(-A15)					F22 = @SUM(F15..F20)			
35	G15 = +C15*(1+$G$6)^(-A15)					G22 = @SUM(G15..G20)			
36	H15 = +F15								
37	H16 = +F16+H15								

Figure 5.A.1 Determining Payback and Discounted Payback for Example 5.A.1

EXAMPLE 5.A.2 *Return on Investment*

Recall Example 5.6. We are considering a $50,000 investment in a project, alternative X, that is expected to generate cash flows as follows:

End of Period:	1	2	3	4	5
Cash flows ($):	5,000	10,000	15,000	20,000	25,000

The project has no expected salvage value. It is further assumed that the entire initial investment ($50,000) will be depreciated over 5 years using the straight line method, thus generating a depreciation expense of $10,000 each year.

For the *original book method*, the return on investment is given by

$$ROI = \frac{\text{average annual accounting profit}}{\text{original book value}}$$

$$= \frac{\$5,000}{\$50,000} = 0.10$$

And, for the *average book method*, the return on investment is given by

$$ROI = \frac{\text{average annual accounting profit}}{0.5(\text{original book value} + \text{salvage value})}$$

$$= \frac{\$5,000}{0.5(\$50,000 + \$0)} = \frac{\$5,000}{\$25,000} = 0.20$$

	A	B	C	D	E	F
1						
2	**EXAMPLE 5.A.2**					
3						
4	**RETURN ON INVESTMENT (ROI)**					
5						
6	Given:	Initial Cost (P)			$50,000	
7		Salvage Value (S)			$0	
8		Depreciable Life (Nd)			5	years
9		Depreciation Method			Straight Line	
10		Guessed value of IRR =			0.30	
11	End of	Cash		Book Value	Profit	
12	year	Flow	Deprec'n	End of year	after	Annual
13	j	Aj	Dj	Bj	Deprec'n	ROI
14	----------------	----------------	----------------	----------------------	----------------------	-------------
15	0	-50,000		50,000		
16	1	5,000	10,000	40,000	-5,000	-0.10
17	2	10,000	10,000	30,000	0	0.00
18	3	15,000	10,000	20,000	5,000	0.17
19	4	20,000	10,000	10,000	10,000	0.50
20	5	25,000	10,000	0	15,000	1.50
21	----------------	----------------	----------------	----------------------	----------------------	-------------
22	TOTAL:	25,000	50,000			
23						
24	AVERAGE:	5,000	10,000			0.413
25						
26	ROI using original book method:				0.10	
27						
28	ROI using average book method:				0.20	
29						
30	Internal Rate of Return (IRR):				0.1201	
31						
32	**Sample Cell Formulas:**					
33	D15 = $E$6			E16 = +B16-C16		
34	C16 = ($E$6-$E$7)/$E$8			F24 = ((@SUM(F16..F20))/A20		
35	B22 = @SUM(B15..B20)			E26 = +B24/$E$6		
36	C22 = @SUM(C15..C20)			E28 = +B24/(0.5*($E$6+$E$7))		
37	B24 = +B22/A20			E30 = @IRR($E$10,B15..B20)		
38	D16 = +D15-C16					

Figure 5.A.2   Determining Return on Investment (ROI) for Example 5.A.2

Both calculations are shown in tabular form in **Figure 5.A.2** on page 193. Note that the year-by-year returns on investment are shown in cells F16–F20, and the average ROI, 0.413, is shown in cell F24.

6

Depreciation and Certain
Other Noncash Expenses

6.1 INTRODUCTION

All relevant methodologies for economic analyses require estimates of the amounts and timing of cash flows over the study period. Cash flows resulting from taxes paid (or avoided) must be included in evaluation models, along with those economic consequences due to investments, maintenance, operations, and so on.

Most individuals and business firms operating in the private sector are directly affected, because income is taxed by the federal government and, in certain instances, by other governmental jurisdictions as well.

Decision makers in the public sector may also be directly concerned with cash flows for taxes, and the associated concept, depreciation. Government agencies, including publicly owned utilities, for example, may purchase goods and services from individuals or firms in the private sector. Thus it may be useful to carry out an economy study from the point of view of the private sector in order to determine the after-tax return to the supplier, the vendor, or the customer. Moreover, regulatory rules in many parts of the United States define a utility's revenue requirements in terms of (1) current operating disbursements, other than interest on debt; (2) allowance for depreciation; (3) income taxes, where appropriate; and (4) a "fair return" on a rate base that usually approximates depreciated book value. Thus depreciation and taxes are of special concern to regulated utilities, whether they be publicly or privately (investor) owned.

In this chapter we will present a variety of depreciation methods permitted at various times by the federal government of the United States as well as state and local taxing authorities. (Certain of these methods, although no longer allowable in the United States, are currently being used by many other countries.) In addition to depreciation, several other allowable noncash expenses are also discussed in this chapter, specifically, amortization, depletion, and the Section 179 expense deduction. All of these expenses have the effect of reducing taxable income, thereby influencing cash flows for income taxes.

6.2 THE RELATIONSHIP BETWEEN NONCASH EXPENSES AND INVESTMENT PROFITABILITY

There are essentially three ways in which the amounts and timing of cash flows for taxes can be affected. First, procedures and rules for determining taxable income can be specified by the taxing authority; second, the rules for determining income taxes can be specified, i.e., the tax rates and taxable income brackets to which these rates apply; and third, tax credits (deduction from income taxes) can be allowed that permit the final tax liability to be reduced. This can be demonstrated by reference to a simple numerical illustration.

Consider the income statement summarized in **Table 6.1**. An income statement is a summary of the taxpayer's income and expenses for a specified period of time, usually a month, quarter, or year. *Income* (or *revenue*) is an accounting concept; revenues generally accrue in the accounting period in which they are earned, which is not necessarily the same as the period in which the cash is received. Similarly, *expense* is also an accrual concept, and cash expenditures do not necessarily occur in the period in which the expense is accrued. In this simplified example, we assume that cash receipts are the same as revenues and cash expenditures are the same as expenses (other than noncash expenses).

The income tax in Table 6.1, an $18,000 negative cash flow from the taxpayer's point of view, is based on $80,000 taxable income, a 30% income tax rate, and a $6,000 tax credit. The taxable income results in part from the $50,000 depreciation expense claimed during the period. Assum-

TABLE 6.1   Income Statement for the Period January 1, 19XX–December 31, 19XX

Revenue (assume all cash receipts)		$200,000
Expenses		
Noncash expenses (e.g., depreciation)	$50,000	
Other (assume all cash expenditures)	70,000	120,000
Taxable income		$ 80,000
Income tax @ 30% of taxable income	$24,000	
Less tax credits (e.g., "energy credit")	6,000	18,000
Net income after taxes		$ 62,000

ing that all income and expense items other than depreciation are cash, the cash flow after taxes is $112,000.

Receipts	$200,000
Less expenditures	70,000
Cash flow before taxes	$130,000
Less cash flow for taxes	18,000
Cash flow after taxes	$112,000

Rules for determining taxable income, tax rates, and tax credits are complex issues. Thus this chapter will focus on the first of these three issues, specifically, various methods for determining the effect on taxable income of allowable depreciation and certain other noncash expenses. The remaining two issues are considered in Chapter 7.

6.3   DEPRECIATION DEFINED

There is a good deal of misunderstanding about the precise meaning of **depreciation**. In economic analysis, depreciation is not a measure of the loss in market value of equipment, land, buildings, and the like. It is not a measure of reduced serviceability. Although assets do decline in value because of obsolescence, deterioration, and other factors, depreciation as it is used here is strictly an *accounting concept*. The American National Standards Institute (ANSI) gives the following definitions:

> (1)(a) Decline in value of a capitalized asset; (b) A form of capital recovery, usually without interest, applicable to property with two or more years' life span in which an appropriate portion of the asset's value periodically is charged to current operations. (2) A loss of value due to physical or economic reasons. (3) In accounting, *depreciation is the allocation of book value of this loss to current operations according to some systematic plan.* Depending on then existing income tax laws, the amount and timing of the charge to current operations for tax purposes may differ from that used to report annual profit and loss.[1]

Definition (3) above is the one which is used throughout the remainder of this chapter.

6.4   FEDERAL REGULATIONS CONCERNING CERTAIN ASPECTS OF DEPRECIATION ACCOUNTING

Rules relating to allowable depreciation expenses and certain other noncash expenses are imposed by all taxing authorities—national govern-

---

[1] *Industrial Engineering Terminology*, rev. ed., ANSI Standard Z94.0-1989. An American National Standard, approved July 10, 1989. Published by the Industrial Engineering and Management Press, Norcross, GA.

ments as well as regional and local governments such as many of the states and some cities. The material that follows is related specifically to the federal government of the United States. For additional reading, especially useful references are:

- **Depreciation**, Publication 534 of the Internal Revenue Service. Updated and published annually. No cost.
- **Tax Guide for Small Business**, Publication 334 of the Internal Revenue Service. Updated and published annually. No cost.

The definitions and procedures summarized in this chapter are not unique to the U.S. federal government. Indeed, they are widely used by many other taxing authorities. Nevertheless, the reader is advised to consult appropriate guidance when evaluating the tax consequences of investments in other environments.

6.4.1 What Are Depreciable Assets? When Is Depreciation Permitted?

Depreciable property may be tangible or intangible. **Tangible property** is any property that can be seen or touched. **Intangible property** is any other property, e.g., a copyright, patent, or franchise; it can be depreciated if it has a determinable life.

Depreciable property may be real or personal. **Real property** is land and generally anything erected on, growing on, or attached to the land. **Personal property** is any other property, e.g., machinery or equipment. (Land is never depreciable, as it has no determinable life.)

To be depreciable, property must meet three requirements: (1) It must be used in business or held for the production of income; (2) it must have a determinable life longer than 1 year; and (3) it must be something that wears out, decays, gets used up, becomes obsolete, or loses value from natural causes.

Depreciation begins when the property is placed in service; it ends when the property is removed from service. In other words, it is the *period of service* that is relevant, not the period of ownership. If the property is not being used, it cannot be depreciated.

6.4.2 Cost Basis, Salvage Value, and Depreciable (Useful) Life

As implied by the definition of depreciation, there are three fundamental parameters of interest: cost basis, salvage value, and depreciable (useful) life.

- The **basis** for determining depreciation is usually the cost of the property, payable either in cash or other property. If property is materially improved—say, through construction of a special foundation or structure to support depreciable equipment—the additional costs are added to the basis. After the property is acquired, the basis may be increased by all items that may be charged to a capital account (with certain exceptions), including the cost of any improvements having a useful life of more than 1 year. The basis may be decreased by any items that represent a return of capital. These changes, if any, result in the **adjusted basis.**

- **Salvage value** is the estimate, at the time of acquisition, of the amount to be realized upon the sale or other disposition of the property after it is no longer useful in the business or in the production of income and is retired from service. Salvage, when reduced by the cost of removal, is called **net salvage**. Either salvage or net salvage may be used to compute depreciation, but the practice must be followed consistently.

- The **useful life** of a piece of property is an estimate of how long it can be expected to be used in a trade or business or of how long it can be expected to produce income for the taxpayer. Useful life is not necessarily the same as physical life or even the period of ownership. Rather, the property is no longer useful when it has been removed from service. Useful life is sometimes known as the **depreciable life** or **recovery period**.

6.4.3 Depreciation and Book Value (Adjusted Basis)

The amount to be depreciated over the useful life is $B - S_d$, where B = cost basis and S_d = salvage value assumed for depreciation purposes. (Note that the actual salvage value, S_a, is not necessarily the same as S_d.) The **depreciation expense** in year j, D_j, depends on which of the various methods is used for this calculation. These will be discussed in the next section.

The **book value** at the end of year j, B_j, is the original cost basis less the accumulated depreciation. (In the sense in which it is used here, book value is also known as the *adjusted basis* of the property.)

$$B_j = B - \sum_{k=1}^{j} D_k \qquad (6.1)$$

For any depreciation method used, the resulting book value cannot be less than the salvage value used for depreciation purposes; that is,

$$B_j \geq S_d \qquad \text{for all } j \qquad (6.2)$$

Once the property has been fully depreciated, i.e., when $\Sigma_j D_j = B - S_d$, the depreciation expenses for the subsequent years the property remains in service must be zero.

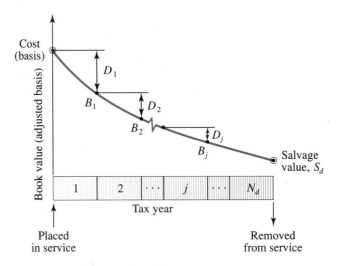

Figure 6.1   Decline in Book Value over Time due to Depreciation (Idealized Case)

Figure 6.1 is an idealized representation of these concepts in that it assumes (1) the property is placed in service at the start of the first tax year[2] and (2) it is removed from service exactly N_d years later.

These assumptions are rarely justified in practice. The more usual cases are illustrated in **Figure 6.2**. In Case A, the actual period the property will remain in service, N_a, is the same as the depreciable life, N_d, but the property is placed in service at some time after the start of the tax year. The depreciation expenses in the first ($j = 1$) and last ($j = N + 1$) tax years must be prorated on the percentage of time the property was in service in those years. In Case B we see that the property is in service longer than the depreciable life ($N_a > N_d$). As the property is fully depreciated in the $N_d + $ 1st tax year, no additional depreciation is permitted for the subsequent years. Case C illustrates *early disposition*, that is, the property is removed from service before the end of its depreciable life ($N_a < N_d$). In this instance, depreciation is prorated for the last tax year the property is in service and no additional depreciation is permitted for subsequent years.

These rules may appear somewhat complicated, but the concepts are really quite simple. Just bear in mind these two principles:

1. If the property is in service for less than one full tax year, the depreciation expense must be prorated on the basis of the time it is actually in service in that year; and

---

[2]A tax year is usually 12 consecutive months. It may either be a *calendar year* (12 months ending on December 31) or a *fiscal year* (12 consecutive months ending on the last day of any month other than December).

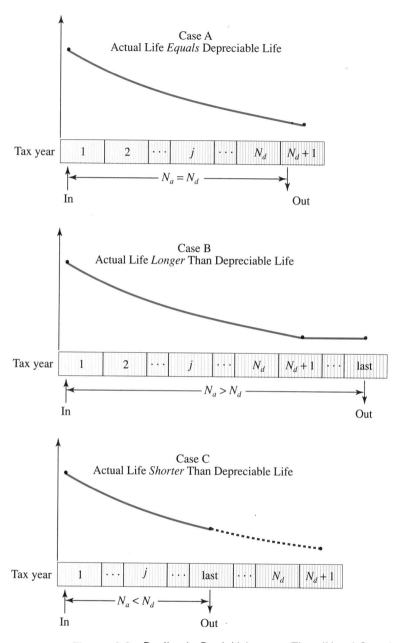

Figure 6.2   Decline in Book Value over Time (Usual Cases)

2. Depreciation expenses must be discontinued once the salvage value (S_d) has been reached or the property has been removed from service.

These principles will be illustrated further in our discussion of several of the depreciation methods.

6.5 THE "CLASSICAL METHODS" OF COMPUTING DEPRECIATION FOR ASSETS PLACED IN SERVICE BEFORE 1981

In this section we will discuss certain depreciation methods that were permitted by the Internal Revenue Code for depreciable property placed in service *before* 1981. These "classical methods" are of more than historical interest, however. There are several important reasons for including them in this book. First, pre-1981 methods are widely used for depreciation accounting in countries other than the United States. Second, they are also specified by tax regulations of a number of other taxing authorities in the United States (states and local governments). Third, as will be shown, the declining balance and straight line methods are imbedded in the depreciation methods introduced subsequently at the federal level. And, fourth, current federal tax law requires that pre-1981 methods be used for property placed in service before 1981 or for property that does not qualify for depreciation by the newer methods. In summary, then, these "classical methods" remain very much in evidence.

6.5.1 The Straight Line Method

For item accounting, that is, for determining the depreciation allowance for a single depreciable asset, the **straight line method** is the simplest method for computing depreciation. The cost or other basis of the asset, less its expected salvage value, is deducted in equal amounts over the period of its estimated useful life. Let

$$B = \text{initial cost (basis)}$$
$$S_d = \text{salvage value}$$
$$N_d = \text{depreciable life}$$

Then the annual depreciation, D, is given by

$$D = \frac{B - S_d}{N_d} \tag{6.3}$$

The **book value**, or **adjusted basis**, of the asset at any point in time is the initial cost less the accumulated depreciation. Generally, the book value, B_j, after j years is given by

$$B_j = B - jD \tag{6.4}$$

EXAMPLE 6.1 *Straight Line Depreciation*

Consider a depreciable asset with cost basis of $80,000, expected salvage value of $8,000, and depreciable life of 5 years. The annual depreciation expenses for years 1–5 are:

$$D = \frac{\$80,000 - \$8,000}{5} = \$14,400$$

The book values at the ends of each year are

$$B_1 = \$80,000 - 1(\$14,400) = \$65,600$$
$$B_2 = \$80,000 - 2(\$14,400) = \$51,200$$
$$\vdots$$
$$B_5 = \$80,000 - 5(\$14,400) = \$8,000$$

As noted earlier, depreciation expenses in the first and last tax year must be prorated on the basis of the period of service for the tax year when it is placed in service and the year in which it is removed from service. If the property in this example had been placed in service at the midpoint of the tax year, then $D_1 = \$14,400(1/2) = \$7,200$ and $B_1 = \$72,800$ ($= \$80,000 - \$7,200$). If the asset is retained in service for at least a full 5 years, then

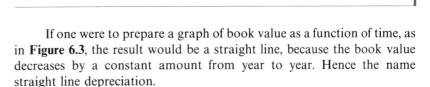

$$\left.\begin{array}{l} D_2 = D_3 = D_4 = D_5 = \$14,400 \\ D_1 = D_6 = \$7,200 \end{array}\right\} \quad \text{total} = \$72,000$$

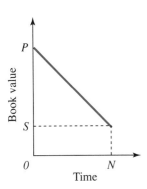

Figure 6.3 Straight Line Depreciation

If one were to prepare a graph of book value as a function of time, as in **Figure 6.3**, the result would be a straight line, because the book value decreases by a constant amount from year to year. Hence the name straight line depreciation.

6.5.2   The Declining Balance Method

In the **declining balance method**, the amount of depreciation taken each year is subtracted from the book value before the following year's depreciation is computed. A constant depreciation rate applies to a smaller, or declining, balance each year. Let

a = depreciation rate

D_j = depreciation in tax year j

B_j = book value at the end of year j, that is, after the depreciation deduction in year j has been recorded

Assuming that the asset is placed in service at the start of the tax year,

$$B_0 = B \qquad\qquad\qquad D_1 = aB_0$$
$$B_1 = B_0 - D_1 = B(1 - a) \qquad D_2 = aB_1 = aB(1 - a)$$
$$B_2 = B_1 - D_2 = B(1 - a)^2 \qquad D_3 = aB_2 = aB(1 - a)^2$$
$$\vdots \qquad\qquad\qquad\qquad \vdots$$

Thus it will be seen that the book value at the end of year j is given by

$$B_j = B(1 - a)^j \tag{6.5}$$

and the depreciation in year j is given by

$$D_j = aB_{j-1} = Ba(1 - a)^{j-1} \tag{6.6}$$

The value selected for the depreciation rate, a, is not entirely arbitrary. Prior to 1981, for example, federal tax regulations limited the rate to $a \leq 2/N$. When $a = 2/N$, the depreciation scheme is known as the **double declining balance method**, or simply **DDB**.

Property depreciated by the declining balance method must have an estimated useful life of at least 3 years, that is, $N_d \geqslant 3$. The reason for this lower bound for N_d is that if $N_d = 2$ under DDB, $a = 100\%$, so all of the depreciation would take place in the first year. There would be no depreciation in the second year, and this result is not permissible.

Salvage value is not deducted from the cost or other basis in determining the annual depreciation allowance, but the asset cannot be depreciated below the expected salvage value. In other words, once book value equals salvage value, no further depreciation may be claimed.

EXAMPLE 6.2 *Declining Balance Depreciation*

Consider the property described in Example 6.1: $B = \$80,000$, $S_d = \$8,000$, and $N_d = 5$ years. If the property is placed in service at the start of the tax year and is depreciated by the 200% declining balance method (DDB), then

$$a = \frac{2.00}{5} = 0.40$$

$$B_0 = \$80,000$$

$$D_1 = 0.40(\$80,000) = \$32,000$$

$$B_1 = \$80,000 - \$32,000 = \$48,000$$

$$D_2 = 0.4(\$48,000) = \$19,200$$

$$B_2 = \$48,000 - \$19,200 = \$28,800$$

$$D_3 = 0.4(\$28,800) = \$11,520$$

$$B_3 = \$28,800 - \$11,520 = \$17,280$$

$$D_4 = 0.4(\$17,280) = \$6,912$$

$$B_4 = \$17,280 - \$6,912 = \$10,368$$

At this point we note that the DDB calculation for year 5 would yield

$$D_5 = 0.4(\$10,368) = \$4,147$$

but this would result in an unacceptable book value:

$$B_5 = \$10,368 - \$4,147 = \$6,221$$

It is unacceptable because we are limited to

$$B_j \geq S_d = \$8,000$$

Thus the depreciation in year 5 must be

$$D_5 = \$10,368 - \$8,000 = \$2,368$$

Equations (6.5) and (6.6) assume that the property is placed in service at the start of the tax year. A more general formulation is to let

$$\pi_1 = \text{portion of the first tax year in which the property}$$
$$\text{is in service } (0 < \pi_1 \leq 1)$$

In this case it can be shown that

$$B_j = B(1-a)^{j-1}(1 - a\pi_1), \quad j = 1, 2, \ldots \tag{6.7}$$

and

$$D_j = \begin{cases} Ba\pi_1 & , \quad j = 1 \\ Ba(1-a)^{j-2}(1 - a\pi_1), & j = 2, 3, \ldots \end{cases} \tag{6.8}$$

For example, suppose that the property in Example 6.2 is placed in service at the start of the fourth month of the tax year. Here, $\pi_1 = 9/12 = 0.75$ and

$$D_1 = \$80,000(0.4)(0.75) = \$24,000$$
$$B_1 = \$80,000 - \$24,000 = \$56,000$$
$$D_2 = 0.4(\$56,000) = \$22,400$$

and so on.

Salvage value is treated only implicitly in the declining balance method; the property cannot be depreciated below the expected salvage value. As shown in **Figure 6.4**, when book value, B_n, equals salvage value, S_d, after n periods, where $n < N_d$, no depreciation may be taken after that time. This condition may occur when the declining balance rate, a, is relatively large and the salvage value as a percentage of the cost basis, S_d/B, is also relatively large. There are two possible approaches in such cases. First, do nothing at all: Simply compute annual depreciation expenses such that resulting book values are reflected as in Figure 6.4. A second approach is to select a declining balance rate such that $B_N = S_d$ after N periods. From Eq. (6.5):

$$S_d = B(1-a)^N$$

Solving for a,

$$a = 1 - \left(\frac{S_d}{B}\right)^{1/N} \tag{6.9}$$

This yields book values as in **Figure 6.5**. Both approaches have the effect of depreciating the entire amount $B - S_d$ over the depreciable life. The second approach is rarely used, however, as the taxing authority most often specifies the depreciation rate to be applied to certain types of property.

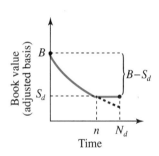

Figure 6.4 Declining Balance Depreciation Limited by Salvage Value

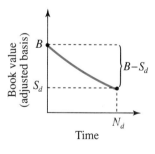

Figure 6.5 Declining Balance Depreciation with Depreciation Rate Set so as to Depreciate $B - S_d$ Over N_d Years

Shift from Declining Balance to the Straight Line Method

When the declining balance rate, a, is relatively small and/or the salvage value percentage, S_d/B, is relatively small, the book value at the end of the depreciable life may be larger than the expected salvage value. See **Figure 6.6**. In such instances we have failed to take full advantage of the opportunity to depreciate the entire amount $B - S_d$ over the depreciable life. Assuming that the rate a is the maximum permitted by law, we can depreciate down to the salvage value by shifting from the declining balance method to the straight line method. In general, it is advantageous to make this switch in the first tax year for which the straight line method when applied to the book value (adjusted basis) at the beginning of the year will yield a larger deduction. That is, make the switch in the first year for which

$$\begin{Bmatrix} \text{Depreciation} \\ \text{under} \\ \text{straight line} \end{Bmatrix} \geq \begin{Bmatrix} \text{Depreciation} \\ \text{under} \\ \text{declining balance} \end{Bmatrix}$$

$$\frac{B_{j-1} - S_d}{N_d - j + 1} \geq aB_{j-1} \tag{6.10}$$

Equation (6.10) assumes that the property is placed in service at the start of the first tax year. Thus, at the start of year j, there remain $N_d - j + 1$ years to be depreciated. This situation is illustrated in **Figure 6.7**.

EXAMPLE 6.3 Declining Balance with Switch to Straight Line

Suppose that the asset described in Examples 6.1 and 6.2 has a zero salvage value ($S_d = \$0$). As before, $B = \$80,000$ and $N_d = 5$ years. The property is placed in service at the start of the tax year and is depreciated initially by the DDB method. Determine the year to switch to the straight line method so as to ensure that the entire amount ($\$80,000 - \$0 = \$80,000$) is depreciated. From Eq. (6.10):

Tax Year j	B_{j-1}	Depreciation in Year j Under:		
		Straight Line[a]		Declining Balance[b]
1	$80,000	$16,000	<	$32,000
2	48,000	12,000	<	19,200
3	28,800	9,600	<	11,520
4	17,280	8,640	>	6,912
5	10,368	10,368	>	4,147

[a] $D_j(\text{SL}) = (B_{j-1} - \$0)/(5 - j + 1)$.
[b] $D_j(\text{DB}) = (\frac{2}{5})(B_{j-1})$.

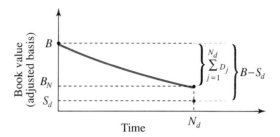

Figure 6.6   Depreciating Less than $B - S_d$ Over N_d Years

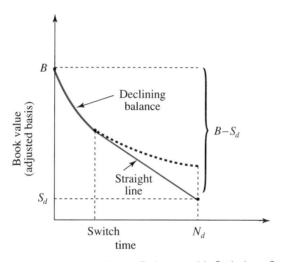

Figure 6.7   Declining Balance with Switch to Straight Line Depreciation

The first year for which the straight line method yields a larger depreciation deduction is the fourth year; thus year 4 is the year the switch should occur. The allowable depreciation expenses would be $8,640 in years 4 and 5.

6.5.3   The Sum of the Years-Digits Method

In the **sum of the years-digits (SYD) method**, the annual depreciation allowance is a declining fraction of the cost or other basis of each single-asset account reduced by the estimated salvage value. In particular, the depreciation in year j is given by

$$D_j = \left(\frac{N_d - j + 1}{\text{SYD}}\right)(B - S_d) \qquad (6.11)$$

where SYD is the sum of the years digits, that is,

$$SYD = 1 + 2 + \cdots + N_d = \frac{N_d(N_d + 1)}{2}$$

As shown in **Figure 6.8**, this method results in a declining book value quite similar to that found by using the declining balance method, except that here we are assured that exactly $B - S_d$ is depreciated over N_d years.

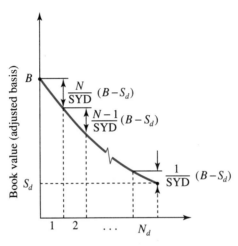

Figure 6.8 Sum of the Years-Digits Method

EXAMPLE 6.4 *SYD Depreciation*

Suppose that the property described in Examples 6.1 and 6.2 is to be depreciated by the sum of the years-digits method. Recall that $B = \$80,000$, $S_d = \$8,000$, and $N_d = 5$ years. We first determine

$$SYD = \frac{5(6)}{2} = 15$$

Then, using Eq. (6.11):

$$D_1 = (5/15)(\$80,000 - \$8,000) = \$24,000$$
$$D_2 = (4/15)(\qquad '' \qquad) = \;\;19,200$$
$$D_3 = (3/15)(\qquad '' \qquad) = \;\;14,400$$
$$D_4 = (2/15)(\qquad '' \qquad) = \;\;\;\;9,600$$
$$D_5 = (1/15)(\qquad '' \qquad) = \;\;\;\;4,800$$

$$\sum_{j=1}^{5} D_j = (15/15)(\qquad '' \qquad) = \$72,000$$

Both the declining balance and the sum of the years-digits methods are called "accelerated" depreciation methods because they result in greater depreciation in the early years and less depreciation in the later years.

6.5.4   Other Depreciation Methods

There are, of course, a variety of other methods that would have the effect of distributing the cost or other basis value of the property over the estimated life in a systematic, rational manner. Generally, there are restrictions or limitations imposed by the taxing authority, however. For example, the Internal Revenue Code of 1954 specifically allowed the straight line, declining balance, and sum of the years-digits methods and, moreover, permitted "any other consistent methods" for federal income tax purposes *if the total depreciation deductions taken during the asset's useful life are not more than the total allowed under the declining balance method.* One cannot depreciate below the salvage value, of course, and "negative depreciation" is not allowed. These conditions remained in effect until the tax code was substantially modified in 1981 with the introduction of the ACRS method.

In this section we introduce two additional methods. One has the effect of taking small depreciation deductions in the early years of ownership and larger deductions in the later years, just the opposite of the DB and SYD methods; the other has the effect of permitting irregular deductions from year to year, depending on the amount of usage of the depreciable property.

The Sinking Fund Method

Whereas the declining balance and the sum of the years-digits methods result in relatively greater depreciation in the early years, the **sinking fund method** results in the reverse situation: Depreciation is lower in the early years and greater in the later years.

In this method, an imaginary sinking fund is established that earns imaginary interest at a predetermined rate, i_s, on a series of imaginary deposits. The sum of these deposits over the life of the asset is equal to the cost or other basis less the expected salvage value. The depreciation charge for the year, D_j, is equal to the imaginary uniform deposit into the imaginary sinking fund, d, plus the accumulated interest on the fund up to that point. That is,

$$D_1 = d = (B - S_d)(A/F, i_s, N_d) \tag{6.12}$$

$$D_2 = d + i_s D_1 = d(1 + i_s)$$

$$D_3 = d + i_s(D_1 + D_2) = d(1 + i_s)^2$$

$$\vdots$$

In general,

$$D_j = d(1 + i_s)^{j-1} \quad \text{for} \quad j = 1, 2, \ldots, N_d \tag{6.13}$$

EXAMPLE 6.5 *Sinking Fund Depreciation*

Suppose that the property described in Example 6.1 is depreciated by the sinking fund method with $i_s = 10\%$. Recall that $B = \$80{,}000$, $S_d = \$8{,}000$, and $N_d = 5$ years. From Eq. (6.12):

$$d = (\$80{,}000 - \$8{,}000)(A/F, \ 10\%, \ 5)$$

$$= \$72{,}000(0.16379) = \$11{,}793$$

and

$$
\begin{aligned}
D_1 &= \$11{,}793 \ (1.10)^0 = \$11{,}793 \\
D_2 &= \quad '' \quad (1.10)^1 = \quad 12{,}973 \\
D_3 &= \quad '' \quad (1.10)^2 = \quad 14{,}270 \\
D_4 &= \quad '' \quad (1.10)^3 = \quad 15{,}697 \\
D_5 &= \quad '' \quad (1.10)^4 = \quad 17{,}267
\end{aligned}
$$

$$\sum_{j=1}^{5} D_j = B - S_d \qquad = \$72{,}000$$

The Units of Production Method

All the depreciation methods presented previously use *time* as the basis for distributing $B - S_d$ over the depreciable life. It is also possible to allocate the depreciable amount on the basis of the expected *output* of the property over its lifetime. This is known as the **units of production method** or the **service output method**. Depreciation in period j is charged in proportion to the number of units produced in that period. That is,

$$D_j = \left(\begin{array}{c} \text{Depreciation charge} \\ \text{per unit of output} \end{array} \right) \times \left(\begin{array}{c} \text{Units of output} \\ \text{in period } j \end{array} \right)$$

$$= \left(\frac{B - S_d}{U_e} \right) U_j \qquad\qquad (6.14)$$

where U_e is the number of units of output expected over the depreciable life of the property and U_j is the number of units of output during period j.

EXAMPLE 6.6 *Units of Production Method*

Suppose that the property described in Example 6.1 is to be depreciated by the units of production method.

Recall that $B = \$80{,}000$, $S_d = \$8{,}000$, and $N_d = 5$ years. It is expected that this property will produce a total of 60,000 units over the 5-year depreciable life, so the depreciation charge per unit of output is

$$\frac{\$80{,}000 - \$8{,}000}{60{,}000 \text{ units}} = \$1.20/\text{unit}$$

The anticipated annual usage is shown in column 2 of the following table; the expected annual depreciation expenses are given in column 3.

Tax Year j (1)	Usage (2)	Depreciation D_j (3)
1	10,000 units	$12,000
2	12,000	14,400
3	15,000	18,000
4	15,000	18,000
5	8,000	9,600
Totals	60,000 units	$72,000

6.6   THE ACCELERATED COST RECOVERY SYSTEM (ACRS)

With certain exceptions, tangible property placed in service after 1980 and before 1987 must be depreciated by the **Accelerated Cost Recovery System (ACRS)** for the purpose of determining federal income taxes. The principal features of this method are summarized below.[3]

6.6.1   Classes of Recovery Property

Perhaps the most striking feature of this new method, introduced by the Economic Recovery Tax Act of 1981, is that depreciation is associated with *classes* of properties rather than the individual assets themselves. Each depreciable property is assigned to a specific class, the depreciation rates for that class are predetermined, and thus the depreciation rate for the property is that of its class.

The **class life** of each property is based on a 1970 study by the U.S. Treasury Department of actual useful lives experienced by a sample of taxpayers. This study led to published guidelines for the ranges and midpoints of useful lives for a large number (~ 100) of classifications of depreciable assets. The **Asset Depreciation Range** midpoint lives subsequently served to associate specific assets with property classes under ACRS.

---

[3]For property placed in service before 1981, taxpayers must continue to use the method of depreciation that was used in the past. Also, ACRS cannot be used for intangible depreciable property or for certain types of real property. Regardless of when property is placed in service, taxpayers may elect to use a method of depreciation that is not based on years of service, such as the units of production method, rather than ACRS.

There are eight ACRS property classes under ACRS. The six principal classes are summarized in **Table 6.2**. (Two other classes, public utility property and low income housing, are not discussed further.)

6.6.2 Determining Depreciation Under ACRS

Under ACRS, the allowable depreciation in year j is given by

$$D_j = B[p_j(k)] \qquad (6.15)$$

where

$B =$ unadjusted basis for ACRS computation

$p_j(k) =$ recovery percentage in tax year j for property class k

These percentages are statutory; that is, the percentages are predetermined and specified in the law.

The recovery percentages for 3-year, 5-year, and 10-year property

TABLE 6.2   ACRS Property Classes Under Former Law
(Other than Public Utility Property and
Low Income Housing)

Property Class	Class Life (N Years)	Additional Description	Examples
3-Year	$N < 4$	Machinery/equipment used in R&D	Autos, light-duty trucks, special mfg. tools
5-Year	$N > 4$	Not 3-year, 10-year, or 15-year public utility property	Computers, copiers, office furniture and fixtures; petroleum storage facilities
10-Year	$N < 12.5$	Certain real property and certain public utility property	Theme park structures, manufactured homes
15-Year real	$N > 12.5$	Placed in service before 3/16/84; all real property other than that designated 5-year or 10-year	Buildings
18-Year real	$N > 12.5$	Placed in service after 3/15/84 and before 5/9/85; all real property other than that designated 5-year, 10-year, 15-year real, or low income housing	
19-Year real	$N > 12.5$	Placed in service after 5/8/85; all real property other than that designated 5-year, 10-year, 15-year, or 18-year real property	

under the former law are summarized in **Table 6.3**. See IRS Publication 534 or similar publications for percentages for the other property classes.

There is an underlying rationale for the statutory recovery percentages. Zero salvage value is assumed in all cases. Further, for the 3-, 5-, and 10-year property classes, it is assumed that all properties placed in service during the taxable year are placed in service at the midpoint of the year. The statutory recovery percentages for the three property classes shown in Table 6.3 are determined assuming 150% declining balance depreciation with switch to the straight line method.

Under ACRS, the entire depreciation expense in the year of acquisition may be included for all property classes other than real property. No depreciation expense may be included in the year of disposal in the event of early disposition of ACRS properties other than real property.

TABLE 6.3   Applicable Percentages for Computing Annual Depreciation Under ACRS

Recovery Year	Type of Property		
	3-Year	5-Year	10-Year
1	25%	15%	8%
2	38	22	14
3	37	21	12
4		21	10
5		21	10
6			10
7			9
8			9
9			9
10			9

EXAMPLE 6.7 *ACRS (Normal Disposal)*

Consider the property introduced in Example 6.1. Recall that $B = \$80,000$ and $N_d = 5$ years. Assume that this asset belongs to the 5-year property class ($k = 5$) and it is to be kept in service for at least 5 years. Using Eq. (6.15) and the depreciation percentages from Table 6.3, allowable depreciation expenses under ACRS are

$$D_1 = \$80,000(0.15) = \$12,000$$
$$D_2 = \$80,000(0.22) = 17,600$$
$$D_3 = \$80,000(0.21) = 16,800$$
$$D_4 = \$80,000(0.21) = 16,800$$
$$D_5 = \$80,000(0.21) = 16,800$$

Total $\qquad \overline{\$80,000}$

EXAMPLE 6.8 *ACRS (Early Disposal)*

Suppose that the property in Example 6.7 is to be kept in service only 3 years; that is, it is placed in service in tax year 1 and removed from service in tax year 4. In that event,

$$D_1 = \$80,000(0.15) = \$12,000$$
$$D_2 = \$80,000(0.22) = 17,600$$
$$D_3 = \$80,000(0.21) = 16,800$$
$$D_4 = \phantom{\$80,000(0.21) = {}}\underline{0}$$
$$\text{Total} \phantom{= \$80,000(0.21) {}} \$46,400$$

No depreciation expense is allowable in tax year 4, the year of disposal.

6.6.3 The Alternate ACRS Method (A/ACRS)

The Economic Recovery Tax Act of 1981 permits taxpayers to elect an alternative recovery percentage determined in a manner similar to that of the straight line method. Using the **alternate ACRS method (A/ACRS)**, for each property class (k), the applicable percentage for each year, $p(k)$, is

$$p(k) = \frac{1}{N(k)} \tag{6.16}$$

where $N(k)$ is the recovery period selected by the taxpayer for that property class. (Note that *recovery period* under ACRS is comparable to *usable life* under the pre-1981 methods.) The three options available to taxpayers for each property class are:

Recovery Property	Optional Recovery Period
3-year	3, 5, or 12 years
5-year	5, 12, or 25 years
10-year	10, 25, or 35 years
15-year real	15, 35, or 45 years

The alternate ACRS method, like the ACRS method, ignores salvage value when computing annual depreciation expenses.

Under the alternate ACRS method, the *half-year convention* must be used for the year in which 3-, 5-, and 10-year property is placed in service. That is, for these classes of property, only half the annual depreciation may be claimed for the initial year of ownership. A full year's deduction may be claimed for subsequent years, and if the property is held for the entire recovery period, $N(k)$, a half-year of depreciation is allowed for the tax year following the end of the recovery period. In general,

Tax Year	Allowable Depreciation
1	One-half
2 through $N(k)$	All
$N(k) + 1$	One-half

The early-disposition rule discussed under ACRS also applies to property for which A/ACRS is elected. That is, no deduction is allowed in the year of disposal or retirement for 3-, 5-, or 10-year property.

EXAMPLE 6.9 *Alternate ACRS*

Suppose that the depreciable property introduced in Example 6.1 is to be depreciated under A/ACRS. Recall that $B = \$80,000$ and $N_a = 5$ years. Assume that this is a 5-year recovery property and the taxpayer elects to use a 5-year recovery period; that is, $N(5) = 5$ years. Because the property has an actual life of 5 years, it will be removed from service in tax year 6.

The applicable percentage is

$$p(5) = 1/5 = 0.20$$

and the allowable depreciation expenses in each tax year are

$$D_1 = 0.20(\$80,000)(\tfrac{1}{2}) = \$\ 8,000$$
$$D_2 = D_3 = D_4 = D_5 = 0.20(\$80,000) = \underline{\quad 16,000}$$
$$\text{Total} = \overline{\$72,000}$$

Note that no depreciation deduction is allowed in the sixth tax year, as that is the year in which the property is removed from service. However, if the actual life is such that removal takes place in the seventh tax year or later, then $D_6 = \$8,000$ would be permissible.

6.7 THE MODIFIED ACCELERATED COST RECOVERY SYSTEM (MACRS)

As described above, prior to 1981 the taxpayer could choose among a number of methods for depreciating assets. In 1981 the Economic Recovery Tax Act eliminated some of these options, leaving the taxpayers with either an accelerated method (ACRS) or, alternatively, a straight line method (A/ACRS). In this section we refer to the modifications in the accelerated depreciation method available under the 1986 Tax Reform Act as **MACRS** (Modified Accelerated Cost Recovery System—pronounced "makers") and the alternate straight line method as **A/MACRS** (Alternate Modified Accelerated Cost Recovery System—pronounced "a-makers").[4] These methods apply to most depreciable property placed in service after December 31, 1986.

---

[4]MACRS is also referred to as the *General Depreciation System* (GDS), and A/MACRS is also known as the *Alternate Depreciation System* (ADS). Recent publications of the Internal Revenue Service seem to be favoring the terms GDS and ADS rather than MACRS and A/MACRS.

6.7.1 Classes of Recovery Property

The 1986 Tax Reform Act describes eight property classes, each with a defined class life. There are six *personal property* classes,

- 3-year property
- 5-year property
- 7-year property
- 10-year property
- 15-year property
- 20-year property

and two *real property* classes,

- Nonresidential real property
- Residential rental property

As was the case with respect to ACRS, the class to which property is assigned is generally determined by its class life. Class lives and recovery periods for most assets are listed in the **Table of Class Lives and Recovery Periods** to be found in *Depreciation*, IRS Publication 534. That reference provides (1) descriptions of the various classes of depreciable assets, (2) a number designator for the asset class, (3) the class life in years, and (4) the recovery period for the asset class when using either the regular MACRS method or the alternate MACRS method. A partial listing is summarized here in **Table 6.4**. The column labeled "Recovery Period, MACRS" indicates the property class.

6.7.2 Determining Depreciation Under MACRS

Unlike the ACRS method, in which the annual depreciation percentages are specified by law, MACRS specifies the procedures to be used to determine the depreciation percentages.

The "Placed in Service" Assumption

For other than nonresidential rental property and residential real property, both MACRS and A/MACRS assume that all property placed in service (or disposed of) during a tax year is placed in service (or disposed of) at the midpoint of that year. This is the **half-year convention**.

There is an exception to the above rule under MACRS. If more than 40% of the cost bases of all depreciable property (other than residential rental property or nonresidential real property) placed in service during the tax year is placed in service during the last 3 months of the year, then a **mid-quarter convention** must be used. That is, properties placed in service during any quarter are assumed to have been placed in service at the midpoint of that quarter.

A **mid-month convention** is used for *nonresidential real* and *residential rental* properties in all situations. These properties are considered placed

TABLE 6.4   Class Lives and Recovery Periods (in Years)
(Partial Listing Abstracted from *Depreciation*,
IRS Publication 534)

Asset Class	Depreciated Asset Used in Business	Class Life	Recovery Period MACRS	Recovery Period A/MACRS
00.11	Office furniture, fixtures, and equipment	10	7	10
00.12	Information systems	6	5	5
00.241	Light general-purpose trucks	4	5	5
00.242	Heavy general-purpose trucks	6	5	6
10.0	Mining	10	7	10
13.3	Petroleum refining	16	10	16
15.0	Construction	6	5	6
22.1	Manufacture of knitted goods	7.5	5	7.5
24.4	Manufacture of wood products and furniture	10	7	10
28.0	Manufacture of chemical and allied products	9.5	5	9.5
30.1	Manufacture of rubber products	14	7	14
31.0	Manufacture of leather and leather products	11	7	11
34.0	Manufacture of fabricated metal products	12	7	12
34.01	Manufacture of special tools	3	3	3
36.1	Any semiconductor mfg. equipment	5	5	5
37.11	Manufacture of motor vehicles	12	7	12
37.2	Manufacture of aerospace products	10	7	10
46.0	Pipeline transportation	22	15	22
49.13	Elec. utility steam production plant	28	20	28
79.0	Recreation	10	7	10

in (removed from) service at the midpoint of the month during which they are actually placed in (removed from) service.

Imbedded Depreciation Methods

The 200% declining balance (DB) method is used for the early years of depreciation for the 3 , 5 , 7 , and 10-year classes; the 150% declining balance method is used for the 15- and 20-year classes. Switch to the straight line (SL) method occurs in the first tax year for which the SL method, when applied to the adjusted basis at the beginning of the year, will yield a larger deduction than had the DB method been continued. (The appropriate switch years are summarized in **Table 6.5.**) Zero salvage value is assumed. For nonresidential real and residential rental property,

TABLE 6.5   Tax Year in Which to Switch from Declining
Balance to the Straight Line Method

Recovery Class k	DB Rate a	Quarter Placed in Service		Half Year		
		1st	2nd	Half Year	3rd	4th
3	200%	3	3	3	4	4
5	200%	4	4	4	5	5
7	200%	5	5	5	5	6
10	200%	7	7	7	7	7
15	150%	7	7	7	7	7
20	150%	8	9	9	9	9

the straight line method is used, assuming the mid-month convention, with recovery periods of 39* and 27.5 years, respectively. These features are summarized in **Table 6.6**.

EXAMPLE 6.10 *Personal Property*

Suppose that the depreciable asset described in Example 6.1 is to be depreciated under regular MACRS as a 5-year recovery property. Recall that $B = \$80,000$ and $N_d = 5$ years. This 5-year property requires the 200% declining balance method. There will be depreciation deductions in each of the first 6 tax years. Assuming the half-year convention,

$$a = \frac{2}{5} = 0.40 \qquad\qquad B_0 = \$80,000$$

$$D_1 = 0.40(\$80,000)(\tfrac{1}{2}) = \$16,000 \qquad B_1 = \$64,000$$
$$D_2 = 0.40(\$64,000) \quad = \$25,600 \qquad B_2 = \$38,400$$
$$D_3 = 0.40(\$38,400) \quad = \$15,360 \qquad B_3 = \$23,040$$

From Table 6.5 we note that, under the half-year convention, we switch to straight line in year 4 when $k = 5$. There are 2.5 years remaining at the time of the switch, the start of year 4. Thus,

$$D_4 = \frac{\$23,040}{2.5} \quad = \$9,216 \qquad B_4 = \$13,824$$

$$D_5 = \frac{\$23,040}{2.5} \quad = \$9,216 \qquad B_5 = \$4,608$$

$$D_6 = \$9,216(0.5) = \$4,608 \qquad B_6 = \$0$$

*The recovery period under MACRS is only 31.5 years for nonresidential real property placed in service prior to May 13, 1993.

TABLE 6.6   Property Classes and Principal Features of MACRS
for Non-Farm Property

Property Class	ADR Class Life (Years)	Examples	Recovery Period (Years)	Primary Depreciation Method	Convention[a]
3-Year	$CL \leq 4$	Over-the-road tractor units	3	200% DB with switch to SL	Half-year
5-Year	$4 < CL < 10$	Heavy trucks, computers and peripheral equip., office machinery, light general-purpose trucks, and R&D property	5	200% DB with switch to SL	Half-year
7-Year	$10 \leq CL < 16$	Office furniture and fixtures; property with no CL and not designated by law as being in any other class	7	200% DB with switch to SL	Half-year
10-Year	$16 \leq CL < 20$	Vessels, barges, tugs, and similar transportation equipment	10	200% DB with switch to SL	Half-year
15-Year	$20 \leq CL < 25$	Municipal wastewater treatment plant	15	150% DB with switch to SL	Half-year
20-Year	$25 \leq CL$	Farm buildings, municipal sewers	20	150% DB with switch to SL	Half-year
Non-residential real	$27.5 \leq CL$		39	SL	Mid-month
Residential rental		Rental bldg. or structure for which at least 80% of the gross rental income for the tax year is from dwelling units	27.5	SL	Mid-month

[a]Except when "40% rule" is invoked.

EXAMPLE 6.11 *Early Disposal*

Suppose that the property in Example 6.10 is removed from service at some time in the third tax year. In this case,

$$D_1 = \$16,000 \qquad \text{(as before)}$$
$$D_2 = \$25,600 \qquad \text{(as before)}$$

but

$$D_3 = \$15,360(\tfrac{1}{2}) = \$7,680$$

because the half-year convention is used in the year of disposal.

EXAMPLE 6.12 *Real Property*

Suppose that a warehouse is to be purchased and placed in service during the fourth month of the tax year. The cost basis is $1,000,000. This is a "nonresidential real" property and must be depreciated under MACRS by the straight line method over 31.5 years using the mid-month convention. Thus,

$$D = \frac{\$1,000,000}{31.5} = \$31,746$$

for all years other than the first and last years. In the first tax year, the warehouse is assumed to be in service 8.5 months ($=12 - 3.5$), so

$$D_1 = \$31,746\left(\frac{8.5}{12}\right) = \$22,487$$

A similar calculation must be performed in the year of disposal.

Although depreciation percentages are not specified by the Tax Reform Act of 1986, the Internal Revenue Service has computed these percentages based on the appropriate methodologies as outlined above. (See IRS Publication 534.) Summarized in **Table 6.7** are the values of $p_j(k)$ for $k = 3, 5, 7, 10, 15$, and 20 years, assuming the half-year convention.

Variations of the MACRS Method

The primary form of MACRS is based on either the declining balance (200% or 150%) or the straight line methods, as described above, using the MACRS recovery period. Several variations are also allowable and may be elected by the taxpayer:

Property Class	Primary MACRS Method	Optional Method
3-5-7-10 year (non-farm)	200% DB over MACRS recovery period	SL over MACRS recovery period or 150% DB over A/MACRS recovery period
15-20 year (non-farm) or Property used in farming business	150% DB over MACRS recovery period	SL over MACRS recovery period
Nonresidential real or residential rental	SL over MACRS recovery period	//

TABLE 6.7   Annual Depreciation Percentages Under MACRS (Half-Year Convention)

Recovery Year	Recovery Period (k)					
	3-Year	5-Year	7-Year	10-Year	15-Year	20-Year
1	33.33	20.00	14.29	10.00	5.00	3.750
2	44.45	32.00	24.49	18.00	9.50	7.219
3	14.81	19.20	17.49	14.40	8.55	6.677
4	7.41	11.52	12.49	11.52	7.70	6.177
5		11.52	8.93	9.22	6.93	5.713
6		5.76	8.92	7.37	6.23	5.285
7			8.93	6.55	5.90	4.888
8			4.46	6.55	5.90	4.522
9				6.56	5.91	4.462
10				6.55	5.90	4.461
11				3.28	5.91	4.462
12					5.90	4.461
13					5.91	4.462
14					5.90	4.461
15					5.91	4.462
16					2.95	4.461
17						4.462
18						4.461
19						4.462
20						4.461
21						2.231

The MACRS and A/MACRS recovery periods are given in the Table of Class Lives and Recovery Periods. For the 150% declining balance method, switch to straight line occurs when that method gives a larger deduction.

EXAMPLE 6.13 *MACRS Variations*

Consider an investment of $40,000 in a heavy general-purpose truck. From Table 6.4, this is asset class 00.242: The class life is 6 years, and the recovery periods are 5 years under MACRS and 6 years under A/MACRS. Assuming the half-year convention and an actual life of at least 6 years for the truck, we will determine the allowable depreciation deductions under the primary MACRS and the two allowable variations.

1. *200% declining balance, MACRS recovery period:* Using the percentages from Table 6.7 for $k = 5$:

$$D_1 = 0.2000(\$40,000) = \$\ 8,000$$
$$D_2 = 0.3200(\$40,000) = \ 12,800$$
$$D_3 = 0.1920(\$40,000) = \ \ \ 7,680$$
$$D_4 = 0.1152(\$40,000) = \ \ \ 4,608$$
$$D_5 = 0.1152(\$40,000) = \ \ \ 4,608$$
$$D_6 = 0.0576(\$40,000) = \ \ \ 2,304$$
$$\text{Total} \qquad\qquad \underline{\quad\quad} \ \$40,000$$

2. *Straight line, MACRS recovery period:* Here,

$$p_j(5) = \begin{cases} \frac{1}{5} = 0.2 & \text{for } j = 2, 3, 4, 5 \\ \frac{1}{2}(0.2) = 0.1 & \text{for } j = 1, 6 \end{cases}$$

Thus,

$$D_2 = D_3 = D_4 = D_5 = 0.2(\$40,000) = \$8,000$$
$$D_1 = D_6 = 0.1(\$40,000) = \$4,000$$

3. *150% declining balance, A/MACRS recovery period:* Here, the recovery period is 6 years, and

$$a = \frac{1.50}{6} = 0.25 \qquad\qquad B_0 = \$40,000$$
$$D_1 = 0.25(\$40,000)(\tfrac{1}{2}) = \$5,000 \qquad B_1 = \$35,000$$
$$D_2 = 0.25(\$35,000) = \$8,750 \qquad B_2 = \$26,250$$
$$D_3 = 0.25(\$26,250) = \$6,563 \qquad B_3 = \$19,687$$

It may be shown that the switch to straight line occurs at the start of the fourth tax year, at which time there are $3.5(=6 - 2.5)$ years remaining.

Thus,

$$D_4 = \frac{\$19,687}{3.5} = \$5,625 \qquad\qquad B_4 = \$14,062$$

$$D_5 = \$5,625 \qquad\qquad\qquad B_5 = \$\ 8,437$$

$$D_6 = \$5,625 \qquad\qquad\qquad B_6 = \$\ 2,812$$

$$D_7 = \$5,625(0.5) = \$2,812 \qquad B_7 = \$\qquad 0$$

6.7.3   The Alternate MACRS Depreciation System (A/MACRS)

The taxpayer may use the **alternate MACRS method** for most property.[5] This method, also known as the **Alternate Depreciation System (ADS)**, uses the straight line method with zero salvage value and with recovery periods taken from the *Table of Class Lives and Recovery Periods*. If the property is not listed in the table, the alternate recovery period is:

- 12 years for personal property, and
- 40 years for real property.

The mid-month convention is used for nonresidential real and residential rental property. The half-year (or mid-month) convention is used for all other property.

EXAMPLE 6.14 *Alternate MACRS*

Consider the heavy general-purpose truck described in Example 6.13. The $40,000 cost basis is to be depreciated under A/MACRS using a recovery period of 6 years. Assuming the half-year convention, and also assuming that the truck will be kept in service at least 6 years, we have

$$D_1 = D_7 = \$40,000(\tfrac{1}{6})(\tfrac{1}{2}) = \$3,333$$

$$D_2 = D_3 = D_4 = D_5 = D_6 = \$40,000(\tfrac{1}{6}) = \$6,667$$

6.8 THE SECTION 179 DEDUCTION

The taxpayer may elect to treat the cost of certain qualifying property as an expense rather than as a capital expenditure in the year the property is placed in service. *Qualifying property* is "Section 38 property"—generally,

---

[5]A/MACRS *must* be used for certain property, e.g., tangible property used during the tax year predominantly outside the United States.

property used in the trade or business with a useful life of 3 years or more for which depreciation or amortization is allowable, with certain limitations—and that is purchased for use in the active conduct of the trade or business.[6]

There are several limitations, however. The total cost that may be deducted for a tax year may not exceed $17,500. The expense deduction is further limited by the taxpayer's total investment during the year in Section 179 property: The $17,500 maximum is reduced by one dollar for each dollar of cost in excess of $200,000. Therefore, no Section 179 expense deduction may be used if total investment in qualifying property during the tax year exceeds $217,500. These total investment limitations are reflected in **Figure 6.9**. The total cost that may be deducted is also limited to the taxable income resulting from the active conduct of any trade or business of the taxpayer during the tax year. See *IRS Publication 534* for more information.

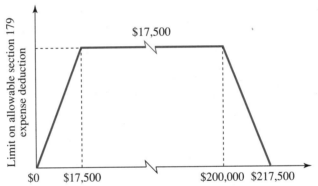

Figure 6.9   Investment Limit on Section 179 Expense Deduction

The Section 179 deduction is, in a sense, a kind of additional depreciation that is allowed in the first tax year, subject to the limitations described. Therefore, before computing regular depreciation, the cost basis of the property must be reduced by the amount of the Section 179 expense deduction, if any.

EXAMPLE 6.15 *The Section 179 Expense Deduction*

Mr. Mason operates a small engineering consulting firm. During the tax year he purchased a total of $20,000 in computer equipment for his firm, all of which is considered qualifying property. His total income, without considering the cost of the Section 179 property, would be $78,000. In this case, neither the investment limit nor the taxable income limit are

---

[6]Qualifying property includes tangible personal property and other tangible property that is used as an integral part of manufacturing. Most buildings and their structural components do *not* qualify. Property held merely for the production of income, such as rental property, does not qualify.

constraining. Therefore, Mr. Mason would be allowed the full $17,500 Section 179 deduction. If he should elect to do so, the cost basis of the computer equipment would be reduced to $2,500 ($=$2,000 - $17,500), and he would depreciate this remaining $2,500 over the depreciable life in the usual manner.

6.9 AMORTIZATION

Amortization permits the taxpayer to recover certain capital expenditures in a way that is like straight line depreciation. Qualifying expenditures include certain costs incurred in setting up a business (e.g., survey of potential markets, analysis of available facilities), the cost of a certified pollution control facility, bond premiums, and the costs of trademarks and trade names. Expenditures are amortized on a straight line basis over a 60-month period or more.[7]

$$\left. \begin{array}{l} \text{Amortization} \\ \text{expense} \\ \text{per month} \end{array} \right\} = \frac{\text{amount to be amortized}}{\text{amortization period (months)}} \qquad (6.17)$$

EXAMPLE 6.16 *Amortization*

In starting up its new business, the firm incurs a cost of $90,000 for the analysis of available facilities and labor supply. These costs are incurred at the start of the fourth month of the tax year. Assuming that these costs are amortizable over 60 months,

$$D_1 = (9 \text{ months}) \left(\frac{\$90,000}{60} \right) \qquad = \$13,500$$

$$D_2 = D_3 = D_4 = D_5 = 12(\$1,500) \ = \ 18,000$$

$$D_6 = (3 \text{ months})(\$1,500) \qquad\qquad = \ \underline{\ \ 4,500}$$

$$\text{Total over 60 months} \qquad\quad = \$90,000$$

6.10 DEPLETION

Depletion is similar to depreciation and amortization. It is a deduction from taxable income applicable to a mineral property, oil and gas wells, natural and geothermal deposits, or standing timber. There are two ways to figure depletion, cost depletion and percentage depletion.

---

[7]The Omnibus Reconciliation Act of 1993 includes a new Section 179 of the Tax Code that provides a single amortization period (15 years) for recovering the cost of a variety of purchased intangible assets. This feature does not apply if the intangible is not created in connection with transactions that involve the acquisition of a trade or business.

Cost depletion, D_{cj}, is determined by dividing the adjusted basis of the mineral property by the total number of recoverable units in the deposit, and multiplying the resulting rate per unit by the number of units sold during the tax year. (Timber depletion may be determined only by the cost method. The cost does not include any part of the cost of the land.) Thus,

$$D_{cj} = U_j \left(\frac{B_j}{U_e}\right) \tag{6.18}$$

where

U_j = the total number of units sold during the interval[8]

U_e = the total number of recoverable units in the deposit (number of tons, barrels, or thousand board-feet, for example)

B_j = adjusted basis of the property; the book value at the start of year j; original cost less depletion previously allowed, whether computed by the percentage or cost depletion method.

Percentage depletion, D_{pj}, is a certain percentage, specified for each mineral, of the taxpayer's gross income from the property during the tax year. Currently, the rates range from 5% to 22%. Some examples are:

Gravel, sand, stone	5%
Coal deposits	10%
Oil and gas wells (domestic production)	15%
Sulfur and uranium deposits	22%

In general, the deduction for depletion under this method cannot be more than 50% of the taxable income from the property, figured without the deduction for the depletion. (This taxable income limit is 100% for domestic oil and gas properties.) *Even if the taxpayer has already recovered the full cost or other basis of the property, the deduction for percentage depletion may be taken.*

If you can use either cost or percentage depletion, you must use cost depletion in any tax year in which it is greater than percentage depletion. That is, if $D_{cj} > D_{pj}$ in any year, then cost depletion must be used for that year.

EXAMPLE 6.17 *Depletion*

Suppose that the owner of a certain mineral deposit estimates, based on engineering studies, that the property contained 100,000 unextracted tons

---

[8] If the cash method of accounting is used, U_j represents the number of units for which payment was received in year j. Otherwise, U_j is the number of units sold in year j.

of ore at the start of the current year. During the year, 15,000 tons were extracted. The original cost of the property was $950,000. A total of $350,000 was allowed for depletion in previous years. Thus

$$D_c = 15{,}000 \text{ tons} \left(\frac{\$950{,}000 - \$350{,}000}{100{,}000 \text{ tons}} \right)$$

$$= \$90{,}000$$

The owner had gross receipts of $500,000 from the sale of the ore during the year. Before deduction of the depletion allowance, the net taxable income is $160,000. Assuming a 15% depletion rate for this mineral,

$$D_p = 0.15(\$500{,}000)$$

$$= \$75{,}000$$

The limit on percentage depletion this year is 50% of net taxable income, or $80,000. Thus D_p as computed above is acceptable.

In this example, cost depletion must be used, since $D_c > D_p$. That is, the taxpayer will declare a depletion expense in the amount of $90,000. The new cost basis for the following year will be $950,000 − $440,000, or $510,000. In the absence of any new engineering data, there are $100,000 − 15,000 = 85,000$ tons of ore remaining in the property at the start of the following year.

SUMMARY

Depreciation expenses are not cash flows, yet they are significant in economic analysis because of their effect on taxable income, which, in turn, leads to cash flows for income taxes. Before 1982, depreciation accounting for federal income tax purposes was generally limited to the straight line, the declining balance, or the sum of the years-digits methods, although certain other methods were also permitted. Beginning in 1982 an entirely new procedure, the ACRS method, was introduced by the federal government, specifying depreciation percentages for classes of recovery properties. The ACRS method was modified by the Tax Reform Act of 1986, and the resulting MACRS (and A/MACRS) is required for most depreciable property placed in service beginning in 1987.

The Section 179 deduction, when applicable, provides an additional deduction in the year in which the property is placed in service. There are significant limitations, however, and the Section 179 deduction is generally of use only to those taxpayers with less than $217,500 in total investment in Section 179 property during the tax year. Of particular note is that the cost basis of depreciable property must be reduced by the Section 179 deduction taken in the tax year.

Amortization is much like straight line depreciation in that it allocates the initial cost or other basis of certain assets uniformly over a given period of time. Generally, assets that may be amortized are certain business start-up costs and organizational expenditures for corporations and partnerships.

Depletion may be thought of as a form of depreciation applicable to mineral deposits, oil and gas wells, and standing timber. Both cost depletion and percentage depletion should be computed, where permitted, and the method that yields the largest deduction should be used. Only cost depletion may be used for timber deposits, however.

Both the absolute and relative economic attractiveness of prospective investment proposals are significantly affected by relevant federal and state tax regulations. It is essential, therefore, that engineering economy studies consider tax rates, scheduling of tax payments, and other matters that relate to the amounts and timing of cash flows for income taxes. These issues are addressed in the next chapter.

PROBLEMS

Federal regulations governing depreciation accounting, investment credit, and the like are complex, so it is difficult to prepare numerical exercises and problems that are reasonably realistic and yet do not unduly burden the student. The objective of the following problems is to reinforce the concepts outlined in the chapter without requiring special expertise in tax accounting. Accordingly, the following assumptions should be used in the problems *unless specifically stated otherwise*:

1. The taxpayer's tax year is the same as the calendar year: January 1–December 31.

2. Depreciable property is placed in service immediately upon acquisition, and the property is sold when it is removed from service.

3. When depreciating a specific capital asset, the same depreciation method is used for all levels of government (local, state, and federal) for determining taxable income.

4. Depreciable property is new (not used) when acquired.

5. Ignore the Section 179 expense deduction when determining the cost basis. That is, the Section 179 expense is either not allowable or, if it is allowable, it is applicable to other of the firm's capital investments in the tax year.

6. The depreciable asset (recovery property) is placed in service at the beginning of the tax year and removed from service at the end of the tax year.

"Classical" Methods

6.1 Consider a depreciable asset, say an industrial robot, having the following characteristics:

Initial cost	$50,000
Depreciable life	8 years
Estimated salvage value after 8 years	$14,000

Find the allowable depreciation expenses and book values for each of the 8 years using (a) the straight line method, (b) the 125% declining balance method, (c) the sum of the years-digits method, and (d) the sinking fund method using an 8% sinking fund.

[*Answers*: a. $4,500/year; b. $7,812,..., $1,222; c. $8,000,..., $1,000; d. $3,385,..., $5,800.]

6.2 The Comma Corporation is planning the purchase of a high-volume, overhead bulk conveyor. If it is purchased, it will have an initial cost of $100,000, an estimated useful life of 10 years, and an expected salvage value of $20,000. During the 10-year period, it is predicted that the conveyor will handle a total of 400,000 tons of bulk raw materials.

Year	Tons Handled	Year	Tons Handled
1	20,000	6	60,000
2	30,000	7	50,000
3	40,000	8	40,000
4	50,000	9	30,000
5	60,000	10	20,000

a. With the above projections, determine the expected depreciation charges for each of the 10 years if the units of production method is used.
b. What is the depreciation charge in the third year if only 35,000 tons are actually handled?

6.3 Consider a special-purpose punch press that costs $100,000 and has an expected salvage value of $16,000 at the end of 6 years. This punch press is expected to produce a total of 600,000 units over its 6-year useful life, or an average of 100,000 units per year. The actual expected production schedule, however, is as follows:

Year	Production
1	80,000
2	100,000
3	120,000
4	120,000
5	100,000
6	80,000

Determine the annual depreciation expenses and end-of-year book values using each of the following methods:
a. Straight line
b. 150% declining balance
c. Sum of the years digits
d. 8% sinking fund
e. Units of production

[Answers: a. $14,000/year; b. $25,000, ..., $5,933; c. $24,000, ..., $4,000; (d). $11,450, ..., $16,825; e. $11,200, ..., $11,200.]

6.4 The XYZ Corporation is considering the pur-

chase of a computer that is currently owned by the ABC Company. ABC paid $200,000 for this computer and depreciated it by the straight line method using a 10-year estimated useful life and an estimated salvage value of $80,000. The computer was purchased by ABC 4 years ago. If the sale takes place, XYZ will pay $100,000 for the computer; it will be depreciated by the 150% declining balance method over a 5-year remaining useful life, assuming a 20% expected salvage value. Assuming that the sale will take place, answer the following questions:
a. What will be the gain (or loss) on disposal to ABC?
b. What will be the depreciation expense in XYZ's first year of ownership?
c. What will be the book value of the computer after 5 years of ownership by XYZ?
d. If the XYZ Company used a rate other than $1.5/N$ for declining balance, it could depreciate the computer down to the expected salvage value after 5 years. What is this rate?

6.5 A manufacturing firm is considering the purchase of an asset having a first cost of $100,000, an 8-year life, and an estimated salvage value of $20,000 at the end of 8 years. If it is purchased, the asset will result in a pretax savings of $30,000 during each year over its 8-year life. Property taxes must be paid at the end of each year at the rate of 2% of the book value of the asset at the start of that year. Insurance premiums of $1,000 must be paid at the start of each year.
   If it is purchased, the asset will be depreciated by the SYD method using a 6-year life for depreciation purposes, assuming $20,000 salvage at the end of 6 years. (The depreciable life need not, of course, be the same as the expected actual life.)
a. Find the pretax (before income taxes) rate of return for this proposed investment.
b. Find the taxable income for each year.
[Answers: a. 23.4%; b. $11,222, ..., $27,778.]

6.6 A certain property was purchased prior to 1981 at a cost of $160,000. The expected service life at the time of acquisition was 15 years, and the expected salvage value after 15 years was estimated to be $40,000. Find the allowable depreciation expenses and book values for each of the 15 years using (a) the straight line method, (b) the double declining balance method, and (c) the sum of the years-digits method.

6.7 The White Trucking Company is considering the installation of an electronically controlled drag line in its Chicago terminal. The initial investment will be $300,000, but it will cause a reduction in annual operating disbursements of $50,000 a year for 12 years (the expected life of the equipment).

The investment will be depreciated for income tax purposes by the straight line method using a 20-year life and zero salvage value. However, the equipment will be useless to the company at the end of 12 years and will be disposed of at a zero net salvage value at that time. Determine the effect on taxable income each year during the study period.

[*Answer:* $35,000, ..., − $85,000.]

6.8 The Black Company is considering a proposal to manufacture a new product. The planning department has been asked to make an economic analysis assuming that production will begin January 3, 19xx (zero date), and will terminate after twelve years.

The original cost of machinery and equipment is $90,000. An estimated life of 12 years with a $12,000 salvage value will be used for tax purposes. The sum of the years-digits depreciation method will be used. In addition, $30,000 of working capital will be required at zero date. It is assumed that the working capital will be fully recovered at the end of 12 years. Estimated receipts from sale of the product are $60,000 in year 1, $90,000 in year 2, and $120,000 a year from years 3 through 12. Estimated operating disbursements are $48,000 in year 1, $65,000 in year 2, and $80,000 a year from years 3 through 12.

Determine the prospective taxable income each year during the study period.

6.9 Suppose that the declining balance method is used with $B = \$1,000$, $S_d = \$200$, and $N_d = 8$ years. Find the declining balance rate such that $B_8 = \$200$.

[*Answer:* 0.1822.]

6.10 Consider a capital asset that is to be depreciated by the 150% declining balance method over a 5-year depreciable life. The initial cost is $10,000;

the expected salvage value for depreciation purposes is $4,000. The asset is acquired and placed in service at the midpoint of the tax year.

a. Determine the allowable depreciation for each of the first 4 tax years.

b. If the asset is sold for $3,000 at the end of the fourth tax year, and if the taxpayer's effective incremental income tax rate is 40%, determine the after-tax residual value of the asset.

6.11 Depreciable property is acquired during the first day of the tax year. The initial cost of the asset is $40,000. It is expected to be retained in service for exactly 5 years and then sold for an expected net salvage value of $10,000. Determine the allowable depreciation each year assuming the:

a. Sum of the years-digits method

b. Double declining balance (DDB) method

[*Answers:* a. **$10,000, $8,000, $6,000, $4,000, $2,000; b. $16,000, $9,600, $4,400, $0, $0.**]

6.12 Consider a capital asset depreciated by the declining balance method based on a 150% DB rate. The asset has a first cost of $1,000, zero salvage value, and a 3-year depreciable life. The actual life is also assumed to be 3 years.

a. Assuming that the asset is placed in service at the beginning of the tax year, determine the allowable depreciation expenses for the first, second, and third tax years. Do not permit switching.

b. Recompute your answers in (a) assuming that a switch to straight line is permitted.

c. Assuming that the asset is placed in service at the middle of the tax year, determine the allowable depreciation expenses for the first, second, third, and fourth tax years. Do not permit switching.

d. Recompute your answers in (c) assuming that a switch to straight line is permitted.

6.13 Consider a depreciable property with

Initial cost	$20,000
Estimated useful life	5 years
Estimated salvage value after 5 years	$5,000

Determine the allowable depreciation expenses if depreciated by the:

a. Straight line method

b. 150% declining balance method

c. Sum of the years-digits method

Assume that the property is placed in service at the start of the first tax year and sold at the end of the fifth tax year.

[*Answers:* a. $3,000/year; b. $6,000, ..., $0; $5,000, ..., $1,000.]

6.14 Tropica, S.A., is a manufacturing firm located in the nation of Andino, a small country in Latin America. Tropica is considering the acquisition of certain equipment that will have an initial cost of 500,000 pesos, will be kept in service for 6 years, then sold for an estimated 100,000 pesos. The equipment will be depreciated by the 150% declining balance method with a 6-year depreciable life. Andino, like the United States, does not permit book value to fall below the expected salvage value. Assuming that the equipment will be acquired and placed in service at the start of the tax year, determine the allowable annual depreciation expenses.

6.15 Certain depreciable property, purchased at the start of the tax year, is to be depreciated by the 200% declining balance method over 10 years. The initial cost is $10,000; the expected salvage value after 10 years is $2,000.
  a. Assuming that there will be no switch to the straight-line method, determine the allowable annual depreciation expenses in the first 6 years.
  b. If the property is removed from service and sold at the midpoint of the seventh tax year, determine the allowable depreciation in year seven.

[*Answers:* a. $2,000, $1,600, $1,280, $1,024, $819, $655; b. $262.]

6.16 Certain depreciable property, purchased at the start of the tax year, is to be depreciated by the SYD method over 10 years. The initial cost is $10,000; the expected salvage value after 10 years is $2,000.
  a. Determine the allowable depreciation expenses in the first 6 years.
  b. If the property is removed from service and sold at the midpoint of the seventh tax year, determine the allowable depreciation in year seven.

6.17 A new milling machine is to be placed in service at the start of the tax year, kept in service for 10 years, then removed from service and sold for an estimated $12,500. The cost basis of the machine when placed in service is $40,000. Determine the allowable annual depreciation expenses when using the:
  a. Straight line method
  b. 150% declining balance method
  c. Double declining balance method
  d. Sum of the years-digits method

[*Answers:* a. $2,750, ..., $2,750; b. $6,000, ..., $0; c. $8,000, ..., $0; d. $5,000, ..., $500.]

6.18 Consider a capital asset that is depreciated by the 150% declining balance method over a 5-year depreciable life. The initial cost is $10,000; the expected salvage value for depreciation purposes is $4,000. The asset is acquired and placed in service at the midpoint of the tax year. Do not shift to the straight line method at any time. Determine the allowable annual depreciation expenses if the asset is kept in service 4 years, that is, until the midpoint of the fifth tax year.

Depreciation of Recovery Property by ACRS and MACRS

Unless otherwise stated, assume that the primary *method will be used under MACRS. For example, for 3, 5, 7 and 10-year non-farm property, use 200% DB over the MACRS recovery period.*

6.19 Consider a depreciable asset purchased for $100,000 at the start of the fourth month of 1985. The asset was retained for exactly 4 years and then sold for $10,000. Determine the allowable depreciation if the asset was depreciated as a 3-year ACRS property.

[*Answer:* $25,000, $38,000, $37,000, $0, $0.]

6.20 Consider a certain Section 1250 class property depreciated under ACRS as 15-year real property. The initial cost is $10,000, the expected salvage value on disposal is $25,000, and the estimated service life is 10 years. Determine the year-by-year ACRS deductions over the 10 year service life.

6.21 Solve Problem 6.7 assuming that the drag line is depreciated by the White Trucking Company as a 15-year recovery property under ACRS.

[*Answer:* $14,000, ..., -$10,000.]

6.22 Solve Problem 6.8 assuming that the machinery and equipment will be depreciated by the Black Company as 10-year recovery property under ACRS.

6.23 Consider a 5-year ACRS recovery property with

Initial cost $20,000
Estimated salvage value after 5 years 5,000

Ignore the Section 179 expense deduction. The property is purchased at the start of the fourth month of the tax year.

a. Assuming that the property will be in service exactly 5 years, use the ACRS method to find the allowable depreciation expenses in years 1, 2,..., 6.

b. Assume that the 5-year recovery property will be depreciated under the alternate ACRS method with $N(5) = 5$ years. If the property will remain in service exactly 5 years, find the allowable depreciation expenses in years 1, 2, ..., 6.

[*Answers:* a. $3,000, $4,400, $4,200, $4,200, $4,200, $0; b. $2,000, $4,000, $4,000, $4,000, $4,000, $2,000.]

6.24 Consider a capital asset to be acquired at a cost of $50,000. It is expected to be placed in service at the midpoint of the tax year, kept in service for 3 years, then removed from service and sold for $25,000.

a. Suppose that this asset is to be depreciated by the DDB method assuming $N_d = 4$ and $S_d = \$20,000$. Determine the allowable annual depreciation expenses.

b. Suppose that the asset is to be treated as a 3-year MACRS recovery property. Regular MACRS is used with a 3-year recovery period. Determine the allowable annual depreciation expenses.

6.25 Consider a depreciable asset with initial cost of $10,000, service life of 6 years, and expected salvage value of $1,000 at the end of 6 years. The asset will be placed in service at the midpoint of the tax year. Determine the annual allowable depreciation expenses using:

a. Regular MACRS over a 3-year recovery period

b. Alternate MACRS over a 3-year recovery period

[*Answers:* a. $3,333, $4,445, $1,481, $741; b. $1,667, $3,333, $3,333, $1,667.]

6.26 A certain depreciable asset (recovery property) has an initial cost (basis) of $50,000. It is expected to remain in service for 6 years, at which time it will be removed from service and sold for an expected $10,000. Assume that the property will be placed in service at the start of the first tax year and sold at the end of the sixth tax year. Determine the allowable annual depreciation expenses under:

a. Regular MACRS with a 5-year recovery period

b. Alternate MACRS with a 5-year recovery period

6.27 A depreciable property is to be acquired and placed in service at the middle of the tax year. The initial cost is $100,000. It will be kept in service for 6 years, then sold for an estimated $20,000. Determine the allowable annual depreciation expenses if the property is depreciated under:

a. Regular MACRS with a 5-year MACRS recovery period

b. Alternate MACRS using a 6-year A/MACRS recovery period

c. 150% DB over the A/MACRS recovery period (6 years)

d. Straight line method over the MACRS recovery period (5 years)

[*Answers:* a. $20,000, $32,000, $19,200, $11,520, $11,520, $5,760; b. $8,333, $16,667, $16,667, $16,667, $16,667, $16,667, $8,333; c. $12,500, $21,875, $16,406, $14,063, $14,063, $14,063, $7,031; d. $10,000, $20,000, $20,000, $20,000, $20,000, $10,000.]

6.28 A depreciable property is to be acquired and placed in service at the beginning of the tax year. The initial cost is $100,000. It will be kept in service for 6 years, then sold for an estimated $20,000. Determine the allowable depreciation expenses in the first and last tax years—that is, in tax years 1 and 6—under each of the following methods:

a. Straight line over 5 years

b. DDB over 5 years

c. SYD over 5 years

d. Regular MACRS over 5 years

e. Alternate MACRS over 5 years

6.29 Certain manufacturing equipment will be acquired at cost (basis) of $100,000. It will be kept in service for 5 years, then sold for an estimated

$20,000. For the depreciation methods shown below, determine the depreciation in the first and second years. Assume that the equipment will be placed in service at the start of the tax year.

a. Straight line
b. DDB
c. SYD
d. Regular MACRS over 5 years
e. Alternate MACRS over 5 years

[*Answers:* a. $16,000, $16,000; b. $40,000, $24,000; c. $26,667, $21,333; d. $20,000, $32,000; e. $10,000, $20,000.]

6.30 A certain depreciable asset (recovery property) is to be placed in service on the first day of the tax year, and then removed from service at the beginning of the sixth tax year. The initial cost basis is $20,000. The expected salvage value when it is removed from service is $8,000.

The taxpayer will use the SYD method with a 4-year life when computing taxable income for the *state* income tax return. The property will be treated as a 3-year recovery property, using the regular MACRS method, when computing taxable income on the *federal* income tax return.

a. Determine the allowable depreciation expenses on the state income tax return.
b. Determine the allowable depreciation expenses on the federal income tax return.
c. Recompute the allowable depreciation expenses on the state income tax return if the property is in fact placed in service at the midpoint of the tax year.

Depletion, Amortization, and the Section 179 Expense Deduction

6.31 The Gusher Corporation is planning to purchase an oil property for $5,000,000. Initial engineering estimates indicate that the property is capable of producing 1,000,000 barrels. It is estimated that deductible operating expenses, other than depletion, will be $24.50 per barrel. The percentage depletion allowance for this small producer during the first year of operation is 20%. If Gusher sells 100,000 barrels during the first year of operation at $30 per barrel, find the maximum allowable depletion expense.

[*Answer:* $500,000.]

6.32 The Dunn Mining Company owns a small mine that produces a certain metallic mineral. The mine, now ending its third year of operation, was originally purchased for $200,000. A total of $20,000 has been charged as depletion expense for the first 2 years. At the beginning of the third year it was estimated that 8,000 tons of ore were still available in the mine. During the third year, 2,000 tons were sold at an average price of $155 per ton, and $230,000 in deductible operating expenses (other than depletion) were incurred. Determine:

a. The depletion deduction
b. Taxable income for the third year of operation

6.33 A small engineering design firm is considering the purchase of computer-aided design (CAD) equipment to be placed in service at the mid-point of the tax year. The initial cost is $20,000. The equipment, purchased new, will be considered as 5-year MACRS recovery property.

a. The firm expects to acquire certain depreciable assets during the tax year, other than the CAD equipment, with a total cost of $3,000. Determine the Section 179 expense deduction due to the acquisition of the new CAD equipment.
b. Assuming that the regular MACRS method will be used, determine the MACRS recovery (depreciation expense) in the first tax year.
c. Assuming that the equipment will be kept in service for 4 years, then sold for an estimated $5,000, determine the gain on disposal at the time of sale.

[*Answers:* a. $7,000; b. $2,600; c. $2,754.]

6.34 Peggy Ho, a recent Computer Science graduate, has a small consulting firm specializing in the development of computer software for the legal profession. She is considering the purchase of a laser printer, to be used for business purposes, which costs $3,000. The printer, purchased new, will be considered a 5-year MACRS recovery property.

a. Ms. Ho expects to acquire certain depreciable assets during the tax year, other than the printer, with a total cost of $9,000. Determine the allowable Section 179 expense deduction due to the acquisition of the new printer.
b. Assuming that the alternate MACRS method

will be used with $N_d = 5$, determine the MACRS recovery (depreciation expense) in the first tax year.

c. Assuming that the equipment will be kept in service for 4 years, then sold for an estimated $1,000, determine the gain on disposal at the time of sale.

6.35 A certain depreciable asset (recovery property) is to be placed in service at the midpoint of the tax year. The initial cost basis is $20,000. The property is expected to remain in service for exactly 4 years, then sold for an estimated $12,000. It

will be treated as a 5-year recovery property, using the regular MACRS method, when computing taxable income on the federal income tax return. This will be the taxpayer's only capital investment during the year, so the Section 179 expense deduction is fully applicable. Determine:

a. The allowable Section 179 expense deduction in the 1st tax year
b. Allowable annual depreciation expenses
c. Book value (adjusted basis) at the time of disposal

[*Answers:* **a. $10,000; b. $2,000, . . . , $576; c. $1,152.**]

APPENDIX 1: *Computer Software Applications*

Recall the depreciable asset first introduced in the text as Example 6.1. The cost basis is $80,000, the depreciable life is 5 years, and the salvage value at the end of 5 years is assumed (for depreciation purposes) to be $8,000. It is also assumed here that the asset is placed in service at the start of the first tax year.

EXAMPLE 6.A.1   *Allowable Depreciation Expenses Under the Straight Line, Declining Balance, and Sum of the Years-Digits Methods*

A computer-generated spreadsheet for determining allowable annual depreciation expenses under the SL, DDB, and SYD methods is shown as **Figure 6.A.1**.

- Columns B and C: straight line method
- Columns D and E: double declining balance method
- Columns F and G: sum of years-digits method

The spreadsheet has been developed using *Lotus 1-2-3*, Release 2.3. Note that cell **E18** contains the book value at the end of the first year under DDB. This calculation takes advantage of a special function available for this software, which creates a conditional test:

@ IF(test, TRUE, FALSE)

As shown at the bottom of Figure 6.A.1 describing the content of cell E18, the test is

$$B_0 - \left(B_0 * \frac{2.00}{N_d} \right) < S_d$$

	A	B	C	D	E	F	G
1	EXAMPLE 6.A.1						
2							
3	METHOD: Straight Line, 200% Declining Balance & Sum of Years Digits						
4	(Assumption: Asset placed in service at the start of the year)						
5							
6	Input Data:						
7	-----	--------					
8	Cost Basis (B)				$80,000.00		
9	Depreciable life in years (Nd)				5		
10	Actual life in years (Na)				5		
11	Salvage Value for depreciation (Sd)				$8,000.00		
12	Declining Balance Rate (%)				200%		
13	-----	----------------------		----------------------		----------------------	
14		STRAIGHT	LINE	200% DECLINING BALANCE		SUM OF YEARS DIGITS	
15	Yr	Depreciation	BV	Depreciation	BV	Depreciation	BV
16	-----	----------------------		----------------------		----------------------	
17	0		$80,000		$80,000		$80,000
18	1	14,400	65,600	32,000	48,000	24,000	56,000
19	2	14,400	51,200	19,200	28,800	19,200	36,800
20	3	14,400	36,800	11,520	17,280	14,400	22,400
21	4	14,400	22,400	6,912	10,368	9,600	12,800
22	5	14,400	8,000	2,368	8,000	4,800	8,000
23	-----	----------------------		----------------------		----------------------	
24		$72,000		$72,000		$72,000	
25	-----	----------------------		----------------------		----------------------	
26							
27	Sample Cell Formulas:						
28							
29	Straight Line Method:				B24 = @SUM(B18..B22)		
30	B18 = ($E$8-$E$11)/$E$9				D24 = @SUM(D18..D22)		
31	C18 = +C17-B18				F24 = @SUM(F18..F22)		
32							
33	200% Declining Balance Method:						
34	E18 = @IF((E17-(E17*$E$12/$E$9))<$E$11,$E$11,+E17-(E17*$E$12/$E$9))						
35	D18 = +E17-E18						
36							
37	Sum of Years Digits Method:						
38	F18 = (($E$9-A18+1)/($E$9*($E$9+1)/2))*($E$8-$E$11)						
39	G18 = +G17-F18						

Figure 6.A.1 *Lotus 1-2-3* Spreadsheet for Example 6.A.1 (SL, DDB, and SYI)

In other words, is the adjusted basis less than the expected salvage value? If the statement is *true*, then the value in cell E18 is the right-hand side of the inequality, i.e., the salvage value. If the statement is *false*, the value in cell E18 is the left-hand side of the inequality, the adjusted basis. Thus the @ IF function is used to ensure that the adjusted basis does not fall below the expected salvage value.

EXAMPLE 6.A.2 **DDB Depreciation with Switch to Straight Line**

For the depreciable asset described above, **Figure 6.A.2** shows the spreadsheet for determining annual depreciation under the DDB method, assuming a switch to the straight line method in the first year for which the switch would be warranted.

	A	B	C	D	E	F	G
1	EXAMPLE 6.A.2						
2							
3	METHOD: 200% Declining Balance, Switch to Straight Line						
4	(Assumption: Asset placed in service at the start of the year)						
5							
6	Input Data:						
7	-------	-------					
8	Cost Basis (B)				$80,000		
9	Depreciable life in years (Nd)				5		
10	Actual life in years (Na)				5		
11	Salvage Value for depreciation (Sd)				$0		
12	Declining Balance Rate (%)				200%		
13							
14							
15		Book Value	Remaining	Dep'n	Dep'n	Actual	Dep'n
16	Yr	End of Yr	Life	if DB	if Switch	Dep'n	Method
17	(j)	(Bj)	(N-j+1)		to SL	Dn	Used
18							
19	0	80,000					
20	1	48,000	5	32,000	16,000	32,000	DB
21	2	28,800	4	19,200	12,000	19,200	DB
22	3	17,280	3	11,520	9,600	11,520	DB
23	4	10,368	2	6,912	8,640	8,640	SL
24	5	6,221	1	4,147	10,368	8,640	SL
25							
26	TOTAL:					$80,000	
27							
28	Sample Cell Formulas:						
29	B20 = @IF((B19-(B19*$E$12/$E$9))<$E$11,$E$11,+B19-(B19*$E$12/$E$9))						
30	C20 = $E$9-A20+1						
31	D20 = +B19-B20						
32	E20 = (B19-$E$11)/C20						
33	F20 = @IF((E20>=D20),E20,D20)						
34	G20 = @IF((E20>=D20),"SL","DB")						
35	F26 = @SUM(F20..F24)						

Figure 6.A.2   *Lotus 1-2-3* Spreadsheet for Example 6.A.2 (DDB with Switch to SL)

Note the use of the @IF function available under Release 2.3 of *Lotus 1-2-3*. At cell **B20**, the conditional statement is used to ensure that the adjusted basis does not fall below the expected salvage value. At cell **F20**, the @IF function tests which is larger, (a) depreciation if the DB method is continued or (b) depreciation if there is a switch to the SL method in that year. The larger of the two values is placed in the cell. As shown in column G, the switch to SL occurs in the fourth year.

EXAMPLE 6.A.3   *Partial Depreciation in First Year*

In the two previous examples we assumed that the depreciable asset is to be placed in service at the start of the tax year, that is,

$$\pi_1 = \text{portion of the first tax year in}$$
$$\text{which the property is in service}$$

$$= 1.00 \quad \text{(assumed)}$$

Now let us assume that the depreciable property described above is placed

	A	B	C	D	E
1	EXAMPLE 6.A.3				
2					
3	METHOD: 200% Declining Balance, Partial Depreciation in 1st Year				
4					
5	Input Data:				
6					
7	Cost Basis (B)				$80,000
8	Salvage Value for depreciation (Sd)				$0
9	Actual Salvage Value (Sa)				$0
10	Depreciable life in years (Nd)				5
11	Actual service life in years (Na)				5
12	Declining Balance Rate (%)				200%
13	Portion of 1st tax year in service (pi sub 1)				0.75
14					
15					
16	Tax	Portion of	Book Value	Allowable	
17	Year	Yr in Service	Start of Yr	Depreciation	
18	(j)	(pi sub j)	(Bj)	(Dj)	
19					
20	0		$80,000		
21	1	0.75	56,000	$24,000	
22	2	1.00	33,600	22,400	
23	3	1.00	20,160	13,440	
24	4	1.00	12,096	8,064	
25	5	1.00	7,258	4,838	
26	6	0.25	4,355	2,903	
27					
28	TOTAL:	5.00		$75,645	
29					
30	Sample Cell Formulas:				
31	C20 = +$E$7				
32	B21 = +$E$13				
33	C21 = +C20-D21				
34	D21 = +C20*$E$12*$B$21/$E$10				
35	C22 = 1+C21-D22				
36	D22 = +C21*$E$12/$E$10				
37	B26 = (1-$B$21)				

Figure 6.A.3   *Lotus 1-2-3* Spreadsheet for Example 6.A.3 ($\pi_1 < 1.0$)

in service at some time during the tax year; specifically, suppose that $\pi_1 = 0.75$. If the property is kept in service 5 years, then $\pi_2 = \pi_3 = \pi_4 = \pi_5 = 1.00$ and $\pi_6 = 0.25$.

The allowable depreciation expenses under the DDB method are determined in spreadsheet format as shown in **Figure 6.A.3** above.

EXAMPLE 6.A.4 *Determining Depreciation Under MACRS*

Consider a depreciable property with cost basis of $40,000, expected service life of 8 years, and expected zero salvage value. This property is to be depreciated under regular MACRS with a 5-year recovery period. The half-year convention is assumed.

The allowable annual depreciation expenses are shown in spreadsheet format in **Figure 6.A.4**. Note that the continuing depreciation under DDB is determined in column C; and the depreciation in year j, assuming that there is a shift to SL in year j, is shown in column F. Thus the actual depreciation in year j is the larger of either the DDB or SL depreciation for that year. For example, the value in cell **F22** is the maximum of the values in **C22** and **E22**. The special function available under Release 2.3 of *Lotus 1-2-3* is

$$F22 = @\text{MAX}(C22, E22)$$

In this example, we computed the depreciation rate each year using the declining balance method, and we determined that the shift to straight line occurs in year 4. An alternative approach would be to use the appropriate table of annual depreciation rates prepared by the Internal Revenue Service. See Table 6.A.1 in Appendix 2 to this chapter. The more complicated approach is used here in order to demonstrate the use of the

	A	B	C	D	E	F	G	H
1	EXAMPLE 6.A.4							
2								
3	METHOD: 200% Declining Balance, MACRS Recovery Period							
4								
5	Input Data:							
6								
7	Asset Class				00.242			
8	Class life, years				6			
9	Recovery Period under MACRS				5			
10	Recovery Period under A/MACRS				6			
11	Cost Basis (B)				$40,000			
12	Actual service life in years (Na)				8			
13	Salvage Value for depreciation (Sd)				$0			
14	Actual Salvage Value (Sa)				$0			
15	Declining Balance Percent				200%			
16	Portion of 1st tax yr in service (pi s				0.50			
17								
18	Tax	<-----DDB	-------->	<----------	SL ------>	Actual	Actual	Dep'n
19	Year	BV	Dep'n	Remaining		Deprec'n	book value	rate
20	(j)	(1/2 Yr convention)		Life at start	Deprec'n	in year j	at end of	
21				of year			year	
22	1	$40,000	$8,000	5.0	$4,000	$8,000	$32,000	20.00%
23	2	32,000	12,800	4.5	7,111	12,800	19,200	32.00%
24	3	19,200	7,680	3.5	5,486	7,680	11,520	19.20%
25	4	11,520	4,608	2.5	4,608	4,608	6,912	11.52%
26	5	6,912	2,765	1.5	4,608	4,608	2,304	11.52%
27	6	4,147	1,659	0.5	2,304	2,304	0	5.76%
28	7	0	0					
29	8	0	0					
30								
31	Total					$40,000		100.00%
32								
33	Sample Cell Formulas:							
34	B21 = +$E$11							
35	C21 = ($B$22*$E$15/$E$9)*0.5							
36	D21 = +E9							
37	E21 = (($E$11-$E$13)/D22)*0.5							
38	G21 = +$E$11-F22							
39	H21 = +F22/$E$11							
40	B22 = @IF(A23<=$E$9+1,+B22-C22,0)							
41	C22 = ($B$22*$E$15/$E$9)*0.5							
42	D22 = +D22-E16							
43	E22 = (($E$11-$E$13)/D22)*0.5							

Figure 6.A.4 *Lotus 1-2-3* Spreadsheet for Example 6.A.4 (MACRS)

computer-generated spreadsheet showing, in general, how these depreciation rates may be calculated.

APPENDIX 2: *Applicable MACRS Percentages*

As indicated in the text, the Internal Revenue Service has determined the applicable MACRS and A/MACRS depreciation percentages under a variety of assumptions related to property class, underlying depreciation method, recovery period, and convention used in the first tax year (half-year, mid-quarter, mid-month). A total of 18 tables are published in *Depreciation*, IRS Publication 534. Reprinted here are six of those tables, namely:

Table No. in Chap. 6	Table No. in Pub. 534	Class	Principal Assumptions		
			Method	Recovery Period*	Convention
6.A.1	A-1	3, 5, 7, 10, 15, 20	DB/SL	MACRS	Half-year
6.A.2	A-2	3, 5, 7, 10, 15, 20	DB/SL	MACRS	Mid-quarter ($q = 1$)
6.A.3	A-3	3, 5, 7, 10, 15, 20	DB/SL	MACRS	Mid-quarter ($q = 2$)
6.A.4	A-4	3, 5, 7, 10, 15, 20	DB/SL	MACRS	Mid-quarter ($q = 3$)
6.A.5	A-5	3, 5, 7, 10, 15, 20	DB/SL	MACRS	Mid-quarter ($q = 4$)
6.A.6	A-6	Res. rental	SL (27.5 yr)	MACRS	Mid-month

*Regular MACRS is also known as GDS.

TABLE 6.A.1 Allowable MACRS Depreciation Rates
(Half-Year Convention)

Year	Depreciation rate for recovery period					
	3-year	5-year	7-year	10-year	15-year	20-year
1	33.33%	20.00%	14.29%	10.00%	5.00%	3.750%
2	44.45	32.00	24.49	18.00	9.50	7.219
3	14.81	19.20	17.49	14.40	8.55	6.677
4	7.41	11.52	12.49	11.52	7.70	6.177
5		11.52	8.93	9.22	6.93	5.713
6		5.76	8.92	7.37	6.23	5.285
7			8.93	6.55	5.90	4.888
8			4.46	6.55	5.90	4.522
9				6.56	5.91	4.462
10				6.55	5.90	4.461
11				3.28	5.91	4.462
12					5.90	4.461
13					5.91	4.462
14					5.90	4.461
15					5.91	4.462
16					2.95	4.461
17						4.462
18						4.461
19						4.462
20						4.461
21						2.231

TABLE 6.A.2 Allowable MACRS Depreciation Rates
(Mid-Quarter Convention: $q = 1$)

Year	Depreciation rate for recovery period					
	3-year	5-year	7-year	10-year	15-year	20-year
1	58.33%	35.00%	25.00%	17.50%	8.75%	6.563%
2	27.78	26.00	21.43	16.50	9.13	7.000
3	12.35	15.60	15.31	13.20	8.21	6.482
4	1.54	11.01	10.93	10.56	7.39	5.996
5		11.01	8.75	8.45	6.65	5.546
6		1.38	8.74	6.76	5.99	5.130
7			8.75	6.55	5.90	4.746
8			1.09	6.55	5.91	4.459
9				6.56	5.90	4.459
10				6.55	5.91	4.459
11				0.82	5.90	4.459
12					5.91	4.460
13					5.90	4.459
14					5.91	4.460
15					5.90	4.459
16					0.74	4.460
17						4.459
18						4.460
19						4.459
20						4.460
21						0.557

TABLE 6.A.3 Allowable MACRS Depreciation Rates
(Mid-Quarter Convention: $q = 2$)

Year	Depreciation rate for recovery period					
	3-year	5-year	7-year	10-year	15-year	20-year
1	41.67%	25.00%	17.85%	12.50%	6.25%	4.688%
2	38.89	30.00	23.47	17.50	9.38	7.148
3	14.14	18.00	16.76	14.00	8.44	6.612
4	5.30	11.37	11.97	11.20	7.59	6.116
5		11.37	8.87	8.96	6.83	5.658
6		4.26	8.87	7.17	6.15	5.233
7			8.87	6.55	5.91	4.841
8			3.33	6.55	5.90	4.478
9				6.56	5.91	4.463
10				6.55	5.90	4.463
11				2.46	5.91	4.463
12					5.90	4.463
13					5.91	4.463
14					5.90	4.463
15					5.91	4.462
16					2.21	4.463
17						4.462
18						4.463
19						4.462
20						4.463
21						1.673

TABLE 6.A.4 Allowable MACRS Depreciation Rates
(Mid-Quarter Convention: $q = 3$)

Year	Depreciation rate for recovery period					
	3-year	5-year	7-year	10-year	15-year	20-year
1	25.00%	15.00%	10.71%	7.50%	3.75%	2.813%
2	50.00	34.00	25.51	18.50	9.63	7.289
3	16.67	20.40	18.22	14.80	8.66	6.742
4	8.33	12.24	13.02	11.84	7.80	6.237
5		11.30	9.30	9.47	7.02	5.769
6		7.06	8.85	7.58	6.31	5.336
7			8.86	6.55	5.90	4.936
8			5.53	6.55	5.90	4.566
9				6.56	5.91	4.460
10				6.55	5.90	4.460
11				4.10	5.91	4.460
12					5.90	4.460
13					5.91	4.461
14					5.90	4.460
15					5.91	4.461
16					3.69	4.460
17						4.461
18						4.460
19						4.461
20						4.460
21						2.788

TABLE 6.A.5 Allowable MACRS Depreciation Rates
(Mid-Quarter Convention: $q = 4$)

Year	Depreciation rate for recovery period					
	3-year	5-year	7-year	10-year	15-year	20-year
1	8.33%	5.00%	3.57%	2.50%	1.25%	0.938%
2	61.11	38.00	27.55	19.50	9.88	7.430
3	20.37	22.80	19.68	15.60	8.89	6.872
4	10.19	13.68	14.06	12.48	8.00	6.357
5		10.94	10.04	9.98	7.20	5.880
6		9.58	8.73	7.99	6.48	5.439
7			8.73	6.55	5.90	5.031
8			7.64	6.55	5.90	4.654
9				6.56	5.90	4.458
10				6.55	5.91	4.458
11				5.74	5.90	4.458
12					5.91	4.458
13					5.90	4.458
14					5.91	4.458
15					5.90	4.458
16					5.17	4.458
17						4.458
18						4.459
19						4.458
20						4.459
21						3.901

TABLE 6.A.6   Allowable MACRS Depreciation Rates
(Residential Rental Property)

Year	\multicolumn{12}{c}{Month property placed in service}											
	1	2	3	4	5	6	7	8	9	10	11	12
1	3.485%	3.182%	2.879%	2.576%	2.273%	1.970%	1.667%	1.364%	1.061%	0.758%	0.455%	0.152%
2-9	3.636	3.636	3.636	3.636	3.636	3.636	3.636	3.636	3.636	3.636	3.636	3.636
10	3.637	3.637	3.637	3.637	3.637	3.637	3.636	3.636	3.636	3.636	3.636	3.636
11	3.636	3.636	3.636	3.636	3.636	3.636	3.637	3.637	3.637	3.637	3.637	3.637
12	3.637	3.637	3.637	3.637	3.637	3.637	3.636	3.636	3.636	3.636	3.636	3.636
13	3.636	3.636	3.636	3.636	3.636	3.636	3.637	3.637	3.637	3.637	3.637	3.637
14	3.637	3.637	3.637	3.637	3.637	3.637	3.636	3.636	3.636	3.636	3.636	3.636
15	3.636	3.636	3.636	3.636	3.636	3.636	3.637	3.637	3.637	3.637	3.637	3.637
16	3.637	3.637	3.637	3.637	3.637	3.637	3.636	3.636	3.636	3.636	3.636	3.636
17	3.636	3.636	3.636	3.636	3.636	3.636	3.637	3.637	3.637	3.637	3.637	3.637
18	3.637	3.637	3.637	3.637	3.637	3.637	3.636	3.636	3.636	3.636	3.636	3.636
19	3.636	3.636	3.636	3.636	3.636	3.636	3.637	3.637	3.637	3.637	3.637	3.637
20	3.637	3.637	3.637	3.637	3.637	3.637	3.636	3.636	3.636	3.636	3.636	3.636
21	3.636	3.636	3.636	3.636	3.636	3.636	3.637	3.637	3.637	3.637	3.637	3.637
22	3.637	3.637	3.637	3.637	3.637	3.637	3.636	3.636	3.636	3.636	3.636	3.636
23	3.636	3.636	3.636	3.636	3.636	3.636	3.637	3.637	3.637	3.637	3.637	3.637
24	3.637	3.637	3.637	3.637	3.637	3.637	3.636	3.636	3.636	3.636	3.636	3.636
25	3.636	3.636	3.636	3.636	3.636	3.636	3.637	3.637	3.637	3.637	3.637	3.637
26	3.637	3.637	3.637	3.637	3.637	3.637	3.636	3.636	3.636	3.636	3.636	3.636
27	3.636	3.636	3.636	3.636	3.636	3.636	3.637	3.637	3.637	3.637	3.637	3.637
28	1.97	2.273	2.576	2.879	3.182	3.485	3.636	3.636	3.636	3.636	3.636	3.636
29							0.152	0.455	0.758	1.061	1.364	1.667

7

Taxes and After-Tax
Economy Studies

7.1 INTRODUCTION

The accounting procedures discussed in Chapter 6—the Section 179 expense deduction, depreciation, depletion, and amortization—all relate to the determination of taxable income in that they are deductible business expenses. This is an essential element, of course, for business entities that are subject to income taxes. In this chapter we will discuss the relationship between taxable income and the amount and timing of cash flows for (income) taxes. A number of numerical examples are provided illustrating the effect of cash flows for income taxes on present worth and (internal) rate of return.

7.2 INCOME TAXES ARE FOR (ALMOST) EVERYONE

In the United States, all individuals and for-profit corporations are required to pay taxes, based on taxable income, to the federal government. (Most not-for-profit entities, such as schools, municipally owned utilities, charitable organizations, church groups, and the like, are not required, with certain exceptions, to pay taxes on their income.) The income of unincorporated business entities, such as *partnerships* and *single proprietorships*, is added to the nonbusiness income of the owners of these firms in order to determine the taxpayer's total taxable income. *Corporations* are also required to pay income taxes, but the tax rates differ somewhat from

Publication 334. It is revised annually, and a current edition may be obtained from any IRS office.

In addition to the federal government, income taxes are also paid to other taxing entities. Most states and, to a lesser extent, certain cities impose taxes on individuals and businesses residing in, or doing business in, these jurisdictions. For example, a corporation headquartered in Los Angeles will pay income taxes to the State of California as well as to the federal government.

It is not feasible here to summarize all relevant tax policies for individuals as well as corporations and for the federal government as well as all states and cities within the United States. Rather, our discussion will be limited to the tax consequences of business activity by *corporations* at the *federal* level. With certain exceptions, the after-tax analysis procedures described in the remainder of this chapter are generally applicable for all forms of business enterprise and for all taxing jurisdictions.

7.3   INCOME TAXES FOR CORPORATIONS

7.3.1   Income Tax Rates

Federal Tax Rates

Income tax rates for corporations are adjusted from time to time by the government, largely in order to affect the level of economic activity. The Omnibus Reconciliation Act of 1993 relates federal income tax rates, t, for corporations to the firm's taxable income, TI, for the tax year as follows:

Range of Taxable Income (TI)	Corporate Income Tax (T)	Marginal Tax Rate (t)
$0 < TI \leq \$50,000$	$0.15(TI)$	0.15
$\$50,000 < TI \leq \$75,000$	$\$7,500 + 0.25(TI - \$50,000)$	0.25
$\$75,000 < TI \leq \$100,000$	$\$13,750 + 0.34(TI - \$75,000)$	0.34
$\$100,000 < TI \leq \$335,000$	$\$22,250 + 0.39(TI - \$100,000)$	0.34 + 0.05
$\$335,000 < TI \leq \$10$ million	$\$113,900 + 0.34(TI - \$335,000)$	0.34
$\$10$ million $< TI \leq \$15$ million	$\$3,400,000 + 0.35(TI - \$10,000,000)$	0.35
$\$15$ million $< TI \leq \$18\frac{1}{3}$ million	$\$5,150,000 + 0.38(TI - \$15,000,000)$	0.35 + 0.03
$\$18\frac{1}{3}$ million $< TI$	$\$6,416,667 + 0.35(TI - \$18,333,333)$	0.35

Note that the nominal *marginal* tax rate is 34% for all taxable income between $75,000 and $10 million. However, for taxable income between $100,000 and $335,000, there is a 5% additional tax, called a *surtax*, which brings the *average* tax rate up to 34% when $TI = \$335,000$. Similarly, the *marginal* tax rate is 35% for taxable income above $10 million, but there is a 3% *surtax* for TI in the range $15,000,000–$18,333,333 so as to increase the average rate to 35% at $TI = 18\frac{1}{3}$ million. Therefore, corporations with taxable income for the tax year of at least $18\frac{1}{3}$ million, are taxed at a flat rate of 35%.

EXAMPLE 7.1   *Determining Income Tax*

Consider a taxpayer, a small corporation, which has a taxable income of $70,000 for the tax year. Determine its tax liability, that is, the required income tax for the year.

Solution:

Tax on first $50,000 of income = 0.15($50,000)	$ 7,500
Tax on next $20,000 of income = 0.25($20,000)	5,000
Total tax on $70,000 of income	$12,500

Note that the *average* tax rate is $12,500/$70,000 = 0.1786.

EXAMPLE 7.2   *Determining Income Tax on Increment of Investment*

Suppose that our taxpayer in Example 7.1 is considering a certain investment that will have the effect of increasing taxable income in this tax year by $20,000, that is, from $70,000 to $90,000. Determine the incremental effect on income taxes.

If the investment is implemented,

Tax on first $50,000 = 0.15($50,000)	$ 7,500
Tax on next $25,000 = 0.25($25,000)	6,250
Tax on next $15,000 = 0.34($15,000)	5,100
Total tax on $90,000	$18,850
Total tax on $70,000 (Example 7.1)	12,500
Incremental income tax	$ 6,350

Note that the *average* tax rate on the incremental investment is $6,350/$20,000 = 0.3175.

Tax Rates Combining Several Levels of Taxing Authorities

When income is taxed by more than one jurisdiction, the appropriate tax rate for economy studies is a combination of the rates imposed by the jurisdictions. If these rates are independent, they may simply be added. But the combinatorial rule is not quite so simple when there is interdependence. Income taxes paid to local and state governments, for example, are deductible from taxable income on federal income tax returns, but the reverse is not true; federal income taxes are not deductible from local returns. Thus, considering only state (t_s) and federal (t_f) income tax rates, the **combined marginal tax rate** (t) for economy studies is given by

$$t = t_s + t_f(1 - t_s) \tag{7.1}$$

EXAMPLE 7.3 **Combining Marginal Tax Rates**

The taxpayer, a large corporation, is located in a state in which its incremental income tax rate is 8%. The relevant federal incremental income tax rate is 35%. Using Eq. (7.1), the combined marginal tax rate is

$$t = 0.08 + 0.35(1 - 0.08) = 0.402$$

This rate can be used as a "shortcut" for determining the combined income taxes.

To illustrate the utility of this shortcut, suppose that the firm is considering an investment that would have the effect of increasing state taxable income by $100,000 during the tax year. The incremental effect on *state* income taxes is

$$0.08(\$100,000) = \$8,000$$

State income taxes are deductible on federal income tax returns, and thus the effect on *federal* income taxes is

$$0.35(\$100,000 - \$8,000) = \$32,200$$

The combined taxes paid is $40,200 (= $8,000 + $32,200), and the combined rate paid on the original $100,000 of increased taxable income is 0.402 (= $40,200/$100,000).

Caution: This formulation for the combined tax rate is valid only in those cases in which, except for state income taxes, federal taxable income is computed in exactly the same way as state taxable income. For example, Eq. (7.1) cannot be used if, say, the straight line depreciation method is used on state returns and MACRS depreciation is used on federal returns. In such instances it will be necessary to determine separately the cash flows for state and federal income taxes.

7.3.2 Tax Treatment of Gains and Losses on Disposal

The value of an asset on disposal is rarely equal to its book value at the time of sale or other disposition. When this inequality occurs, a **gain or loss on disposal** is established and the transaction has certain tax consequences.[1]

---

[1]Depreciable property used in a trade or business are *not* capital assets, and therefore, with certain exceptions, any gains or losses are *not capital* gains or losses. There is some confusion about this, because one normally thinks of investments in plant and equipment as capital investments. From the standpoint of tax law, however, gains and losses resulting from the disposal of depreciable property are treated as *ordinary* gains and losses. (Examples of capital assets include stocks or bonds; gold, silver, or any other metal.)

In general, the *gain* on disposition of depreciable property is the net salvage value minus the adjusted basis of the property (its book value) at the time of disposal. The **adjusted basis** is the original cost basis less any accumulated depreciation, amortization, Section 179 expense deduction, and, where appropriate, any basis adjustments due to investment credit claimed on the property. A negative gain is considered a loss on disposal. Thus, in general,

$$G = \text{gain (loss) on disposal}$$

$$= S_a - \left(B - \sum_{j=1}^{N_a} D_j \right) \tag{7.2}$$

where

$$S_a = \text{net salvage value on disposal}$$
$$B = \text{initial cost basis}$$
$$D_j = \text{depreciation in year } j, j = 1, 2, \ldots, N_a$$
$$N_a = \text{year of disposal}$$

Note that $(B - \sum D_j)$ is the book value at the time of disposal.

As a general rule, with some important exceptions, any gains or losses on the disposition of tangible personal property used in a taxpayer's trade or business are treated as ordinary income, limited to the total amount of depreciation taken up to the time of disposal. That is:

Gain	Ordinary Income	Capital Gain
Less than depreciation	Entire gain	None
Greater than depreciation	Accumulated depreciation	Excess of gain over accumulated depreciation

Currently (1993), the tax rate for capital gains is the same as that for ordinary income.

EXAMPLE 7.4 *Tax on Gain on Disposal*

Certain manufacturing equipment was purchased at a cost of $62,000. This was the cost basis. Four years later the equipment was removed from service and sold for $16,000. At the time of disposal, the book value was $12,000, resulting from $50,000 accumulated depreciation over 4 years. If the firm's incremental tax rate is 40%, determine the effect of the sale on cash flow for taxes.

Solution:

$$G = \text{(salvage value)} - \text{(book value at disposal)}$$

$$= \$16{,}000 - (\$62{,}000 - \$50{,}000)$$

$$= \$4{,}000$$

$$\text{Tax on gain} = tG = 0.40(\$4{,}000) = \$1{,}600$$

In this example, the net salvage value after taxes is

$$\$14{,}400 \; (=\$16{,}000 - \$1{,}600).$$

7.3.3 Timing of Tax Payments

In some corporations, the tax year (for accounting purposes) coincides with the calendar year. Such firms must file tax returns with the federal government not later than March 15 following the end of the calendar year. If the corporation's tax year is not the same as the calendar year, however, the due date is the 15th day of the third month following the close of the tax year. Local and state regulations generally, but not necessarily, parallel those of the federal government.

Because of the time value of money, it is normally advantageous for taxpayers to delay tax payments as long as possible. Conversely, it is to the government's advantage to collect tax receipts as quickly as possible, preferably as soon as income is earned by the taxpayer. To compromise, the government requires corporations to compute their estimated tax liability in advance of the tax year, and if this estimated tax can reasonably be expected to be $500 or more, periodic installments must be paid. For example, for firms that meet the "$500 requirement" prior to the fourth month of the tax year, 25% of the estimated tax must be paid by the first day of the 4th, 6th, 9th, and 12th months of the tax year. The corporation must pay the balance due in full by the due date of the return, that is, before the 15th day of the third month following the close of the tax year. This sequence of payments is shown in **Figure 7.1**.

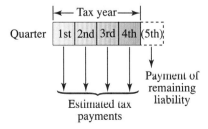

Figure 7.1   Timing of Tax Payments

There may be considerable time lag between the pretax cash flow (for example, the investment) and the resulting effect on cash flow for income taxes. Tax consequences that reduce tax liability are generally claimed as quickly as possible; consequences that increase taxes are generally delayed as long as possible. The general rule is to accelerate positive cash flows and defer negative cash flows. Authors of economic analysis textbooks usually adopt the following conventions in their illustrations and problems:

Consequence	Timing of Tax Effect
Initial investment	Beginning of year
Depreciation and other tax-deductible expenses	End of year
Disposal of asset	End of year

It should be stressed, however, that these are merely simplified textbook assumptions. They are probably appropriate to the vast majority of situations in which the dollar amounts are relatively small and the discount rate is relatively low. In the few cases where these assumptions do not hold, the analyst should identify, insofar as possible, the precise time lag between the pretax events and the cash flows for taxes.

7.4 TAX CREDITS

Federal tax law permits the taxpayer to claim **tax credits** for certain types of investments. These are *deductions from income taxes*, not taxable income. Currently (1993), here is a partial listing of investments that may qualify for credit against taxes:

- Disabled (permanent and total disability) access
- Foreign tax (taxes paid to a foreign country)
- Jobs credit (for employers who hire people who are members of special targeted groups)
- Alcohol used as fuel
- Research and experimental expenditures
- Low income housing

An additional general business credit, the **investment credit**, has been generally repealed for property placed in service after 1985. It is not now currently in effect.

7.5   AFTER-TAX ECONOMY STUDIES

After-tax analyses of investment alternatives can be quite complex. Not all pretax cash flows affect taxable income. Moreover, the timing of the various pretax cash flows may differ substantially from the timing of the cash flows for income taxes appropriate to the proposed investment. These complexities notwithstanding, it may be useful to adopt certain simplifying assumptions that enable us to develop approximate solutions relatively easily and with a level of precision that may be tolerable. In this section, then, both a general procedure and a simplified procedure are presented. Numerical examples will illustrate the two analytical approaches.

7.5.1   A General Procedure

The *general procedure* consists of seven steps as follows:

Step 1. Specify the assumptions and principal parameter values, including:
   a. Tax rates (federal and other taxing jurisdictions, as appropriate).
   b. Relevant methods related to depreciation, amortization, depletion, investment tax credit, and Section 179 expense deduction.
   c. Length of planning horizon (study period).
   d. Minimum attractive rate of return—the interest rate to be used for discounting cash flows. Call this $\hat{i}$, pronounced "eye hat." *Caution:* This rate should represent the after-tax opportunity cost to the taxpayer; it will almost always be lower than the pre-tax MARR. The same discounting rate should *not* be used for both before-tax and after-tax analyses.

Step 2. Estimate the amounts and timing of cash flows other than income taxes. It will be useful to separate these cash flows into three categories:
   a. *Cash flows that have a direct effect on taxable income,* either as "income" or "expense." Examples include sales receipts, direct labor savings, material costs, property taxes, interest payments, and state and local income taxes (on federal returns).
   b. *Cash flows that have an indirect effect on taxable income* through depreciation, amortization, depletion, Section 179 expense deduction, and gain or loss on disposal. Examples include initial cost of depreciable property and salvage value.
   c. *Cash flows that do not affect taxable income.* Examples include working capital and that portion of loan repayments that represents payment of principal.

Step 3. Determine the amounts and timing of cash flows for income taxes.

Step 4. Find the equivalent present value of cash flows for income taxes at the beginning of the first tax year. Call this $P(T)$.

Step 5. Find the equivalent present value of the cash flows for taxes, where "present" is defined as the start of the planning horizon. For example, if the property is placed in service at the end of the third month of the tax year, the present value adjustment would be $P(T) \times (1 + \hat{i})^{3/12}$.

Step 6. Find the equivalent present value of all other cash flows estimated in Step 2. Use the after-tax MARR, $\hat{i}$.

Step 7. Combine Steps 5 and 6 to yield the total present worth (PW).

Note: If it is desired to determine the after-tax rate of return rather than PW (or FW, EUAC, etc.), Steps 4–7 must be modified. With the appropriate present worth equation for all cash flows, set the equation equal to zero, and find the value of the interest rate such that PW = 0. This is the after-tax IRR for the proposed investment, $\hat{i}^*$

EXAMPLE 7.5 *After-Tax Present Worth*

Consider the possible acquisition of certain manufacturing equipment with initial cost of $400,000. The equipment is expected to be kept in service for 6 years and then sold for an estimated $40,000. Working capital of $50,000 will be required at the start of the 6-year period; the working capital will be recovered intact at the end of 6 years. If it is acquired, this equipment is expected to result in savings of $90,000 each year. The timing of these savings is such that the continuous cash flow assumption will be adopted throughout each year. The firm's after-tax MARR is 10% per year.

Assume that there is no Section 179 expense deduction. The equipment will be placed in service at the middle of the tax year and depreciated under MACRS as a 5-year recovery property using the half-year convention. The incremental federal income tax rate is 0.34; there are no other relevant income taxes affected by this proposed investment.

We now turn to our "general procedure" to determine the after-tax present worth of this proposed investment.

Step 1. Assumptions and principal parameter values
Tax rate $(t) = 0.34$
Depreciation by MACRS with recovery period = 5 years
No Section 179 expense deduction
Cost basis = $400,000
Planning horizon = 6 years
MARR$(\hat{i}) = 10\%$ (after-tax)
Property placed in service at middle of tax year

Step 2. Cash flows other than income taxes
Affecting taxable income:

$$\text{Annual savings} = \$90,000 \text{(continuous cash flow)}$$

Affecting income taxes:

$$\text{Initial cost} = \$400{,}000$$

$$\text{Salvage value} = \$\ 40{,}000$$

Other cash flows:

$$\text{Working capital} = \$50{,}000 \text{ (to be recovered at end of 6 years)}$$

Step 3. Cash flows for income taxes: See columns 1–7 of **Table 7.1.**

Step 4. PW of T_j's at start of first tax year: See columns 8 and 9 of Table 7.1.

Step 5. PW of T_j's at start of study period:

$$\$29{,}028(1.10)^{0.5} = \$30{,}445$$

Step 6. PW of all other cash flows:

Initial cost	$= -\$400{,}000$
Salvage value $= \$40{,}000(P/F,\ 10\%,\ 6)$	$=\qquad 22{,}579$
Working capital	$= -\qquad 50{,}000$
Working capital recovered $= \$50{,}000(P/F,\ 10\%,\ 6) =$	$28{,}224$
Annual savings $= \$90{,}000(P/\bar{A},\ 10\%,\ 6)$	$=\qquad 411{,}261$
	$\overline{\$\ 12{,}064}$

Step 7. Total present worth:

Cash flows other than income taxes	$\$12{,}064$
Less PW of additional income taxes	$-\ 30{,}445$
Net after-tax present worth	$\overline{-\$18{,}381}$

TABLE 7.1   Cash Flows for Income Taxes—Example 7.5

Tax Year j (1)	Depr. Rate $p_j(5)$ (2)	Depr. D_j (3)	Gain G (4)	Other Revenue R_j (5)	Taxable Income TI_j (6)	Income Taxes T_j (7)	PW Factor $(1.10)^{-j}$ (8)	PW @10% P_j (9)
1	0.2000	$ 80,000		$ 45,000	$-\$ 35,000$	$-\$11,900$	0.9091	$-\$10,818$
2	0.3200	128,000		90,000	$-38,000$	$-12,920$	0.8264	$-10,678$
3	0.1920	76,800		90,000	13,200	4,488	0.7513	3,372
4	0.1152	46,080		90,000	43,920	14,933	0.6830	10,199
5	0.1152	46,080		90,000	43,920	14,933	0.6209	9,272
6	0.0576	23,040		90,000	66,960	22,766	0.5645	12,851
7	0	0	$40,000	45,000	85,000	28,900	0.5132	14,830
Total	1.0000	$400,000	$40,000	$540,000	$180,000	$61,200		$29,028

PW measured at start of first tax year	$29,028
Adjustment factor (to move to the middle of the first tax year)	$\times (1.10)^{0.5}$
PW measured at start of planning horizon	$30,445

Here, since the after-tax PW < \$0, the "do nothing" alternative is preferred.

EXAMPLE 7.6 *After-Tax Internal Rate of Return*

Suppose that our problem now is to find the after-tax IRR, $\hat{i}^*$, of the proposed investment described in Example 7.5. (The cash flow diagrams are shown in **Figure 7.2**. The pretax cash flows are displayed in the lower part of the figure; the cash flows resulting from income taxes are in the upper part.) In this case, $\hat{i}^*$ is the solution to the equation:

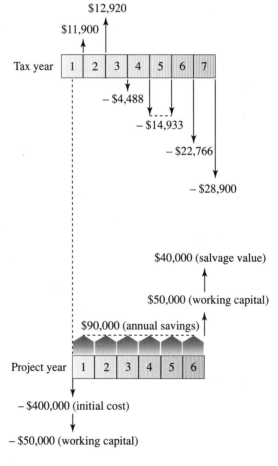

Figure 7.2   Cash Flow Diagrams for Examples 7.5 and 7.6

$$\mathrm{PW}(\hat{i}^*) = 0$$

$$= [\text{PW of cash flows other than income taxes}]$$
$$- [\text{PW of income taxes}]$$

$$= [-\$400{,}000 - \$50{,}000$$
$$+ (\$40{,}000 + \$50{,}000)(P/F, \hat{i}^*, 6)$$
$$+ \$90{,}000(P/\bar{A}, \hat{i}^*, 6)]$$
$$+ [\$11{,}900(1 + \hat{i}^*)^{-0.5} + \cdots - \$28{,}900(1 + \hat{i}^*)^{-6.5}]$$

The trial-and-error solution yields $\hat{i}^* = 8.35\%$. This IRR is less than the after-tax MARR (10%), and thus the proposed project appears to be less desirable than the do-nothing alternative.

7.5.2 A Simplified Procedure

In the general solution described above we were careful to distinguish between the timing of the cash flows for income taxes (Steps 4 and 5) and the timing of all other cash flows (Step 6). The analytical procedure is greatly simplified if we assume that cash flows for taxes occur at the end of each tax year and, further, that the tax year and the project year are coincident. This simplifying assumption is illustrated in **Figure 7.3**. This "offset" in timing does introduce an error, of course, but the approximation is probably reasonable in those cases where great precision is unwarranted.

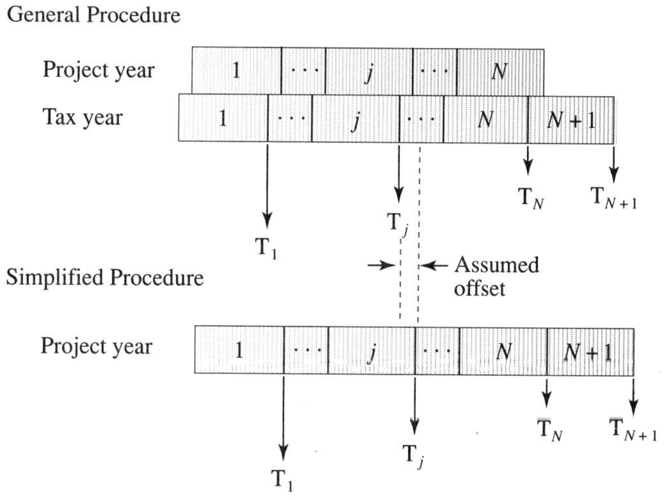

Figure 7.3 Assumed Timing of Cash Flows for Income Taxes

Assuming that the cash flows for income taxes occur at the end of each project year, the simplified procedure for after-tax analyses is as follows:

Step 1.
Step 2. } As described previously ("general procedure")
Step 3.

Step 4. Prepare an after-tax cash flow table with these columns:

1. End of year j.
2. Cash flows before taxes occurring at the end of year j. (If cash flows actually occur within the year, e.g., monthly or "continuous" cash flows, the equivalent end-of-year values should be calculated using the appropriate after-tax MARR as the interest rate for compounding.) Denote the values in this column as A_j.
3. Noncash deductions from taxable income: the Section 179 expense, depreciation, depletion, and amortization. This will usually be the depreciation expense, D_j. [In the last row of the table, the row in which the disposal of the asset is described, show the book value at the time of the disposal, B_j, in parentheses. This will help in determining the gain or loss on disposal.]
4. The total effect of the project on taxable income in year j, TI_j.
5. Cash flow for income taxes in year j, TI_j. For each row j, this is the product of the incremental tax rate, t, times the taxable income:

$$T_j = t(TI_j) \qquad (7.3)$$

6. Cash flow after taxes at the end of year j, $\hat{A}_j$. For each row j, this is column 2 minus column 5.

$$\hat{A}_j = A_j - T_j \qquad (7.4)$$

If all the pretax cash flows in column 2 affect taxable income, then

$$TI_j = A_j - D_j \qquad (7.5)$$

Combining Eqs. (7.3), (7.4), and (7.5),

$$\hat{A}_j = A_j - t(A_j - D_j)$$
$$= A_j(1 - t) + tD_j \qquad (7.6)$$

It is useful to display two rows for the final year: one to represent the effects of the disposal of depreciable assets and the second to show the effects of all other cash flows. This provides a simple mechanism for computing the after-tax salvage value, that is, the pretax salvage value less any tax consequences due to gain or loss on disposal.

Step 5. Determine the PW (or FW, AC, etc.) using the after-tax cash flows $(\hat{A}_j)$ and the after-tax discount rate $(\hat{i})$:

$$PW = \sum_{j=0}^{N} \hat{A}_j(1 + \hat{i})^{-j}$$

If it is desired to find the after-tax internal rate of return $(\hat{i}^*)$, solve the equation

$$\sum_{j=0}^{N} \hat{A}_j(1 + \hat{i}^*)^{-j} = 0$$

in the usual way.

EXAMPLE 7.7 *After-Tax PW Using the Simplified Procedure*

Consider the numerical example introduced in Example 7.5. Our analysis will proceed as before, except that here we assume that the cash flows for taxes occur at the end of each project year. The analysis is summarized in **Table 7.2**. Note that in

$$\text{column 2:}\quad A_j = \bar{A}_j \left[\frac{i}{\ln(1 + i)} \right]$$

$$= \$90,000 \left[\frac{0.10}{\ln 1.10} \right]$$

$$= \$94,429 \qquad \text{for } j = 1, 2, \ldots, 6$$

TABLE 7.2   After-Tax Cash Flows—Example 7.7

End of Year j	Effect on Cash Flow Before Taxes A_j	Depreciation Expense D_j	Effect on Taxable Income TI_j	Effect on Cash Flow for Taxes T_j	Effect on Cash Flow After Taxes $\bar{A}_j$	Present Value at 10% P_j
(1)	(2)	(3)	(4)	(5)	(6)	(7)
0	− $400,000				− $400,000	− $400,000
0	− 50,000				− 50,000	− 50,000
1	94,429	$ 80,000	$ 10,000	$ 3,400	91,029	82,754
2	94,429	128,000	− 38,000	− 12,920	107,349	88,718
3	94,429	76,800	13,200	4,488	89,941	67,574
4	94,429	46,080	43,920	14,933	79,496	54,297
5	94,429	46,080	43,920	14,933	79,496	49,361
6	94,429	23,040	66,960	22,766	71,663	40,452
6	50,000				50,000	28,224
6	40,000	⟨0⟩	40,000	13,600	26,400	14,902
Totals	$206,574	$400,000	$180,000	$61,200	$145,374	−$23,718

$$\hat{i}^* = 8.25\%$$

column 3: MACRS depreciation using a 5-year recovery period.

column 4: $TI_j = \$90,000 - D_j$ for $j = 1, 2, \ldots, 6$

The gain (G) is

$G = S_a - B_a = \$40,000 - \$0 = \$40,000$

column 5: $T_j = t \times TI_j$

column 6: $\hat{A}_j = A_j - T_j$

column 7: $P_j = \hat{A}_j(1.10)^{-j}$

The total present worth in this case ($-\$23,718$) is somewhat less than that determined in Example 7.5 ($-\$18,381$). The difference is due to our assumption concerning the timing of cash flows for income taxes. By assuming that the T_j's occur at the end of each project year, we have postponed the beneficial effects by one-half year on the average. This penalizes the project and results in a smaller present worth.

7.5.3 After-Tax Analyses: "Classical" Depreciation Methods

To further illustrate the simplified procedures for after-tax analyses, several examples are presented in which the "classical" methods of depreciation are used by the investor: straight line, sum of the years-digits, and declining balance.

EXAMPLE 7.8 *Straight Line Depreciation*

The firm is considering an investment of $80,000 in an asset that is to be kept in service for 5 years and then sold for an estimated zero salvage value. (Allowable depreciation for this asset was described in Chapter 6 under a variety of methods.) If it is acquired, the asset will result in annual savings of $24,000. It will be placed in service at the beginning of the tax year and depreciated using the straight line method. The taxpayer's incremental tax rate is 40%. Our problem is to determine the after-tax internal rate of return on this proposed investment.

The analysis is summarized in **Table 7.3**. Of particular note,

$$G = S_a - B_5 = \$0 - \$0 = \$0$$

$$A_j = \$24,000$$

$$D_j = \frac{\$80,000 - \$0}{5} = \$16,000$$

$$TI_j = A_j - D_j = \$24,000 - \$16,000 = \$8,000$$

$$T_j = 0.40(TI_j)$$

$$\hat{A}_j = A_j - T_j$$

The after-tax IRR $(\hat{i}^*)$ for this project is 9.43%.

TABLE 7.3   After-Tax Analyses Under "Classical" Depreciation Methods[a]

	End of Year j	Effect on Cash Flow Before Taxes A_j	Depreciation Expense D_j	Effect on Taxable Income TI_j	Effect on Cash Flow for Taxes T_j	Effect on Cash Flow After Taxes $\hat{A}_j$	After-Tax Rate of Return (IRR) $\hat{i}^*$
Example 7.8	0	− $80,000				− $80,000	
Straight line	1	24,000	$16,000	$8,000	$3,200	20,800	
depreciation	2	24,000	16,000	8,000	3,200	20,800	
$P = B = \$80,000$	3	24,000	16,000	8,000	3,200	20,800	
$S_a = S_d = \$0$	4	24,000	16,000	8,000	3,200	20,800	
$N_a = N_d = 5$ years	5	24,000	16,000	8,000	3,200	20,800	
	5	0	(0)[b]	0	0	0	
	Total	$40,000	$80,000	$40,000	$16,000	$24,000	9.43%
Example 7.9	0	− $80,000				− $80,000	
Straight line	1	24,000	$20,000	$4,000	$1,600	22,400	
depreciation with	2	24,000	20,000	4,000	1,600	22,400	
depreciable life shorter	3	24,000	20,000	4,000	1,600	22,400	
than actual life	4	24,000	20,000	4,000	1,600	22,400	
$P = B = \$80,000$	5	24,000	0	24,000	9,600	14,400	
$S_a = S_d = \$0$	5	0	(0)	0	0	0	
$N_d = 4$ years							
$N_a = 5$ years							
	Total	$40,000	$80,000	$40,000	$16,000	$24,000	9.97%
Example 7.10	0	− $80,000				− $80,000	
Sum of years-	1	24,000	$26,667	− $ 2,667	− $ 1,067	25,067	
digits depreciation	2	24,000	21,333	2,667	1,067	22,933	
$P = B = \$80,000$	3	24,000	16,000	8,000	3,200	20,800	
$S_a = S_d = \$0$	4	24,000	10,667	13,333	5,333	18,667	
$N_d = N_a = 5$ years	5	24,000	5,333	18,667	7,467	16,533	
	5	0	(0)	0	0	0	
	Total	$40,000	$80,000	$40,000	$16,000	$24,000	10.20%
Example 7.11	0	− $80,000				− $80,000	
Double declining	1	24,000	$32,000	− $ 8,000	− $ 3,200	27,200	
balance depreciation	2	24,000	19,200	4,800	1,920	22,080	
$P = B = \$80,000$	3	24,000	11,520	12,480	4,992	19,008	
$S_a = S_d = \$0$	4	24,000	6,912	17,088	6,835	17,165	
$N_d = N_a = 5$ years	5	24,000	4,147	19,853	7,941	16,059	
	5	0	(6,221)	−6,221	−2,488	2,488	
	Total	$40,000	$80,000	$40,000	$16,000	$24,000	10.26%
Example 7.12	0	− $80,000				− $80,000	
DB depreciation with	1	24,000	$32,000	− $ 8,000	− $ 3,200	27,200	
shift to straight line	2	24,000	19,200	4,800	1,920	22,080	
$P = B = \$80,000$	3	24,000	11,520	12,480	4,992	19,008	
$S_a = S_d = \$0$	4	24,000	8,640	15,360	6,144	17,856	
$N_d = N_a = 5$ years	5	24,000	8,640	15,360	6,144	17,856	
	5	0	(0)	0	0	0	
	Total	$40,000	$80,000	$40,000	$16,000	$24,000	10.29%

[a]Incremental income tax rate (t) is 40%. See text discussion for other general assumptions.
[b]Book value at time of disposal is shown in parentheses.

EXAMPLE 7.9 *Straight Line Depreciation, with Depreciable Life Shorter than Actual Life*

Consider the project as described in the previous example, but we will now assume that the taxpayer is allowed to depreciate the asset over 4 years ($N_d = 4$), even though the actual life is 5 years ($N_a = 5$).

The analysis is summarized in Table 7.3. Although the total of the after-tax cash flows ($24,000) is the same as determined previously, the after-tax IRR has increased to 9.97%. The increase is due entirely to the change in timing of the cash flows for taxes.

EXAMPLE 7.10 *Sum of Years-Digits Depreciation*

Consider the project described above, but assume here that depreciation will be by the sum of the years-digits method. As in Example 7.8, we will continue to assume that $N_d = 5$ years and S_d is zero. The analysis, summarized in Table 7.3, indicates that the after-tax IRR is 10.20%.

EXAMPLE 7.11 *Declining Balance Depreciation*

Continuing with the project described above, assume that depreciation will be by the declining balance method with $a = 200\%/N_d = 2/5 = 0.40$. The analysis, summarized in Table 7.3, indicates that the book value after 5 years is $6,221, and thus

$$G = S_a - B_5 = \$0 - \$6,221 = -\$6,221$$

That is, there is a loss on disposal of $6,221, and, with a 40% tax rate, the effect of this loss is to reduce the tax liability by $2,488 ($= 0.40 \times \$6,221$). This calculation is described in the final row for $j = 5$. The after-tax IRR is 10.26%.

EXAMPLE 7.12 *Declining Balance Method with Switch to Straight Line*

In the previous example, the "unrecovered" depreciation ($6,221) was "recaptured" in the form of a loss on disposal at the end of year 5. The taxpayer can avoid this situation by switching from DB to SL at any time using the book value and remaining life at the time of the switch. (See Section 6.5.2.) In this example, the optimal time to make the switch is in year 4, resulting in $D_4 = D_5 = \$8,640$. The effect on cash flows for taxes is shown in Table 7.3. The resulting after-tax IRR is 10.29%.

The amounts and timing of the before-tax cash flows are constant, of course, in each of the five examples described above. However, changes in depreciation method lead to changes in the *timing* of cash flows for income taxes. (The total amount of cash flows for taxes, $16,000, is the same in each case.) Summarizing our results:

Cash Flows for Taxes (T_j)

End of Year	Straight Line		SYD	Declining Balance	
j	$N_d = N_a$	$N_d < N_a$		No Shift	Shift to SL
1	$ 3,200	$ 1,600	−$ 1,067	−$ 3,200	−$ 3,200
2	3,200	1,600	1,067	1,920	1,920
3	3,200	1,600	3,200	4,992	4,992
4	3,200	1,600	5,333	6,835	6,144
5	3,200	9,600	7,467	5,453	6,144
Total	$16,000	$16,000	$16,000	$16,000	$16,000
IRR	9.43%	9.97%	10.20%	10.26%	10.29%

If the tax rate is constant or decreasing over time, there is an advantage in taking depreciation expenses as early as possible in order to postpone the payment of income taxes as long as possible. The precise effect can be measured as outlined in the five examples above.

7.5.4   After-Tax Analyses: ACRS and MACRS

As indicated in Chapter 6, the Accelerated Cost Recovery System (ACRS) was introduced by federal tax law for most assets placed in service beginning in 1981. ACRS and its alternate method, A/ACRS, remained in effect until the Tax Reform Act of 1986 introduced the Modified Accelerated Cost Recovery System (MACRS) for most assets placed in service beginning in 1987. MACRS and its alternate method, A/MACRS, are currently required by federal tax law. Some numerical examples illustrating these methods are included in this section.

EXAMPLE 7.13 *Regular ACRS*

Recall the proposed investment presented initially in Example 7.8. The initial cost is $80,000. It is to be kept in service for 5 years, during which time the asset is expected to result in pretax savings of $24,000 annually. The net salvage value at the end of 5 years is assumed to be zero.

We will assume here that the asset was placed in service between 1981 and 1986, during which time the Accelerated Cost Recovery System was mandated by federal law. We are "postauditing" the investment. We now would like to determine the internal rate of return, after taxes, of this project.

The solution is summarized in **Table 7.4**. The annual depreciation expenses are based on the assumption that the asset was depreciated under regular ACRS as a 5-year recovery property. As noted in Section 6.6.2, the statutory depreciation rates for a 5-year recovery property under ACRS are

Year j	Rate $p_j(5)$
1	0.15
2	0.22
3, 4, 5	0.21

Thus the depreciation expenses each year are found by

$$p_j(5) \times \$80,000$$

Using the resulting after-tax cash flows, $\hat{A}_j$, as shown in the table, the internal rate of return is found to be 9.31%.

TABLE 7.4   After-Tax Analyses Assuming ACRS and A/ACRS Depreciation[a]

	End of Year j	Effect on Cash Flow Before Taxes A_j	Depreciation Expense (Recovery) D_j	Effect on Taxable Income TI_j	Effect on Cash Flow for Taxes T_j	Effect on Cash Flow After Taxes $\hat{A}_j$	After-Tax Rate of Return (IRR) $\hat{i}^*$
Example 7.13	0	$-\$80,000$					$-\$80,000$	
Regular ACRS	1	24,000	$12,000	$12,000	$ 4,800	19,200	
$k = 5$	2	24,000	17,600	6,400	2,560	21,440	
$N_a = 5$ years	3	24,000	16,800	7,200	2,880	21,120	
	4	24,000	16,800	7,200	2,880	21,120	
	5	24,000	16,800	7,200	2,880	21,120	
	5	0	$(0)^b$	0	0	0	
	Total	$40,000	$80,000	$40,000	$16,000	$24,000	9.31%
Example 7.14	0	$-\$80,000$					$-\$80,000$	
Alternate ACRS	1	24,000	$ 8,000	$16,000	$ 6,400	17,600	
$k = 5$ with	2	24,000	16,000	8,000	3,200	20,800	
$N(5) = 5$ years	3	24,000	16,000	8,000	3,200	20,800	
$N_a = 5$ years	4	24,000	16,000	8,000	3,200	20,800	
	5	24,000	16,000	8,000	3,200	20,800	
	6	0	0	0	0	0	
	6^c	0	$(8,000)^b$	$-8,000$	$-3,200$	3,200	
	Total	$40,000	$80,000	$40,000	$16,000	$24,000	8.94%

[a]Incremental tax rate (t) is 40%. See text discussion for other general assumptions.
[b]Book value at time of disposal shown in parentheses.
[c]It is assumed here that the property is sold at the *start* of the sixth tax year, thus permitting depreciation in the fifth year. The tax effect of the loss on disposal occurs at the *end* of the sixth year.

EXAMPLE 7.14 *Alternate ACRS*

Assume that the investment in the prior example had been depreciated using the alternate ACRS method (A/ACRS) using an optional recovery period of 5 years. (Recall that under A/ACRS the taxpayer may select a recovery period of either 5, 12, or 25 years.) The analysis is summarized in Table 7.4.

Under A/ACRS, with $N(5) = 5$ years,

$$p_1(5) = \left(\frac{1.00}{5}\right)\left(\frac{1}{2}\right) = 0.10$$

$$p_j(5) = \frac{1.00}{5} = 0.20 \qquad \text{for } j = 2, 3, 4, 5$$

A full year of depreciation is allowed for the fifth year only if the asset is not sold until the start of the sixth year (or later). We will assume that that was the case.

Recall that, under ACRS and A/ACRS, no depreciation is allowed in the year of disposition or retirement of 3-, 5-, or 10-year recovery property. Thus, when the asset is sold in the sixth year, the book value at the time of disposal (B_6) is \$8,000. The actual salvage value is zero, so there is an \$8,000 loss on disposal at the end of year 6 which, at a 40% tax rate, results in tax savings of \$3,200.

The after-tax cash flows result in an internal rate of return of 8.94%.

EXAMPLE 7.15 *Regular MACRS*

We continue with our proposed investment as described in Example 7.8, except that now we assume that the asset is to be depreciated under regular MACRS as a 5-year recovery property. As before, we assume that the cost basis is the initial cost ($P = B = \$80,000$) and there is no Section 179 expense deduction.

The 200% declining balance method is used for a 5-year property, so the depreciation rate is $a = 2.00/5 = 0.40$. Assuming the half-year convention, it may be shown that

$$p_1(5) = 0.2000 \qquad p_3(5) = 0.1920 \qquad p_5(5) = 0.1152$$

$$p_2(5) = 0.3200 \qquad p_4(5) = 0.1152 \qquad p_6(5) = 0.0576$$

The resulting annual depreciation expenses, shown in **Table 7.5**, are the result of

$$p_j(5) \times \$80,000 \qquad \text{for } j = 1, 2, \ldots, 5$$

Note that depreciation is allowed under MACRS in the year of disposal. There is no gain or loss on disposal in this case, because $B_6 = S_6 = \$0$.

The resulting after-tax internal rate of return is 9.66%.

TABLE 7.5   After-Tax Analyses Assuming MACRS and A/MACRS[a]

	End of Year j	Effect on Cash Flow Before Taxes A_j	Depreciation Expense (Recovery) D_j	Effect on Taxable Income TI_j	Effect on Cash Flow for Taxes T_j	Effect on Cash Flow After Taxes $\hat{A}_j$	After-Tax Rate of Return (IRR) $\hat{i}*$
Example 7.15	0	−$80,000				−$80,000	
Regular MACRS	1	24,000	$16,000	$ 8,000	$ 3,200	20,800	
(DDB/SL)	2	24,000	25,600	−1,600	−640	24,640	
N_d = 5 years	3	24,000	15,360	8,640	3,456	20,544	
$P = B = \$80,000$	4	24,000	9,216	14,784	5,914	18,086	
$S_a = \$0$	5	24,000	9,216	14,784	5,914	18,086	
N_a = 5 years	6[b]	—	4,608	−4,608	−1,843	1,843	
	6[b]	0	(0)	0	0	0	
	Total	$40,000	$80,000	$40,000	$16,000	$24,000	9.66%
Example 7.16	0	−$80,000				−$80,000	
Alternate MACRS	1	24,000	$ 6,667	$17,333	$ 6,933	17,067	
(straight line over	2	24,000	13,333	10,667	4,267	19,733	
the A/MACRS	3	24,000	13,333	10,667	4,267	19,733	
recovery period)	4	24,000	13,333	10,667	4,267	19,733	
N_d = 6 years	5	24,000	13,333	10,667	4,267	19,733	
$P = B = \$80,000$	6[b]	—	6,667	−6,667	−2,667	2,667	
$S_a = \$0$	6[b]	0	(13,333)	−13,333	−5,333	5,333	
N_a = 5 years							
	Total	$40,000	$80,000	$40,000	$16,000	$24,000	8.59%
Example 7.17	0	−$80,000				−$80,000	
(straight line over	1	24,000	$ 8,000	$16,000	$ 6,400	17,600	
the MACRS	2	24,000	16,000	8,000	3,200	20,800	
recovery period)	3	24,000	16,000	8,000	3,200	20,800	
N_d = 5 years	4	24,000	16,000	8,000	3,200	20,800	
$P = B = \$80,000$	5	24,000	16,000	8,000	3,200	20,800	
$S_a = \$0$	6[b]	0	8,000	−8,000	−3,200	3,200	
N_a = 5 years	6[b]	0	(0)	0	0	0	
	Total	$40,000	$80,000	$40,000	$16,000	$24,000	8.94%
Example 7.18	0	−$80,000				−$80,000	
(150% DB over	1	24,000	$10,000	$14,000	$ 5,600	18,400	
the A/MACRS	2	24,000	17,504	6,496	2,598	21,402	
recovery period)	3	24,000	13,128	10,872	4,349	19,651	
N_d = 6 years	4	24,000	11,248	12,752	5,101	18,899	
$P = B = \$80,000$	5	24,000	11,248	12,752	5,101	18,899	
$S_a = \$0$	6[b]	—	5,624	−5,624	−2,250	2,250	
N_a = 5 years	6[b]	0	(11,248)	−11,248	−4,499	4,499	
	Total	$40,000	$80,000	$40,000	$16,000	$24,000	8.89%

[a]Marginal tax rate (t) is 40%. See text discussion for other general assumptions.
[b]Assumes property is sold at *start* of sixth tax year, thereby allowing one-half year of depreciation in year 6.

EXAMPLE 7.16 *Alternate MACRS*

As mentioned in Section 6.7.3, the taxpayer may elect to depreciate the asset by the straight line method over the alternate MACRS recovery period. Continuing the previous example, suppose that the firm chooses to exercise this option. The A/MACRS recovery period for this asset class is 6 years, and so

$$
p_j(6) =
\begin{cases}
\left(\frac{1}{6}\right)\left(\frac{1}{2}\right) = \frac{1}{12} & \text{for } j = 1 \\[2mm]
\qquad\quad = \frac{1}{6} & \text{for } j = 2, 3, \ldots, 6 \\[2mm]
\left(\frac{1}{6}\right)\left(\frac{1}{2}\right) = \frac{1}{12} & \text{for } j = 7
\end{cases}
$$

Assuming that the asset is sold at the start of the sixth tax year, only one-half year of depreciation ($\frac{1}{2} \times \frac{1}{6} \times \$80{,}000 = \$6{,}667$) is allowable in that year. Thus there is a loss on disposal of $13,333, the tax consequences of which occur at the end of year 6.

$$
B_6 = \$80{,}000 - \sum_{j=1}^{6} D_j
$$
$$
= \$80{,}000 - \$66{,}667 = \$13{,}333
$$
$$
G = S_a - B_6 = \$0 - \$13{,}333 = -\$13{,}333
$$

The after-tax cash flows, summarized in Table 7.5, yield an internal rate of return of 8.59%.

EXAMPLE 7.17 *Straight Line over the MACRS Recovery Period*

Current tax law permits the taxpayer to depreciate the asset by the straight line method over the regular MACRS recovery period. Using this approach for the asset described in Example 7.15, we have

$$
p_j(5) =
\begin{cases}
\left(\frac{1}{5}\right)\left(\frac{1}{2}\right) = 0.10 & \text{for } j = 1 \\[2mm]
\frac{1}{5} \qquad\quad = 0.20 & \text{for } j = 2, 3, 4, 5 \\[2mm]
\left(\frac{1}{5}\right)\left(\frac{1}{2}\right) = 0.10 & \text{for } j = 6
\end{cases}
$$

The annual depreciation expenses and the after-tax cash flows are summarized in Table 7.5. Note that there is no gain or loss on disposal, because

$$
B_6 = \$80{,}000 - \$80{,}000 = \$0
$$
$$
G = S_a - B_6 = \$0 - \$0 = \$0
$$

The resulting after-tax internal rate of return is 8.94%.

EXAMPLE 7.18   *150% Declining Balance over the A/MACRS Recovery Period*

Another MACRS option permitted by current tax law is to depreciate the property by the *150% declining balance method,* with switch to straight line, over the alternate MACRS recovery period. Suppose that the taxpayer elects this option with respect to the asset described above. Recall that the A/MACRS recovery period is 6 years. The switch year is year 4, and the resulting depreciation percentages are

$$p_1(6) = 0.1250$$

$$p_2(6) = 0.2188$$

$$p_3(6) = 0.1641$$

$$p_4(6) = p_5(6) = p_6(6) = 0.1406$$

$$p_7(6) = 0.0703$$

With a cost basis of $80,000, the allowable depreciation expenses are as shown in Table 7.5. Note that there is a premature disposal in that the asset is sold in the sixth year. The allowable depreciation in that year is only one-half the normal depreciation:

$$D_6 = 0.1406(\$80,000)\left(\frac{1}{2}\right) = \$5,624$$

There is a loss of $11,248 on disposal, because

$$B_6 = \$80,000 - \sum_{j=1}^{6} D_j$$

$$= \$80,000 - \$68,752 = \$11,248$$

$$G = S_a - B_6 = \$0 - \$11,248 = -\$11,248$$

The resulting after-tax internal rate of return is 8.89%.

Summarizing the results of this section, Examples 7.13–7.18, the cash flows for income taxes (T_j) and internal rates of return have been shown to be as follows:

End of Year j	(7.13) Regular ACRS (5 yr)	(7.14) Alternate ACRS (5 yr)	(7.15) Regular MACRS (5 yr)	(7.16) Alternate MACRS (6 yr)	(7.17) SL MACRS (5 yr)	(7.18) 150% DB MACRS (6 yr)
1	$ 4,800	$ 6,400	$ 3,200	$ 6,933	$ 6,400	$ 5,600
2	2,560	3,200	−640	4,267	3,200	2,598
3	2,880	3,200	3,456	4,267	3,200	4,349
4	2,880	3,200	5,914	4,267	3,200	5,101
5	2,880	3,200	5,914	4,267	3,200	5,101
6		−3,200	−1,843	−8,000	−3,200	−6,749
Total	$16,000	$16,000	$16,000	$16,000	$16,000	$16,000
IRR	9.31%	8.94%	9.66%	8.59%	8.94%	8.89%

The total cash flows for taxes is the same in each case ($16,000). Differences in after-tax rates of return are due entirely to differences in the *timing* of these tax payments. As was noted previously, if the tax rate remains constant over time, there is an advantage in postponing income tax payments as long as possible.

7.6   SOME ADDITIONAL EXAMPLES OF THE EFFECTS OF INCOME TAXES

The concept of after-tax analysis may be extended, of course, to *any* proposed investment in which there are income tax consequences. These effects are caused by pretax cash flows affecting taxable income. Several additional examples of after-tax economy studies are given in this section.

7.6.1   Loans

In addition to the return of the original loan, the *principal*, the borrower generally pays *interest* as a cost of obtaining the loan. The amounts of the timing of the interest payments will vary, depending on the terms of the loan agreement. In any event, interest paid in connection with a loan for business purposes is considered to be a business expense and is allowable as a deduction from taxable income.[2] This will affect cash flows for income taxes, of course, as illustrated in the following examples.

EXAMPLE   7.19   *"Interest Only" Payments*

In order to obtain additional funds for the acquisition of new equipment, the firm obtains a $100,000 loan at a pretax cost, i_d, of 10% (per year). "Interest only" payments will be made at the end of each year, and the entire principal will be repaid at the end of 5 years. If the firm's incremental income tax rate is expected to be 34% at that time, determine the after-tax cost of this loan. (*Note:* The cost of a loan is usually expressed as an effective interest rate, per annum. The cost to the borrower is the rate of return to the lender.)

As before, let A_j = pretax cash flow and $\hat{A}_j$ = after-tax cash flow at the end of period j. Then

$$\hat{A}_j = A_j - T_j$$
$$= A_j - tI_j \tag{7.7}$$

where

$$t = \text{incremental tax rate}$$

$$I_j = \text{interest expense in period } j$$

---

[2]With the exception of interest on home mortgages, *nonbusiness* interest expenses (e.g., consumer credit) cannot be deducted on federal income tax returns.

The "period" here is one year. In this example, the interest to be paid at the end of each year is

$$I_j = \$100,000(0.10) = \$10,000$$

and the after-tax cash flows are

$$\hat{A}_0 = \$100,000$$

$$\hat{A}_j = -\$10,000 + 0.34(\$10,000)$$

$$= -\$6,600 \quad \text{for } j = 1, 2, 3, 4$$

$$\hat{A}_5 = -\$100,000 - \$6,600$$

$$= -\$106,600$$

The after-tax cost of this loan, $\hat{i}_d$, is found from

$$\$100,000 - \$6,600(P/A, \hat{i}_d, 5) + \$100,000(P/F, \hat{i}_d, 5) = 0$$

Solving, $\hat{i}_d = 6.60\%$. The subscript "d" signifies "debt."

EXAMPLE 7.20 *Uniform Series of Annual Payments*

Suppose that the repayment terms for the $100,000 loan in the previous example are changed somewhat: The loan will be repaid by five equal end-of-year payments assuming a pretax loan rate, i_d, of 10%. (This is known as **loan amortization**; the loan is to be fully amortized over 5 years.)

For an initial loan P, each loan payment, A, is given by

$$A = P(A/P, i_d, N)$$

$$= \$100,000(A/P, 10\%, 5) = \$26,380 \quad \text{(to the nearest dollar)}$$

Our next step is to determine the portion of each loan payment that represents interest, because this will be the effect on taxable income. In general,

$$I_j = i_j P_{j-1} \tag{7.8}$$

where P_{j-1} is the principal owed at the end of period $j - 1$ (the beginning of period j).

Determination of the after-tax cash flows for this loan is summarized in **Table 7.6**. The after-tax cost of the loan is found from

$$\sum_{j=0}^{5} \hat{A}_j (1 + \hat{i}_d)^{-j} = 0$$

Solving using the $\hat{A}_j$ values from Table 7.6, $\hat{i}_d = 6.60\%$.

TABLE 7.6   Determining the After-Tax Cash Flows for a Uniform Series of Annual Loan Payments (Example 7.20)

End of Year j	Principal Owed at Start of Year P_{j-1}	Interest Owed I_j	Payment A	Portion of Payment that Is Principal $A - I_j$	Effect on Income Taxes T_j	Cash Flow After Taxes $\hat{A}_j$
0						$100,000
1	$100,000	$10,000	$ 26,380	$ 16,380	$ 3,400	−22,980
2	83,620	8,362	26,380	18,018	2,843	−23,537
3	65,602	6,560	26,380	19,820	2,230	−24,150
4	45,782	4,578	26,380	21,802	1,557	−24,823
5	23,980[a]	2,398	26,380	23,982[a]	815	−25,565
Total		$31,900	$131,900	$100,000	$10,845	−$ 21,055

[a]Differences due to rounding.
Notes: $I_j = i_d P_{j-1} = 0.10 P_{j-1}$
$T_j = t I_j = 0.34 I_j$
$\hat{A}_j = -(A - T_j)$ for $j = 1, 2, \ldots, 5$

The solutions to the previous two examples, 7.19 and 7.20, could have been determined by using the relationship

$$\hat{i}_d = (1 - t)i_d \tag{7.9}$$

In both examples,

$$\hat{i}_d = (1 - 0.34)(0.10) = 0.066$$

This simplified approach is appropriate in only a limited number of situations:

- When the entire principal of the loan is repaid at the end of N periods, and "interest only" payments are made at the end of each and every period
- When the loan is fully amortized by a series of uniform payments consisting of principal and interest

Because the conditions necessary to use the simplified approach, Eq. (7.9), are not always present for a given problem, it is preferable to develop a table of after-tax cash flows as illustrated in Example 7.20. This point is further demonstrated in the next example.

EXAMPLE 7.21   *Uniform Series of Quarterly Payments*

Suppose that our firm has the opportunity to borrow $100,000 at an effective rate of 12.55% (per year). Repayment is to be made by eight equal payments at the end of each 3-month period. The effective quarterly rate corresponding to the annual rate of 12.55% is

$$i_d = (1.1255)^{1/4} - 1 = 0.03 \quad \text{(per quarter)}$$

Each quarterly payment, before adjustment for income taxes, is

$$\$100,000(P/A, 3\%, 8) = \$14,246 \quad \text{(to the nearest dollar)}$$

Assuming a 40% tax rate, determination of the cash flows for income taxes is summarized in **Table 7.7**.

At this point the question arises as to the *timing* of the cash flows for income taxes. There are (at least) two possibilities: We can assume that the tax consequences occur when the interest expense is incurred, that is, at the end of each quarter; or, alternatively, we can assume that the cash flows for taxes occur only at the end of each year.

Assuming that the cash flows for taxes occur quarterly, the solution procedure parallels that of the previous example. Using the cash flow table approach, the after-tax cost of this loan can be obtained by solving

$$\sum_{j=0}^{8} \hat{A}_j (1 + \hat{i}_d)^{-j} = 0$$

where

$$A_0 = \$100,000$$

$$A_j = \$14,246 + (T_j \text{ from Table 7.7}) \quad \text{for } j = 1, 2, \ldots, 8$$

TABLE 7.7   Determining the After-Tax Cash Flows for a $100,000 Loan, Quarterly Payments at 3% Paid over 8 Quarters (Example 7.21)

End of Quarter j	Principal Owed at Start of Quarter P_{j-1}	Interest Owed I_j	Payment A	Portion of Payment that Is Principal $A - I_j$	Effect on Income Taxes $(t=0.40)$ T_j	Total Annual Income Taxes
1	$100,000	$ 3,000	$ 14,246	$ 11,246	$1,200	
2	88,754	2,663	14,246	11,583	1,065	
3	77,171	2,315	14,246	11,931	926	
4	65,240	1,957	14,246	12,288	783	$3,974
5	52,952	1,586	14,246	12,657	634	
6	40,295	1,209	14,246	13,037	484	
7	27,258	818	14,246	13,428	327	
8	13,830[a]	415	14,246	13,831[a]	166	1,611
Total		$13,965	$113,965	$100,000	$5,585	$5,585

[a]Difference due to rounding.
Notes: $I_j = 0.03P_{j-1}$
$T_j = 0.40I_j$

Solving, $\hat{i}_d = 1.8\%$ per quarter. This is the same rate as that which would have resulted using Eq. (7.9):

$$\hat{i}_d = 0.03(1 - 0.40) = 0.018$$

The equivalent effective annual rate is

$$(1.018)^4 - 1 = 7.40\%$$

As shown in Table 7.7, the total cash flows for taxes are $3,974 in year 1 and $1,611 in year 2. Assuming that these tax savings due to loan interest occur at the end of each year, that is, quarters 4 and 8, the after-tax cost of this loan, per quarter, is given by

$$\$100,000 - \$14,246(P/A, \hat{i}_d, 8)$$

$$+ \$3,974(P/F, \hat{i}_d, 4) + \$1,611(P/F, \hat{i}_d, 8)$$

$$- \$100,000 = 0$$

Solving, $\hat{i}_d = 1.84\%$. The effective annual rate (cost) corresponding to this rate is

$$(1.0184)^4 - 1 = 7.56\%$$

The simplified approach, Eq. (7.9), cannot be used in this instance because of the timing of the cash flows for taxes.

7.6.2 Leases

A **lease** is a type of rental agreement for which the "rental period" is generally longer than one year. Units to be leased may be buildings, automobiles, office or storage space, computers, trucks, or any other equipment and facilities for which there may be an interest. The renter is known as the **lessor**; the **lessee** is the individual or organization to whom the unit is leased.

There are a number of apparent advantages to the lessee, including the following:

- Equity capital is freed up for other purposes.
- If equity capital is simply not available, leasing may provide an opportunity to acquire needed assets over a limited time period.
- In most cases, lease payments (for business-related assets) are fully deductible as a business expense.

For these and other reasons, leasing is a growing, multibillion dollar industry.

The **lease agreement**, the contractual agreement between lessor and lessee, specifies the terms and conditions of the lease. Among the features generally included in a lease agreement are:

- Amount of the deposit, usually refunded at the end of the lease

- Frequency (timing) of lease payments; usually monthly over a period of 36–60 months
- Amount of each lease payment
- Down payment, usually the equivalent of two or three lease payments
- Whether the lease is *open ended* (no residual value is specified) or *close ended* (the residual value at the end of the lease period is fixed)
- Who is responsible, lessor or lessee, for additional costs such as property taxes and licenses, insurance, maintenance, etc. (normally these are the responsibility of the lessee)
- Prepayment penalties to be assessed if the lease is terminated prematurely by the lessee
- Excessive use penalties, that is, penalties to be paid by the lessee if usage is greater than some agreed-upon level

Appropriate economic analysis can assist the decision maker in choosing among alternative leasing options or between leasing and other forms of financing (e.g., 100% equity, partial equity, 100% debt). The following example illustrates how this analysis might be effected.

EXAMPLE 7.22 *An Equipment Lease*

A small engineering design firm is exploring the possibility of leasing certain computer equipment for its engineering staff. The equipment may be purchased outright for $20,000. Management believes that this equipment can be used for only two years.

As an alternative to purchase, the identical equipment can be leased over a 24-month period with monthly payments of $1,000 at the beginning of each month. A $2,500 deposit is required, to be returned at the end of 24 months. This is a closed-end lease, and the residual value is $6,000 at the end of 24 months. However, the firm does not expect to exercise the option to buy at that time. The firm, the lessee, will be responsible for all property taxes, insurance, equipment maintenance, and other operating costs.

The firm's after-tax MARR is 12.7% (per year). The marginal income tax rate is 30%. Our problem is to determine the after-tax present worth of the cost of the lease. The result will be compared to the after-tax cost of the alternative, outright purchase.

Our first task is to determine the pretax cash flows, A_j, for each of the 24 months.

$$A_0 = (\text{deposit}) + (\text{first month lease payment})$$

$$= -\$2,500 - \$1,000 = -\$3,500$$

$$A_j = -\$1,000 \qquad \text{for } j = 1, 2, \ldots, 23$$

$$A_{24} = (\text{return of deposit}) = \$2,500$$

The after-tax cash flows are determined as in **Table 7.8**. Note that only the lease payments affect the taxable income in this case.

TABLE 7.8   Determining After-Tax Cash Flows for Equipment Lease
(Example 7.22)

End of Month j	Cash Flow Before Taxes A_j	Lease Payments L_j	Effect on Taxable Income TI_j	Effect on Income Taxes T_j	Cash Flow After Taxes $\hat{A}_j$
0	−$ 3,500	−$ 1,000	−$ 1,000	−$ 300	−$ 3,200
1–23	− 1,000	− 1,000	− 1,000	− 300	− 700
24	2,500	0	0	0	2,500
Total	−$24,000	−$24,000	−$24,000	−$7,200	−$16,800

Our next step is to determine the appropriate discount rate per month.

$$\hat{i}_m = (1 + \hat{i}_m)^{1/12} - 1$$

$$= (1.127)^{1/12} - 1 = 0.010 \text{ per month}$$

The after-tax present worth of the cost of this lease is given by

$$\text{PWOC } (1.0\%) = \$3,200 + \$700(P/A, 1\%, 23)$$

$$- \$2,500(P/F, 1\%, 24)$$

$$= \$3,200 + \$14,319 - \$1,969$$

$$= \$15,550$$

This solution assumes that cash flows for taxes occur at the end of each month.

If we should assume that the cash flows for taxes occur at the end of each year, i.e., at the end of months 12 and 24, then the lease payments within each year must be aggregated to determine the annual taxable income.

$$T_{12} = T_{24} = 0.3(12)(-\$1,000) = -\$3,600$$

The after-tax present worth of costs is now

$$\text{PWOC}(1\%) = (\text{PWOC of pretax cash flows})$$
$$- (\text{PWOC of cash flows for taxes})$$

$$- [\$3,500 + \$1,000(P/A, 1\%, 23) \quad \$2,500(P/F, 1\%, 24)]$$

$$- [\$3,600(P/F, 1\%, 12) + \$3,600(P/F, 1\%, 24)]$$

$$= (\$3,500 + \$20,456 - \$1,969) - (\$3,195 + \$2,835)$$

$$= -\$21,987 - \$6,030 = \$15,957$$

This latter result is more costly ($15,957 − $15,550 = $407) because the beneficial effects of the tax savings are postponed somewhat—to the end of each year rather than the end of each month.

7.6.3 After-Tax Analysis: Equity Viewpoint

Throughout the preceding discussion we have assumed that the minimum attractive rate of return, either before or after income taxes, reflects any and all opportunities foregone if the proposed investment is undertaken. The firm's capital can be used "elsewhere," and therefore the cost of capital is an appropriate concern. Since capital consists of the owners' own funds (**equity**) as well as borrowed funds (**debt**), we have assumed that the MARR reflects the combined costs of equity and debt.

A somewhat different viewpoint is proposed by some analysts—evaluation based solely on the equity funds. (In a corporate form of business enterprise, the stockholders provide the equity.) With this perspective, the MARR is defined by the cost of equity, that is, the cost of capital provided by the investors. Similarly, all cash flows should be measured from the investor's point of view. This issue will arise when a particular investment consists of a mix of debt and equity capital.

The following example illustrates the appropriate analysis for return on equity.

EXAMPLE 7.23 *Present Worth of Equity Investment*

A construction company is considering the purchase of certain equipment at a cost of $100,000. If it is acquired, the equipment will be kept in service for 5 years and then sold for an expected $20,000. It is believed that the equipment will result in pretax savings in operating costs of $24,000 each year over the 5-year period.

There is no applicable Section 179 expense, and the cost basis will be $100,000. The equipment will be depreciated by the regular MACRS method over 3 years; the half-year convention will be used. The firm's marginal income tax rate is expected to remain at 40% over the next 5 years.

The firm will borrow $25,000 to partially finance this investment. The loan is to be repaid by four equal end-of-year payments based on a pretax rate of 10% per year. Each annual payment, then, is

$$\$25,000(A/P, 10\%, 4) = \$7,887 \quad \text{(to the nearest dollar)}$$

The firm's owners want to achieve at least a 15% annual rate of return, after income taxes. The present worth method will be used to determine if this investment appears justified.

The complete analysis is summarized in **Table 7.9**. Note that the pretax cash flows, A_j, are as follows:

$$A_0 = \text{(funds borrowed)} - \text{(investment)}$$

$$= \$25,000 - \$100,000 = -\$75,000$$

$$A_j = \text{(operating cost savings)} - \text{(debt service)}$$

$$= \$24,000 - \$7,887 = \$16,113, \quad j = 1, 2, 3, 4$$

There are two effects in year 5: the $24,000 savings in operating costs and the $20,000 salvage value when the equipment is sold.

Next we must determine the interest costs each year, I_j.

$$I_j = \text{(interest rate, 10\%)} \times \text{(unpaid principal at start of year)}$$

For regular MACRS with recovery over 3 years,

$$p_1 = 0.33333 \qquad p_3 = 0.14815$$

$$p_2 = 0.44445 \qquad p_4 = 0.07407$$

The allowable depreciation expenses in year j are given by

$$D_j = p_j B = p_j \times \$100,000$$

The *entire* investment can be depreciated, not just the portion funded by equity capital.

The effect of the proposed investment on taxable income each year is given by

$$TI_j = \text{(cost savings)} - \text{(interest)} - \text{(depreciation)}$$

$$= \$24,000 - I_j - D_j, \quad j = 1, 2, \dots, 5$$

Note that the book value will be zero when the equipment will be sold, and thus a $20,000 gain on disposal will result at the end of year 5:

$$G = S_a - B_5 = \$20,000 - \$0 = \$20,000$$

TABLE 7.9   Determining After-Tax Cash Flows for Equity Investment—Example 7.23

Year j	Cash Flow Before Tax A_j	Unpaid Principal at Start of Year	Portion of Payment that Is Principal E_j	Portion of Payment that Is Interest I_j	Depreciation D_j	Taxable Income TI_j	Income Tax T_j	Cash Flow After Tax $\hat{A}_j$	PW @ 15% P_j
0	−$75,000							−$75,000	−$75,000
1	16,113	$25,000	$ 5,387	$2,500	$ 33,333	−$11,833	−$ 4,733	20,846	18,127
2	16,113	19,613	5,925	1,961	44,445	−22,407	−8,963	25,076	18,961
3	16,113	13,688	6,518	1,369	14,815	7,816	3,126	12,987	8,539
4	16,113	7,170	7,170	717	7,407	15,876	6,350	9,763	5,582
5	24,000	0	0	0	0	24,000	9,600	14,400	7,159
5	20,000				(0)	20,000	8,000	12,000	5,966
Total	$33,452		$25,000	$6,547	$100,000	$33,452	$13,380	$20,072	−$10,666

Income taxes are equal to the tax rate times the taxable income:

$$T_j = t(TI_j) = 0.40 TI_j$$

and the total after-tax cash flows are given by

$$\hat{A}_j = A_j - T_j$$

The present values of these after-tax cash flows, discounted at the MARR on equity, 15%, are given in the last column of Table 7.9. The total present worth ($-\$10,666$) is negative, indicating that the proposed project fails to provide a 15% return on the equity investment.

SUMMARY

Income taxes are paid by individuals and corporations, sometimes at various levels of government (e.g., city, state, and federal). These taxes, representing negative cash flows to the taxpayer, are based on (1) the amount of taxable income and (2) the appropriate tax rate at that level. The tax rate to be used in economy studies is that rate or rates relevant to the incremental effect of the project on the taxpayer's taxable income.

The disposal of a depreciable asset could result in a taxable gain or loss at the time of the disposal. (A loss on disposal is a reduction from taxable income, and hence there is a savings in income taxes.) In general, a gain is taxed at the same rate as ordinary income.

The timing of cash flows for taxes presents certain complexities in the analysis. A convenient approximation is to assume that the cash flows for taxes occur at the end of each project year.

A *general procedure* is presented in Section 7.5.1 for evaluating investments on an after-tax basis. Here, the present worths of the pretax cash flows and the cash flows for income taxes are evaluated separately, then combined. Alternatively, a *simplified procedure* (Section 7.5.2) is presented for use in those cases in which complexities of timing of tax payments may be relatively insignificant. This latter approach is followed throughout the remainder of the chapter.

Numerical examples are provided to illustrate the appropriate economic analysis under a variety of situations:

- *Classical depreciation methods*
  - Straight line, $N_d = N_a$ (Example 7.8)
  - Straight line, $N_d < N_a$ (Example 7.9)
  - Sum of the years-digits (Example 7.10)
  - Double declining balance (Example 7.11)
  - Declining balance, shift to straight line (Example 7.12)

- *Accelerated Cost Recovery System*
  - ○ Regular ACRS (Example 7.13)
  - ○ Alternate ACRS (Example 7.14)
- *Modified Accelerated Cost Recovery System*
  - ○ Regular MACRS, 5-year property (Example 7.15)
  - ○ A/MACRS, straight line over A/MACRS recovery period (Example 7.16)
  - ○ MACRS, straight line over MACRS recovery period (Example 7.17)
  - ○ MACRS, 150% declining balance over A/MACRS recovery period (Example 7.18)

After-tax analyses were also discussed in the context of three additional topics: (1) loans, (2) leases, and (3) returns on equity investment.

PROBLEMS

Unless specifically directed otherwise, assume that:

1. Depreciable assets are purchased and placed in service at the start of the first tax year.
2. Depreciable assets are removed from service and sold at the end of the final tax year.
3. The tax year is a 12-month period coincident with the calendar year (January 1–December 31).
4. Income taxes are paid at the end of the tax year.
5. The Section 179 expense deduction is not used.
6. Regular MACRS (not the alternate MACRS method) is used for depreciating recovery property, that is, DB with switch to SL for 3, 5, 7, 10, 15, and 20-year property classes, MACRS recovery period, and half-year convention.
7. There are no allowable tax credits.
8. Federal income tax rates are as specified in the text.

Income Tax Rates and Cash Flows for Taxes

7.1 A corporation is considering the purchase of a certain asset that, if acquired, will increase the firm's taxable income during the first year of ownership from \$40,000 to \$70,000.
   a. Determine the increase in federal income taxes attributable to this proposed investment during this tax year.
   b. If the asset is acquired, determine the average income tax rate paid by the corporation during the year.
   c. Suppose that the corporation's incremental federal income tax rate is 0.30 and the incremental state income tax rate is 0.10. What is the combined incremental income tax rate?

[*Answers:* a. \$6,500; b. 0.1786; c. 0.37.]

7.2 The Fortune Corporation is considering the acquisition of a certain asset. The company pays corporate income taxes both to the state in which it is located and to the federal government. Without this asset, the company's taxable income as reported to the state tax collector would be \$40,000 next year; with the asset, taxable income reported to the state would be increased to \$60,000. Assume that the state's corporate income tax rate is 11% for all taxable income.
   a. If the asset is not acquired, determine the amount of next year's combined state and federal income taxes.
   b. Determine next year's combined state and

federal income taxes if the asset is acquired.

c. What is the total combined incremental income tax rate appropriate to this analysis?

7.3 Consider a small U.S. corporation doing business in a state in which there is no state income tax. Investment in certain manufacturing equipment will have the effect of increasing taxable income from $40,000 (without the equipment) to $55,000 (with the equipment). Determine the additional federal income taxes to be paid due to this $15,000 increase in taxable income.
[*Answer: $2,750.*]

7.4 A certain U.S. corporation's taxable income for a given tax year is expected to be $72,000 on its federal income tax return. The firm is considering a certain investment proposal that will have the effect of increasing taxable income by $18,000 (to $90,000 total). The firm's after-tax MARR is 15%.

a. If the proposal *is not* adopted, determine the corporation's federal income tax liability for the tax year.

b. If the proposal *is* adopted, determine the incremental effect of the investment on the corporation's cash flow for federal income taxes.

7.5 The Belgrade Corporation is considering a plant expansion that will have the effect of increasing taxable income (on state tax returns) from $80,000 to $120,000. The state tax rate is 8% on all taxable income.

a. Determine the federal corporate income taxes paid by Belgrade without and with expansion.

b. Determine the total (state + federal) average tax rate paid on the additional $40,000 in taxable income.
[*Answers: a. $13,400 and $26,306; b. 0.4027.*]

7.6 A certain proposed investment will have the effect of increasing a corporation's taxable income on its state income tax return by $10,000. The corporation's taxable income on its federal tax return is $46,000 without the proposed investment. The corporation's state income tax is 8% on all taxable income.

a. Determine the federal corporate income taxes without and with the new investment.

b. Determine the total (state + federal) average tax rate paid if the proposed investment is adopted.

7.7 Consider a firm expecting to earn taxable income for the year of $60,000. The firm is considering an investment that, if made, would increase taxable income to $80,000. Determine the average incremental tax rate on the additional $20,000 in taxable income.
[*Answer: 27.25%.*]

7.8 Consider a small U.S. corporation doing business in a state in which there is no state income tax. Investment in certain manufacturing equipment will have the effect of increasing taxable income from $70,000 (without the equipment) to $85,000 (with the equipment). Determine the additional federal income taxes to be paid due to this $15,000 increase in taxable income.

7.9 A U.S. corporation currently has taxable income of $80,000.

a. What is its marginal tax rate?

b. What is the federal income tax liability?

c. What is the average income tax rate?

d. If the corporation increases its taxable income to $100,000, what will be the new average income tax rate?
[*Answers: a. 0.34; b. $15,450; c. 0.1931; d. 0.2225.*]

7.10 The Little Corporation is a supplier of special purpose jigs and fixtures to a large local manufacturer. It is considering the purchase of computer equipment that, if acquired, will be the only recovery property acquired during this current tax year. Without this new equipment, the Little Corporation's taxable income this year will be $160,000. If it is acquired, this new equipment will have the effect of reducing operating costs by $13,000 this year. The computer equipment will cost $20,000, be kept in service 6 years, and then be sold for an estimated $2,000. It will be depreciated as a 5-year recovery property.

a. What is the allowable Section 179 expense?

b. If the equipment *is not* acquired, determine the federal income tax liability this tax year.

c. If the equipment *is* acquired, determine the federal income tax liability this tax year.

7.11 A California-based corporation is considering a certain investment that will have the effect of increasing the firm's taxable income on the state income tax return by $100,000 in the tax year.

The marginal state tax rate is 9.3%; the marginal federal tax rate for this corporation is 34.0%. Assume that depreciation expense and other deductions are handled identically on both the state and federal tax returns. Remember that state income taxes paid are deductible expenses on federal income tax returns.

   a. Determine the state income taxes paid in the tax year.

   b. Determine the federal income taxes paid in the tax year.

   c. Determine the combined (state and federal) income tax rate.

   d. Revise the above problem slightly. Suppose that the Section 179 expense in the tax year is $10,000, and this is fully deductible on the federal (but not the state) return. In this case, determine the federal income tax paid in the tax year and the combined (state and federal) income tax rate.

[*Answers:* a. $9,300; b. $30,838; c. 40.138%; d. $27,438 and 36.738%.]

7.12 Consider a 5-year property depreciated under MACRS. The initial cost is $10,000, it has an expected 6-year physical life, and the firm expects to sell it after 39 months for $3,000. Assume that the asset is purchased at the midpoint of its tax year, that is, on the first day of the seventh month of the tax year. Also assume that cash flows for taxes occur at the end of each tax year. The firm's after-tax MARR is 20%.

   a. Determine the allowable depreciation expense for the first, second, third, and fourth tax years.

   b. Determine the gain or loss on disposal at the time of disposal.

   c. If the taxpayer's marginal tax rate is 40%, determine the present value of tax savings due to depreciation. *Hint:* Of relevance here are the annual depreciation expenses as well as the gain or loss on disposal. Note that "the present" is the midpoint of the first tax year.

7.13 In an attempt to lower communications costs, the Fletcher Company is considering the acquisition of certain automatic switching equipment for its telephone system. The initial cost of the equipment is $80,000. If it is purchased, it will be kept in service for an expected 6 years, then sold for an estimated $25,000.

   The equipment will be depreciated using the DDB method when determining taxable income on Fletcher's state income tax return. Depreciable life will be 4 years; salvage value for depreciation will be 10% of initial cost, or $8,000.

   If it is acquired, the new equipment will result in savings of $25,000 per year. The firm's state tax rate is 10%, its pretax MARR is 25%, and its after-tax MARR is 15%. Assume that the purchase of the equipment, if it is acquired, will take place at the midpoint of the tax year. Also assume that savings will occur continuously and uniformly throughout the year. Thus savings will be $12,500 the first year, $25,000 the second year, and so forth. The equipment, if it is acquired, is assumed to be removed from service and sold at the midpoint in the seventh tax year.

   If it is acquired, the equipment will be treated as a 5-year MACRS property on the federal income tax return, using the regular MACRS method. Fletcher's incremental federal tax rate is 34%. Complete the cash flow table showing the effect of the proposed investment on cash flows for federal taxes.

   a. Determine the present worth of cash flows for state income taxes.

   b. Determine the present worth of cash flows for federal income taxes. *Hint:* "Cash flow before taxes" must reflect the state income taxes computed in part (a).

[*Answers:* a. $4,333; b. $10,642.]

7.14 The Really Big Manufacturing Company (RBM) is considering the purchase of certain manufacturing equipment that will cost $45,000, will be placed in service in the third month of the tax year, kept in service for 3 years, and then sold in the third month of the fourth tax year for an estimated $15,000. The equipment will be depreciated as 7-year recovery property under MACRS. RBM's marginal income tax rate is 40%, and its after-tax MARR is 10%.

   a. Determine the present worth of the cash flows for income taxes due to depreciation. Include the effect of the gain or loss on disposal. Assume that all cash flows for taxes occur at the ends of the tax year, and that the "present" is the start of the first tax year.

   b. Modify your answer to part (a) by defining the "present" as the beginning of the planning horizon, that is, the start of the third month of the tax year. Stated somewhat differently, what is the equivalent worth of these tax savings in part (a) just 2 months later?

7.15 The firm plans to acquire a truck on or about September 15. (The calendar year is the same as the tax year.) The truck will be depreciated by the alternate MACRS method, A/MACRS, with a 5-year recovery period. The original cost is $17,300, and the entire Section 179 expense will be applicable for this investment. The truck, if it is acquired, will be sold about three years later, that is, toward the end of the fourth tax year, for an expected $2,500.

Assuming that the firm's marginal income tax rate will be 25% throughout the period of ownership, determine the savings in cash flows for income taxes for each of the four tax years.

[*Answer:* **$2,683, $365, $365, $288.**]

After-Tax Economy Studies

7.16 Missile and Space Electronics, Inc., has just received a 4-year contract to manufacture a small, high-frequency pulse generator for the Air Force. Their manufacturing expense estimates at the time MSE submitted its bid for the contract are shown below as Alternative I. Since submitting the bid, the company's engineering division has developed what is possibly a better proposal that uses extensive mechanization, shown below as Alternative II. What alternative is preferable? The after-tax minimum attractive rate of return is 15%.

For this project, assume negligible salvage values. Capitalize the machines and tooling as one group, using regular MACRS depreciation over the 3-year project life. Assume that all other items may be expensed in the year in which they are purchased. The incremental income tax rate is 40%. Other appropriate input data are:

	Alternative I	Alternative II
Initial product development	$ 40,000	$ 50,000
Preproduction engineering	90,000	122,000
Machine tools and tooling	12,000	40,000
Annual labor cost	35,000	15,000
Annual operating expenses for this contract (other than labor)	20,000	20,000
Annual sales revenue from this contract	115,000	115,000

7.17 The General Corporation is considering the purchase of an industrial robot for a plant operation that is currently performed by human labor. The firm's minimum attractive rate of return is 20% before income taxes and 12% after income taxes. The firm's effective incremental income tax rate is 60%.

The initial cost of the robot is $150,000. If it is purchased, the robot will be depreciated under regular MACRS as a 5-year recovery property. However, the operation for which this robot will be used will last for 7 years, at the end of which time the firm expects to sell the robot for $25,000. In other words, even though the robot will be depreciated over 5 years, it will actually be used for 7 years.

Property taxes for the robot, computed at 2% of the initial cost, are paid at the end of each year. Operating costs, assumed to occur uniformly and continuously throughout each year, are estimated to be $5,000 annually.

a. Prepare a table of after-tax cash flows for the 7-year study period.

b. Determine the present worth before income taxes.

c. Determine the equivalent uniform annual cost after income taxes.

d. Let $\bar{x}$ = current cost before income taxes of this operation, expressed as a cash flow occurring continuously and uniformly during each year. (This would be an appropriate assumption in the case of human labor paid, say, bi-weekly.) At what pretax value of $\bar{x}$ would the proposed robot be economically attractive?

[*Answers:* **b.** −**$180,584; c. $23,000; d. $45,670.**]

7.18 Tammi Trojan, a recent engineering school graduate, is considering the purchase of a minicomputer for her wholly owned consulting firm, TTA. Tammi bought all her office furniture and other necessary equipment previously, thus the minicomputer will be her only business purchase during the tax year. TTA is not incorporated; Tammi is not married. Her combined incremental income tax rate is 0.40. If it is purchased, the minicomputer will be depreciated under MACRS as 5-year recovery property. The initial cost (cost basis) of the minicomputer is $5,000. If Tammi purchases the minicomputer, she plans to keep it for 4 years and then sell it for an estimated $2,000. Determine the after-tax equivalent uniform annual cost for this prospective purchase if her after-tax MARR is 8%.

7.19 A manufacturing firm, with a facility in a Latin American country, is considering the purchase of a certain machine that costs $80,000. It will be depreciated over a 5-year life to an estimated $20,000 salvage value using the SYD method. If the machine is purchased, the firm will keep it for 6 years, however, and will probably be able to receive $30,000 for it if it is sold at the end of 6 years. That is, although the machine will be depreciated over 5 years, it will be kept for 6 years.

Annual maintenance costs are expected to be $5,000 the first year and will increase by $1,000 per year over the machine's 6-year life. Annual operating costs are expected to remain constant at $20,000 per year over the 6-year life. Assume that both of these classes of expenses occur at end-of-year.

Property taxes and insurance for this machine are estimated to be $1,000 in the first year and will decrease by $100 per year. These costs are expected to occur at the start of years 1 through 6. The firm's incremental income tax rate is 60%.

If the firm's MARR is 10% after income taxes, find the equivalent uniform annual cost for this proposed investment.

[*Answer:* **$24,250.**]

7.20 The Brown Company is considering the purchase of some special equipment that costs $60,000. If the equipment is purchased, it will cause an expected excess of receipts over disbursements of $16,000 a year for 5 years; at the end of that time the equipment will be worthless. It is expected that the equipment will be purchased with a down payment of $10,000, and the remainder will be paid off at the rate of $10,000 per year plus 6% on the unpaid balance. Determine the after-tax rate of return on the original investment ($10,000) assuming a 50% effective tax rate and sum of the years-digits depreciation based on a 5-year life and zero salvage value.

7.21 Bigditch Construction Company is considering the acquisition of a mobile van to serve as a field office during a 3-year period of construction. The relevant data are as follows:

- The firm's after-tax MARR = 20%.
- Combined incremental corporate income tax rate = 40%.
- Initial cost = $50,000.
- Service life = 3 years (sold when removed from service).
- Salvage value after 3 years = $20,000.
- Property taxes (paid end of year) = $1,500 each year.
- Insurance (paid start of year) = $500 each year.

The van, if it is purchased, will be depreciated under regular MACRS as 3-year recovery property. Assume that the van will be purchased at the start of the first tax year and removed from service (sold) at the start of the fourth tax year.

Find the equivalent present value, after tax, of the costs of the proposed investment.

[*Answer:* **$30,377.**]

7.22 An industrial engineer is developing a facilities layout for a chemical processing plant in a European country. Under his "labor-intensive" plan, 10 people will be required to operate the plant. However, he can reduce the number of operators to only four under a "capital-intensive" plan requiring the expenditure of $800,000 for additional instrumentation and controls. The plant is expected to be in operation for 5 years, at the end of which time the additional instrumentation and controls would add $300,000 to the selling price.

The plant manager feels that the additional expenditure is probably warranted. Since the cost per worker-year is $25,000, a savings of 30 worker-years (6 people per year × 5 years) results in a savings of $750,000, which more than offsets the $500,000 net cost of additional instrumentation and controls ($800,000 − $300,000). In any case, she asks the industrial engineer to complete an after-tax economy study.

If it is purchased, the new instrumentation will be depreciated by the 200% declining balance (DDB) method over the first 4 years, with a switch to straight line depreciation over the next 6 years. The salvage value for depreciation purposes is $100,000 at the end of 10 years. The firm's incremental income tax rate is 0.60.

Determine the after-tax rate of return on the proposed investment.

7.23 Consider a capital asset that, under the state income tax regulations, will be depreciated by the 125% declining balance method. Shift to the straight line method is permitted at any time. The initial cost of the asset is $240,000. For depreciation purposes, the depreciable life is 5 years and the expected salvage value is 10%, or $24,000.

The asset will be placed in service at the start of the first tax year and then sold after 3 years for an anticipated $100,000.

If it is acquired, this asset will result in pretax savings of $50,000 each year that it is in service. (Assume that these savings are end-of-year cash flows.) The firm's marginal state income tax rate is 10%; its after-tax MARR is 15%.

Determine the after-tax PW of the cost of this proposed investment.

[*Answer:* $126,500.]

7.24 Consider a capital asset that has an initial cost of $10,000, a service life of 3.5 years, and expected salvage value of $3,000. The asset is acquired and placed in service at the midpoint of the tax year. The taxpayer's marginal income tax rate is 40%.
a. Suppose that the asset is to be depreciated by the 150% declining balance method over a 5-year depreciable life. The expected salvage value for depreciation purposes is $4,000. Determine the after-tax residual value of the asset when it is sold at the end of the fourth tax year.
b. Suppose that the asset is to be depreciated as 5-year recovery property using the regular MACRS method. Determine the after-tax residual value of the asset when it is sold at the end of the fourth tax year.

7.25 A firm plans to purchase a mechanical device for $9,000. The device is expected to remain in service for 3 years and will reduce labor costs by $5,000 annually. Its expected salvage value at the end of 3 years is $3,000. For tax purposes, the device is classified as 3-year recovery property.

The firm will depreciate the device using the alternate MACRS method with a 3-year recovery period. The firm's marginal income tax rate is 15%. Assume that the device will be placed in service at the middle of the first tax year and sold at the middle of the fourth tax year. Also assume that all cash flows for taxes occur in the *middle* of the tax year.
a. Determine the before-tax rate of return.
b. Determine the after-tax rate of return.

[*Answers:* **a. 40.3%; b. 36.1%.**]

7.26 In November, Ann Fletcher, a manufacturing engineering consultant, is reviewing her tax liability for the year. (Her tax year is the calendar year.) If she purchases no new capital equipment, she anticipates a taxable income of $40,000 for the year. Ms. Fletcher is considering the purchase of a microcomputer and word processing system at a cost of $10,500, to be acquired and placed in service immediately.
a. If Ms. Fletcher has no other Section 179 expenses to claim for the tax year, determine the Section 179 expense deduction if the computer system is purchased.
b. The computer system is to be depreciated for income tax purposes as a 3-year recovery property using the regular MACRS method. If it is acquired, the system will be kept in service for 5 years, then sold for an expected $1,000. Ms. Fletcher's marginal income tax rate is 25%; her after-tax MARR is 10%. Determine the after-tax present worth of the cost of this system.

7.27 The General Manufacturing Company is considering the purchase of certain machine tools to be used in connection with its research and development effort. These tools, costing $30,000, will be depreciated as 3-year MACRS property. It is expected that the tools will be used for 4 years, then sold for an expected $6,000. The GMC effective incremental tax rate is 40%. The tools will be placed in service at the midpoint of the first tax year. If GMC's after-tax MARR is 15%, find the after-tax present worth of costs.

[*Answer:* **$18,000.**]

7.28 Easy Electronics has an overseas manufacturing plant located in the African nation of Gumby. The on-site manufacturing engineer at the Gumby plant has recommended the acquisition and installation of certain machinery that has an initial cost of $350,000, a useful life of 6 years, and salvage value of $50,000 at the end of 6 years. If it is acquired, additional operating costs are estimated to be $20,000 per year, but savings in direct labor are estimated to be $90,000 annually. The net savings ($70,000) are assumed to flow continuously and uniformly throughout the year. The firm's pretax MARR for investments of this type is 25%.

a. Determine the pretax PW of costs of this proposed investment.

b. Local tax laws in Gumby require that this machinery be depreciated by the straight line method over 5 years to a 10% salvage value. Any gains/losses on disposal are treated as ordinary income/losses. The relevant local tax rate is 20%. The firm's after-tax MARR for investments of this type is 15%. Determine the after-tax PW of costs of this proposed investment.

7.29 Consider a depreciable asset purchased for $100,000 at the start of the first tax year. It is expected that this property will be kept in service for almost 4 years, then sold for an estimated $10,000 at the end of the fourth tax year. The firm's after-tax MARR is 12%.

a. The property will be depreciated by the SYD method for state income tax purposes. Assume that $N_d = 4$ years and $S_d = \$10,000$. If the firm's state income tax rate is 10%, determine the present worth of cash flows for state income taxes due to depreciation. Assume that cash flows for taxes occur at the end of each tax year.

b. For federal income tax purposes, the asset will be depreciated as a 3-year recovery property under regular MACRS. If the firm's federal income tax rate is 34%, determine the present worth of cash flows for federal income taxes due to depreciation and any gain or loss on disposal. Assume that cash flows for taxes occur at the end of each tax year.

c. Find the total present worth of costs, after state and federal income taxes, of this capital investment.

d. Suppose that this capital investment has the effect of reducing operating costs in the firm by $2,000 per month over 48 months. Find the present worth of these savings. *Hint:*

These savings will increase taxable income by $24,000 per year.

[*Answers:* a. $7,200; b. $25,200; c. $70,000; d. $47,200.]

7.30 The Micro Company, a small industrial laboratory, is considering the purchase of an electron microscope at a cost of $20,000. This will be the firm's only acquisition during the tax year, so the Section 179 expense will be fully applicable. Although the microscope will be depreciated as 5-year recovery property, the microscope will be used for only 3 years, at the end of which time it will be sold for an expected $4,000. The microscope will be acquired and placed in service of the midpoint of the tax year. The regular MACRS method will be used for determining annual depreciation. The firm's incremental income tax rate is 40%.

If the firm's after-tax MARR is 15%, determine the after-tax present worth of the cost of this proposed investment.

7.31 A proposal to acquire an auto loader is being evaluated. The initial cost is $360,000 and, if it is purchased, the auto loader will be kept in service for 3 years. It will have zero salvage value at the end of 3 years.

The auto loader will be depreciated under regular MACRS as a 5-year recovery property. The firm's marginal tax rate is 0.34. Its after-tax MARR is 0.15.

If it is acquired, the auto loader will have the following effects on pretax cash flows:

	Year 1	Year 2	Year 3
Direct labor savings	$300,000	$300,000	$300,000
Engineering (expense)	−20,000	0	0
Maintenance (expense)	−40,000	−40,000	−40,000
Spare parts (expense)	−15,000	−15,000	−15,000
Total	$225,000	$245,000	$245,000

Assume here that the auto loader will actually be sold at the start of the fourth year, thereby allowing a full year of depreciation to be claimed in the third year. Also assume that the resulting tax effect will be at the start of the fourth year.

a. Assuming the after-tax MARR is 15%, what is the after-tax present worth of this proposal?

b. In the original problem statement, it is assumed that the salvage value at the end of 3 years is zero. This is questionable, of course. So let us assume that the auto loader will in fact be sold for its book value, resulting in no gain or loss on disposal. Determine the after-tax PW of the proposal based on this new assumption.

[*Answers:* a. $87,300; b. $132,200.]

7.32 A small engineering design firm is considering the purchase of computer-aided design (CAD) equipment to be placed in service at the mid-point of the tax year. The initial cost is $20,000. The equipment will be considered as 5-year MACRS recovery property.

a. The firm expects to acquire other depreciable assets during the tax year, besides the CAD equipment, with a total cost of $4,000. Determine the Section 179 expense deduction due to the acquisition of the new CAD equipment.

b. Assuming that the regular MACRS method will be used, and assuming that the equipment will be kept in service for 4 years, then sold for an estimated $6,000, determine the gain (loss) on disposal at the time of sale.

c. If it is acquired, the CAD equipment will have the effect of saving $8,000 each year during the 4 years that it will be in use. These savings may be assumed to be distributed continuously and uniformly during each year. Determine the before-tax rate of return (i^*) of this proposed investment.

d. If the firm's marginal income tax rate is 40%, determine the after-tax rate of return ($\hat{i}^*$) of this proposed investment.

7.33 A trucking company is considering the acquisition of a tractor unit for use in over-the-road, intercity hauling. If it is acquired, it will be purchased for $80,000, kept in service for 5 years, then removed from service and sold for an expected $20,000. The tractor will be purchased and placed in service at the beginning of the tax year. Cash flows for income taxes are assumed to occur at the end of the tax year. Assume that the tax consequence of the sale of the tractor will occur at the time of the sale, that is, at the start of the sixth tax year.

The firm will depreciate the tractor using the SYD method for purposes of state income taxes. The depreciable life will be 3 years with a $26,000 salvage value at the end of 3 years. The firm's marginal state income tax rate is 10%.

For purposes of federal income taxes, the tractor will be depreciated under regular MACRS as 3-year recovery property. The marginal federal income tax rate is 34%.

The pretax costs of operating the tractor are expected to be $100,000 per year. These costs are expected to flow continuously and uniformly throughout each of the 5 years during which the tractor will be operated.

The firm's after-tax MARR is 15%.

a. Determine the present worth of tax savings due to depreciation on the state and federal income tax returns. Include the effect, if any, of the gain or loss on disposal.

b. Determine the present worth of tax savings due to operating costs on the state and federal income tax returns.

c. Determine the total after-tax present worth of costs of the tractor unit.

[*Answers:* a. $20,481; b. $136,097; c. $3,249.]

7.34 Southwest Manufacturing, Inc. (SMI), located in central Texas, is a major aerospace industry subcontractor. The firm's marginal income tax rate is 40%.

The SMI plant engineering department is considering the purchase of certain materials handling equipment that, if acquired, will result in labor savings of $10,000 per month, assumed to flow at end-of-month. Preventive maintenance for this equipment will take place at the end of each year except the final year, and will cost $10,000 per occurrence. The initial cost of the equipment is $400,000. The equipment will be depreciated as 5-year MACRS property. The equipment will be kept in service for 6 years, then sold for an estimated $20,000. SMI's after-tax MARR is 15%.

a. Determine the present worth of the monthly savings.

b. Determine the present worth of the annual maintenance expenses.

c. Determine the present worth of tax consequences.

d. Determine the after-tax PW of this proposed investment.

7.35 Certain manufacturing equipment has an initial cost of $30,000. It will be placed in service at the start of the first tax year, kept in service for 3 years, and then sold at the end of the third tax year for net zero salvage value. Pretax "income" (before depreciation), assumed to be all cash flows, are $20,000 each year for 3 years. Summarizing,

Year	Cash Flow Before Taxes	Taxable Depreciation	Taxable Income
0	− $30,000	—	—
1	20,000	$15,000	$ 5,000
2	20,000	10,000	10,000
3	20,000	5,000	15,000

(The depreciation method used in this case is the SYD method.)

a. Defining each year's return on investment (ROI) as the ratio of annual profit (after taxes) to that year's average book value, and assuming a tax rate of 40%, determine the ROI each year.

b. What is the average ROI?

c. Determine the true after-tax internal rate of return of this investment.

[*Answers:* **a. 13.3%, 60%, 360%; b. 144%; c. 29.2%.**]

Loans and Leases

7.36 GAF Contractors is a medium-sized general construction company. The firm is considering the acquisition of certain earth-moving equipment to be used in various projects over the next 4 years. The initial cost of the equipment is $90,000; expected salvage value at the end of 4 years is $15,000. The firm's pretax MARR is 20%. Insurance for the equipment, payable at the start of each year, is $1,000 per year.

a. The equipment will be depreciated on the state income tax return by the SYD method over 5 years. The firm's effective tax rate for state income taxes is 10%. Assuming that the equipment will be purchased and placed in service at the start of the first tax year and sold at the end of the fourth tax year, determine the cash flows for state income taxes for each of the 4 tax years.

b. The equipment will be depreciated on the federal income tax return as 3-year recovery

property using MACRS. GAF Contractors is in the 34% tax bracket.

c. If the firm's after-tax MARR is 12%, determine the total after-tax present worth of costs. Include the effects of both state and federal income taxes.

7.37 GAF Contractors has an opportunity to lease the earth-moving equipment described in Problem 7.36. The leasing company will provide insurance at no additional cost. The lease payments will be $4,000 per month, payable at the start of each month for 48 months. Determine the after-tax PWOC of this lease. *Hint:* Lease payments are deductible expenses for both state and federal income taxes. As above, assume that the cash flows for taxes occur at the end of each taxable year.

[*Answer:* **$95,920.**]

7.38 Boyum Products, Inc. (BPI), can purchase certain manufacturing equipment for $10,000. The equipment will be needed for 5 years. Alternatively, BPI can lease the equipment from an equipment rental agency. Lease payments are made quarterly, at the beginning of each quarter, in the amount of $700 each. That is, rather than spend $10,000 for the equipment, the firm can lease the asset by making 20 lease payments of $700 each at the beginning of each quarter. In a sense, then, the firm can "free up" $10,000 by leasing rather than purchasing the equipment. If it is owned outright, the equipment can be depreciated under regular MACRS; if it is leased, the lease payments are deductible from taxable income. Assume that the firm pays both federal and state income taxes such that $t_f = 0.34$ and $t_s = 0.10$. The firm's after-tax MARR is 12%.

a. Determine the after-tax cost, expressed as a percentage rate, of capital obtained in this manner.

b. What is the maximum lease payment such that it would be worthwhile to lease, rather than purchase, the equipment?

7.39 The Allstar Products Company is considering the lease of a photocopy machine for use in the sales office. The terms of the lease are as follows:

● Monthly lease payments of $200 each.

● Lease payments to be paid at the beginning of each month.

● Total period of the lease is 36 months.

The firm's pretax MARR is 20% per year; its after-tax MARR is 12%. The marginal income tax rate is 0.40.

   a. Determine the present worth of the cost of this lease before income taxes.

   b. Determine the present worth of the cost of this lease after income taxes.

[*Answers:* **a. $5,590; b. $3,830.**]

7.40 In order to obtain investment funds, the firm can borrow $15,000 to be repaid in 20 equal quarterly payments. The interest rate paid on this debt, before considering the effect of income taxes, is 9%. If the firm's state income tax rate is 10%, and if the federal income tax rate is 34%, determine the after-tax cost of this increment of debt.

7.41 In order to fund a certain investment, the Universal Services Corporation (USC) borrows $20,000 to be repaid in two equal end-of-year payments of $12,000 each. USC's state tax rate is 10%; its taxable income is approximately $500,000 annually.

   a. What is USC's marginal federal income tax rate?

   b. What is the combined (state and federal) income tax rate? (Assume that interest on debt is deductible from taxable income in both state and federal tax returns.)

   c. Determine the pretax cost of this loan.

   d. Determine the after-tax cost of this loan.

[*Answers:* **a. 0.34; b. 0.406; c. 13.07%; d. 7.76%.**]

7.42 In order to fund a plant expansion program, Easy Electronics has an opportunity to borrow $200,000 at an effective rate of 12% per year. Repayment of the loan is to be in three equal end-of-year payments. Easy Electronics has an effective incremental state tax rate of 8% and an effective incremental federal tax rate of 34%. Determine the after-tax effective annual cost of this increment of debt. (Assume that cash flows for taxes are concurrent with the loan repayments.)

7.43 In order to fund the acquisition of certain manufacturing equipment, the firm can borrow $50,000 to be repaid in four semiannual payments (at the end of months 6, 12, 18, and 24). The nominal interest rate on the loan is 14% per year. Assume that the firm's incremental income tax rate is 40%, the loan is made at the middle of the tax year, and cash flows for taxes occur at the end of each tax year. Determine the after-tax cost of this loan. *Hint:* the second and third payments will have tax effects in the second tax year.

[*Answer:* **8.7%.**]

7.44 The Really Big Corporation (RBC) is considering the purchase of a computer workstation for its engineering design department. The initial cost of a Moon system is $90,000. The vendor requires a 20% down payment and will finance the balance at an interest rate of 12% per year for a 3-year period with three equal end-of-year payments. Thus, if RBC wants to buy the Moon system on these terms, it would have to repay $72,000 (= 0.80 × $90,000) in three equal payments at the end of years 1, 2, and 3.

   Assume that RBC's state income tax rate is 10%, its federal income tax rate is 34%, and income taxes paid to the state are deductible from taxable income on federal income tax returns. Also assume that cash flows for taxes are paid at the end of each year. Determine the after-tax cost of the "loan," expressed as a percentage or decimal fraction.

7.45 In order to finance the purchase of an industrial robot, the Needham Corporation is offered a $100,000 loan. The loan is to be repaid in five equal payments of $25,000 each, with payments to be made at the beginning of months 1, 4, 7, 10, and 13. The tax consequences of each loan payment—the interest portion of each payment is deductible—will take place at the end of months 3, 6, 9, 12, and 15, respectively. The firm's tax rate is 40%. Determine the after-tax cost of this loan (stated as a percentage) per year.

[*Answer:* **35.7%.**]

APPENDIX: *Computer Software Applications*

EXAMPLE 7.A.1 *After-Tax Present Worth: General Procedure*

Certain manufacturing equipment has an initial cost of $400,000. The equipment is expected to be kept in service for 6 years, then sold for an estimated $40,000 salvage value. Working capital of $50,000 is required at the start of the 6-year period; it will be recovered intact at the end of 6 years. If it is acquired, this equipment is expected to result in savings of $90,000 each year. The timing of these savings is such that the continuous cash flow assumption is appropriate. The firm's after-tax MARR is 10% per year. There is no Section 179 expense deduction. The equipment will be placed in service at the middle of the tax year and depreciated under MACRS as a 5-year recovery property using the half-year convention. The incremental federal income tax rate is 0.34. Our problem is to determine the after-tax present worth of this proposed investment using the "general procedure." (This is Example 7.5 in the text.)

Figure 7.A.1 on page 288 is a spreadsheet approach to the general procedure. The spreadsheet has been generated using **Excel** software. Note that the spreadsheet is divided into three sections: The top section (rows 5–17) summarizes all the relevant input data; the middle section (rows 19–30) includes the calculations to determine the year-by-year cash flows for income taxes; and the bottom section (rows 32–50) calculates the present worth at 10%, −$18,382. That is, the net after-tax PW of costs is $18,382.

Because of the complexity of this spreadsheet, the relevant cell formulas are displayed in **Figure 7.A.2** on page 289.

	A	B	C	D	E	F	G
1	EXAMPLE 7.A.1						
2							
3	AFTER-TAX PRESENT WORTH ▓ GENERAL PROCEDURE (Example 7.5)						
4							
5	Assumptions:	Portion of 1st tax year property in service (pi) =				0.50	
6		Initial cost of recovery propoerty (P) =				$400,000	
7		Expected life, years (Na) =				6	
8		Expected salvage value (Sa) =				$40,000	
9		Depreciation by regular MACRS method					
10		MACRS revovery period, years (Nd) =				5	
11		Depreciation percentage from Table 6.7					
12		Section 179 expense deduction to be taken =				$0	
13		Working capital to be recovered end of project life =				$50,000	
14		Annual savings, CONTINUOUS cash flows =				$90,000	
15		Marginal income tax rate (t) =				0.34	
16		After-tax MARR =				0.10	
17		Cash flows for taxes occur at end of tax year.					
18							
19		Tax	Deprec'n	Deprec'n	Sec. 179	Other	Taxable	Cash Flow
20		Year	Percent.	Expense	$ Gain	Income	Income	For Taxes
21								
22		1	0.2000	$80,000	$0	$45,000	($35,000)	($11,900)
23		2	0.3200	$128,000		$90,000	($38,000)	($12,920)
24		3	0.1920	$76,800		$90,000	$13,200	$4,488
25		4	0.1152	$46,080		$90,000	$43,920	$14,933
26		5	0.1152	$46,080		$90,000	$43,920	$14,933
27		6	0.0576	$23,040		$90,000	$66,960	$22,766
28		7			$40,000	$45,000	$85,000	$28,900
29								
30	Totals		1.0000	$400,000	$40,000	$540,000	$180,000	$61,200
31								
32	PW of taxes @ 10%, START of 1st tax year =						$29,029
33								
34		End of			Other	Total	Present	
35		Project	Recovery	Working	Pre-tax	Cash Flow	Worth at	
36		Year	Property	Capital	Cash Flows*	Before Tax	0.10	
37								
38		0	($400,000)	($50,000)		($450,000)	($450,000)	
39		1			$94,429	$94,429	$85,844	
40		2			$94,429	$94,429	$78,040	
41		3			$94,429	$94,429	$70,946	
42		4			$94,429	$94,429	$64,496	
43		5			$94,429	$94,429	$58,633	
44		6	$40,000	$50,000	$94,429	$184,429	$104,105	
45								
46	Totals		($360,000)	$0	$566,571	$206,571	$12,064	
47								
48	Less PW of taxes, MIDDLE of 1st tax year =						$30,446
49								
50	NET PRESENT WORTH =						($18,382)
51								
52	* Aj = $90,000 * [0.10 / Ln (1.10)] =					$94,429	(equiv.)

Figure 7.A.1   Excel Spreadsheet for Example 7.A.1 (7.5 in Text) Illustrating the General Procedure

	B	C	D	E	F	G	H
2	**EXAMPLE 7.A.1**						
4	**AFTER-TAX PRESENT WORTH GENERAL PROCEDURE (Example 7.5)**						
6	Assumptions	Portion of 1st tax year property in service (pi) =				0.50	
7		Initial cost of recovery property (P) =				$400,000	
8		Expected life, years (Na) =				6	
9		Expected salvage value (Sa) =				$40,000	
10		Depreciation by regular MACRS method					
11		MACRS recovery period, years (Nd) =				5	
12		Depreciation percentage from Table 6.7					
13		Section 179 expense deduction to be taken =				$0	
14		Working capital to be recovered end of project life =				$50,000	
15		Annual savings, CONTINUOUS cash flows =				$90,000	
16		Marginal income tax rate (t) =				0.34	
17		After-tax MARR =				0.10	
18		Cash flows for taxes occur at end of tax year.					
20	Tax	Deprec'n	Deprec'n	Sec. 179	Other	Taxable	Cash Flow
21	Year	Percent.	Expense	$ Gain	Income	Income	For Taxes
23	1	0.2000	+C23*($G$7-$G$13)	+$G13	+$G$6*$G$15	-D23+E23+F23	+G23*$G$16
24	2	0.3200	+C24*($G$7-$G$13)		+$G$15	-D24+E24+F24	+G24*$G$16
25	3	0.1920	+C25*($G$7-$G$13)		+$G$15	-D25+E25+F25	+G25*$G$16
26	4	0.1152	+C26*($G$7-$G$13)		+$G$15	-D26+E26+F26	+G26*$G$16
27	5	0.1152	+C27*($G$7-$G$13)		+$G$15	-D27+E27+F27	+G27*$G$16
28	6	0.0576	+C28*($G$7-$G$13)		+$G$15	-D28+E28+F28	+G28*$G$16
29	7			+$G$9-($G$7-D31-$G$13)	(1-$G$6)*$G$15	-D29+E29+F29	+G29*$G$16
31	Totals	@SUM(C23..C29)	@SUM(D23..D29)	@SUM(E23..E29)	@SUM(F23..F29)	@SUM(G23..G29)	@SUM(H23..H29)
33	PW of taxes @ 10%, START of 1st tax year =						@NPV($G$17,H23..H29)
35	End of			Other	Total	Present	
36	Project	Recovery	Working	Pre-tax	Cash Flow	Worth at	
37	Year	Property	Capital	Cash Flows*	Before Tax	0.10	
39	0	-$G$7	-$G$14		+C39+D39+E39	+F39*((1+$G$17)^-B39)	
40	1			+$G15*($G$17/@LN(1+$G$17))	+C40+D40+E40	+F40*((1+$G$17)^-B40)	
41	2			+$G15*($G$17/@LN(1+$G$17))	+C41+D41+E41	+F41*((1+$G$17)^-B41)	
42	3			+$G15*($G$17/@LN(1+$G$17))	+C42+D42+E42	+F42*((1+$G$17)^-B42)	
43	4			+$G15*($G$17/@LN(1+$G$17))	+C43+D43+E43	+F43*((1+$G$17)^-B43)	
44	5			+$G15*($G$17/@LN(1+$G$17))	+C44+D44+E44	+F44*((1+$G$17)^-B44)	
45	6	+$G$9	+$G$14	+$G15*($G$17/@LN(1+$G$17))	+C45+D45+E45	+F45*((1+$G$17)^-B45)	
47	Totals	@SUM(C39..C45)	@SUM(D39..D45)	@SUM(E39..E45)	@SUM(F39..F45)	@SUM(G39..G45)	
49	Less PW of taxes, MIDDLE of 1st tax year =					+H33*((1+$G$17)^-G6)	
51	NET PRESENT WORTH =					+G47-G49	
53	* Aj = $90,000 * [0.10/Ln (1.10)] =			+$G$15*($G$17/@LN(1+$G$17))		(equiv.)	

Figure 7.A.2   Excel Cell Formulas for Spreadsheet in Figure 7.A.1

	A	B	C	D	E	F	G
1	**EXAMPLE 7.A.2**						
2							
3	**AFTER-TAX ANALYSIS-SIMPLIFIED PROCEDURE (Example 7.15)**						
4							
5							
6	Assumptions:	Tax year corresponds to project year.					
7		Cash flows for taxes occur at end of project years.					
8		Depreciation using regular MARCS method.					
9		MARCS recovery period (Nd) is				5 years	
10		Depreciation percentages from Table 6.7.					
11		All pre-tax cash flows are end-of-year.					
12		Marginal tax rate =				0.40	
13		After-tax discount rate =				0.10	
14		Cost basis of depreciable asset =				$80,000	
15		Actual salvage value =				$0	
16		Actual service life =				5 years	
17							
18							
19	End of	Deprec'n		Cash Flow			
20	year	Percent	Deprec'n	Before Taxes			
21	j	Pj	Dj	Aj	Ij	Tj	Aj^
22							
23	0			-80,000			-80,000
24							
25	1	0.2	$16,000	$24,000	$8,000	$3,200	$20,800
26	2	0.32	$25,600	$24,000	-1600	-640	$24,640
27	3	0.192	$15,360	$24,000	$8,640	$3,456	$20,544
28	4	0.1152	$9,216	$24,000	$14,784	$5,914	$18,086
29	5	0.1152	$9,216	$24,000	$14,784	$5,914	$18,086
30	6	0.0576	$4,608	$0	-4608	-1843.2	$1,843
31							
32	6		$0*	$0	$0	$0	$0
33	TOTALS:	1.00	$80,000	$40,000	$40,000	$16,000	$24,000
34							
35	* Book value (adjusted basis) when removed from service.						
36							
37							
38	IRR =9.66%						
39							

Figure 7.A.3   Excel Spreadsheet for Example 7.A.2 (7.15 in Text) illustrating the Simplified Procedure

EXAMPLE 7.A.2 *After-Tax IRR: Simplified Procedure*

The "simplified procedure" is described in the text in Section 7.5.2. Certain simplifying assumptions concerning the timing of cash flows permit a much less complex analysis as illustrated in Example 7.15. Here, the $80,000 asset is depreciated by regular MACRS over a 5-year recovery period with the half-year convention assumed. The actual service life is 5 years, and there is no salvage value. The marginal income tax rate is 0.40; the after-tax MARR is 0.10. If it is acquired, this asset will result in savings of $24,000 each year.

The after-tax analysis in spreadsheet format is displayed in **Figure 7.A.3**. The spreadsheet has been generated using Excel software, and the underlying cell formulas are summarized in **Figure 7.A.4**. The resulting after-tax IRR is 9.66%.

Note the similarity between the spreadsheet of Figure 7.A.3 and the tabular representation given in the text in Table 7.5.

	A	B	C	D	E	F	G
1							
2	EXAMPLE 7.A.2						
3							
4	AFTER-TAX	ANALYSIS-SIMPLIFIED PROCEDURE	(Example7.15)				
5							
6	Assumptions:	Tax year corresponds to project year.					
7		Cash flows for taxes occur at end of project y					
8		Depreciation using regular MARCS method.					
9		MARCS recovery period (Nd) is				5 years	
10		Depreciation percentages from Table 6.7.					
11		All pre-tax cash flows are end-of-year.					
12		Marginal tax rate =				0.4	
13		After-tax discount rate =				0.1	
14		Cost basis of depreciable asset =				80000	
15		Actual salvage value =				0	
16		Actual service life =				5 years	
17							
18							
19	End of	Deprec'n		Cash Flow			
20	year	Percent	Deprec'n	Before Taxes			
21	j	Pj	Dj	Aj	Ij	Tj	Aj^
22							
23	0			=-F14			=D23-F23
24							
25	1	0.2	=B25*$F$14	24000	=D25-C25	=E25*$F$12	=D25-F25
26	2	0.32	=B26*$F$14	24000	=D26-C26	=E26*$F$12	=D26-F26
27	3	0.192	=B27*$F$14	24000	=D27-C27	=E27*$F$12	=D27-F27
28	4	0.1152	=B28*$F$14	24000	=D28-C28	=E28*$F$12	=D28-F28
29	5	0.1152	=B29*$F$14	24000	=D29-C29	=E29*$F$12	=D29-F29
30	6	0.0576	=B30*$F$14	0	=D30-C30	=E30*$F$12	=D30-F30
31							
32	6		$0*	0	=D32-C32	=E32*$F$12	=D32-F32
33	TOTALS	=SUM(B23:B32)	=SUM(C23:C32)	=SUM(D23:D32)	=SUM(E23:E32)	=SUM(F23:F32)	=SUM(G23:G32)
34							
35							
36		* Book valu (adjusted basis) when removed from service.					
37							
38	IRR =	=IRR(G23:G30)					
39							

Figure 7.A.4 Excel Cell Formulas for Spreadsheet in Figure 7.A.3

	A	B	C	D	E	F	G
1	EXAMPLE 7.A.3						
2							
3	AFTER-TAX ANALYSIS: EQUIPMENT LEASE (Example 7.22)						
4	Assumption:						
5		Monthly lease payment					$1,000
6		Lease Payment at beginning of each month.					
7		Total number of lease payments =					24
8		Refundable deposit =					$2,500
9		Cash flows for taxes at end of each month.					
10		Marginal income tax rate =					0.3
11		After-tax MARR per year =					0.127
12							
13	End of	Cash Flow	Lease	Effect on	Effect on	Cash Flow	PW @
14	Month	Before Tax	Payments	Taxable Income	Income Tax	After Taxes	0.01
15	j	Aj	Lj	TIj	Tj	Aj^	per month
16	---------	-----------------	---------------	--------------------------	---------------------	--------------------	-------------------
17	0	($3,500)	($1,000)	($1,000)	($300)	($3,200)	($3,200.00)
18	1	($1,000)	($1,000)	($1,000)	($300)	($700)	($693.06)
19	2	($1,000)	($1,000)	($1,000)	($300)	($700)	($686.19)
20	3	($1,000)	($1,000)	($1,000)	($300)	($700)	($679.39)
21	4	($1,000)	($1,000)	($1,000)	($300)	($700)	($672.65)
22	5	($1,000)	($1,000)	($1,000)	($300)	($700)	($665.98)
23	6	($1,000)	($1,000)	($1,000)	($300)	($700)	($659.38)
24	7	($1,000)	($1,000)	($1,000)	($300)	($700)	($652.84)
25	8	($1,000)	($1,000)	($1,000)	($300)	($700)	($646.37)
26	9	($1,000)	($1,000)	($1,000)	($300)	($700)	($639.96)
27	10	($1,000)	($1,000)	($1,000)	($300)	($700)	($633.62)
28	11	($1,000)	($1,000)	($1,000)	($300)	($700)	($627.34)
29	12	($1,000)	($1,000)	($1,000)	($300)	($700)	($621.12)
30	13	($1,000)	($1,000)	($1,000)	($300)	($700)	($614.96)
31	14	($1,000)	($1,000)	($1,000)	($300)	($700)	($608.86)
32	15	($1,000)	($1,000)	($1,000)	($300)	($700)	($602.83)
33	16	($1,000)	($1,000)	($1,000)	($300)	($700)	($596.85)
34	17	($1,000)	($1,000)	($1,000)	($300)	($700)	($590.93)
35	18	($1,000)	($1,000)	($1,000)	($300)	($700)	($585.08)
36	19	($1,000)	($1,000)	($1,000)	($300)	($700)	($579.28)
37	20	($1,000)	($1,000)	($1,000)	($300)	($700)	($573.53)
38	21	($1,000)	($1,000)	($1,000)	($300)	($700)	($567.85)
39	22	($1,000)	($1,000)	($1,000)	($300)	($700)	($562.22)
40	23	($1,000)	($1,000)	($1,000)	($300)	($700)	($556.64)
41	24	$2,500	$0	$0	$0	$2,500	$1,968.30
42	---------	-----------------	---------------	--------------------------	---------------------	--------------------	-------------------
43	Totals	($24,000)	($24,000)	($24,000)	($7,200)	($16,800)	($15,549)

Figure 7.A.5 Excel Spreadsheet for Example 7.A.3 (7.22 in Text)
Illustrating After-Tax Analysis of Equipment Lease

EXAMPLE 7.A.3 *After-Tax Analysis: Lease Payments*

Consider an equipment lease over a 24-month period of $1,000 at the beginning of each month. A $2,500 deposit is required, refundable at the end of 24 months. The marginal income tax rate is 30%. The firm's after-tax MARR is 12.7% per year, which, when compounded monthly, is equivalent to 1.0% per month. Our problem is to determine the after-tax present worth of the cost of the lease. (This is Example 7.22 in the text.)

The after-tax analysis in spreadsheet format is shown in **Figure 7.A.5**. The spreadsheet has been generated with Excel software, and the underlying cell formulas are summarized in **Figure 7.A.6**. The after-tax present worth of costs is $15,549.

	A	B	C	D	E	F	G
1	EXAMPL						
2							
3	AFTER-T						
4	Assumpt						
5		Monthly lease pa					1000
6		Lease Payment a					
7		Total number of l					24
8		Refundable depo					2500
9		Cash flows for ta					
10		Marginal income					0.3
11		After-tax MARR					0.127
12							
13	End of	Cash Flow	Lease	Effect on	Effect on	Cash Flow	PW @
14	Month	Before Tax	Payments	Taxable Income	Income Tax	After Taxes	=(1+G11)^(1/12)-1
15	j	Aj	Lj	Tlj	Tj	Aj^	per month
16	----------	--------------------	--------------------	--------------------	--------------------	--------------------	-------------------------------
17	0	=-$G$5-$G$8	=-$G$5	=-$G$5	=D17*$G$10	=B17-E17	=F17*((1+$G$14)^(-A17))
18	1	=-$G$5	=-$G$5	=-$G$5	=D18*$G$10	=B18-E18	=F18*((1+$G$14)^(-A18))
19	2	=-$G$5	=-$G$5	=-$G$5	=D19*$G$10	=B19-E19	=F19*((1+$G$14)^(-A19))
20	3	=-$G$5	=-$G$5	=-$G$5	=D20*$G$10	=B20-E20	=F20*((1+$G$14)^(-A20))
21	4	=-$G$5	=-$G$5	=-$G$5	=D21*$G$10	=B21-E21	=F21*((1+$G$14)^(-A21))
22	5	=-$G$5	=-$G$5	=-$G$5	=D22*$G$10	=B22-E22	=F22*((1+$G$14)^(-A22))
23	6	=-$G$5	=-$G$5	=-$G$5	=D23*$G$10	=B23-E23	=F23*((1+$G$14)^(-A23))
24	7	=-$G$5	=-$G$5	=-$G$5	=D24*$G$10	=B24-E24	=F24*((1+$G$14)^(-A24))
25	8	=-$G$5	=-$G$5	=-$G$5	=D25*$G$10	=B25-E25	=F25*((1+$G$14)^(-A25))
26	9	=-$G$5	=-$G$5	=-$G$5	=D26*$G$10	=B26-E26	=F26*((1+$G$14)^(-A26))
27	10	=-$G$5	=-$G$5	=-$G$5	=D27*$G$10	=B27-E27	=F27*((1+$G$14)^(-A27))
28	11	=-$G$5	=-$G$5	=-$G$5	=D28*$G$10	=B28-E28	=F28*((1+$G$14)^(-A28))
29	12	=-$G$5	=-$G$5	=-$G$5	=D29*$G$10	=B29-E29	=F29*((1+$G$14)^(-A29))
30	13	=-$G$5	=-$G$5	=-$G$5	=D30*$G$10	=B30-E30	=F30*((1+$G$14)^(-A30))
31	14	=-$G$5	=-$G$5	=-$G$5	=D31*$G$10	=B31-E31	=F31*((1+$G$14)^(-A31))
32	15	=-$G$5	=-$G$5	=-$G$5	=D32*$G$10	=B32-E32	=F32*((1+$G$14)^(-A32))
33	16	=-$G$5	=-$G$5	=-$G$5	=D33*$G$10	=B33-E33	=F33*((1+$G$14)^(-A33))
34	17	=-$G$5	=-$G$5	=-$G$5	=D34*$G$10	=B34-E34	=F34*((1+$G$14)^(-A34))
35	18	=-$G$5	=-$G$5	=-$G$5	=D35*$G$10	=B35-E35	=F35*((1+$G$14)^(-A35))
36	19	=-$G$5	=-$G$5	=-$G$5	=D36*$G$10	=B36-E36	=F36*((1+$G$14)^(-A36))
37	20	=-$G$5	=-$G$5	=-$G$5	=D37*$G$10	=B37-E37	=F37*((1+$G$14)^(-A37))
38	21	=-$G$5	=-$G$5	=-$G$5	=D38*$G$10	=B38-E38	=F38*((1+$G$14)^(-A38))
39	22	=-$G$5	=-$G$5	=-$G$5	=D39*$G$10	=B39-E39	=F39*((1+$G$14)^(-A39))
40	23	=-$G$5	=-$G$5	=-$G$5	=D40*$G$10	=B40-E40	=F40*((1+$G$14)^(-A40))
41	24	=$G$8	0	0	=D41*$G$10	=B41-E41	=F41*((1+$G$14)^(-A41))
42	----------	--------------------	--------------------	--------------------	--------------------	--------------------	-------------------------------
43	Totals	=SUM(B17:B41)	=SUM(C17:C41)	=SUM(D17:D41)	=SUM(E17:E41)	=SUM(F17:F41)	=SUM(G17:G41)

Figure 7.A.6 Excel Cell Formula for Spreadsheet in Figure 7.A.5

EXAMPLE 7.A.4 *After-Tax Analysis: Equity Investment*

Recall Example 7.23 in the text. Certain depreciable equipment will cost $100,000 and will result in savings in operating costs of $24,000 each year over a 5-year period. The equipment will be kept in service for 5 years and then sold for an estimated $20,000.

The equipment will be depreciated under regular MACRS with a 3-year recovery period. The half-year convention will be used.

The firm will borrow $25,000 to partially finance this investment. The loan is to be repaid by four equal end-of-year payments based on a pretax cost of 10% per year.

The firm's after-tax MARR is 15%. The relevant marginal income tax rate is 40%. Our problem is to determine the after-tax present worth resulting from the equity investment.

The spreadsheet analysis of this problem is shown in **Figure 7.A.7**. The spreadsheet is prepared using Excel, and the underlying cell formulas are summarized in **Figure 7.A.8** on page 296.

Note that the spreadsheet is in three sections: the top section (rows 5–16) summarizes the input data for the analysis; the middle section (rows 18–26) shows the necessary calculations to determine the allowable year-by-year depreciation expenses; and the bottom section (rows 28–40) summarizes the equity cash flows before and after income taxes. As shown in cell G39, the after-tax PW on the equity investment, based on a 15% MARR, is −$10,665. The after-tax rate of return on equity, cell D40, is 8.65%.

	A	B	C	D	E	F	G
1	EXAMPLE 7.A.4						
2							
3	AFTER-TAX ANALYSIS: EQUITY INVESTMENT (Example 7.23)						
4	Assumptions:						
5		Initial investment, total =				$100,000	
6		Amount borrowed =				$25,000	
7		Annual pre-tax savings =				$24,000	
8		Loan to be repaid by equal end-of-year payments					
9		Number of equal payments =				4	
10		Before-tax cost of loan				0.1	
11		Marginal income tax rate =				0.4	
12		MARR on equity investment =				0.15	
13		Expected actual salvage value =				$20,000	
14		Depreciation using the regular MACRS method					
15		MACRS recovery period =				5	Years
16		Depreciation percentage from Table 6.7.					
17							
18	Year	Unpaid	Total	Amount of Payment, That Is:		Deprec'n	Deprec'n
19		Principal at	Payment	Interest	Principal	Percentage	Expense
20	j	Start of year		Ij	Ej	Pj	Dj
21	1	$25,000	$7,887	$2,500	$5,387	0.33333	$33,333
22	2	$19,613	$7,887	$1,961	$5,925	0.44445	$44,445
23	3	$13,688	$7,887	$1,369	$6,518	0.14815	$14,815
24	4	$7,170	$7,887	$717	$7,170	0.07407	$7,407
25							
26	Totals:		$31,547	$6,547	$25,000	$1	$100,000
27							
28	End of	Operating	Equity Cash	Taxable	Income	Equity Cash Flow	Present
29	Year	savings	Flow Before	Income	Taxes	After Taxes	Worth
30	j		Tax Aj	TIj	Tj	Aj - Tj	Pj
31	0		($75,000)			($75,000)	($75,000)
32	1	$24,000	$16,113	($11,833)	($4,733)	$20,846	$18,127
33	2	$24,000	$16,113	($22,406)	($8,963)	$25,076	$18,961
34	3	$24,000	$16,113	$7,816	$3,126	$12,987	$8,539
35	4	$24,000	$16,113	$15,876	$6,350	$9,763	$5,582
36	5	$24,000	$24,000	$24,000	$9,600	$26,400	$13,125
37	5		$20,000	$20,000	$8,000		
38							
39	Totals:	$120,000	$33,453	$33,453	$13,381	$20,072	($10,665)
40		Rate of return on equity =		8.65%			

Figure 7.A.7   Excel Spreadsheet for Example 7.A.4 (7.23 in Text) Illustrating
After-Tax Analysis of Equity Investment

	A	B	C	D	E	F	G
1	EXAM						
2							
3	AFTE						
4	Assu						
5		Initial investment,				100000	
6		Amount borrowe				25000	
7		Annual pre-tax sa				24000	
8		Loan to be repaid					
9		Number of equal				4	
10		Before-tax cost o				0.1	
11		Marginal income				0.4	
12		MARR on equity i				0.15	
13		Expected actual s				20000	
14		Depreciation usin					
15		MACRS recovery				5	Years
16		Depreciation perc					
17							
18	Year	Unpaid	Total	Amount of Payment, T		Deprec'n	Deprec'n
19		Principal at	Payment	Interest	Principal	Percentage	Expense
20	j	Start of year		Ij	Ej	Pj	Dj
21	1	=$F$6	=$F$6*(($F$10*(1+$F$10)^$F$9)/((1+$F$10)^$F$9-1))	=$F$10*B21	=C21-D21	0.33333	=F21*$F$5
22	2	=B21-E21	=$F$6*(($F$10*(1+$F$10)^$F$9)/((1+$F$10)^$F$9-1))	=$F$10*B22	=C22-D22	0.44445	=F22*$F$5
23	3	=B22-E22	=$F$6*(($F$10*(1+$F$10)^$F$9)/((1+$F$10)^$F$9-1))	=$F$10*B23	=C23-D23	0.14815	=F23*$F$5
24	4	=B23-E23	=$F$6*(($F$10*(1+$F$10)^$F$9)/((1+$F$10)^$F$9-1))	=$F$10*B24	=C24-D24	0.07407	=F24*$F$5
25							
26	Totals		=SUM(C21:C24)	=SUM(D21:D24)	=SUM(E21:E24)	=SUM(F21:F24)	=SUM(G21:G24)
27							
28	End of	Operating	Equity Cash	Taxable	Income	Equity Cash Flow	Present
29	Year	savings	Flow Before	Income	Taxes	After Taxes	Worth
30	j		Tax Aj	Tij	Tj	Ai - Tj	Pj
31	0		=$F$6-$F$5			=C31-E31	=F31*((1+$F$12)^(-A31))
32	1	=$F$7	=B32-$C$21	=$F$7-D21-G21	=$F$11*D32	=C32-E32	=F32*((1+$F$12)^(-A32))
33	2	=$F$7	=B33-$C$21	=$F$7-D22-G22	=$F$11*D33	=C33-E33	=F33*((1+$F$12)^(-A33))
34	3	=$F$7	=B34-$C$21	=$F$7-D23-G23	=$F$11*D34	=C34-E34	=F34*((1+$F$12)^(-A34))
35	4	=$F$7	=B35-$C$21	=$F$7-D24-G24	=$F$11*D35	=C35-E35	=F35*((1+$F$12)^(-A35))
36	5	=$F$7	=B35	=$F$7	=$F$11*D36	=C36-E36+C37-E37	=F36*((1+$F$12)^(-A36))
37	5	=$F$13	=$F$13	=$F$13-($F$5-G26)	=$F$11*D37		
38							
39	Totals	=SUM(B31:B37)	=SUM(C31:C37)	=SUM(D31:D37)	=SUM(E31:E37)	=SUM(F31:F37)	=SUM(G31:G37)
40				=IRR(F31:F36)			

Figure 7.A.8   Excel Cell Formulas for Spreadsheet in Figure 7.A.6

8

Retirement and Replacement

8.1 INTRODUCTION

As noted in preceding chapters, the methodology presented in this book can be used to solve a wide variety of investment problems: which administrative procedure to adopt, which route to select, which equipment to purchase, which new product design is preferable, whether to lease or purchase, when to implement a new program or initiate construction, and so on. Perhaps the most common of these problems is that of **retirement** and **replacement** of existing plant and equipment. Because the retirement/replacement problem occurs so frequently and because of the special analytical techniques appropriate to it, this separate chapter has been developed.

8.1.1 Causes of Retirement from Service

In modern industry, existing assets are not retired merely because they are physically incapable of performing their original function. Rather, retirement generally occurs or is "encouraged" because of changes in economics or the operating environment, such as, for example, when:

1. An existing asset was originally purchased to meet a certain demand, but demand has risen and the equipment can no longer satisfy current or anticipated demand.
2. New, improved equipment is available that is less expensive to operate, say, as a result of lower maintenance costs.
3. New, improved equipment is available that is more efficient than the existing asset.

4. The demand that prompted the purchase of the original equipment no longer exists, so the equipment no longer fills a useful purpose.

5. Existing equipment has become a casualty as a result of, say, fire, accident, or major breakdown.

These are only examples. In most practical situations, retirement (and possible replacement) is considered because of a combination of causes.

8.1.2 Factors to Be Considered in Replacement Analyses

There are three fundamental characteristics of all economic analyses: (1) the *amounts* of the various cash flows anticipated over the planning horizon, (2) the *timing* of those cash flows within the study period, and (3) the *minimum attractive rate of return*, that is, a measure of the cost of foregone opportunities resulting from investment in the proposed project. Given these three elements, calculation of the appropriate measure of worth (present worth, rate of return, etc.) is relatively straightforward. It should be emphasized, however, that the complexities of the real world are such that reasonable estimates of the amount and timing of the cash flows and/or the MARR present significant challenges to the analyst. This observation is especially relevant with respect to problems of retirement/replacement.

It will be helpful at this point to consider some of the variations in assumptions that, when taken together, describe the context underlying a specific problem. This will give us some appreciation for the range of issues to be addressed by the analyst.[1]

- If a unit fails, must it be removed permanently from service?
- Are standby units available if the system should fail?
- Do components or units fail independently of the failure of other components?
- Is there a budget constraint?
- In the event that a unit can be repaired after failure, is there a constraint on the capacity of the repair facility?
- Is there only one replacement allowed over the planning horizon? Are subsequent replacements allowed at any time during the study period?
- Is there more than one replacement unit (price and quality combination) available at a given point in time?

---

[1] This summary is adapted from James T. Luxhoj and Marilyn S. Jones, "A Framework for Replacement Modeling Assumptions," *The Engineering Economist*, Vol. 32, No. 1, Fall 1986, pp. 39–49.

- Do future replacement units differ over time? Are technological improvements considered?
- Is preventive maintenance included in the model?
- Are periodic operating and maintenance costs constant or variable over time?
- Is the planning horizon finite or infinite?
- Are consequences other than economic inpacts, i.e., sociotechnical issues, considered?
- Are income tax consequences considered?
- Is "inflation" (i.e., relative price changes) considered?
- Does replacement occur simultaneously with retirement, or are there nonzero lead times?
- Are cash flow estimates deterministic or stochastic?

This listing of issues is not intended to be exhaustive, but merely illustrative of the many complexities that must be addressed to define the context for any specific analysis. The remaining discussion in this chapter necessarily focuses on a limited set of factors.

8.1.3   The Terminology of Replacement Theory

Every discipline and subdiscipline has its own technical terms. Some are useful; others appear to have little value. The literature of replacement theory is no exception, so before going on, you will find it useful to understand the following terms:

Defender: the existing asset; the one under consideration for retirement[2]

Challenger: the proposed replacement asset

Economic life: that asset life resulting in minimum (equivalent uniform) annual cost

(Of course, the economic life must be equal to or less than the asset's maximum physical life. Physical life is affected by the design of the equipment itself and the physical environment in which it operates. Economic life is a reflection of the economic consequences generated by owning and operating the challenger or defender.)

8.1.4   Plan of Development for Chapter 8

There is a logical structure to the development of the following sections. Section 8.2 examines simple retirement (abandonment), that is, retirement

---

[2]The terms *defender* and *challenger* originated in George Terborgh, *Dynamic Equipment Policy* (New York: McGraw-Hill, 1949).

without replacement. Section 8.3 considers the problem of replacement where the currently available challenger is identical to the defender. Section 8.4 develops an evaluation procedure for the situation in which the current challenger is different from the defender but all future challengers are identical to the current challenger. And Section 8.5 briefly discusses the general case where all future challengers are different. This development plan is shown in **Figure 8.1**.

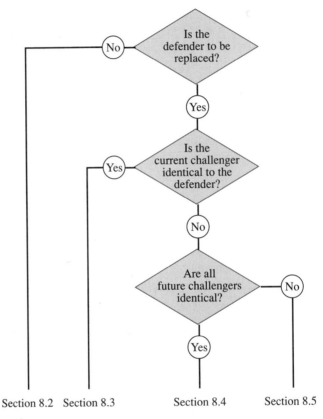

Section 8.2   Section 8.3   Section 8.4   Section 8.5

Figure 8.1   Plan of Development for Chapter 8

8.2   RETIREMENT WITHOUT REPLACEMENT (ABANDONMENT)

As with all retirement problems, simple retirement concerns only two immediate alternatives: (1) *keep* the existing asset, the defender; or (2) *retire* it. But simple retirement, unlike other retirement problems, concerns only the retirement of the asset, not its replacement. Thus, the analysis is straightforward. If the asset is to be kept in service for N more periods, the present worth of the resulting cash flows is

$$PW(N) = \sum_{j=1}^{N} A_j(1 + i)^{-j} \qquad (8.1)$$

where A_j is the cash flow at the end of period j and i is the appropriate discount rate. [Note that Eq. (8.1) reflects the end-of-period assumptions for cash flows and discounting.]

The alternative is to retire (*abandon*) the asset and to invest any funds received from its sale at interest rate i. Letting P represent the current market value of the asset, it follows that retirement is justified if there exists no possible remaining life such that the present worth of the "keep" alternative is less than the present worth of the "retire" alternative. That is, *retire the asset* if

$$PW(N) < P \qquad \text{for any } 1 \leq N \leq N_{max} \qquad (8.2)$$

where N_{max} is the maximum possible remaining life. Keep the asset in service if $PW(N) > P$; and if $PW(N) = P$, we are indifferent between the alternatives.

Noting that $P = -A_0$, where A_0 is the cash flow at the start of the first period (end of period 0), and combining Eqs. (8.1) and (8.2), our decision rule is to *keep the asset* if

$$A_0 + \sum_{j=1}^{N} A_j(1 + i)^{-j} > 0 \qquad \text{for any } 1 \leq N \leq N_{max}. \qquad (8.3)$$

In other words, keep the asset if the present worth of the cash flows for any remaining life is positive.

EXAMPLE 8.1

To illustrate this procedure, consider an asset that has a current market value of $1,000. If it is retained, it will generate net positive cash flows of $300 per year each year for the next 3 years. Moreover, at the end of 3 years it can be sold for $200. The appropriate discount rate is assumed to be 6%. To determine whether the asset should be retained, compute the present worth:

End of Period	Cash Flows	Discount Factor	Present Worth
0	− $1,000	1.0000	− $1,000
1–3	300	2.673	802
3	200	0.8396	168
Total	$ 100		− $ 30

Inasmuch as the PW is negative at $i = 6\%$, the asset should be retired. It is economically preferable to dispose of the asset and invest the $1,000 elsewhere at 6%.

In this example, the current market value, $1,000, represents a cost to be charged against the asset should it be retained in service. In general, the **disposal value**, or **salvage value**, is not necessarily the current market value of the asset. Rather, the value used in the analysis should reflect the value to the firm if the asset is retired from its current activity. If it is retired, it may be sold. But it may also remain inside the firm to be employed in some other activity, in which case the disposal value is represented by the equivalent present worth if the asset were to be used elsewhere, say, in another operating unit. Put somewhat differently, the disposal value should represent the cost of the foregone opportunity of continuing to employ the asset in its current function—a measure of the monetary cost of the opportunity foregone.

It should be noted that the analysis in Example 8.1 is incomplete. It evaluates only one scenario, namely, *keep the asset exactly 3 more years*. If the only alternative action would be to sell the asset now and invest the proceeds elsewhere at the MARR, 6%, then that action would be justified. However, suppose that one were to ask: "Shall we keep the asset in service only 1 more year? Or 2 more years?" A complete analysis requires evaluation of all possible combinations of remaining lives and associated cash flows, as is demonstrated in the next example.

EXAMPLE 8.2

Consider an asset that has a current market value of $1,000. Estimates for the next 2 years are as follows:

End of Year	Market Value	Net Revenue
1	$800	$400
2	600	200

Assuming an effective annual discount rate of 10%,

$$PW(N) = \text{present worth if asset is kept in service } N \text{ more years}$$

$$PW(1) = -\$1,000 + \$800(1.10)^{-1} + \$400(1.10)^{-1}$$

$$= \$91$$

$$PW(2) = -\$1,000 + \$600(1.10)^{-2} + \$400(1.10)^{-1} + 200(1.10)^{-2}$$

$$= \$25$$

Note that there are three mutually exclusive alternatives at this point in time: (1) Do nothing, retire the asset from service now, resulting in a present worth of exactly zero; (2) keep the asset in service one more year, with PW = $91; or (3) keep the asset 2 more years, with PW = $25. Of the three, the preferred alternative is to keep the asset 1 more year. The worst alternative is to abandon the asset now.

Suppose that the asset is kept 1 more year and, at that point in time, the decision to keep the asset a second year is to be considered. The present worth at the end of year 1 (the beginning of year 2) is

$$PW = -\$800 + \$600(1.10)^{-1} + \$200(1.10)^{-1}$$
$$= -\$73$$

The present worth is negative, indicating that the asset should not be retained the second year. (Note that the present worth measured at the start of year 1 is $\$91 - \$73(1.10)^{-1} = \$25$, the result previously obtained.) In general, although the present worth of an N-year remaining life is positive, this does not necessarily imply that the worth measured at any future point in time is also positive.

EXAMPLE 8.3

Another example may be helpful in understanding the problem of simple retirement without replacement. Consider the following cost and revenue estimates for an asset that, at the time of the analysis, is 3 years old and has a remaining physical life of 2 years:

Age	End of Year	Revenue	Operating Costs	Net Revenue	End-of-Year Salvage Value
4	1	$1,200	$800	$400	$300
5	2	1,100	900	200	0

Assume that the asset may currently be sold for $600 and that the pretax minimum attractive rate of return for the firm is 10%.

If the asset is kept 1 more year, the relevant consequences are (1) a foregone opportunity to receive $600 "today," (2) net receipts of $400 during the next year, and (3) a salvage value of $300 at the end of the year. Assuming end-of-year cash flows and discounting, the present worth of keeping the asset 1 more year is

$$PW(1) = -\$600 + (\$400 + \$300)(1.10)^{-1}$$
$$= \$36$$

Thus, as the analysis indicates, the asset should not be retired immediately; it should be kept for *at least* 1 more year.

Although the retirement question has been answered for the moment, it may be of some interest to determine whether, based on current estimates, the asset should be kept for 2 years, that is, through the end of its physical life. The expected consequences are (1) an immediate cost (opportunity forgone) of $600, (2) net receipts from operations of $400 and $200 at the end of years 1 and 2, respectively, and (3) zero net salvage value.

$$PW(2) = -\$600 + \$400(1.10)^{-1} + \$200(1.10)^{-2}$$

$$= -\$71$$

It appears at this point that, unless there are some changes in the estimates for the second year, the asset should be retired from service at the end of the first year. Thus we have answered the question concerning retirement and we have also determined *how long* the asset should be retained.

8.3   RETIREMENT WITH IDENTICAL REPLACEMENT

Consider an asset that has been in service, say, for N_0 years. Its remaining maximum service life is N_{max} years, so it has a total service life of $N_0 + N_{max}$ years. Assume that there currently exists a replacement asset, a challenger, that is identical to the defender in every respect—its total expected service life is the same, the expected cash flows in each and every year of service are identical to those of the defender, and so on. Our problem is to determine if the defender should be retired from service and replaced with the challenger.

(It is difficult to conceive of a challenger that is identical in *every* respect to a given defender. However, an approximate situation occurs when the asset under consideration is relatively unchanged by technological advances and prices remain reasonably constant, as is the case with certain hand tools, basic construction equipment, and many pipes and cables.)

8.3.1   Determining Economic Life

We should focus our attention first on the challenger. Let $AC(N)$ represent the (equivalent uniform) *annual cost* of the challenger if it is purchased and kept in service for exactly N periods. Here, annual cost is the negative of annual worth as defined in Section 3.4. The $AC(N)$ is computed for all values of $N = 1, 2, \ldots, N_{max}$. The project life for which the AC is minimized is the economic life, N^*, as illustrated in **Figure 8.2**. Note that the total annual cost consists of two elements: operating costs, which generally increase with the age of the asset; and capital recovery costs, which generally decrease with N because the capital costs are spread over a longer time period.

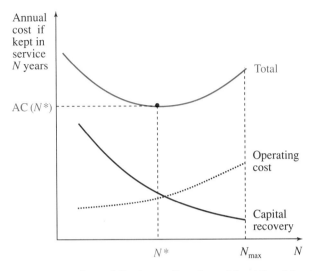

Figure 8.2   Annual Cost as a Function of the Life of the Asset

EXAMPLE   8.4

To illustrate the procedure for determining economic life, consider a challenger that is currently available at a cost of $10,000. The challenger, identical to the defender, has a maximum service life of 6 years. The firm's pre-tax MARR is 10%. The expected year-by-year salvage values, S_N, are shown in column 2 of **Table 8.1**, and the capital recovery calculations,

TABLE 8.1   Determining Capital Recovery for a $10,000 Asset if It Is Retained for Varying Periods from 1 Through 6 Years, Assuming 10% Interest Rate (Example 8.4)

Year N (1)	Salvage Value at End of Year S_N (2)	Initial Cost Less Salvage Value (3)	Capital Recovery Factor (4)	Interest on Salvage Value (5)	Capital Recovery if Kept N Years (6)
1	$7,000	$ 3,000	1.10000	$700	$4,000
2	4,500	5,500	0.57619	450	3,619
3	2,500	7,500	0.40211	250	3,266
4	1,000	9,000	0.31547	100	2,939
5	0	10,000	0.26380	0	2,638
6	0	10,000	0.22961	0	2,296

Col. 3: $10,000 - S_N$
Col. 4: $(A/P, 10\%, N)$
Col. 5: $0.10 S_N$
Col. 6: (col. 3) × (col. 4) + (col. 5)

CR(N), are reflected in the remaining columns of the table. The underlying equation is

$$CR(N) = (P - S_N)(A/P, i, N) + iS_N \quad (8.4)$$
$$= (\$10{,}000 - S_N)(A/P, 10\%, N) + 0.10S_N$$

Note that, for all N in this example, the capital recovery decreases as the life of the asset increases. This will not always be the case, however. The capital recovery factor, $(A/P, i, N)$, always decreases as N increases, but it is not always true that the terms $(P - S_N)$ and iS_N also decrease. Suppose, for example, that S_2 is only $1,000 rather than the $4,500 originally stated. Then the cost of capital recovery if the asset is kept through 2 years would be

$$CR(2) = (\$10{,}000 - \$1{,}000)(0.57619) + 0.10(\$1{,}000)$$
$$= \$5{,}285$$

This is larger than AC(1). If the year-to-year salvage values remain constant, then the cost of capital recovery will steadily decrease. Otherwise, the direction of change depends on the salvage values as well as on the interest rate.

The remaining calculations necessary to the computation of the equivalent uniform annual costs, AC(N), are presented in **Table 8.2**. The

TABLE 8.2 Determining Equivalent Uniform Annual Costs

			Present Worth of Operating Costs			Equivalent Uniform Annual Costs if Retired After N Years		
Year j, N (1)	Operating Costs During jth Year (2)	Present Worth Factor (3)	During jth Year (4)	Through N Years (5)	Capital Recovery Factor (6)	Operating Costs (7)	Capital Recovery (8)	Total Annual Cost, AC(N) (9)
1	$ 5,000	0.9538	$4,769	$ 4,769	1.10000	$5,246	$4,000	$9,246
2	5,400	0.8671	4,682	9,451	0.57619	5,446	3,619	9,065
3	6,100	0.7883	4,809	14,260	0.40211	5,734	3,266	9,000
4	7,100	0.7166	5,088	19,348	0.31547	6,104	2,939	9,043
5	8,400	0.6515	5,473	24,821	0.26380	6,548	2,638	9,186
6	11,000	0.5922	6,514	31,335	0.22961	7,195	2,296	9,491

Col. 2: Given $\bar{C}_j, j = 1, 2, \ldots, N_{max}$
Col. 3: $(P/F, 10\%, N)$
Col. 4: $P_j = $ (col. 2) × (col. 3)
Col. 5: $\text{PWOC}(N) = \sum_{j=1}^{N} P_j$
Col. 6: $(A/P, 10\%, N)$
Col. 7: (col. 5) × (col. 6)
Col. 8: From col. 6 of Table 8.1
Col. 9: (col. 7) + (col. 8)

estimated operating costs during years 1 through 6, $\bar{C}_j$, are given in column 2, and the total annual cost if the asset is kept through N_{max} years is calculated. The results are given in column 9 of the table and shown in **Figure 8.3**.

This pattern is typical of most replacement problems: Capital recovery costs tend to decrease over time, whereas operating costs tend to increase. (The cost histories of most private autos are excellent examples of this characteristic.) The minimum total equivalent uniform annual cost occurs at 3 years, thus 3 years is the economic life of the challenger. At this life, AC(3) = $9,000.

As shown by both column 9 of Table 8.2 and Figure 8.3, the annual costs for lives other than the economic life are only slightly higher than the minimum value. The total cost curve is relatively flat, suggesting that the economic life is rather insensitive to values assumed for input data. This phenomenon appears to occur quite frequently in actual replacement studies.

8.3.2 The Keep/Replace Decision

The procedure for arriving at a decision as to whether to keep or replace the defender with an identical challenger can be summarized as follows.

1. Determine the economic life of the challenger, N^*, and the annual cost at that life, $AC(N^*)$, as described previously.

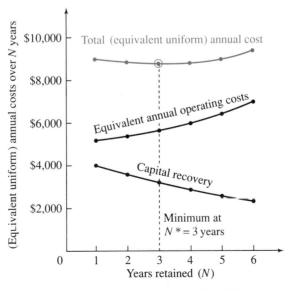

Figure 8.3   Determining Economic Life (N^*)

2. Determine the current age of the defender, N_0.
3. If $N_0 < N^*$, keep the defender in service. Otherwise continue.
4. If $N_0 = N^*$, the defender should be replaced. There is no economic advantage in continuing it in service, because there can be no smaller AC(N) at any $N \neq N^*$.
5. If $N_0 > N^*$, determine AC(Δ) for each Δ, where Δ = remaining life.
6. Keep the asset in service if there is any value of AC(Δ) < AC(N^*). Otherwise, replace the asset with the challenger.

EXAMPLE 8.5

We are considering the replacement of a certain piece of equipment, the defender, with an identical challenger. The initial cost of the challenger is $50,000 and, once purchased, the net salvage value is expected to be zero at all times. This is a before-tax analysis, and the pretax MARR is 10%. A partial analysis is summarized in **Table 8.3**. The values in column 6 represent the equivalent uniform annual cost if the challenger is acquired and kept in service N years. The minimum annual cost occurs when $N^* = 2$ years: AC(2) = $51,810.

If the defender, identical to the challenger, is now 1 year old, we should keep it in service another year. To do so would incur a cost of only $12,000. This compares favorably with the average annual cost of the replacement, a challenger to be kept 2 years.

TABLE 8.3   Determining Annual Cost of the Challenger (and Defender) for Example 8.5

End of Year j j (1)	Costs in Year j C_j (2)	PW of Costs at $i=10\%$ P_j (3)	PW if Kept N Years PW(N) (4)	Capital Recovery Factor (5)	EUAC if Kept N Years AC(CN) (6)
0	$50,000	$50,000	—	—	—
1	33,000	30,000	$ 80,000	1.1000	$88,000
2	12,000	9,917	89,917	0.5762	51,810
3	67,000	50,338	140,255	0.4021	56,396
4	40,000	27,321	167,576	0.3155	52,870
5	60,000	37,255	204,831	0.2638	54,034

Col. 2: Given
Col. 3: $P_j = C_j(1 + i)^{-j}$
Col. 4: $\text{PW}(N) = \sum_{j=0}^{N} P_j$
Col. 5: $(A/P, 10\%, N), j = N$
Col. 6: AC(CN) = PW(N) × $(A/P, 10\%, N)$

If the defender is now 2 years old, it should be replaced now, as there is no economic advantage to delay. It is already at its economic life.

But suppose that the defender is now 3 years old, 1 year past its economic life. We now determine the (equivalent uniform) annual costs resulting from prolonging the life of the defender until age 4 or 5, that is, for $\Delta = 1$ and 2 years. From Table 8.3, note that $C_4 = \$40,000$ and $C_5 = \$60,000$. Thus,

$$AC(\Delta = 1) = C_4(1 + i)^{-1}(A/P, i, 1)$$

$$= \$40,000(1.10)^{-1}(A/P, 10\%, 1)$$

$$= \$40,000$$

$$AC(\Delta = 2) = [C_4(1 + i)^{-1} + C_5(1 + i)^{-2}](A/P, 10\%, 2)$$

$$= [\$40,000(1.10)^{-1} + \$60,000(1.10)^{-2}](0.5762)$$

$$= \$49,525$$

Our 3-year-old defender should be kept in service 1 more year, i.e., until age 4, because $AC(\Delta = 1) = \$40,000$ is less than the $AC(N^*) = \$51,810$. We must take care in interpreting the significance of $AC(\Delta = 2)$, however. Note that the marginal cost for year 6 is \$60,000, somewhat greater than the $AC(N^*)$ of the challenger. Thus, if the defender is now 3 years old, it should be retained in service 1 more year, then replaced with an identical challenger.

To summarize our results,

If the Age of the Defender Is:	The Optimal Policy Is:
1 year	Keep defender 1 year
2 year	Replace with challenger
3 year	Keep defender 1 year
4 year	Replace with challenger

It should be emphasized that there are two assumptions that are critical to our development of the solution procedure discussed in this section. First, it is assumed that the current challenger is not only identical to the defender now in service, but this challenger will also be available at any time throughout the planning horizon. Future challengers will be identical to the current challenger. Second, it is assumed that the length of the planning horizon is unlimited or, if finite, that it is equal to the least common multiple (LCM) of the alternative lives under consideration. This assumption is important, because otherwise we would be unjustified in comparing directly the annual worths of alternatives with different lives. (See Section 3.5.) Stated somewhat differently, it is this assumption that permits us to compare alternative asset lives of N_1 and N_2, say, when $N_1 \neq N_2$.

8.4 RETIREMENT WITH UNLIKE REPLACEMENT: ALL CHALLENGERS IDENTICAL

Consider the case in which an available challenger is under consideration to replace a certain defender. The current challenger is unlike the defender—perhaps it is more or less costly to operate or it is more efficient or it has a different physical life, and so on. Assume that all future challengers are identical to the current challenger, so if replacement is forgone at this time, there will be future opportunities to replace the defender with new challengers. Here, however, we are assuming that all future challengers do not differ significantly in any respect from the current challenger.

We will also assume here that the planning horizon is very long; that is, it approaches infinity. As in the case of identical challengers and defenders, this assumption permits the use of the annual worth method to compare directly the annual equivalents of the various alternatives.

The appropriate solution procedure is closely related to the one described in Section 8.3.

1. Determine the (equivalent uniform) annual cost for the challenger at its economic life, CN^*. (See **Figure 8.4**.) The procedure for determining the AC(CN^*) is precisely the same as the one presented in Section 8.3.

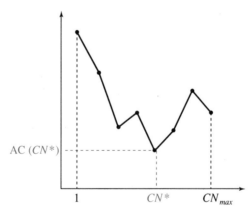

Figure 8.4   Determining AC for Challenger at Its Economic Life

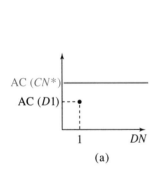

(a)

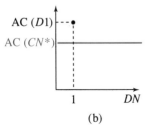

(b)

Figure 8.5 Cost to Keep Defender 1 More Year

2. Determine the annual cost for the defender if it is kept 1 more year. That is, determine AC($D1$).

   a. If AC($D1$) < AC(CN^*), it is preferable to keep the defender for at least 1 more year; thus it should not be replaced at this time. (See **Figure 8.5a**.)

   b. If AC($D1$) > AC(CN^*), it is not preferable to keep the defender exactly 1 more year. (See **Figure 8.5b**.) However, this does not signal replacement. Indeed, the pattern of future cash flows may be such that the annual cost may be reduced if the defender is kept 2 years or more. Step 3 explores this possibility.

3. Determine AC(D2).

   a. If AC(D2) < AC(CN*), as in **Figure 8.6a**, it is preferable to keep the defender for at least 2 more years.

   b. If AC(D2) > AC(CN*), using the same reasoning as in Step 2b, it is necessary to determine whether the defender should be kept for at least 3 additional years. (See **Figure 8.6b**.)

4. The above procedure should be continued until one of two conditions occurs:

   a. Examine the defender if it is kept DN more years. If the AC(DN) < AC(CN*), as in **Figure 8.7a**, the defender should be kept at least DN more years.

   b. Examine the defender throughout its remaining physical life, that is, $DN = 1, 2, \ldots, DN_{max}$. If there is no life such that AC(DN) < AC(CN*), as in **Figure 8.7b**, then replacement is indicated.

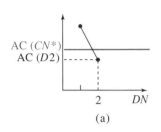

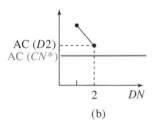

Figure 8.6 Cost to Keep Defender 2 More Years

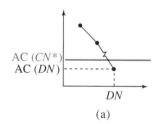

Figure 8.7 Cost to Keep Defender DN More Years

EXAMPLE 8.6

To illustrate this procedure, consider the challenger described in Example 8.4. Recall that AC(CN*) = \$9,000. Suppose that the alternative, the defender, has a current market value of \$8,000. If it is retained for 1 more year, it may be sold for \$5,000 and it will incur operating costs of \$5,100 during the year. In general,

AC(DN) = (equivalent uniform) annual cost for the defender if kept DN more years

= (capital recovery) + (annualized operating costs)

$$= [(P - S_N)(A/P, i, N) + iS]$$
$$+ [(\text{PW of operating costs})(A/P, i, N)] \tag{8.5}$$

Here, when DN = 1 for the defender,

$$AC(D1) = (\$8,000 - \$5,000)(A/P, 10\%, 1) + \$5,000(0.10)$$
$$+ \$5,100(P/\bar{F}, 10\%, 1)(F/P, 10\%, 1)$$
$$= \$3,000(1.10) + \$500 + \$5,100(1.049)$$
$$= \$9,150$$

As AC(D1) > AC(CN*), we conclude that the defender should *not* be kept 1 more year.

Perhaps the defender should be kept 2 more years or more. Suppose that the expected salvage value at the end of 2 years is \$4,000 and operating expenses during the second year are expected to continue at $5,100. Then,

$$AC(D2) = (\$8,000 - \$4,000)(A/P, 10\%, 2) + \$4,000(0.10)$$
$$+ \$5,100(P/\bar{A}, 10\%, 2)(A/P, 10\%, 2)$$
$$= \$4,000(0.57619) + \$400 + \$5,100(1.049)$$
$$= \$8,055$$

Inasmuch as this value is less than $9,000, it is preferable to keep the defender for at least 2 more years.

In problems of this type, it may be useful to think of the challenger and the defender as sets of mutually exclusive alternatives, as shown by:

$$C1 = \text{challenger with life of 1 year}$$
$$C2 = \text{challenger with life of 2 years}$$
$$\vdots$$
$$CN_{max} = \text{challenger with life of } N_{max} \text{ years}$$

Thus the first step is to determine which alternative represents the optimal challenger. The resulting $AC(CN^*)$ then serves as the standard against which all defender alternatives are measured. If there are no defender alternatives such that $AC(DN) < AC(CN^*)$, then replacement is signaled.

Note that the economic life of the defender is not necessarily determined by this procedure. The previous example showed that replacement should not take place because the AC for the-2-year defender life is less than the AC of the challenger. But it is possible that the AC will be still lower if the defender is kept for 3 more years. This additional information is irrelevant, however, since the question at hand deals only with current replacement, not with the economic life of the defender.

Once again, we emphasize that the procedure outlined above is valid only when the need for the service provided by the defender and/or its successive challengers is indeterminate, that is, when the planning horizon is essentially infinite. Problems that specify relatively short finite planning horizons generally result in complications that are beyond the scope of this discussion.

The assumption of "infinite planning horizon" is unnecessary in those problems for which only two alternatives are specified: (1) keep the defender in service for DN years or (2) replace the defender with a challenger that, if acquired, will also be kept in service for $CN = DN$ years. There is a common planning horizon, namely, $CN = DN$ years. This assumption is not very realistic, however. It implies that there are no lives for the defender other than DN years, no challenger lives other than CN years can be considered, and, finally, that $CN = DN$. Retirement/replacement problems rarely, if ever, are so simplistic.

8.5   THE GENERALIZED REPLACEMENT MODEL

The preceding sections postulated certain assumptions concerning the character of current and future challengers. Section 8.2 examined retirement with no replacement. Section 8.3 examined the situation where the current challenger (and all future challengers) is identical to the defender, and Section 8.4 examined the situation where the current challenger is unlike the defender but all future challengers are identical to the current challenger. Now the assumptions are relaxed in order to establish the most general, and probably the most realistic, case: *The current challenger is unlike the defender, and all future challengers are different from one another.*

It is generally expected that future challengers will be superior to those presently available in terms of initial cost, maintainability, useful physical life, and so on. The **generalized replacement model** provides for the broadest possible range of assumptions.

8.5.1   Exhaustive Search

Consider the stream of cash flows generated by the replacement of a defender, the subsequent replacement of the second replacement, and so on, throughout the planning horizon. There are a total of K such replacements over the entire planning horizon, as shown in **Figure 8.8**.

Let N_k represent the life of the kth replacement, $k = 1, 2, \ldots, K$. For the sake of symmetry and completeness, let $k = 0$ specify the defender. The problem, then, becomes one of determining the economic lives (N_k^*) of all $K + 1$ assets in the replacement sequence. If $N_0^* = 0$ as the result of these calculations, then the defender should be replaced now. That is, the value of N_0^* represents the economic life of the defender.

In general, the economic lives of the defender and its chain of replacements can be determined by finding those values of N_k such that the present worth of all cash flows over the planning horizon is maximized. The present worth, PW, is given by

$$\text{PW}(N_0, N_1, \ldots, N_K) = \sum_{k=0}^{K} [\text{PW}(k)](1 + i)^{-t} \qquad (8.6)$$

where

$\text{PW}(k)$ = net present value of the kth replacement in the sequence, computed at the time the kth replacement is acquired

i = discount rate (per period)

t = number of periods that will have elapsed until the kth replacement is made

$\quad = N_0 + N_1 + N_2 + \cdots + N_{k-1}$

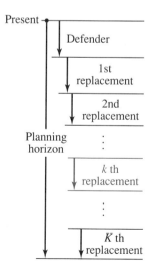

Figure 8.8   *K* Successive Replacements over a Finite Planning Horizon

The present worth of the kth replacement, computed at time t, is found from

$$\text{PW}(k) = \sum_{j=0}^{N_k} A_{k_j}(1+i)^{-j} \qquad (8.7)$$

where A_{k_j} is the anticipated cash flow in relative period j if the kth replacement is acquired at the end of t periods. Note that the PW formulation in Eq. (8.7) assumes end-of-period cash flows, as shown in **Figure 8.9**.

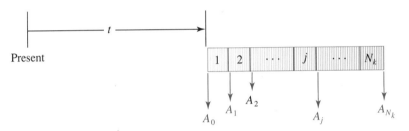

Figure 8.9   Cash Flows for the kth Replacement

In summary, the analyst must find the set $(N_0, N_1, \ldots, N_K)$ such that the total present worth is maximized. The optimal values of N_k are indicated by asterisks (*). The values of N_k^* represent the optimal replacement strategy over the planning horizon. In particular, immediate replacement of the current defender is signaled if $N_0^* = 0$.

Unfortunately, this procedure can be tedious if the cost and revenue functions are such that the present worths must be computed for all possible values of k. This will be illustrated in the following example.

EXAMPLE 8.7

Consider a planning horizon of only 3 periods. A total of eight separate present worths must be calculated for the eight combinations of lives of the defender and subsequent challengers, as displayed in **Table 8.4**. In the first combination, $N_0 = 3$. That is, the defender is retained throughout the 3-period planning horizon. In the second combination, the defender is retained for 2 periods and the first replacement, the challenger available at the end of 2 periods hence, is retained for 1 period. And so on. Finally, in the eighth combination, the defender is replaced immediately ($N_0 = 0$) and subsequent replacements are then effected at the start of the second ($N_1 = 1$) and third ($N_2 = 1$) periods.

Next, the anticipated cash flows for each combination, A_{k_j}, must be estimated and present worths computed using Eqs. (8.6) and (8.7). For the sake of brevity we will assume that this has been done. The resulting PW(N_0, N_1, N_2, N_3) are shown in the last column of Table 8.4.

Inspection of the results indicates that PW is maximized with combination 5. The optimal policy is to replace the defender now

TABLE 8.4   Present Worth of All Possible Combinations of
Defender and Replacement(s) Lives Given a
3-Period Planning Horizon (Example 8.7)

Combination	Defender Life (N_0)	Lives of Replacements			Present Worth (PW)
		1st (N_1)	2nd (N_2)	3rd (N_3)	
1	3	0	0	0	$50,000
2	2	1	0	0	45,000
3	1	2	0	0	55,000
4	1	1	1	0	60,000
5	0	3	0	0	(65,000) max
6	0	2	1	0	60,000
7	0	1	2	0	50,000
8	0	1	1	1	40,000

$(N_0^* = 0)$ with a challenger that itself will not be replaced throughout the planning horizon $(N_1^* = 3)$. Note that "replacement" is indicated for all combinations for which $N_0 = 0$.

8.5.2   Efficient Algorithms

In general, given a planning horizon of N periods, there are 2^N possible sets of assumptions, or separate values of PW, that must be calculated. If $N = 10$, for example, 1,024 values of PW must be calculated. Assuming a 20-period planning horizon, the number of possible values of PW exceeds a million! Clearly, the burden of calculation quickly becomes excessive for the general case unless an efficient algorithm can be used. Here, an "efficient algorithm" is one that leads to the optimal solution, that is, the set of N_k's such that PW is maximized, without an exhaustive search of all possible combinations. The generalized replacement model may be cast into a convenient solution framework by using **dynamic programming**, although the procedure is beyond the scope of this book.[3]

A number of simplified mathematical models have been developed to solve the equipment replacement problem when future challengers are unlike the current challenger. All such models depend on critical assump-

---

[3]One of the earliest articles that discusses this procedure and includes a numerical example is S. E. Drayfus, "A Generalized Equipment Replacement Study," *Journal of the Society for Industrial and Applied Mathematics*, Vol. 8, No. 3, September 1960, pp. 425–435.

tions concerning the characteristics of these unlike future challengers. Some of the more interesting of these models were developed by the research staff of the Machinery and Allied Products Institute (MAPI), under the direction of George Terborgh. The first appeared in *Dynamic Equipment Policy* (1949), the second in the *MAPI Replacement Manual* (1950), and the third in *Business Investment Policy* (1958). Although the MAPI models attracted a good deal of interest when they were first published, there is little indication that these models are used in industry today. Students interested in in-depth discussions of the MAPI methods should consult the sources directly. Other references to MAPI techniques appear in the Bibliography.

SUMMARY

Of those problems of interest in the capital allocation context, questions of retirement and replacement probably occur most frequently in the real world. Moreover, retirement/replacement problems present special conceptual and computational difficulties.

The simplest retirement case occurs when an existing asset is to be retired but not replaced. In this instance the discounted present value (present worth) of all prospective consequences during the remaining life of the asset should be evaluated. If the present worth of these consequences is greater than the current disposal value, then the asset should be retained; otherwise it should be terminated. This procedure may also be used to determine the economic life of the asset.

An additional complexity is generated when the problem of replacement is added to that of simple retirement. If the existing asset, the defender, is to be replaced by an *identical* asset, the challenger, and if the challenger that will become available in future years is also identical to the defender, the appropriate replacement decision is indicated by determining whether the age of the defender, N_0, is less than the economic life, N^*. If so, keep the defender in service. If $N_0 = N^*$, replace the asset now. If $N_0 > N^*$, determine the (equivalent uniform) annual cost, AC, for all possible remaining lives and keep the asset if there is any remaining life such that the AC at that life is less than AC(N^*). (Replacement must occur, of course, if the physical life of the asset has been met.)

In the event that the current challenger is *different* from the defender but identical or superior to all challengers that may become available in the future, the first step in resolving the replacement problem is to determine the economic life of the challenger. The AC of the challenger at this life is then compared to the cost of keeping the defender in service 1 more year. If the AC of the defender is lower, then keep the defender; if it is higher than the AC of the challenger, then the AC of the defender if kept 2 more years must be determined. The comparison is then repeated. If the AC of the defender with a 2-year life is less than the AC of the challenger, the defender should not be replaced; if it is greater, then the AC of the defender if kept 3 more years should be computed. This procedure is repeated until retention is indicated or the physical life of the defender has been met. If there is no defender life resulting in an AC less than that of the best challenger, the defender should be replaced.

The most difficult problems arise from cases in which the current challenger is different from the defender and succeeding challengers are different from the existing challenger. It is possible to structure a mathematical model that describes the general situation, but optimal solutions for this model are extremely difficult to determine. (For the generalized replacement model, if an improved challenger becomes available each year, there are 2^N possible solutions for a planning horizon of N years.) Mathematical programming procedures may be used to search for the optimal solution more efficiently—dynamic programming, for example, can be employed under certain conditions—but these procedures are beyond the scope of this book. References are included in the Bibliography for those who may be interested in additional discussion of this important topic.

| PROBLEMS

Unless otherwise indicated, assume that all cash flows are end-of-period and all discount rates are effective, per annum. Also assume that all future challengers will not differ significantly from the current challenger and, when the lives of the defender and the replacement assets are unequal, the planning horizon will be very long (approaching infinity).

Determining Economic Life

8.1 Machines of a given class have the cost characteristics shown below. The cost of a new machine is $100,000.

Machine Age (years)	Maintenance and Operating Costs	Salvage Value at End of Year
1	$12,000	$60,000
2	17,000	40,000
3	19,000	31,000
4	20,000	24,000
5	22,000	15,000
6	24,000	10,000
7	33,000	2,000
8	37,000	0

If the pretax discount rate is 12%, find the economic life of this machine and the (equivalent uniform) annual cost at the economic life. Ignore the effect of income taxes.

[*Answer:* AC($N^* = 6$) = **$42,514**.]

8.2 Consider the acquisition of a new asset (the challenger) having a 5-year physical life. The relevant economic data are:

	Estimated Market Value		Operating Costs (at End of Year)
Year	Start of Year	End of Year	
1	$10,000	$7,000	$1,000
2	7,000	5,000	2,000
3	5,000	4,000	3,000
4	4,000	3,200	4,000
5	3,200	2,500	5,000

The firm's pretax minimum attractive rate of return is 10%.
a. What is the (equivalent uniform) annual cost for this proposed asset if it is purchased and kept for the full 5 years?

b. Determine the economic life for this alternative and the (equivalent uniform) annual cost at the economic life.

8.3 [This problem was adapted from L. T Plank and A. J. Tarquin, *Engineering Economy*, ... ed. (New York: McGraw-Hill, 1983), p. 216.] A prospective acquisition, a capital investment, is currently available at a cost of $13,000. Anticipated salvage values and operating costs are:

End of Year	Salvage Value	Operating Costs
1	$9,000	$2,500
2	8,000	2,700
3	6,000	3,000
4	2,000	3,500
5	0	4,500

Assuming a 10% discount rate, determine the economic life for this challenger and the (equivalent uniform) annual cost at the economic life.
[*Answer*: $AC(N^* = 3) = \$6,132.$]

8.4 Certain manufacturing equipment with a current cost of $10,000 is being considered for a production process expected to continue for another 4 years. The end-of-year market values (salvage value) for this equipment are assumed as follows:

End of Year:	1	2	3	4
Market Value:	$8,000	$6,500	$5,500	$5,000

The other annual costs associated with this acquisition are equal from year to year. Determine the economic life for this proposed acquisition and the EUAC at that life.

8.5 The plant engineering department is considering the purchase of a new air compressor at a cost of $20,000. Its salvage value at the end of 5 years, the physical life of the equipment, is expected to be $5,000. The salvage values at the end of years 1, 2, 3, and 4 are approximated by the book values resulting from SYD depreciation. Annual costs of operation, $1,000 per year, are assumed to be distributed uniformly and continuously during each year. If the firm's

pretax MARR is expected to be 15%, determine the AC of this "challenger" at its economic life.
[*Answer*: $AC(N^* = 5) = \$6,298.$]

Simple Retirement (Abandonment)

8.6 Consider the retirement of a certain machine, currently in service, which has a remaining physical life of 2 years. If it is retired, the machine will not be replaced. The firm's effective annual discount rate is 10% per year. The machine has a current market value of $1,000. Estimates for the next 2 years are as follows:

End of Year	Market Value	Net Revenue
1	$800	$400
2	600	200

Should the machine be abandoned now or retained in service?

8.7 A company uses a compactor to press scrap materials, which are then sold to a salvage yard. The issue is whether to keep the compactor in service. If it is removed from service, the compactor will not be replaced. Estimates are as follows:

End of Year	Salvage Value	Net Value
0	$100,000	—
1	80,000	$ 5,000
2	70,000	20,000
3	60,000	35,000
4	60,000	45,000

For simplicity, assume that the firm's MARR is 0%. Shall the compactor be kept in service? Explain.
[*Answer*: Yes, because $PW(N) > 0$ for all $N = 3, 4.$]

8.8 The firm owns a photocopy machine, which it uses for volume copying of technical reports. The current market value of the machine is $10,000. It has a remaining physical life of 4 years. Estimated salvage values at the end of each of the next 4 years are:

End of Year:	1	2	3	4
Market Value:	$7,500	$5,500	$4,000	$3,000

If the machine is sold, the reports will be copied by an outside vendor at a cost to the firm of $2,000 each year. Thus the firm is considering two alternatives: Either continue the machine in service (not more than 4 more years) or retire the machine from service now.

Assume here that the current machine will not be replaced under any circumstances. The firm's pretax MARR is 12%.
a. Should the machine be sold now or kept in service (1 or more years)?
b. In terms of present worth, what is the value to the firm of keeping the machine in service exactly 4 more years?

8.9 Certain manufacturing equipment has a current salvage value of $10,000. It has a remaining physical life of 4 years. If it is continued in service, the equipment will result in operating savings as summarized below. Also shown are end-of-year salvage values. If the firm's MARR is 15%, determine the optimal policy. That is, should the equipment be retired from service?

Year n	Operating Savings in Year n	Salvage Value at End of Year n
1	$5,000	$6,000
2	4,000	4,000
3	3,000	2,000
4	2,000	0

[*Answer*: Do not abandon now.]

8.10 [From Thuesen and Fabrycky, *Engineering Economy*, 7th ed., Prentice-Hall, 1989, p. 299, Prob. 8.36.] A company is considering the retirement of an existing machine. (If it is retired, the machine will not be replaced.) This machine is expected to produce receipts of $8,000 in year 1, $9,000 in year 2, and $10,000 in year 3. Subsequent receipts will be $11,500 per year. Other relevant economic data are:

End of Year	Operating and Maintenance Costs	Salvage Value
0	—	$5,000
1	$ 4,000	4,000
2	5,000	3,000
3	6,000	2,000
4	7,000	1,000
5	9,000	700
6	11,000	0

If the MARR is 20%, determine the retirement age of the machine. Note that an answer of "zero" implies that the machine should be retired immediately.

Replacement with Identical Challenger

8.11 Consider a challenger with the following prospective consequences:

End of Year	Market Value	Operating Expenses
0	$10,000	—
1	8,000	$2,000
2	7,000	2,500
3	6,500	3,000
4	2,000	0

The initial cost of the challenger, as shown above, is $10,000. It has a maximum physical life of 4 years. Assume that $i = 8\%$ and all future challengers are identical to the current challenger.
a. Find the economic life of the challenger (CN^*).
b. Find the (equivalent uniform) annual cost of the challenger at its economic life.
c. Suppose that the defender is identical to the challenger and that the defender is now exactly 1 year old. Find the remaining defender life before the defender should be replaced.

[*Answers*: a. $CN^* = 3$; b. $4,450; c. 2 years.]

8.12 A manufacturing firm is considering the purchase of certain production machinery with a maximum physical life of 4 years. The current cost of this machinery is $10,000. End-of-year salvage (market) values, as well as annual operating costs, are summarized in the following table.

End of Year	Salvage Value	Operating Costs
1	$8,000	$6,000
2	6,500	6,000
3	5,500	6,500
4	5,000	7,500

The firm's pretax MARR is 10%.

a. Determine the equivalent uniform annual cost of this challenger if it is kept in service 4 years.

b. Determine the AC of the challenger at its economic life.

c. Suppose this challenger is now considered as a replacement for a defender that is identical to the challenger. All future challengers are also assumed to be identical. The defender is now 2 years old. Should the defender be replaced now? Why?

8.13 A manufacturing firm is considering the acquisition of certain materials handling equipment at a current cost of $20,000. The equipment will be used in connection with a certain product that will be discontinued after 4 years. Estimated end-of-year market values for this equipment are as follows:

End of Year:	1	2	3	4
Market Value:	$15,000	$11,000	$8,000	$5,000

The annual operating costs are expected to be constant at $18,000 per year. The firm's pretax MARR is 15%. It is assumed that future models of this equipment (i.e., future challengers) will not be less costly than this current challenger.

a. Determine the AC of this challenger at its economic life.

b. The equipment now being used by the firm, the defender, is identical to the challenger. It is now 3 years old. Should the defender be replaced now, or kept in service for 1 more year? Explain.

[*Answers*: a. $24,000; b. keep defender 1 more year.]

8.14 A 2-year old forklift truck is currently in service, and we are considering replacing it with an identical piece of equipment. Based on a 12% minimum attractive rate of return, preliminary analysis of the challenger indicates:

If Kept in Service N years	Capital Recovery if Kept in Service N years	Equivalent Uniform Annual Operating Costs if Kept in Service N years
1	$4,000	$2,500
2	3,400	3,000
3	3,000	3,500
4	2,800	4,000
5	2,700	4,500

a. Determine the EUAC of the challenger at its economic life.

b. Should the defender be replaced now? Why?

8.15 Consider the problem of replacing a certain 3-year-old asset (the defender) with an identical asset (the challenger). Assume that all future challengers will also be identical to the current challenger and the defender.

The challenger may be purchased for $10,000. Its maximum physical life is 5 years. End-of-year market values and annual operating costs are:

End of Year	Market Value	Operating Costs
1	$8,000	$10,000
2	4,000	10,000
3	2,000	10,000
4	0	10,000
5	0	10,000

The firm's before-tax minimum attractive rate of return is 15%.

a. Find the economic life for the challenger and the (equivalent uniform) annual cost at the economic life.

b. Decide whether the defender should be replaced. Explain why or why not.

[*Answers*: a. AC($N^* = 5$) = **$12,983; b. do not replace now.**]

8.16 A certain asset, now 2 years old, is a candidate for replacement by an identical challenger. Cash flow estimates for the challenger are as follows:

End of Year	Salvage Value	Operating Costs
0	$50,000	—
1	35,000	$15,000
2	25,000	15,000
3	17,000	18,000
4	10,000	22,000
5	5,000	24,000

For simplicity, assume that the firm's MARR is 0%.
a. Determine the EUAC of the challenger at its economic life.
b. Shall the defender be replaced now? Explain.

8.17 Consider the problem of replacing a certain defender with an identical challenger. The relevant data for the challenger are as described below. The original cost of the challenger is $10,000.

Year n	Operating Costs in Year n	Market Value at End of Year n
1	$7,000	$7,000
2	7,000	5,000
3	7,500	4,000
4	7,450	3,400
5	8,300	3,000
6	9,300	2,800
7	8,850	2,650

The maximum physical life of the challenger is 7 years. For ease of calculation, assume a 0% discount rate. (This is an artificial assumption, of course, but it does serve to expedite calculations without sacrificing the essential character of the problem.)
a. Determine the AC at the challenger's economic life.
b. The defender is now 3 years old. Should it be retained in service? If so, for how long?
[*Answers*: a. $8,900; b. keep for 2 more years.]

Challenger Unlike Defender, But All Future Challengers Identical to (or No Worse than) Current Challenger

8.18 Suppose that the production machinery in Problem 8.12 is under consideration to replace a defender that has no current market value. This is not expected to change in the future. The defender has a remaining physical life of 4 years. Operating costs will be $7,500 next year and will increase by $1,000 per year. That is, costs will be $7,500, $8,500, $9,500 and $10,500 at the end of each of the next 4 years.
a. Determine the AC of the defender if it is kept in service for 4 more years.
b. Should the defender be replaced now with the challenger described in the Problem 8.12? Explain.

8.19 The Blue Company owns a machine that cost $26,000 ten years ago. A new machine is available that costs $11,000 and will save $2,000 annually.

If the new machine is purchased, the old machine will be sold for $8,000. If the old machine is retained, it will be scrapped in 5 years for zero salvage value. Assume that the estimated salvage value of the new machine will be $1,000 at the end of 5 years. The minimum attractive rate of return before income taxes is 20%. Should the existing machine be replaced with the challenger? Why or why not?
[*Answer*: Yes, replace the old machine now.]

8.20 General Hospital is considering the purchase of certain new kitchen equipment to replace its present 3-year-old equipment. The total physical life of the present equipment is 5 years. Cost history for the past 3 years and estimates for the next 2 years are:

Year	Operations	Maintenance	End-of-Year Salvage Value
	Past Disbursements		
1	$ 6,000	$1,000	$23,000
2	8,000	3,000	18,000
3	10,000	5,000	14,000
	Estimated Future Disbursements		
4	12,000	7,000	11,000
5	14,000	9,000	9,000

The kitchen superintendent has suggested that the present equipment be replaced by a new model costing $60,000. The manufacturer claims that the new equipment will require no maintenance and will reduce operating costs by $1,000 per year.

Assuming a 5-year physical life for the challenger with a $10,000 terminal salvage value, make a comparison of (equivalent uniform) annual costs using a 10% minimum attractive rate of return.

8.21 The manager of the Western Division of the Brown Manufacturing Company has received a request from the superintendent of the Los Angeles plant for a new heavy-duty generator to replace the existing generator. The manager has forwarded the request to corporate headquarters for approval. The existing generator was purchased 5 years ago for $5,000. It is believed that it may remain in operation for another 8 years before it must be scrapped. The existing generator is specialized equipment but, fortunately, there is a buyer who is willing to pay $600 for it. However, this will probably be the last chance to sell it at any price.

The proposed replacement will cost $6,000, and it has an expected useful life of 10 years. It is expected that the actual market value of the proposed equipment will be:

End of Year	Market Value
1	$4,000
2	2,400
3	1,200
4	400
5–10	0

The estimated annual operating costs for the new equipment are $600. Operating costs for the existing equipment are expected to be $1,800 per year. The company plans to sell the plant and all its equipment in 5 years. The sale price is independent of the present decision concerning the generator.

Assuming a before-tax minimum attractive rate of return of 15%, determine whether the new equipment should be purchased. Explain why or why not.

[*Answer*: **Keep the defender 5 more years.**]

8.22 The Core Corporation is considering replacing or remodeling a small commercial structure. A new building will cost $600,000 and will have an estimated useful life of 20 years. It is expected that the present structure, if it is repaired, will have an additional useful life of 7 years. Remodeling costs are estimated to be $180,000. The existing building may be sold for $80,000 in its present condition, but if it is remodeled and retained for another 7 years, it will have no disposal value at that time. The terminal salvage value of a replacement structure after 20 years is estimated to be $1,000,000. All other costs are unaffected by the replacement decision.

If the before-tax minimum attractive rate of return is 15%, determine the (equivalent uniform) annual costs of the two alternatives.

8.23 A high-speed, special-purpose, automatic strip-feed punch press costing $150,000 has been proposed to replace three hand-fed presses now in use. The life of this automatic press has been estimated to be 5 years. Expenditures for labor, maintenance, and so forth, are estimated to be $30,000 per year.

The general-purpose hand-fed punch presses each cost $20,000 ten years ago and were estimated to have a life of 20 years, with a salvage value of $2,000 each at the end of that time. Their present net disposal value is $7,500 each. Operating expenditures for labor, and so forth, will be about $28,500 per machine per year. The net salvage value is expected to decrease by about $1,000 per press per year. It is expected that the required service will continue for only 5 more years and that the salvage value of the challenger will be $40,000 at the end of that time.

If the company's before-tax minimum attractive rate of return is 20%, determine whether the defenders should be replaced.

[*Answer*: **Replace the defenders now.**]

8.24 A firm is considering the replacement of a certain machine (the defender) with a certain new machine (the challenger). The relevant economic data are:

	Defender		Challenger	
End of Year	Market Value	Operating Costs	Market Value	Operating Costs
−4	$8,000	$2,000	—	—
−3	6,000	2,200	—	—
−2	4,500	2,400	—	—
−1	3,200	2,600	—	—
0 (now)	2,500	2,800	$6,500	—
1	2,000	3,000	4,000	$3,000
2	1,600	3,200	4,000	3,000
3	1,300	3,400	4,000	3,000
4	1,100	3,600	4,000	3,000
5	1,000	3,800	4,000	3,000
6	—	—	1,000	3,000
7	—	—	1,000	3,000
8	—	—	1,000	3,000
9	—	—	1,000	3,000
10	—	—	1,000	3,000

The firm's minimum attractive rate of return is 10%. The present time (now) is the end of year 0. Assume that the operating costs at the end of year 0, $2,800, are sunk costs, that is, they were incurred prior to the beginning of the planning horizon. The defender will reach its maximum physical life at age 10.

a. Determine the economic life of the challenger and its (equivalent uniform) annual cost at its economic life.

b. Assume that the need for this type of equipment, whether defender or challenger, will cease at the end of year 5. Given this 5-year planning horizon and assuming that all future challengers are identical to the current challenger, determine whether the defender should be replaced. Explain your answer.

8.25 The General Bottling Company is considering the replacement of its 20-year-old capping machine, which was originally purchased for $120,000. The current net salvage value of the machine is zero because the cost to haul it away is about equal to its scrap value. Operating costs are currently $200,000 per year and are expected to increase at the rate of $20,000 per year. That is, next year's costs will be $200,000, the following year's costs will be $220,000, and so on. The remaining physical life of the machine is 5 years.

A new capping machine is available for $400,000, but because it is a special-purpose machine, its salvage value will drop sharply:

Age (years)	Salvage Value
1	$240,000
2	120,000
3	40,000
More than 3	0

Operating costs for this new machine should hold steady at $160,000 per year over its 20-year physical life. Assume that all future challengers will be identical to the current challenger. The firm's pretax minimum attractive rate of return is 20%.

a. Find the (equivalent uniform) annual cost for the challenger at its economic life.

b. Find the (equivalent uniform) annual cost for the defender if it is kept for 5 more years.

c. Should the defender be replaced now? Should it be replaced at any time over the next 5 years? Explain your answers.

[*Answers*: a. AC($CN^* = 20$) = **$257,657**;
b. AC($DN = 5$) = **$255,384**;
c. **Keep defender 2 more years, then replace.**]

8.26 Assume that the challenger in Problem 8.3 is being compared to a defender with a current market value of $5,000. Anticipated salvage values and operating costs for the defender are:

End of Year	Salvage Value	Operating Costs
1	$4,000	$3,000
2	2,000	3,500
3	0	4,000

The defender was originally purchased 1 year ago for $8,000. Its remaining physical life is 3 years. There is a perpetual need for this type of equipment. The discount rate is 10%.

Determine the optimal strategy. How many more years, if any, should the defender be retained before it is replaced?

8.27 The Bruin Company is considering replacement

of certain manufacturing equipment. It has been determined that the economic life of the challenger is 5 years, and the EUAC over 5 years is $4,000. Bruin's pretax MARR is 12%. Economic estimates for the defender are as follows:

End of Year	Salvage Value	Operating Costs
0	$5,000	—
1	3,000	$1,800
2	1,000	2,000
3	0	2,500

It is assumed that there will be an indefinite need for this equipment—at least 5 years—and that future challengers will be identical to the challenger. The maximum remaining physical life for the defender is 3 years. Should the defender be replaced at any time? Why?

[*Answer*: **Replace the defender now.**]

8.28 [From E. P. DeGarmo, W. G. Sullivan, and J. A. Bontadelli, *Engineering Economics*, 8th ed., Macmillan: New York, 1988, p. 454.] Suppose that it is desired to determine how much longer an old forklift truck should remain in service before it is replaced by a new truck. The expected salvage value and operating expenses for the defender are shown below:

End of Year	Salvage Value	Operating Expenses
1	$4,000	$5,500
2	3,000	6,600
3	2,000	7,800
4	1,000	8,800

The defender is now 2 years old, originally cost $13,000, and has a present realizable market value of $5,000. The firm's pretax cost of capital is 10%.

a. Determine the (equivalent uniform) annual costs of this defender if kept 1, 2, 3, or 4 more years.

b. Suppose that the AC for the challenger is $8,600 at its economic life. Assume that all future challengers are identical to the current

challenger. For how much longer should the defender be kept in service before it is replaced with the challenger? Explain.

8.29 The Specialty Products Co. is currently using certain equipment that it purchased 20 years ago for $25,000. Management had judged the equipment to have no market value, but an offer has just been received from a similar company in Latin America, offering $10,000 for the equipment. This is probably Specialty Product's last chance to sell the equipment at any price.

When the equipment was first acquired, operating costs were averaging about $35,000 annually. Next year, operating costs are expected to be $40,000, increasing each year by $2,000. If it is retained, Specialty products plans to keep the equipment for at most 5 more years, at which time operating costs will be $48,000. The firm's pretax MARR is 15%.

a. Assume that, if the equipment is sold now, Specialty Products will have to lease comparable equipment at a cost of $48,000 annually over the next 5 years. (Assume that lease payments are end-of-year.) Should Specialty Products retain the equipment? If so, for how long?

b. Assume that replacement equipment, the challenger, is currently available. What is the maximum AC for the challenger that would justify replacement at this time, assuming that all future challengers are identical to the current challenger?

[*Answers*: **Keep the defender 4 more years; b. $46,155.**]

8.30 Consider a replacement problem in which it is assumed that all future challengers will be identical to the current challenger. Assume a zero interest rate ($i = 0\%$). The defender's maximum remaining physical life is 4 years. The challenger's maximum physical life is 5 years.

The cost of the challenger is $10,000; its market value will be zero at any time in the future. Operating costs are expected to be $4,000 per year.

The relevant data for the defender are as shown:

End of Year	0	1	2	3	4
Market Value:	$7,000	$3,000	$2,000	$1,000	0
Operating Cost:	—	$4,800	$4,800	$4,800	$4,800

a. Determine the EUAC of the challenger at its economic life.

b. Should the defender be replaced now? Why?

[*Answers*: a. **$6,000 at 5 years; b. yes.**]

A certain piece of production equipment was purchased several years ago at a cost of $100,000. Its current market value is $6,000, but the market value is expected to be negligible in the future. The remaining physical life of this defender is 2 years. Operating costs are expected to be $2,000 per year over the next 2 years.

A challenger is currently under consideration that will cost $9,000 and has an expected service life of 3 years. Operating costs are $1,000 per year over the next 3 years. In addition, there will be a $1,500 negative salvage value at the end of the third year due to the cost of dismantling the equipment.

Additional assumptions: infinite planning horizon (indefinite need for this equipment); identical replacement (all future challengers identical to the one currently available); zero discount rate (MARR = 0%); end-of-period cash flows.

a. First, consider the current market value of the defender to be its initial "cost." Then the appropriate cash flow table for this problem is as shown.

j	$A_j(D)$	$A_j(C)$
0	− $6,000	− $9,000
1	− 2,000	− 1,000
2	− 2,000	− 1,000
3		− 2,500

Based on comparisons of AC, what is the optimal strategy (keep defender or buy challenger)?

b. Second, consider the current market value of the defender to be a reduction in the purchase price of the challenger. Then the appropriate cash flow table for this problem is as shown.

j	$A_j(D)$	$A_j(C)$
0		− $3,000
1	− $2,000	− 1,000
2	− 2,000	− 1,000
3		− 2,500

Based on comparisons of AC, what is the optimal strategy (keep defender or buy challenger)? Why?

c. Which of the above solutions is correct? Why?

[*Answers*: a. **replace now, AC(C3) = $4,500;** b. **keep defender 2 years, AC(D2) = $2,000; c. first approach is correct.**]

8.32 A forklift truck is being considered for acquisition. The current cost is $6,000. It will be kept in service for at most 6 years. The firm's pretax MARR is 10%. Other relevant data are as follows:

End of Year	Salvage Value	Operating Costs
0	$6,000	—
1	4,500	$1,700
2	3,400	1,900
3	2,400	2,500
4	2,000	3,000
5	1,500	3,500
6	1,000	4,000

a. Determine the EUAC of this "challenger" at its economic life.

b. Assume that all future challengers are identical to the current challenger. As an alternative to purchasing the forklift truck, the company can rent a forklift truck for $3,400 per year, with rental payments at the start of each year. There will be a need for a forklift truck of this type for 6 years. What is the optimal policy? That is, should the truck be purchased or leased? If purchased, for how long should it be kept in service? Keep in mind that the lease opportunity is available at any time.

8.33 A prospective acquisition, a capital investment, is currently available at a cost of $13,000. Anticipated salvage values and operating costs of this "challenger" are as shown:

End of Year:	1	2	3	4	5
Salvage Value:	$9,000	$8,000	$6,000	$2,000	$0
Operating Costs:	$2,500	$2,700	$3,000	$3,500	$4,500

The equipment currently in service, the defender, has a current market value of $5,000. Anticipated salvage values and operating costs are as shown:

End of Year:	1	2	3
Salvage Value:	$4,000	$2,000	$0
Operating Costs:	$3,000	$3,500	$4,000

The defender was originally purchased 1 year ago at a cost of $8,000. Its remaining physical life is 3 years. Assume that all future challengers will be identical to the current challenger. The firm's pretax MARR is 15%.

a. Determine the AC of the challenger at its economic life.

b. Should the defender be replaced now? If not, when?

[*Answers*: a. AC($CN^* = 3$) = $6,676; b. keep defender 3 more years.]

8.34 A $10,000 "challenger" is being considered to replace existing equipment. The cash flows associated with the challenger are as follows:

End of Year	Market Value	Operating Costs
0	$10,000	—
1	6,000	$2,000
2	3,000	2,000
3	0	2,000

The maximum physical life for the challenger is 3 years. The firm's MARR is 20% before income taxes.

a. Determine the AC of the challenger at its economic life.

b. If the defender is *identical* to the challenger, and if the defender is now 2 years old, should the defender be replaced now?

c. Suppose that the defender is *not identical* to the challenger, and its remaining physical life is 4 years. Analysis of the defender indicates that:

AC if kept 1 more year = $7,000

AC if kept 2 more years = $6,800

AC if kept 3 more years = $7,000

AC if kept 4 more years = $7,200

In this case, should the defender be kept in service?

8.35 Testing equipment whose initial cost is $4,000 has the following characteristics:

End of Year:	1	2	3	4
Salvage Value:	$3,000	$2,000	$1,600	$1,000
Operating Cost:	600	700	900	1,200

Disregard the cost of capital for this problem; that is, assume that $i = 0\%$. There is a perpetual need for the services of the equipment.

a. If the equipment is to be replaced by an identical one, when should this replacement take place and what would be the (equivalent uniform) annual cost for the years it is kept in use?

b. Suppose the equipment has been in use for 2 years now and a new challenger has recently become available. The economic life of the challenger is 5 years and the corresponding equivalent annual cost is $1,400. Determine if the challenger should replace the old equipment and, if so, determine when the replacement should take place.

[*Answers*: a. **$1,533 at 3 years; b. keep defender 1 more year, then replace.**]

Retirement/Replacement After Income Taxes

8.36 An automatic lathe was purchased 2 years ago for $20,000. It has been depreciated using the straight line method over 10 years using an estimated salvage value of $4,000 at the end of the 10th year.

The machine can now be sold for $15,000. If it is kept for another 8 years, as originally planned, it appears that the sale price at disposal will be $5,000. The company has an effective income tax rate of 40%. Assume that long-term gains (and losses) of this depreciable asset are taxed at the ordinary income rate. The firm's minimum attractive rate of return is 10% after taxes.

Determine the (equivalent uniform) annual cost after taxes if the machine is kept for 8 more years.

8.37 The Cardinal Company owns a machine that cost $26,000 ten years ago. It is being depreciated by the straight line method over 12 years to an estimated $2,000 salvage value. A new machine is available that costs $11,000 and, if purchased, will save $2,000 annually. This new machine will be depreciated by the sum of the years-digits method over 4 years to an estimated $1,000 salvage value.

If the new machine is purchased, the old machine will be sold for $8,000. If the old machine is retained, it will be scrapped in 5 years for zero salvage value. Assume that the estimated salvage value of the new machine will be $1,000 at the end of 5 years, that the minimum attractive rate of return after taxes is 12%, and that the incremental income tax rate is 40%. Should the existing machine be replaced with the challenger described above? Why or why not?

[*Answer*: **Replace now.**]

8.38 Refer to Problem 8.37. Repeat the analysis assuming that the challenger is to be depreciated under regular MACRS as a 5-year recovery property. All other assumptions are the same.

8.39 The manager of the Western Division of the Brown Manufacturing Company has received a request from the superintendent of the Detroit plant for a new heavy-duty generator to replace existing equipment. The manager has forwarded the request to corporate headquarters for approval. The existing generator was purchased 5 years ago for $50,000. It was depreciated by the straight line method over 5 years to zero salvage value.

Although its present book value is zero, it is believed that it may remain in operation for another 8 years before it must be scrapped. The existing generator is specialized equipment, but fortunately there is a buyer who is willing to pay $6,000 for the equipment. This will probably be the last chance to sell it at any price.

The proposed equipment will cost $60,000 and has an expected useful life of 10 years. If it is purchased, the regular MACRS depreciation method will be used with a 5-year recovery period. It is expected that the actual market value of the equipment at any point in time will be the same as its book value at that time. The estimated annual operating costs for the new equipment are $6,000; operating costs for the existing equipment are expected to be $18,000 per year.

The company plans to sell the plant and all its equipment in 5 years The sale price is independent of the present decision concerning the generator.

Assuming an after-tax minimum attractive rate of return of 10% and an incremental tax rate of 40%, determine whether the new equipment should be purchased. Expain why or why not.

[*Answer*: **Keep the defender**].

8.40 Refer to Problem 8.39. Repeat the analysis assuming that the defender has been depreciated using the alternate MACRS (A/MACRS) method with a 5-year recovery period.

APPENDIX: *Computer Software Application*

As was shown in previous chapters, generic computer software may be used in the analysis of the economic consequences of retirement/replacement problems. In this appendix we show how a *Lotus 1-2-3* spreadsheet is developed to determine the (equivalent uniform) annual costs for a challenger in order to identify the challenger's economic life.

EXAMPLE 8.A.1

Consider a challenger that has an original cost of $6,000, maximum service life of 6 years, and year-by-year salvage values, S_j, as shown in cells B15–B21 of **Figure 8.A.1**. The year-by-year operating costs, C_j, are shown in cells C16–C21. Summarizing,

Cells A15–A21: Year j, $j = 0, 1, 2,...,6$
Cells B15–B21: End-of-year salvage values, S_j
Cells C16–C21: End-of-year operating costs, C_j

	A	B	C	D	E	F	G
1	**Retirement and Replacement Analysis:**						
2							
3	**EXAMPLE 8.A.1**						
4	(Determine EUAC of the Challanger)						
5							
6	Current Cost					$6,000	
7	Service Life (years)					6	
8	Pre-tax MARR					0.10	
9							
10							
11	End of	Salvage	Operating	Marginal	Total Margina	Cum.	EUAC if kept
12	year	Value	Costs	Cost of CR	Cost	PW	in service
13	j	Sj	Cj	CRj	Cj + CRj	@ 10%	j years
14							
15	0	6,000					
16	1	4,500	1,700	2,100	3,800	3,455	3,800
17	2	3,400	1,900	1,550	3,450	6,306	3,633
18	3	2,400	2,500	1,340	3,840	9,191	3,696
19	4	2,000	3,000	640	3,640	11,677	3,684
20	5	1,500	3,500	700	4,200	14,285	3,768
21	6	1,000	4,000	650	4,650	16,910	3,883
22							
23							
24	The Economic Life of the Challenger is the year in						
25	which the EUAC is minimum.						
26							
27	**Sample Cell Formulas:**						
28	D16 = (B15-B16) + $F$8*B15						
29	E16 = +C16+D16						
30	F16 = +E16*(1+$F$8)^(-A16)						
31	G16 = ($F$8/(1-(1+$F$8)^(-A16)))*F16						

Figure 8.A.1 Spreadsheet to Determine the EUAC of the Challenger, Example 8.A.1

The firm's pretax MARR is 10%.

The marginal cost of capital recovery in year j, CR_j, is given by

$$CR_j = \text{(decline in salvage value in year } j)$$
$$+ \text{ (interest foregone on the capital invested}$$
$$\text{at the start of the year)}$$
$$= (S_{j-1} - S_j) + iS_{j-1}$$

The total marginal cost in year j is the sum of the marginal cost of capital recovery plus the operating costs in year j, $CR_j + C_j$. The cumulative present worth if the challenger is kept through N years is given by

$$\sum_{j=0}^{N} (CR_j + C_j)(1 + i)^{-j}$$

and the (equivalent uniform) annual cost is given by

$$AC(N) = \left[\sum_{j=0}^{N} (CR_j + C_j)(1 + i)^{-j} \right](A/P, i, N)$$

For $i = 10\%$, the values of AC(N) are shown in cells G16–G21 for $N = 1, 2, \ldots, 6$. The minimum AC occurs at $N = 2$. That is,

$$AC(N^* = 2) = \$3{,}633$$

9

The Revenue Requirement Method

9.1 INTRODUCTION

All industries in the United States are regulated to some extent. Local, state, and federal agencies exercise control in varying degrees over a wide spectrum of business activities. To list but a few examples: city and county building codes impose restrictions on the design of a firm's new manufacturing plant and warehouse; state highway departments issue regulations controlling the types of vehicles permitted on certain roads in the state as well as the license fees to be paid by highway users; and the federal Environmental Protection Agency issues policies concerning the disposal of toxic waste materials. In some instances the prices that member firms in an industry can charge their customers are regulated by government agencies. Tariffs on many goods transported interstate are regulated, as are the rates charged by public utilities, especially those firms that provide gas, water, or electricity to the community.

The regulation of these public utilities has given rise to an evaluation methodology known as the **revenue requirement method**, an approach that differs markedly from the other methodologies presented earlier. In this chapter we will present the revenue requirement method in some detail, provide numerical examples to illustrate its application, and, finally, show the equivalence between the revenue requirement and the other conventional approaches used in "nonregulated" industries.

9.1.1 Some Important Differences Between Regulated Public Utilities and Nonregulated Industries

Nonregulated business firms are generally free to price goods and services so as to meet the owners' objectives. Revenues are affected by the interaction of market forces and the firm's investment strategies. How-

330

ever, an important exception arises in the case of regulated utilities, whether they are owned by private shareholders or public entities.[1] Because utilities hold monopolistic or oligopolistic positions within a given community, the community protects its interests by regulating the behavior of the utility through a *utilities commission* or a similar regulatory body. Prices, or rates, charged by the utility are normally subject to review and approval by the regulatory agency.

With few exceptions, public utilities require very large amounts of capital for necessary plant and equipment. Capital costs are relatively higher, and operating costs are relatively lower, than those experienced in competitive industries. The required funds for these capital investments are usually raised by extensive borrowing (debt) rather than through the sale of additional ownership shares (equity). Thus the ratio of debt to equity capital is somewhat higher for regulated public utilities.

9.1.2 Basic Philosophy

The philosophy underlying the regulation of prices (rates) is that utilities are entitled to price their products and services so that all costs are recovered, including a *fair return on the rate base.* (There is no guarantee that a utility will in fact earn a fair return under a permitted rate structure. Operational experience that occurs after the authorization may well prove to be different from what was predicted, or forecasted, when authorization to charge certain rates was originally sought.) Because regulatory agencies act on behalf of the consumers of the utility's products and services, investment decisions should be made in such a way that revenue requirements, while meeting the costs and providing a fair return to the utility, are minimized.

9.2 ELEMENTS OF PERMITTED REVENUE

As illustrated in **Figure 9.1**, the permitted revenue, or **revenue requirement** (RR), may, in general, be characterized as the sum of six elements:[2]

K = current operating disbursements (labor and material costs associated directly with the investment, but also other expenses such as property taxes and insurance)

---

[1]Regulated utilities are "public" because their customers are, generally, all the members of the community. Some public utilities are owned by the community; others are owned by individual shareholders.

[2]Inasmuch as the notation introduced in this chapter differs somewhat from the notation used in other chapters, a special glossary has been included as Appendix 1 at the end of the chapter.

D = depreciation expense, the return *of* investment

I = interest on that portion of the investment representing debt capital, the return to lenders

T = income taxes paid on taxable income

P = dividends paid to *preferred stockholders* based on the portion of the investment that represents funds raised through preferred stock

S = return to owners, that is, return to the utility's *common stock holders*

The return *on* investment is the sum of the return to the lenders (I) plus the returns to the preferred and common stockholders ($Q = P + S$). **Carrying charges** (CC) are all permitted cost elements except the operating disbursements, or $\text{CC} = R - K$.

It should be noted that there are two types of depreciation. The first, **book depreciation** (D') represents the decline in the amount of invested capital over the course of the year. This is a measure of the

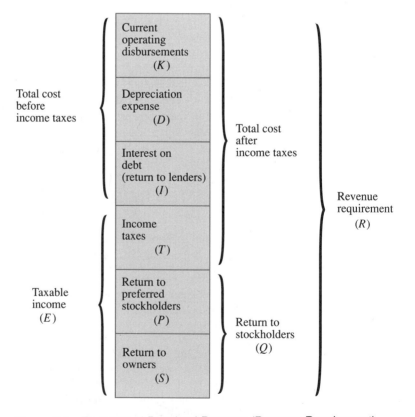

Figure 9.1   Elements of Permitted Revenue (Revenue Requirement)

decline in the **rate base**. Book depreciation is a form of capital recovery in the sense that it allows the initial investment to be returned to the investors in a systematic manner over the life of the investment.

The second type of depreciation is used in the firm's income statement. This is an allowable expense that reduces taxable income. (See Chapter 6.) This is known as **tax depreciation** (*D*).

9.3   DETERMINING THE "FAIR (RATE OF) RETURN"

At this point it will be helpful to review the concepts underlying the "costs" of the various types of capital invested by the utility. These concepts, as you will see, are essential to the development of the revenue requirement models.

Broadly speaking, a utility's investment funds consist of debt and equity capital. **Debt** is simply borrowed money, the cost of which is reflected by interest payments to lenders. **Equity capital** is obtained through investments by the owners—taxpayers, in the case of publicly owned utilities—and through retained earnings. There is, of course, an opportunity cost associated with equity capital in the sense that these funds could be employed elsewhere if they were not used for investment by the utility. In addition to loans and equity obtained through the sale of **common stock**, utilities (and other firms) frequently obtain needed investment funds by issuing **preferred stock**. Dividends on preferred stock are paid at a specified rate after all other expenses, including interest on debt, are paid, but before returns to common stockholders are paid.[3] In summary, then, if investment funds are made available by lenders as well as by common and preferred stockholders, the investment must provide sufficient revenues to reward lenders as well as common and preferred stockholders.

(Some authors categorize preferred stock as a type of equity capital. We prefer to treat preferred stock separately, although the distinction is not important in this context.)

The **weighted average cost of capital** (*i*) after income taxes is given by

$$i = w_b k_b + w_p k_p + w_e k_e \tag{9.1}$$

where

w_b = proportion of debt k_b = cost of debt before tax

w_p = proportion of preferred stock k_p = cost of preferred stock

w_e = proportion of equity k_e = cost of equity

---

[3]Dividends to preferred stockholders are not guaranteed, of course. These dividends can be paid only after more senior obligations (debt, for example) have been met.

and

$$w_b + w_p + w_e = 1.0$$

The cost of debt (k_b) is the average interest rate paid for the borrowed capital committed for investment. Similarly, the cost of preferred stock (k_p) is the average dividend rate paid on investment funds raised through issuing preferred stock. The cost of equity (k_e) is that **fair return** necessary to attract investors (in the case of investor-owned utilities) or to compensate taxpayers (in the case of publicly owned utilities).

EXAMPLE 9.1

Suppose that a utility's investment capital consists of the following:

Source	Proportion (w)	Cost (k)
Debt	0.55	0.10
Preferred stock	0.10	0.10
Common stock	0.35	0.14

The weighted-average cost of capital is

$$i = 0.55(0.10) + 0.10(0.10) + 0.35(0.14)$$

$$= 0.055 + 0.010 + 0.049$$

$$= 0.114$$

9.4 DETERMINING INCOME TAXES

The income taxes paid by the utility in a given year $j(T_j)$ are found by multiplying the effective tax rate for that year (t_j) by the taxable income in that year (E_j).

$$\begin{pmatrix} \text{Income} \\ \text{tax} \end{pmatrix} = \begin{pmatrix} \text{taxable} \\ \text{income} \end{pmatrix} \times \begin{pmatrix} \text{tax} \\ \text{rate} \end{pmatrix}$$

$$T_j = E_j t_j \qquad (9.2)$$

where

$$\begin{pmatrix} \text{Taxable} \\ \text{income} \end{pmatrix} = (\text{revenue}) - \left[\begin{pmatrix} \text{recurring} \\ \text{expenses} \end{pmatrix} + \begin{pmatrix} \text{interest} \\ \text{on debt} \end{pmatrix} + \begin{pmatrix} \text{tax} \\ \text{depreciation} \end{pmatrix} \right]$$

$$E_j = R_j - (K_j + I_j + D_j) \qquad (9.3)$$

where

$$R_j = \text{Revenue} = \begin{pmatrix} \text{carrying} \\ \text{charges} \end{pmatrix} + \begin{pmatrix} \text{recurring} \\ \text{expenses} \end{pmatrix}$$

$$R_j = CC_j + K_j \qquad (9.4)$$

where

$$\begin{pmatrix} \text{Carrying} \\ \text{charges} \end{pmatrix} = \begin{pmatrix} \text{return on} \\ \text{investment} \end{pmatrix} + \begin{pmatrix} \text{book} \\ \text{depreciation} \end{pmatrix} + \begin{pmatrix} \text{income} \\ \text{taxes} \end{pmatrix}$$

$$CC_j = (ROI)_j + D'_j + T_j \qquad (9.5)$$

where

$$\begin{pmatrix} \text{Return on} \\ \text{investment} \end{pmatrix} = \begin{pmatrix} \text{return to} \\ \text{lenders} \end{pmatrix} + \begin{pmatrix} \text{return to} \\ \text{stockholders} \end{pmatrix}$$

$$(ROI)_j = I_j + Q_j \qquad (9.6)$$

where

$$\begin{pmatrix} \text{Return to} \\ \text{lenders} \end{pmatrix} = \begin{pmatrix} \text{proportion of} \\ \text{investment that is debt} \end{pmatrix} \times \begin{pmatrix} \text{cost of} \\ \text{debt} \end{pmatrix} \times \begin{pmatrix} \text{rate base at} \\ \text{start of year} \end{pmatrix}$$

$$I_j = w_{bj} k_{bj} B'_j \qquad (9.7)$$

and

$$\begin{pmatrix} \text{Return to} \\ \text{stockholders} \end{pmatrix} = \begin{pmatrix} \text{proportion of invest-} \\ \text{ment that is equity} \end{pmatrix} \times \begin{pmatrix} \text{cost of} \\ \text{equity} \end{pmatrix} \times \begin{pmatrix} \text{rate base at} \\ \text{start of year} \end{pmatrix}$$

$$Q_j = (w_{pj} k_{pj} + w_{ej} k_{ej}) B'_j$$

$$= (i_j - w_{bj} k_{bj}) B'_j \qquad (9.8)$$

Combining Eqs. (9.3) through (9.8), inserting into (9.2) and simplifying,

$$T_j = \left(\frac{t_j}{1 - t_j} \right) (Q_j + D'_j - D_j) \qquad (9.9)$$

$$= \left(\frac{t_j}{1 - t_j} \right) [(i_j - w_{bj} k_{bj}) B'_j + D'_j - D_j] \qquad (9.10)$$

EXAMPLE 9.2

Suppose that the utility with the capital structure given in Example 9.1 is considering investment in certain equipment requiring an initial cost of $90,000. This equipment will be depreciated for *tax* purposes as a 5-year property under regular MACRS. However, the *rate base* will be decremented by the SYD depreciation method over 5 years and assuming a zero salvage value. The tax rate, t, is 40%. The capital structure and the tax rate are assumed to remain constant over the life of the investment.

TABLE 9.1   Determining Annual Income Taxes (Example 9.2)

Year j (1)	Unrecovered Investment at Start of Year (Rate Base) B'_j (2)	Reduction in the Rate Base (Book Depr.) D'_j (3)	Tax Depreciation D_j (4)	Income Taxes T_j (5)
1	$90,000	$30,000	$18,000	$11,540
2	60,000	24,000	28,800	−840
3	36,000	18,000	17,280	1,896
4	18,000	12,000	10,368	1,796
5	6,000	6,000	10,368	−2,676
6	0	0	5,184	−3,456
Total		$90,000	$90,000	$ 8,260

Notes: (2) $B'_j = \$90,000$ for $j = 1$

$\qquad = B'_{j-1} - D'_{j-1}$ for $j = 2, 3, \ldots, 6$

(3) D'_j based on SYD method, $N_d = 5$ and zero salvage value

(4) D_j based on regular MACRS method, $N_d = 5$

(5) $T_j = \left(\dfrac{0.40}{1-0.40}\right)\{[0.114 - 0.55(0.10)]B'_j + D'_j - D_j\}$

$\qquad = \left(\dfrac{2}{3}\right)\left(0.059B'_j + D'_j - D_j\right)$

The income tax consequences are summarized in **Table 9.1**. A negative value in year j indicates that the investment will have the effect of reducing income taxes by that amount in year j.

9.5 THE LEVELIZED REVENUE REQUIREMENT

Having derived the mathematical formulation for the elements of the revenue required to meet all "costs" in year j, we can summarize the process as follows (the subscripts j have been omitted for brevity).

	Return to lenders	$w_b k_b B' = I$
Plus	Return to stockholders	$w_p k_p B' + w_e k_e B' = P + S = Q$
Equals	Return on investment	ROI
Plus	Book depreciation	D'
Equals	Capital recovery cost	$I + Q + D'$
Plus	Income taxes	$T = $ Eq. (9.8) or (9.11)
Equals	Carrying charges	CC
Plus	Operating expenses	K
Equals	Revenue requirement	R

The revenue requirement, R_j, is found for each year j over the planning horizon, $j = 1, 2, \ldots, N$. We want a single statistic, the **levelized revenue requirement**, RR, which is the uniform annual equivalent of the R_j's.

$$RR = \left(\begin{array}{c} \text{PW of annual} \\ \text{revenue requirements} \end{array} \right) \times \left(\begin{array}{c} \text{capital recovery} \\ \text{factor} \end{array} \right)$$

$$= \left[\sum_{j=1}^{N} R_j (1 + i)^{-j} \right] (A/P, i, N) \tag{9.11}$$

The statistic RR is equivalent to the AC as discussed in Chapter 3. Just as investment alternatives in competitive enterprises can be rank-ordered on the basis of (equivalent uniform) annual costs, so can investments in regulated public utilities. Under the revenue requirement method, the preferred alternative is that which minimizes the revenue requirement.

EXAMPLE 9.3

Suppose that the capital investment described in Example 9.2 is equipment that requires $35,000 each year for operation and maintenance. The equipment will be placed in service at the start of the tax year and removed from service at the end of the sixth tax year. The net salvage value at the end of 6 years is expected to be negligible. All other assumptions concerning the firm's capital structure and marginal income tax rate are as described previously.

Determination of the year-by-year revenue requirements is summarized in **Table 9.2**.

Values in the table are based on the following:

Col. 2: Rate base at start of year, B'_j=(Table 9.1, col. 2)
Col. 3: Book depreciation, D'_j=(Table 9.1, col. 3)

TABLE 9.2   Determining Annual Revenue Requirements (Example 9.3)

Year j	Rate Base at Start of Year B'_j	Book Depreciation D'_j	Return to Lenders I_j	Return to Stockholders Q_j	Income Taxes T_j	Operating Expenses K_j	Revenue Requirements R_j	PW of Revenue Requirements $R_j(1+i)^{-j}$
(1)	(2)	(3)	(4)	(5)	(6)	(7)	(8)	(9)
1	$90,000	$30,000	$4,950	$5,310	$11,540	$ 35,000	$ 86,800	$ 77,917
2	60,000	24,000	3,300	3,540	−840	35,000	65,000	52,377
3	36,000	18,000	1,980	2,124	1,896	35,000	59,000	42,677
4	18,000	12,000	990	1,062	1,796	35,000	50,848	33,017
5	6,000	6,000	330	354	−2,676	35,000	39,008	22,737
6	0	0	0	0	−3,456	35,000	31,544	16,505
Total		$90,000	$11,550	$12,390	$8,260	$210,000	$332,200	$245,230

Col. 4. Return to lenders, $I_j = 0.55(0.10)B'_j = 0.055B'_j$

Col. 5: Return to stockholders, $Q_j = [0.10(0.10) + 0.35(0.14)]$
$$B'_j = 0.059B'_j$$

Col. 6: Income taxes, $T_j = $ (Table 9.1, col. 5)

Col. 7: Operating expenses, $K_j = \$35,000/\text{year}$

Col. 8: Revenue requirements, $R_j = D'_j + I_j + Q_j + T_j + K_j$

Finally, the levelized revenue requirement is determined using Eq. (9.11):

$$\text{RR} = \left[\sum_{j=1}^{6} R_j (1.114)^{-j} \right] (A/P, 11.4\%, 6)$$

$$= \$245,230(0.23911)$$

$$= \$58,637$$

In other words, this $90,000 investment would yield returns of 10% to lenders, 10% to preferred stockholders, and 14% to common stockholders, *if* it resulted in increased revenues of the equivalent of $58,637 each year over the 6-year planning horizon.

9.6 A BASIC MODEL ASSUMING STRAIGHT LINE DEPRECIATION FOR "BOOK" AND "TAXES"

A number of engineering economy textbooks introduce the minimum revenue requirement method based on a set of simplifying assumptions. As before, we will assume that:

1. The tax rate (t_j) remains constant throughout the planning horizon $(j = 1, 2, \ldots, N)$.
2. The costs of debt, preferred stock, and equity $(k_b, k_p, \text{and } k_e)$ remain constant throughout the life of the investment. Furthermore, the proportions of debt, preferred stock, and equity (w_b, w_p, w_e) also remain constant for all values of j.
3. The amount of unrecovered invested capital remaining at the start of year j (B'_j) is equal to the book value of the investment at the start of year j (B_j).

In addition, we adopt the following simplifying assumptions:

4. The annual reduction in unrecovered invested capital, book depreciation (D'_j), is equal to the depreciation affecting taxable income in year j, tax depreciation (D_j). Put somewhat differently, we assume that $B_j = B'_j$ at any point in time. See **Figure 9.2**.
5. The straight line method is used for computing annual depreciation expenses. Thus the depreciation expense (D) remains constant over the depreciable life of the investment, and

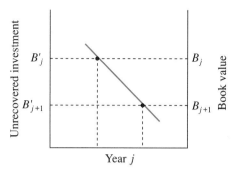

Figure 9.2 Simplifying Assumption: Identical Depreciation Methods for Book and Tax Purposes

$$D = \frac{C - L}{N_d} \tag{9.12}$$

where C is the amount of the initial investment (cost basis), L is the expected salvage value, and N_d is the depreciable life. It also follows that the book value at the start of year j is given by

$$B_j = C - (j - 1)D$$

$$= C + D - jD \tag{9.13}$$

9.6.1   Income Taxes

Given these assumptions, Eq. (9.10) can be rewritten as

$$T_j = \left(\frac{t}{1 - t}\right)(i - w_b k_b)(C + D - jD)$$

$$= \alpha(C + D - jD)$$

$$= \alpha(C + D) - j(\alpha D) \tag{9.14}$$

where

$$\alpha = \left(\frac{t}{1 - t}\right)(i - w_b k_b)$$

As we examine the year-by-year taxes, it may be seen that the taxes decrease by a uniform amount, αD, each year:

$$\left.\begin{array}{l} T_1 = \alpha(C + D) - \alpha D = \alpha C \\[4pt] T_2 = \alpha(C + D) - 2\alpha D = \alpha C - \alpha D \\[4pt] T_3 = \alpha(C + D) - 3\alpha D = \alpha C - 2\alpha D \\[4pt] \quad\vdots \\[4pt] T_N = \alpha(C + D) - N\alpha D = \alpha C - (N - 1)\alpha D \end{array}\right\} \quad \text{gradient} = \alpha D$$

Thus the uniform annual series, T, equivalent to the series of income taxes $(T_1, T_2, \ldots, T_N)$, can be found from

$$T = \alpha C - \alpha D(A/G, i, N)$$

$$= \alpha[C - D(A/G, i, N)] \tag{9.15}$$

(The annualized income taxes are generally known as **levelized taxes** in the literature of utility economics.)

9.6.2 Fixed Charge Rate

It is often useful to express the fixed costs of a proposed investment as a constant percentage of the initial investment, C. This is known as the **fixed charge rate**, FCR. We will now show how this can be done within the context of our basic model.

Income Taxes

Let τ be defined as the ratio of (levelized) income taxes to the initial investment: $\tau = T/C$. Thus, dividing Eq. (9.15) by C,

$$\tau = \alpha \left[1 - \frac{D}{C}(A/G, i, N) \right] \tag{9.16}$$

But $D/C = (1 - c)/N$, where $c = L/C$. Thus

$$\tau = \alpha \left[1 - \left(\frac{1-c}{N} \right)(A/G, i, N) \right] \tag{9.17}$$

After additional algebraic manipulation, it may be shown that

$$\tau = \left(\frac{t}{1-t} \right)\left(1 - \frac{w_b k_b}{i} \right)\left[i - \left(\frac{1-c}{N} \right) + (1-c)(A/F, i, N) \right] \tag{9.18}$$

Both Eqs. (9.17) and (9.18) may be found in the literature of utility economics. Of course, they give identical results.

Capital Recovery

The cost of capital recovery is an important element in evaluating proposed investments by public utilities. Given an initial investment (C), expected terminal salvage value (L), and service life (N), the cost of capital recovery (CR) is the equivalent uniform annual series of the two cash flows:

$$CR = Ci + (C - L)(A/F, i, N) \tag{9.19}$$

Let χ be defined as the ratio of capital recovery to the initial investment. Thus, dividing Eq. (9.19) by C,

$$\chi = i + (1 - c)(A/F, i, N) \tag{9.20}$$

where $c = L/C$ as before.

Other fixed charges, such as property taxes and maintenance expenses, can also be expressed as percentages of the initial cost, as will be demonstrated in the following example.[4]

EXAMPLE 9.4

Consider the utility described in the previous examples. Recall that $i = 0.114$ and $t = 0.40$. Let us now suppose that our utility is considering an investment that has a depreciable life (N) of 28 years and an expected salvage value (L) of zero. (The factor $c = L/C = 0$ because $L = 0$.) Property taxes are assumed to be 1.0% of the initial cost each year; annual maintenance costs are assumed to be 2.0% of the initial cost.

The annual **fixed charge rate** (FCR) is now determined as follows.[5]

Income taxes [Eq. (9.18)]:

$$\tau = \left(\frac{0.40}{1-0.40}\right)\left[1 - \frac{0.55(0.10)}{0.114}\right]$$

$$\times \left[0.114 - \left(\frac{1-0}{28}\right) + (1-0)(A/F,\, 11.4\%,\, 28)\right] = 0.02902$$

Capital recovery [Eq. 9.20)]:

$$\chi = 0.114 + (1-0)(A/F,\, 11.4\%,\, 28) = 0.11983$$

$$\text{Property taxes} = 0.01000$$

$$\text{Maintenance} = \underline{0.02000}$$

$$\text{Total fixed charge rate} = 0.17885$$

That is, for each dollar invested, the equivalent of 17.88 cents of revenue is needed each year for 28 years to cover these fixed charges.

Now, suppose that our utility is considering the construction and operation of a steam-powered generating plant that will produce 300,000 kilowatts (kw) of electric energy. The initial investment cost is $400/kw; operating costs, other than fixed charges, are anticipated to be $20/kw per year. If the plant is constructed, it is expected that it will be operated for 28 years with no residual salvage value. Thus the total (equivalent uniform) annual cost per kilowatt for the proposed investment is

$$(\$400 \times 0.1788) + \$20.00 = \$\ 91.52/\text{kw}$$
$$\times\ 300,000\,\text{kw}$$
$$\text{Total cost for } 300,000\,\text{kw} = \overline{\$27,456,000}$$

---

[4]The numerical illustration in this chapter is based on an example in J. R. Canada and J. A. White, *Capital Investment Decision Analysis for Management and Engineering* (Englewood Cliffs, NJ: Prentice-Hall, 1980).

[5]It may be helpful to note that $(A/F,\, 11.4\%,\, 28) = 0.00583$. See Appendix 2 at the end of the chapter for additional values.

This is the equivalent annual revenue required if the utility is to recover all relevant costs of the investment.

9.7 COMPARING ALTERNATIVES

Once we recognize that the levelized annual revenue requirement for a given investment proposal is the (equivalent uniform) annual cost, it follows that alternatives may be rank-ordered by simply minimizing the revenue requirement. Thus the annual worth method or the present worth method may be applied as outlined in Chapter 3.

EXAMPLE 9.5

Suppose that our utility is considering an alternative to the steam-powered plant described above. The alternative plant is powered by *internal combustion*, has an initial cost of $150/kw, and has an operating life of 14 years. The residual value at the end of its service life is expected to be negligible (zero). The fixed charge rate is

Income taxes (τ)	0.0258
Capital recovery (χ)	0.1463
Property taxes	0.0100
Maintenance	0.0200
Total FCR	0.2021

If it is assumed that total operating costs, other than fixed charges, will be $70/kw each year over the 14-year service life, then the total equivalent uniform annual cost, or revenue requirement, is

$$(\$150 \times 0.2021) + \$70 = \$100.31/\text{kw}$$

If the plant is sized to generate 300,000 kw annually, then the total annual revenue requirement would be $30,093,000.

It should be emphasized that direct comparison of the annualized (levelized) revenue requirements — $27.46 million for the steam plant versus $30.09 million for the internal combustion plant — is valid only under two critical assumptions. First, it must be assumed that there will be a need for this power output for at least 28 years, the least common multiple of the two alternatives. Second, it must be assumed that the cash flow consequences during the first life cycle of the internal combustion plant will also occur during the second life cycle. (See **Figure 9.3**.) If these assumptions hold, then annual worths (costs) may be compared directly. Otherwise, it is necessary to specify the likely consequences after the shorter-lived alternative completes its first life cycle.

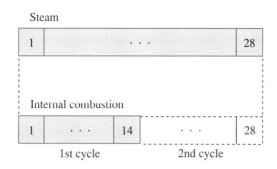

Figure 9.3   Alternatives with Unequal Lives

9.8   RELATING THE REVENUE REQUIREMENT METHOD TO "CONVENTIONAL" ANALYSES

9.8.1   Analyses Based on After-Tax Cash Flows

It may be shown that the revenue requirement method is related directly to the more traditional approaches used widely in nonutility industries: the present worth method, the (internal) rate of return method, and the like. Recall that these more "conventional" techniques depend on the determination of after-tax cash flows, from which the appropriate figure of merit (PW, AW, IRR, and so on) may be computed. We will illustrate these relationships using the data from Example 9.3.

EXAMPLE   9.6

The relevant calculations for this example are summarized in **Table 9.3**. Values in the table are based on the following:

Col. 2: Annual revenue, $R_j = $ (Table 9.2, col. 8)

TABLE 9.3   Conventional Analysis Based on Cash Flows (Example 9.6)

End of Year j (1)	Revenue R_j (2)	Cash Flow due to Investment C, L (3)	Operating Expenses K_j (4)	Interest Expense I_j (5)	Tax Depreciation D_j (6)	Effect on Taxable Income (7)	Effect on Income Taxes T_j (8)	Cash Flow After Taxes $\hat{A}_j$ (9)
0	—	− $90,000	—	—	—	—	—	− $90,000
1	$86,800		$ 35,000	$ 4,950	$18,000	$28,850	$11,540	40,260
2	65,000		35,000	3,300	28,800	− 2,100	− 840	30,840
3	59,000		35,000	1,980	17,280	4,740	1,896	22,104
4	50,848		35,000	990	10,368	4,490	1,796	14,052
5	39,008		35,000	330	10,368	− 6,690	− 2,676	6,684
6	31,544	0	35,000	0	5,184	− 8,640	− 3,456	0
Total		− $90,000	$210,000	$11,550	$90,000	$20,650	$ 8,260	$23,940

Col. 3: Cash flows due to investment, $C = \$90,000$ and $L = \$0$

Col. 4: Annual operating expenses, $K = \$35,000$

Col. 5: Interest expense, $I_j = $ (Table 9.2, col. 4)

Col. 6: Tax depreciation, $D_j = $ (Table 9.1, col. 4)

Col. 7: Effect on taxable income $= R_j - K_j - I_j - D_j$ for $j = 1, 2, \ldots, 6$

Col. 8: Effect on income taxes,

$$T_j = (\text{tax rate}) \times (\text{taxable income}) = 0.40 \text{ (col. 7)}$$

Col. 9: Cash flow after taxes, $\hat{A}_j = R_j - K_j - T_j$

Example 9.6 illustrates three points that are especially noteworthy. First, the cash flows for income taxes, column 8, are generated in this table in the "conventional" way: We first determine the effects on taxable income, column 7, and then multiply these values by the marginal income tax rate. The results are identical to those values of T_j found using Eq. (9.10) and shown in column 5 of Table 9.1.

Second, the after-tax cash flows, $\hat{A}_j$, are those net funds available as returns to the contributors of capital — the lenders, preferred stockholders and common stockholders. To illustrate, consider $\hat{A}_1 = \$40,260$.

Return to lenders
$$= w_b k_b B'_1 + w_b D'_1$$
$$= 0.55(0.10)(\$90,000) + 0.55(\$30,000)$$
$$= \$4,950 + \$16,500$$
$$= \$21,450$$

Return to preferred stockholders $= w_p k_p B'_1 + w_p D'_1$
$$= 0.10(0.10)(\$90,000) + 0.10(\$30,000)$$
$$= \$900 + \$3,000$$
$$= \$3,900$$

Return to common stockholders $= w_e k_e B'_1 + w_e D'_1$
$$= 0.35(0.14)(\$90,000) + 0.35(\$30,000)$$
$$= \$4,410 + \$10,500$$
$$= \$14,910$$

Total return to participants
$$= iB'_1 + D'_1$$
$$= 0.114(\$90,000) + \$30,000$$
$$= \$40,260$$

Note that the reduction in the rate base, D'_1, is shared by the participants in proportion to their relative capital contributions to the rate base.

Third, this "conventional" analysis based on after-tax cash flows, $\hat{A}_j$, produces an internal rate of return that is exactly equal to the weighted average cost of capital, i. Recall that the IRR is the value, $i*$, such that

$$PW(i*) = 0$$

From the example,

$$PW = \sum_{j=0}^{6} \hat{A}_j (1 + i*)^{-j} = 0$$

from which $i* = 0.114$. That is, the after-tax rate of return is the weighted average cost of capital, or $i* = i$.

9.8.2 IRR Based on Equity Cash Flows Only

The relationship between the revenue requirement method and "conventional" analyses can be viewed somewhat differently from the perspective of the *common stockholders only*. Using the previous example for illustration, we determine the net cash flows available to the owners of the utility after returning the lenders' contribution (with interest), after returning the preferred stockholders' contribution (with dividends paid on the preferred stock), and after paying the costs of operations and maintenance as well as relevant income taxes. These cash flows are summarized in **Table 9.4** on page 346.

Col. 2:   Annual revenue, $R_j = $ (Table 9.2, col. 8)

Col. 3:   Return of principal to lenders, $U_j = w_b D'_j = 0.55 D'_j$, where the D'_j values are from Table 9.2, col. 3

Col. 4:   Interest paid to lenders, $I_j = w_b k_b D'_j = $ (Table 9.3, col. 5)

Col. 5:   Return of capital to preferred stockholders,
$$V_j = w_p D'_j = 0.10 D'_j$$

Col. 6:   Dividends paid to preferred stockholders $=$
$$w_p k_p B'_j = 0.10(0.10) B'_j,$$
where the B'_j values are from Table 9.2, col. 2

Col. 7:   Cost of asset $=$ $90,000 with zero salvage value

Col. 8:   Miscellaneous pretax cash flows.
$$K = \$35,000 \text{ each year}$$

Col. 9:   Income taxes, $I_j = $ Table 9.2, col. 6

Col. 10:   Net cash flow after taxes, $\hat{A}_j = $ sum of cols. 2 through 9

The cash flows in column 10 of the table, $\hat{A}_j$, produce an internal rate of return, i_e^*, of 14.0%, the cost of equity capital. That is, the present worth of all future positive cash flows, when discounted at $k_e = 0.14$, is equal to the initial imvestment by owners. This is the expected result, of course, inasmuch as the permitted revenues have been determined so as to provide the 14% return to owners.

TABLE 9.4 Cash Flow Summary from the Perspective of the Owners (Example 9.6)

End of Year j (1)	Revenue R_j (2)	Borrowed Capital		Preferred Stock		Cost of Asset C, L (7)	Miscellaneous Pretax Cash Flows K_j (8)	Income Taxes T_j (9)	Net Cash Flow After Taxes $\hat{A}_j$ (10)
		Principal U_j (3)	Interest I_j (4)	Payments V_j (5)	Dividends P_j (6)				
0	—	$49,500	—	$9,000	—	-$90,000	—	—	-$31,500
1	$ 86,800	-16,500	-$ 4,950	-3,000	-$ 900		-$ 35,000	-$11,540	14,910
2	65,000	-13,200	-3,300	-2,400	-600		-35,000	840	11,340
3	59,000	-9,900	-1,980	-1,800	-360		-35,000	-1,896	8,064
4	50,848	-6,600	-990	-1,200	-180		-35,000	-1,796	5,082
5	39,008	-3,300	-330	-600	-60		-35,000	2,676	2,394
6	31,544	0	0	0	0	0	-35,000	3,456	0
Total	$332,200	$0	-$11,550	$0	-$2,100	-$90,000	-$210,000	-$8,260	$10,290

SUMMARY

Because of the relatively large capital investment in plant and equipment required of companies that furnish utility services to the general public, and because most of these investments are incurred well in advance of the period during which the costs are to be recovered through revenues, public utility companies are often allowed to operate as a monopoly within a political jurisdiction. In return for this special treatment, certain aspects of the firm's behavior are regulated by some regulatory body, generally a utilities commission or the like. Of particular concern are the rates, or prices, that the firm is permitted to charge customers for its services.

A fundamental principle underlying the regulation of rates is that they be sufficient to cover all costs, including a "fair return" on invested capital. Consequently, decisions between alternative investment opportunities may be judged on the basis of their respective effects on revenue requirements, or the rates charged to customers. The preferable alternative is the one that minimizes revenue requirements.

If revenue requirements include the cost of capital, then minimizing revenue requirements is equivalent to maximizing present worth or annual worth. The choices resulting from these methods of analysis are consistent; only the perspective is different. Nevertheless, it follows that decisions taken from the viewpoint of stockholders (the present worth or annual worth method) will be consistent with the interests of ratepayers (the revenue requirement method) when rates are regulated.

PROBLEMS

9.1 Recall Example 9.4. We assumed that straight line depreciation would be used for tax purposes as well as for determining the unrecovered invested capital. Using Eq. (9.13), we determined that the fixed charge for income taxers, τ, is 0.02902 when $i = 0.114$, $t = 0.40$, $w_b = 0.55$, $k_b = 0.10$, $N = 28$, and $c = 0$. Now, we will change the assumptions in only one respect: Assume that the property will be depreciated under MACRS as a 15-year public utility property, with annual depreciation percentages as given in Table 6.A.1 in the Appendix to Chapter 6. Compute the annual fixed charge rate for income taxes, τ, with this modification.
[*Answer:* 0.0215.]

9.2 Recall Example 9.4. Here, all assumptions are as originally stated, except that we now assume

that the capital structure consists of 60% debt with $k_b = 0.11$ and 40% common stock with $k_e = 0.15$.
a. Find the weighted average cost of capital (i).
b. Determine the sinking fund factor for this value of i and $N = 28$; that is, find $(A/F, i, 28)$.
c. Determine the revised total fixed charge rate with these revised data.

9.3 Recall the steam-powered generating plant described in Example 9.4. Construct a table, similar to Table 9.2, showing the year-by-year revenue requirements. Determine the levelized revenue requirements.
[*Answer:* $23,863,000.]

9.4 Recall the internal combustion plant described in Example 9.5. Construct a table similar to

Table 9.2, showing the year-by-year revenue requirements. Determine the levelized revenue requirements.

9.5 Recall the steam-powered plant described in Example 9.4. Construct a table, similar to Table 9.2, showing the year-by-year revenue requirements assuming that the miscellaneous pretax cash flows, K_j, are expected to be $15,000 the first year, increasing by $1,000 each year over the 28-year life of the plant. All other assumptions are as originally stated. Determine the levelized annual revenue requirements.
[*Answer:* **$17,885,000.**]

9.6 (Adapted from R. R. Mayer, "Finding Your Minimum Revenue Requirements," *Industrial Engineering*, April 1977, pp. 16–22.) A public utility is considering investment in an asset that has a first cost of $130,000, an expected service life of 3 years, and an estimated salvage value of $40,000. Recurring costs of operation and administration are expected to be $52,000 the first year, $59,000 the second year, and $66,000 the third year. The company's before-tax rate of return requirement is 20% per year. Determine the (pretax) levelized annual revenue requirement. (This pretax analysis would be suitable for publicly owned utilities that do not pay income taxes.)

9.7 (This is an after-tax analysis of the investment described in Problem 9.6.) The company's combined federal and state income tax rate is 40%, financing will be by 30% debt and 70% equity, the cost of debt financing is 8% per year, and the after-tax rate of return requirement, the overall cost of capital, is 10% per year. The asset will be depreciated by the sum of years-digits method with a 3-year life and $40,000 salvage value. Assume that the unrecovered investment at any time is equal to book value and that the amount of debt at any time (30%) remains constant throughout the asset life.
a. Determine the levelized annual income taxes.
b. Determine the (after-tax) levelized annual revenue requirements.
[*Answers:* **a. $4,682; b. $103,400.**]

9.8 A publicly owned gas company is considering the construction of a small warehouse on land that it currently owns. The need for this warehouse is not pressing; it may be built now or,

alternatively, 3 years from now. The issue is whether it is economically desirable to delay construction.

The cost of the warehouse is $1,000,000; it has an expected useful life of 30 years. These estimates are independent of when the warehouse is built.

The company plans to sell the land and the warehouse in 20 years. If the warehouse is built now, it will be 20 years old at the time of sale, at which point its residual value is estimated to be $200,000; if the company waits 3 years to build, the warehouse will be 17 years old at the time of sale, at which point its residual value is estimated to be $225,000.

If construction of the warehouse is delayed, the company will rent space nearby at a cost of $10,000 per month over the 3-year period.

The company uses 30% debt at a cost of 7%. The regulatory body permits a "fair return" of 9% on the remaining 70% equity capital. This publicly owned company pays no income taxes, but straight line depreciation is used for book purposes. (Here, use a 30-year life with zero salvage value.)

Determine the levelized annual revenue requirements for both alternatives: (a) build now; and (b) delay construction 3 years.

9.9 (Adapted from T. L. Ward and W. G. Sullivan, "Equivalence of the Present Worth and Revenue Requirement Methods of Capital Investment Analysis," *AIIE Transactions*, March 1981, pp. 29–40.) An investor-owned power company is considering the acquisition of a small coal-fired boiler. The initial cost is $2,500,000, the expected useful life is 4 years, and estimated salvage value at the end of 4 years is $500,000. The boiler will be depreciated by the straight line method for book purposes; the regular MACRS method will be used for tax purposes. A 4-year life and a $500,000 salvage value will be used for both book and tax purposes. (Note here that the value of unrecovered investment at any point in time is based on straight line depreciation; income taxes are based on MACRS depreciation.)

Operating costs are expected to be $300,000 per year over the 4-year useful life. The thermal output of the boiler is 300×10^9 Btu per year.

The boiler will be purchased entirely with equity funds that cost 20% per year. The effective income tax rate is 40%.

a. Determine the levelized equivalent revenue requirement for the proposed investment.
b. Determine the revenue requirement per 10^6 Btu.

[*Answers:* **a. $1,373,000; b. $4.58.**]

9.10 In Example 9.5 we found that the levelized annual revenue requirements were $27,456,000 for the steam-powered plant (28-year life) and $30,093,000 for the internal combustion plant (14-year life). We concluded that the steam-powered plant is preferred, based on the assumption that the replacement for the internal combustion plant would be identical in every respect to the first one, thereby resulting in a revenue requirement of exactly $30,093,000 per year during years 15–28. Let us relax that assumption and assume only that the replacement will have a life of 14 years: It will be in use during years 15–28. For the two alternatives (steam versus internal combustion) to result in equal levelized revenue requirements over the 28-year planning horizon, what must be the levelized revenue requirement for the replacement plant over years 15–28?

9.11 A publicly owned utility company is considering investment in a compressor motor that has initial cost of $40,000, an expected service life of 5 years, and an estimated salvage value of $5,000 at the end of 5 years. Annual cost of operation is expected to be $1,500 per year. Determine the levelized annual revenue requirement if the company's rate of return requirement is 12% per year. Note that income taxes are not relevant in this problem.

[*Answer:* **$11,800.**]

9.12 Recall the internal combustion plant discussed as Example 9.5.

Annual capacity	= 300,000 kw
Investment cost	= $150/kw
Annual operating cost	= $70/kw
Life of plant	= 14 years

In the original problem statement, we assumed zero salvage value after 14 years. Let us now change that assumption: Assume a 20% salvage value ($30/kw) after 14 years. All other assumptions remain as before. Determine:

a. Annual depreciation expense (D)
b. Levelized taxes (τ)
c. Charge rate: cost of capital recovery (χ)
d. Total annual fixed charge rate (FCR)
e. Levelized annual revenue requirement (RR)

9.13 Recall the steam-powered generating plant discussed in Example 9.4. Let us now assume that the utility's income tax rates have been revised:

$$\text{State tax rate } (t_s) = 0.093$$
$$\text{Federal tax rate } (t_f) = 0.340$$

Assume that all other assumptions for the 28-year plant are as described in the text.

a. Determine the combined incremental income tax rate.
b. Determine the fixed charge rate for income taxes.
c. Determine the total fixed charge rate.
d. Determine the total levelized revenue requirement per kilowatt for the proposed investment in the steam-powered generating plant.
e. Determine the income taxes per kilowatt in the first year (T_1).

[*Answers:* **a. 0.4014; b. 0.0292; c. 0.1790; d. $91.61; e. $15.82.**]

APPENDIX 1: *Summary of Symbols Used in Chapter 9*

$\hat{A}_j$ = cash flow after income taxes
B_j = book value for tax purposes at start of year j
B'_j = value of the unrecovered investment at the start of year j

c = ratio of salvage value to initial investment
C = initial investment (cost basis)
CC_j = carrying charges in year j

continues

D_j = depreciation for tax purposes in year j (affecting taxable income)

D'_j = decline in rate base (book depreciation) in year j

E_j = taxable income in year j

I_j = interest expense (return to lenders) in year j

i = weighted-average cost of capital

i^* = internal rate of return

i_e^* = internal rate of return on equity investment

j = index for year (shown as subscript for appropriate cost/revenue element)

k_b = cost of debt (interest rate paid on borrowed capital)

k_e = cost of equity (return to common stockholders)

k_p = cost of preferred stock (return to preferred stockholders)

K_j = total pretax cash flows, other than debt service and preferred stock dividends; current operating disbursements in year j

L = terminal salvage value of investment

N = life of investment

N_d = recovery period used for depreciation

P_j = preferred stock dividends paid in year j

Q_j = net profit after income taxes in year j

R_j = operating revenue (permitted revenue) in year j

RR = levelized revenue requirement

S_j = return to owners (common stockholders) in year j

t_j = marginal income tax rate in year j

T_j = income taxes paid in year j

U_j = repayment of loan principal in year j

V_j = retirement of preferred stock in year j

w_b = proportion of debt (borrowing)

w_e = proportion of equity (common stock)

w_p = proportion of preferred stock

τ = ratio of income taxes to the original investment

χ = ratio of capital recovery cost to the original investment

APPENDIX 2: *Some Compound Interest Factors for i = 0.114*

	$N=6$	$N=14$	$N=28$
$(A/P, 11.4\%, N) = \dfrac{0.114(1.114)^N}{(1.114)^N - 1}$	0.23911	0.14627	0.11983
$(P/A, 11.4\%, N) = (A/P, 11.4\%, N)^{-1}$	4.1822	6.8368	8.3451
$(A/F, 11.4\%, N) = \dfrac{0.114}{(1.114)^N - 1}$	0.12511	0.03227	0.00583
$(A\,G, 11.4\%, N) = (0.114)^{-1} - N[(1.114)^N - 1]^{-1}$	2.1874	4.8094	7.3396

10

Economic Risk Analysis[1]

10.1 INTRODUCTION

The various methods of analysis for economic justification discussed previously are based in the assumption that all of the component cash flows for the proposed investment are known and certain. However, in most cases the amount and timing of these cash flows are estimated, and uncertainties exist in the estimation process. Furthermore, there is usually more uncertainty about some component cash flows than about others, and some of these component flows affect the economic criteria more than others. Thus additional methodologies and concepts are needed for economic analysis when explicit information on the effects of uncertainties in the timing and amounts of the cash flows is important. These methodologies and concepts are the focus of this chapter.

Numerous factors contribute to the uncertainties in the estimates of the amount and timing of component cash flows. Delivery or construction delays, unexpected bottlenecks in new projects, inflationary or recessionary pressures, labor negotiations, and problems in R&D are only a few examples of changes that can and do occur to alter the amounts and timing of disbursements and receipts of moneys. Although these possibilities are usually recognized during the early planning phases of a project, the actual cash flows are uncertain, and there is a risk

---

[1]Portions of this chapter are adopted, substantially unchanged, from G. A. Fleischer, "Economic Risk Analysis," Chapter 52 in the *Handbook of Industrial Engineering*, 2nd ed., edited by Gavriel Salvendy (New York: Wiley Interscience and the Institute of Industrial Engineers, 1992), pp. 1343–1376.

associated with the resulting project's present worth, benefit–cost ratio, or other measure of economic merit being used. Since this "economic risk" is as important to the decision maker as the other aspects of economic analysis, explicit information regarding the noncertain estimates should be developed as part of the analysis. Approaches to this form of analysis and some of the relevant techniques are described in this chapter.

A variety of measures have been proposed for dealing with a noncertain operating environment, that is, where the relevant parameters of the analytical model cannot be assumed with certainty. The relevant literature is very extensive, and an encyclopedic treatment is beyond the scope of this chapter. Our discussion will be confined, therefore, to a limited number of concepts selected for their popularity among practitioners and because they are representative of the spectrum of possible approaches to this issue. We begin with *sensitivity analysis*, a technique that, surveys show, appears to be most commonly used in industry. Also discussed here are analytical procedures that can broadly be described as *risk analysis* and decision theory applications.

10.2   SENSITIVITY ANALYSIS

Sensitivity analysis is the process whereby one or more system input variables are changed and corresponding changes in the system output, or figure of merit, are observed. If a decision is changed as a certain input is varied over a reasonable range of possible values, the decision is said to be *sensitive* to that input; otherwise it is *insensitive*.

The term **breakeven analysis** is often used to express the same concept for a single input variable. Here, the value of the input variable at which the decision is changed is determined. If the **breakeven point** lies within the range of expected values, the decision is said to be sensitive to that parameter. Thus sensitivity and the breakeven point are directly related.

10.2.1   Numerical Example: Certainty Analysis

A manufacturing firm is considering the introduction of a new product to be produced and sold over a 15-year period. The initial cost of capital facilities is $100,000; the anticipated net salvage value at the end of 15 years is $20,000. It is expected that 7,000 units will be produced each year at a cost of $10.00 per unit and sold at $12.00 per unit. The firm's minimum attractive rate of return (MARR) is 10% per year.

The anticipated "profitability" of this proposed investment can be measured by present worth (PW) as follows.

$$PW = Q(r - c)(P/A, i, N) - P + S(P/F, i, N) \qquad (10.1)$$

where

$$Q = \text{quantity sold per year}$$
$$r = \text{revenue per unit}$$
$$c = \text{cost per unit}$$
$$P = \text{initial cost of capital facilities}$$
$$S = \text{net salvage value of capital facilities}$$
$$N = \text{project life, in years}$$
$$i = \text{MARR, the discount rate per year}$$

Assuming the "certainty estimates" for these seven parameters as described in the preceding paragraph, the solution is

$$PW = 7,000(\$12 - \$10)(P/A, 10\%, 15) - \$100,000$$
$$+ \$20,000(P/F, 10\%, 15)$$
$$= \$14,000(7.606) - \$100,000 + \$20,000(0.2394)$$
$$= \$11,273$$

Since the PW is positive, we conclude that the proposal appears to be economically attractive. This result, of course, is based on the presumption that all of the parameter values assumed for the analysis will in fact occur as anticipated.

10.2.2 Classical Sensitivity Analysis: Single Variable

Algebraic Solution

Suppose that there is some reason to question the validity of the assumption concerning the number of units produced and sold annually. Additional investigation, for example, may suggest that the "certainty estimate" of 7,000 units per year is questionable; it now appears that this parameter value could occur anywhere over the range of 6,000 to 7,500 units. With this new information the resulting range of values for the present worth is

$$\text{Min PW} = 6,000(\$2)(7.606) - \$95,212 = -\$3,940$$
$$\text{Max PW} = 7,500(\$2)(7.606) - \$95,212 = \$18,878$$

The breakeven point can be determined by determining that value of $Q = Q_0$ such that $PW = 0$.

$$PW = 0 = Q_0(\$2)(7.606) - \$95,212$$

Solving,

$$Q_0 = \frac{\$95,212}{\$15.212} = 6,259 \text{ units}$$

Since the breakeven point lies within the range $(6{,}000 < 6{,}259 < 7{,}500)$, the decision is sensitive to the estimate for Q.

Graphical Presentation

Sensitivity analyses are usually presented in graphical format. Indeed, it is this "power of pictures" that probably accounts for its widespread popularity. The graphical portrayal of sensitivity of PW to the variable Q in our example is illustrated in **Figure 10.1**. The linear function in the figure is the graph of

$$\text{PW} = \$15.212Q - \$95{,}212, \qquad 0 \le Q \le 9{,}000$$

Note the breakeven point at $Q_0 = 6{,}259$. Also note that the range for Q is highlighted at $Q(\text{min}) = 6{,}000$ and $Q(\text{max}) = 7{,}500$.

Percent Deviation Graph

An alternative approach is a plot of the measure of worth—here, the present worth (PW)—as a function of the *percent deviation* of the variable of interest. In the example, let $p_Q =$ the percent deviation of the number

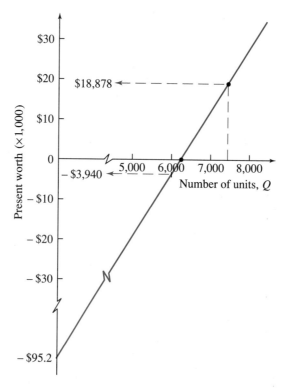

Figure 10.1 Present Worth as a Function of Number of Units Produced and Sold Annually (Breakeven = 6,259 units)

of units sold annually, Q, such that

$$PW = \$15.212(7{,}000)(1 - p_Q) - \$95{,}212$$
$$= \$11{,}272 + \$106{,}485p_Q \qquad (10.2)$$

The function is graphed in **Figure 10.2**. Also shown in the figure are similar graphs for percent deviation of revenue per unit (p_r), cost per unit (p_c), and the project life (p_N).

Although percent deviation graphs for one or more variables may be shown in a single illustration, it should be emphasized that sensitivity to only one variable at a time is being examined. The graph of PW as a function of P_Q, for example, is based on the assumptions that *all* other variables (r, c, P, S, N, i) are held constant at their "certainty estimates." For example, when sensitivity to p_r is being examined, we set $Q = 7{,}000$. And so on.

One notable advantage of the percent deviation graph is that it makes apparent the relative degree of sensitivity for the various parameters. The greater the slope (steepness of the function), the more likely

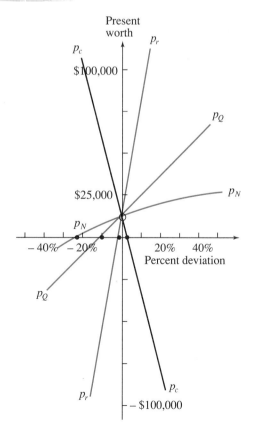

Figure 10.2   Present Worth as a Function of Percent Deviation in Estimates for r, c, Q, and N

is the decision to be sensitive to that parameter; that is, the breakeven point for percent deviation will be relatively small. In Figure 10.2 it is apparent that the decision is somewhat more sensitive to per unit revenue (r) and cost (c) and is relatively insensitive to number of years of service (N) and annual sales (Q). This conclusion may be misleading, however, because it is based on the presumption of equal likelihoods of deviation for the various parameters. To illustrate, we found that the breakeven percent deviations are about -23% for p_N and -11% for p_Q. But suppose that there is evidence to suggest that:

Parameter	Certainty Estimate	Expected Range	Corresponding Deviation
Quantity	7,000 units	6,000–7,500	-14% to $+7\%$
Life	15 years	10–18	-33% to $+20\%$

Thus it would appear that the decision maker would be well advised to give careful attention to the assumption concerning service life (N) as well as quantity produced (Q). The point here is that the *range* of interest for percent deviation may be different for different parameters.

10.2.3 Sensitivity to Two Variables Considered Simultaneously

Suppose that our decision maker in this example is concerned about the sensitivity to the revenue per unit (r) as well as the number of units produced and sold (Q). Considering these two parameters, now variables, simultaneously:

$$PW = Q(r - \$10)(7.606) - \$95,212 \qquad (10.3)$$

As before, assume that $6,000 \le Q \le 7,500$, and assume further that $\$11.25 \le r \le \$12.50$.

One approach to sensitivity analysis for two variables considered simultaneously is to construct a three-dimensional graph with the x and y axes representing the two variables and the z axis serving as the measure of worth. The combined function is now a *surface*, and we now have a breakeven *line*. But three-dimensional graphs are difficult to construct and generally harder to interpret. A useful alternative is a variant of the two-dimensional graph as illustrated in **Figure 10.3**. One of the two variables is represented along the x axis. The second variable is reflected by a family of curves, specifically, curves based on the maximum and minimum values of the variable.

The two functions plotted in Figure 10.3 are

$$PW = Q(\$12.50 - \$10)(7.606) - \$95,212$$

and

$$PW = Q(\$11.25 - \$10)(7.606) - \$95,212$$

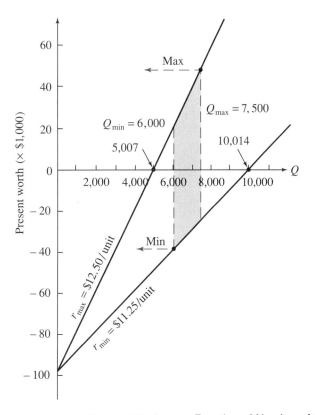

Figure 10.3   Present Worth as a Function of Number of Units Produced (Q) and Revenue per Unit (r)

These represent the upper and lower bounds of the r variable, respectively. Two additional vertical lines are drawn at the lower and upper bounds of the Q variable, at 6,000 and 7,500 units. The polygon thus formed contains all possible combinations of r and Q, and the maximum and minimum values of the figure of merit (PW) can be readily determined. The decision is insensitive if the polygon lies either wholly above the x axis (PW = $0) or wholly below the x axis.

One problem in the interpretation of sensitivity graphs can be illustrated by this numerical example. It would appear from Figure 10.3 that, since the area of the polygon lying above the x axis is roughly the same as the area lying below the line, the likelihood of making money on this project (PW > $0) is about the same as the likelihood of losing money. Implicit in this conclusion is the assumption that all points in the polygon are equally probable. But this is not necessarily the case. Indeed, it would be reasonable to assume that there is an inverse relationship between price per unit and quantity sold, so that Q would decrease as r increases. This dependency is not reflected in the graph.

The simultaneous consideration of sensitivity to two variables can also be displayed in a percent deviation format. In **Figure 10.4**, the

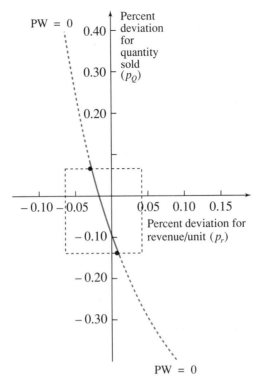

Figure 10.4 Present Deviation Graph for Two Variables: Quantity Sold and Revenue per Unit

percent deviations for each of the variables are shown on the x and y axes. The function is

$$PW = 7,000(1 + p_Q)[\$12(1 + p_r) - \$10](7.606) - \$95,212 = 0$$

This is an **indifference curve**, the locus of all points (p_r, p_Q) such that $PW = \$0$, which is in fact the breakeven line. The solid portion of the line represents the set of possible outcomes: $(-6\% \leq p_r < +4\%)$ and $(-14\% \leq p_Q \leq +7\%)$.

10.2.4 Sensitivity to More than Two Variables

Using the previous example, suppose that, in addition to uncertainty about quantity sold (Q) and revenue per unit (r), there is also uncertainty as to the cost per unit (c). Suppose that the following range of values is possible:

Parameter	Minimum	Most Likely	Maximum
Quantity	6,000	7,000	7,500
Revenue/unit	$11.25	$12.00	$12.50
Cost/unit	$9.00	$10.00	$11.00

As mentioned previously, a percent deviation graph, as in Figure 10.2, permits the plotting of the measure of worth (PW) as function of the percent deviation from the most likely value for any number of parameters. However, the *interactive* effects of the parameter are ignored.

It is possible, of course, to reduce the original problem to a series of two-dimensional graphs. Here, for example, consider: (1) PW as a function of quantity, assuming $c = \$9.00$, and a family of curves for $r = \$11.25$ and $\$12.50$; and (2) PW as a function of quantity, assuming $c = \$11.00$, and a family of curves for $r = \$11.25$ and $\$12.50$. This approach suffers from two defects. First, although a series of smaller problems is solved, we are not testing for the sensitivity of *all* parameters *simultaneously*. Second, the number of graphs required grows exponentially as the number of uncertain parameters increases arithmetically.

A second approach is based on the **a fortiori** ("strength of the argument") **principle**. If it can be shown that a certain course of action is indicated regardless of the input assumptions, then it has been proven, *a fortiori*, that there can be no other possible outcome. To illustrate, the following is a computation of both the minimum and maximum possible values for PW, given the ranges for the input assumptions:

$$\text{Min PW} = 6,000(\$11.25 - \$11.00)(7.606) - \$95,212 = -\$83,803$$
$$\text{Max PW} = 7,500(\$12.50 - \$9.00)(7.606) - \$95,212 = \$104,446$$

If both present worths had been negative, we would have proven, *a fortiori*, that the proposal should be rejected on economic grounds. Conversely, if both PW values had been positive, an "accept" decision would have been indicated.

Unfortunately, this test of extreme values rarely yields a clear result, and the *a fortiori* argument cannot be used. Nevertheless, analysts would be well advised to try this approach before proceeding further. The calculations can be completed relatively easily, and the few cases for which a clear signal is indicated more than justify the time involved.

10.2.5   The Equal-Likelihood Assumption

What can be said about instances in which a specific outcome is strictly sensitive, that is, the breakeven point does lie within the expected range but is not *very* sensitive? **Figure 10.5** on page 360 illustrates this case. Here the breakeven point lies within the range, albeit very close to the minimum anticipated value. Given the two alternatives implied in Figure 10.5, (1) accept the proposal or (2) do nothing, it would appear that the proposal should be accepted. Why? Because, apparently, the likelihood of obtaining a favorable outcome (for example, positive PW) is substantially greater than the likelihood of obtaining an adverse outcome (negative PW, for example). This conclusion stems from the **equal-likelihood assumption**, the assumption that the relative likelihood that these two outcomes will occur is approximately equal to the relative

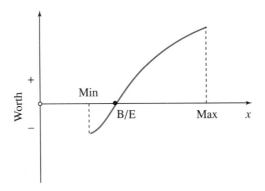

Figure 10.5

distances between the breakeven point and the minimum and maximum possible values.

Also implicit in this argument is the assumption that specific values for the parameter in question are equally likely through the range of possible values. Let us suppose, for example, that there is supplementary information, as illustrated in **Figure 10.6**. Note that the future value of the parameter in question (annual saving, in this case) is much more likely to occur at the lower end of the range. If such is the case, then the decision maker may be well advised to reject this particular proposal. It is not slightly sensitive; it is in fact very sensitive.

In real-world applications, there is no reason to believe that the relative likelihoods, or probabilities, of future events are distributed symmetrically around some most likely value. Indeed, there are many applications for which highly skewed distributions are quite common, including performance characteristics of certain electrical components, dimensions of manufactured parts after "substandard" units have been rejected, and meteorological data such as rainfall. Thus decision makers and analysts must be cautious about arriving at judgments concerning degree of sensitivity.

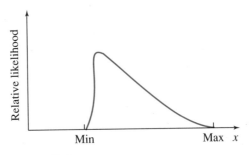

Figure 10.6

| 10.3   **RISK ANALYSIS**

The material in this section assumes that the reader is familiar with the fundamental concepts of probability theory, especially the notions of *probability*, expectation, and the *probability density function*. For those readers who are unfamiliar with these concepts, or for those for whom a review may be in order at this point, see Appendix 1 at the end of the chapter. Alternatively, the reader may wish to skip this material entirely and proceed directly to Section 10.4, "Decision Theory Applications."

10.3.1   **Alternative Risk Measures**

A number of different statistics have been proposed for the measure of "riskiness" of proposed plans, programs, and projects. Perhaps the most widely used measure is the **variance** (or **standard deviation**) **of the prospective return**, where return is generally the present worth, internal rate of return, et cetera. The variance (σ^2) of the distribution for a continuous random variable x is given by

$$\sigma^2 = \int_{-\infty}^{\infty} (x - \mu)^2 f(x)\, dx \qquad \textbf{(10.4)}$$

where μ is the mean of the distribution and $f(x)$ is the probability density function. Large variance signifies large risk; relatively small variance indicates relatively small risk. In general, everything else being equal, risk is to be minimized.

An alternative view is that the **semivariance** is a preferred statistic, because it focuses on the variability in negative return, that is, on the reduction of losses. The semivariance (S_h) of a distribution for random variable X is given by

$$S_h = \int_{-\infty}^{h} (h - x)^2 f(x)\, dx \qquad \textbf{(10.5)}$$

Still another measure of risk is the **probability of loss**, a statistic that measures the probability that the return will lie below some predetermined critical level, h. The probability of loss (L) for a continuous random variable is given by

$$L = \int_{-\infty}^{h} f(x)\, dx \qquad \textbf{(10.6)}$$

Limited space precludes a full discussion of these (and other) risk measures. Therefore, in the remainder of this section we will limit our remarks to the variance, a statistic that has proven most popular in use as well as in the literature of engineering economy.

10.3.2 Determining the Probability Distribution for Present Worth

Expected Present Worth

Consider an uncertain stream of cash flows, A_j, occurring at the end of periods $1, 2, \ldots, j, \ldots, N$. If the project life, N, the discount rate, i, and the amounts and timing of the cash flows are known with certainty, then

$$\text{PW} = \sum_{j=0}^{N} A_j(1 + i)^{-j} \tag{10.7}$$

Now, suppose that the cash flows are random variables with associated probability or density functions $f(A_j)$. The PW is a function of random variables, so it is itself a random variable with mean μ_p,

$$\mu_p = \text{Exp}[\text{PW}] = \sum_{j=0}^{N} \mu_j(1 + i)^{-j} \tag{10.8}$$

where $\mu_j = \text{Exp}(A_j)$ for $j = 0, 1, \ldots, N$.

Variance of Present Worth

The variance of the PW distribution depends on the degree of correlation between the individual cash flows. In general,

$$\sigma_p^2 = \text{Var}[\text{PW}] = \sum_{j=0}^{N} \sigma_j^2(1 + i)^{-2j}$$

$$+ 2\sum_{j=0}^{N-1}\sum_{k=j+1}^{N} \rho_{jk}\sigma_j\sigma_k(1 + i)^{j+k} \tag{10.9}$$

where ρ_{jk} is the correlation coefficient between two cash flows, A_j and A_k; and σ_j and σ_k are the standard deviations of the distribution of A_j and A_k, respectively. This formulation is intractable in practice because of the difficulty, if not impossibility, of estimating the correlation coefficients. However, formulations of the variance under the two extreme cases—independent cash flows ($\rho_{jk} = 0$) and perfectly correlated cash flows ($\rho_{jk} = 1$)—is helpful, as will be shown.

Independent Cash Flows

If there is no causative or consequential relationship between the cash flows, they are said to be *independent* and

$$\sigma_p^2 = \sum_{j=0}^{N} \sigma_j^2(1 + i)^{-2j} \tag{10.10}$$

EXAMPLE 10.1

A numerical example, summarized in **Table 10.1**, illustrates Eqs. (10.9) and (10.10). This is a 5-period project life with means (μ_j) and variances

TABLE 10.1   Numerical Example: Determining the Mean
and Variance of PW Given
Probabilistic Cash Flows and 10%
Discount Rate—Independent Cash Flows

End of Period j	Cash Flow Estimates		Present Worth @ 10%	
	Mean μ_j	Variance σ_j^2	$\mu_j(1.10)^{-j}$	$\sigma_j^2(1.10)^{-2j}$
0	− $400	$$\ 0^2	− $400.00	$$\ \ 0.0000
1	100	10^2	90.91	82.6446
2	130	15^2	107.44	153.6780
3	160	20^2	120.21	225.7896
4	130	20^2	88.79	186.6030
5	100	20^2	62.09	154.2173
Total			$ 69.44	$$802.9325

(σ_j^2) of the cash flows as shown. The results: $\mu_p = \$69.44$ and
$\sigma_p = \sqrt{\$\$802.9325} = \$28.34$.

Perfectly Correlated Cash Flows

Cash flows in any two periods, X and Y, are perfectly correlated if, given
that A_x is the actual value of $\mu_x + d\sigma_x$, then

$$A_y = \mu_y + d\sigma_y$$

In words, if random factors cause A_x to deviate from its mean value by
d standard deviations, the same factors will cause A_y to deviate from its
mean in the same direction by d standard deviations. Under these
conditions,

$$\sigma_p = \sum_{j=0}^{N} \sigma_j(1 + i)^{-j} \qquad (10.11)$$

EXAMPLE 10.2

Consider the problem summarized in **Table 10.2**. Assuming a 10%
discount rate, the expected value of the PW of the five cash flows is
$625.92, and the standard deviation of the PW is $62.60. Note that the
expected value of PW is given by Eq. (10.8) and is independent of the
degree of correlation.

Combining Independent and Perfectly Correlated Cash Flows. Suppose
that it is feasible, in a given problem situation, to identify two types of

TABLE 10.2   Numerical Example.   termining the
Mean and Variance oı PW Given
Probabilistic Cash Flows and 10% Discount
Rate—Perfectly Correlated Cash Flows

End of Period j	Cash Flows		Present Worth @ 10%	
	Mean μ_j	Standard Deviation σ_j	$\mu_j(1.10)^{-j}$	$\sigma_j(1.10)^{-j}$
1	100	$10	$ 90.91	$ 9.09
2	ı50	15	123.97	12.40
3	200	20	150.26	15.03
4	200	20	136.60	13.66
5	200	20	124.18	12.42
Total			$625.92	$62.60

cash flows: those that are statistically independent and those that are
perfectly correlated. In this case the variance of the PW distribution is
the sum of (1) the sum of the variances of the independent cash flows,
discounted, and (2) the sum of the variance of each of the subsets of
perfectly correlated cash flows, where the variance of each subset is the
square of the sum of the standard deviations of the cash flows in that
subset. That is,

$$\sigma_p^2 = \sum_{j=0}^{N} \sigma_j^2 (1+i)^{-2j} + \sum_{k=1}^{M} \left\{ \sum_{j=0}^{N} [\sigma_{jk}(1+i)^{-j}] \right\}^2 \qquad (10.12)$$

where

σ_j^2 = variance of the distribution of the independent A_j's

σ_{jk} = standard deviation of the distribution of the perfectly
correlated cash flows in subset k, $k = 1, 2, \ldots, M$

EXAMPLE 10.3

Returning to the previous example (Table 10.2), suppose that there is a
cash flow A_0 such that $\mu_0 = -\$500$ and $\sigma_0 = \$10$, and A_0 is independent
of the positive cash flows in periods 1 through 5. All cash flows for the
proposal are now completely specified, and

$$\mu_p = \sum_{j=0}^{5} \mu_j(1.10)^{-j}$$

σ_p^2 = (variance of A_0) + (variance of the sum of the perfectly
correlated cash flows $A_1, A_2, \ldots, A_5$)

$= (\$10)^2 + (\$62.60)^2 = \$\$4{,}018.76$

$$\sigma_p = \sqrt{\$\$4018.76} = \$63.39$$

Note in this example that $M = 1$; there is only one subset of perfectly correlated cash flows.

10.3.3 Analysis Based on the Probability Distribution for Present Worth

As before, the mean and variance of the probability distribution for the present worth statistic (PW) are denoted by μ_p and σ_p^2, respectively. These are measures of central tendency and variability, or dispersion, of the PW distribution. Under certain conditions, the underlying probability distribution may be fully or partially characterized. When such is the case, it may be useful to describe the riskiness of the figure of merit in terms other than the variance of the distribution, for example, the probability that the PW will exceed some specified critical level.

Discrete Distribution for Present Worth

Consider a 2-period problem as summarized in **Figure 10.7**.[2] A cash outlay of $100 occurs at the start of period 1 $(j = 0)$. There are two

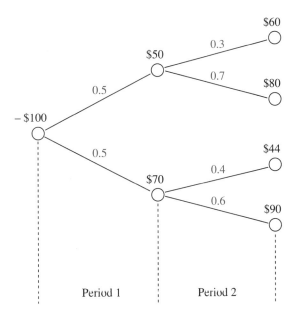

Figure 10.7   Cash Flows and Their Probabilities

[2]Example problem from C. S. Park and G. P. Sharp-Bette, *Advanced Engineering Economy* (New York: John Wiley, 1990), p. 419.

possible discrete cash flows at end of period 1: $A_1 = \$50$ with probability 0.5, or $A_1 = \$70$ with probability 0.5. If $A_1 = \$50$, there are two possibilities for the cash flow at end of period 2: Either $A_2 = \$60$ with probability 0.3, or $A_2 = \$80$ with probability 0.7. If $A_1 = \$70$, then either $A_2 = \$44$ with probability 0.4, or $A_2 = \$90$ with probability 0.6. The diagram of the possible outcomes shown in Figure 10.7 is sometimes known as a **probability tree**.

There are four possible present worths (outcomes), each with an associated joint probability. Assuming a 10% discount rate:

Outcome	A_0	A_1	A_2	$\sum\limits_{j=0}^{2} A_j(1.10)^{-j}$	Joint Probability
1	$-\$100$	$\$50$	$\$60$	$-\$\ 4.96$	$0.5 \times 0.3 = 0.15$
2	-100	50	80	11.57	$0.5 \times 0.7 = 0.35$
3	-100	70	44	0	$0.5 \times 0.4 = 0.20$
4	-100	70	90	38.02	$0.5 \times 0.6 = 0.30$

The remainder of the analysis is summarized in **Table 10.3**. Note that column 4 reflects the calculation of Exp [PW] and column (5) reflects the calculation of Exp [PW^2]. Moreover

$$\sigma_p^2 = \text{Exp}[(PW)^2] - [\text{Exp}(PW)]^2 \tag{10.13}$$

as discussed previously.

Now suppose that it is of interest to determine the probability that this investment will be profitable, that is, PW > $0. Only two of the possible outcomes, 2 and 4, meet this requirement, and, as they are independent events, the sum of their probabilities is

$$\text{Prob}[PW > \$0] = \text{Prob}[PW = \$11.57] + \text{Prob}[PW = \$38.02]$$
$$= 0.35 + 0.30 = 0.65$$

TABLE 10.3   Determining the Expected Present Worth

Outcome (1)	PW @ 10% (2)	Joint Probability (3)	(4) = (2) × (3)	(5) = (2)^2 × (3)
#1	$-\$\ 4.96$	0.15	$-\$\ 0.744$	$\$\$\ \ 3.690$
#2	11.57	0.35	4.050	46.853
#3	0	0.20	0	0
#4	38.02	0.30	11.406	433.656
		$\overline{1.00}$	$\overline{\$14.712}$	$\overline{\$\$484.199}$

Exp[PW] = $14.712
Var[PW] = $$484.199 − ($14.712)^2 = $$267.756
$\sigma_p = \sqrt{\text{Var}[PW]} = \$16.363$

Using Only the Mean and Variance
of the PW Distribution

Tchebycheff's (sometimes written Chebyshev's) **inequality** states that

$$\text{Prob}[\mu - k\sigma < X < \mu + k\sigma] \geq 1 - \frac{1}{k^2} \qquad (10.14)$$

where X is any random variable having mean μ and variance σ^2, and k is a positive constant. This is a useful relationship when only the mean and variance of the distribution are known. In terms of an unknown PW distribution with known mean μ_p and variance σ_p^2,

$$\text{Prob}[\mu_p - k\sigma_p < \text{PW} < \mu_p + k\sigma_p] \geq 1 - \frac{1}{k^2} \qquad (10.15)$$

To illustrate, suppose that the mean and variance of the PW distribution have been determined to be \$800 and (\$50)^2, respectively. The analyst has been asked to determine the probability that the PW lies between two values—say, between \$600 and \$1,000. Note here that

$$\mu_p - k\sigma_p = \$800 - k(\$50) = \$600$$

and

$$\mu_p + k\sigma_p = \$800 + k(\$50) = \$1,000$$

from which it is apparent that $k = 4$. Thus,

$$\text{Prob}[\$600 < \text{PW} < \$1,000] \geq 1 - \frac{1}{16} \quad \text{or} \quad 0.9375$$

Put somewhat differently, in the absence of any knowledge as to the shape of the distribution, the probability is *at least* 0.9375 that the random variable lies within $\pm 4\sigma$ of the mean.

When the Normal Distribution
Can Be Assumed

Consider a stream of risky cash flows A_j occurring at the ends of periods $1, 2, \ldots, j, \ldots, N$. The project life ($N$) and the discount rate (i) are known with certainty. The only stochastic variable here is the amount of the cash flow. The resulting PW is a random variable with mean given by Eq. (10.8) and, assuming independent cash flows, with variance given by Eq. (10.10). Under some general conditions, application of the Central Limit Theorem leads to the result that

$$Z_N = \frac{(\text{PW}) - [\Sigma_{j=0}^{N} \mu_j (1 + i)^{-j}]}{\sqrt{\Sigma_{j=0}^{N} \sigma_j^2}} \qquad (10.16)$$

is approximately normally distributed, with $\mu = 0$ and $\sigma = 1$, as N approaches infinity. The "general condition" may be summarized as follows: The terms A_j, taken individually, contribute a negligible amount to the variance of the sum, and it is unlikely that any single A_j makes a relatively large contribution to the sum.

The terms A_j may have essentially any distribution. As a general rule of thumb, if the A_j's are approximately normally distributed, then the Central Limit Theorem is a very good approximation when $N \geq 4$. If the distribution of the A_j's has no prominent mode(s), that is, approximately uniform distribution, then $N \geq 12$ is a reasonable rule of thumb for applicability of the Central Limit Theorem.

To illustrate the application of Eq. (10.16), consider the numerical example given in Table 10.1. It was determined that $\mu_p = \$69.44$ and $\sigma_p = \sqrt{\$\$802.9325} = \$28.34$. The probability distribution for PW is shown in **Figure 10.8a**; the equivalent standardized normal distribution is shown in **Figure 10.8b**.

Consider the question: What is the probability that this proposal will result in a present worth greater than $50?

$$\text{Prob}[PW > \$50] = 1 - \text{Prob}[PW < \$50]$$

$$= 1 - \text{Prob}\left[Z < \frac{\$50.00 - \$69.44}{\$28.34}\right]$$

$$= 1 - \text{Prob}[Z < -0.686] = 0.50 - 0.25 = 0.75$$

(a) Distribution of random variable, PW

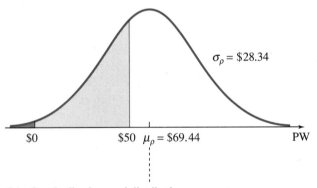

(b) Standardized normal distribution

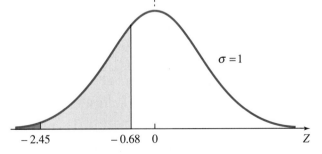

Figure 10.8   Probability Distribution for Present Worth

(Tables for the standardized normal distribution are given in Appendix 2 at the end of the chapter.)

Consider a second question: What is the probability that this proposal will result in a loss?

$$\text{Prob}[PW < \$0] = \text{Prob}\left[Z < \frac{\$0 - \$69.44}{\$28.34}\right]$$

$$= \text{Prob}[Z < -2.45] \cong 0.01$$

Note that the probability of a loss is identical here to the probability that the proposal's internal rate of return will be less than the minimum attractive rate of return. That is,

$$\text{Prob}[PW < \$0] = \text{Prob}[IRR < MARR]$$

10.3.4 Comparing Risky Proposals

As indicated previously, decision makers are generally risk avoiders. Additional risk, as measured by the variance of the figure of merit, is to be avoided whenever possible. Thus there are two criteria to be considered simultaneously: the figure of merit, e.g., present worth, as measured by the expected value (μ) of the distribution; and the riskiness of the outcome as measured by the variance (σ^2) of the distribution. The former is to be maximized, and the latter is to be minimized.

Consider two mutually exclusive alternatives. Let (μ_1, σ_1^2) and (μ_2, σ_2^2) represent the mean and variance of alternatives I and II, respectively. Alternative II is preferred to Alternative I if

Case A: $\mu_1 = \mu_2$ and $\sigma_1 > \sigma_2$

Case B: $\mu_1 < \mu_2$ and $\sigma_1 = \sigma_2$

In case A the two alternatives have equal worth, but alternative II has lower risk. In case B the measures of risk are equal but alternative II has the larger measure of worth. (See **Figure 10.9** on page 370.)

With similar argument, alternative I is preferred to alternative II if

$$\mu_1 = \mu_2 \quad \text{and} \quad \sigma_1 < \sigma_2$$

or

$$\mu_1 > \mu_2 \quad \text{and} \quad \sigma_1 = \sigma_2$$

There is a third possibility, of course. The conclusion is ambiguous if

Case C: $\mu_1 < \mu_2$ and $\sigma_1 < \sigma_2$

or

$$\mu_1 > \mu_2 \quad \text{and} \quad \sigma_1 > \sigma_2$$

When this situation arises, trade-offs must be made between risk and return.

Case A: $\mu_1 = \mu_2$ and $\sigma_1 > \sigma_2$

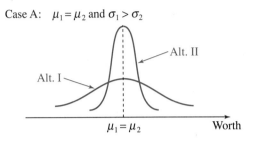

Case B: $\mu_1 < \mu_2$ and $\sigma_1 = \sigma_2$

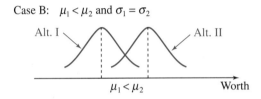

Case C: $\mu_1 < \mu_2$ and $\sigma_1 < \sigma_2$

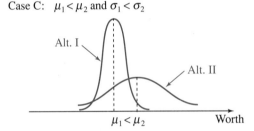

Figure 10.9 Alternative II Preferred to Alternative I in Cases A and B; Ambiguous Conclusion in Case C

10.3.5   Other Models

There are a variety of other analytical models for assessing risky investments. The randomness ("riskiness") of cash flow amounts, timing, project life, and discount rate are considered singly and/or in combination. The complexity of the analytical procedure is roughly a function of the number of variables considered, as well as the assumptions concerning mutual independence between random variables. In almost all cases the mean and variance of the distribution of the measure of worth are of primary concern. In some instances it is also possible to approximate the statistical distribution as well. Space limitations preclude an exhaustive review of the extant literature. For further reading, consult the Bibliography for this chapter.

10.4   DECISION THEORY APPLICATIONS

The approach to risk analysis outlined in the previous section is based on the premise that the decision maker desires to (1) maximize expected return and (2) minimize risk. This section presents some additional principles of choice that may be appealing under certain conditions. A simple numerical example is used as a basis for the discussion.

10.4.1   Problem Statement

The International Manufacturing Company (IMC) is considering five mutually exclusive alternatives for constructing a new manufacturing plant in a certain Asian country. The costs of each alternative, stated in terms of (equivalent uniform) annual costs, depend on the outcome of negotiations that are currently underway between IMC, lending agencies, and the government of the host country. IMC analysts have concluded that four specific mutually exclusive outcomes are possible, and they have computed the equivalent uniform annual cost for each alternative–outcome combination. These are shown as cell values in the **cost matrix** in **Table 10.4**.

 If the future is known with certainty, then the least costly alternative may be selected by any of the methods presented in Chapter 3 or 4. For example, if it is known that outcome s_3 will definitely occur, then a_3 should be selected because it will result in the lowest (equivalent uniform) annual cost. On the other hand, if a_3 is selected and s_2 perversely occurs, choosing a_3 will have resulted in the most costly event.

 Assume that sufficient information exists to warrant statements about the relative probabilities of the possible future outcomes. Specifically, these probabilities (expected relative frequencies) are

$$\text{Prob}[s_1] = 0.3 \qquad \text{Prob}[s_3] = 0.2$$
$$\text{Prob}[s_2] = 0.4 \qquad \text{Prob}[s_4] = 0.1$$

TABLE 10.4   Cost Matrix for Illustrative Problem (Cell Entries in Units of $100,000)

		Possible Outcomes			
		s_1	s_2	s_3	s_4
	a_1	18	11	11	10
	a_2	16	16	16	16
Alternatives	a_3	17	20	8	17
	a_4	9	12	17	16
	a_5	10	13	17	18

Given this additional information, which alternative should be selected? A number of principles that may be applied in this situation are discussed below.

A problem statement of this type is known as a **decision under risk** because the underlying probability distribution for the future scenarios, or **states of nature**, is known or can be assumed. "Risk," in the previous section, was used in a more general sense to characterize the absence of certainty. The term was used analogously to "randomness" or uncertainty. Here, in a more limited sense, a problem statement in which the underlying distribution for the s_j's is *not* known or assumed is a **decision under uncertainty**.

10.4.2 Dominance

Before applying any of the principles of choice, it is first desirable (although not absolutely necessary) to apply the **dominance principle** to determine which alternatives, if any, are inferior to one or more of the other alternatives. If, of two alternatives, one would never be preferred no matter what future occurs, it is said to be dominated and may be removed from further consideration. From the example, consider the costs associated with a_4 and a_5:

	s_1	s_2	s_3	s_4
a_4	9	12	17	16
a_5	10	13	17	18

Since a_5 is always at least as costly as a_4, regardless of which future outcome occurs, a_5 may be ignored in the remaining discussion.

If one alternative dominates all others, it is said to be **globally dominant**, and the decision maker need look no further; the optimal solution has been found. Unfortunately, globally dominant alternatives are rare. But in any event, the dominance principle is frequently effective in reducing the number of alternatives to be considered.

10.4.3 Principles for Decisions Under Risk

The Principle of Expectation

The **principle of expectation** states that the alternative to be selected is the one that has the minimum expected cost (or maximum expected profit or revenue). In general,

$$\text{Min } E[C(a_i)] = \sum_j C(a_i \mid s_j) \times p_j \qquad \textbf{(10.17a)}$$

or

$$\text{Max } E[R(a_i)] = \sum_j R(a_i \mid s_j) \times p_j \qquad \textbf{(10.17b)}$$

where

$$C(a_i \mid s_j) = \text{total cost of alternative } a_i \text{ given that state of}$$
$$\text{nature } s_j \text{ occurs}$$

$$R(a_i \mid s_j) = \text{total net return of alternative } a_i \text{ given that}$$
$$\text{state of nature } s_j \text{ occurs}$$

$$p_j = \text{probability that } s_j \text{ will occur}$$

From the example it may be shown that

$$E[C(a_1)] = \$1{,}300{,}000 \qquad E[C(a_3)] = \$1{,}640{,}000$$
$$E[C(a_2)] = \$1{,}600{,}000 \qquad E[C(a_4)] = \$1{,}250{,}000$$

Here, a_4 should be selected because it yields the minimum expected cost.

Principles that depend on determination of expected values by the mathematics of probability theory are frequently criticized on the grounds that the theory holds only when trials are repeated many times. It is argued that, for certain types of decisions—for example, whether to finance a major expansion—expectation is meaningless, since this type of decision is not made very often. According to the counterargument, even if the firm is not faced with a large number of repetitive decisions, it should apply the principle to many different decisions and thus realize the long-run effects. Moreover, even if the decision is unique, the only way to approach decisions for which probabilities are known is to behave as if the decision were a repetitive one and thus minimize expected cost or maximize expected revenue or profit.

The Principle of Most Probable Future

Assume that the future event to expect is the most likely event. Thus, observing that s_2 has the highest probability of occurring, assume that it will in fact occur. In this case, a_1 (with AC = $1,100,000) is the least costly of the four available alternatives.

This principle is particularly appealing in cases in which one future is significantly more probable than all other possibilities.

The Aspiration-Level Principle

The **aspiration-level principle** requires the establishment of a goal, or "level of aspiration." Thus, the alternative that maximizes the probability that the goal will be met or exceeded should be selected. To illustrate, suppose that the management of IMC wishes to minimize the probability that equivalent uniform annual costs will exceed $1,500,000. (This is identical to the requirement that it maximize the probability that costs will *not* exceed $1,500,000.) The probabilities are

$$\text{Prob}[C(a_1) > \$1{,}500{,}000] = 0.3 \qquad\qquad = \boxed{0.3}$$
$$\text{Prob}[C(a_2) > \$1{,}500{,}000] = 0.3 + 0.4 + 0.2 + 0.1 = 1.0$$

$$\text{Prob}[C(a_3) > \$1,500,000] = 0.3 + 0.4 \qquad + 0.1 = 0.8$$
$$\text{Prob}[C(a_4) > \$1,500,000] = \qquad + 0.2 + 0.1 = 0.3$$

Thus the aspiration level will be met if either a_1 or a_4 is selected.

10.4.4 Principles for Decisions Under Uncertainty

This section examines a number of principles of choice that may be used when the relative likelihoods of future states of nature *cannot* be estimated. These principles will be demonstrated by using the example problem introduced above.

The Minimax (or Maximin) Principle

The **minimax principle** is pessimistic in the extreme. It assumes that, if any alternative is selected, the worst possible outcome will occur. The maximum cost associated with each alternative is examined, and the alternative that *minimizes* the *maximum* cost is selected. In general, the mathematical formulation of the minimax principle is

$$\underset{i}{\text{Min}} \left[\underset{j}{\text{Max}} \, C_{ij} \right] \tag{10.18a}$$

From the example,

Alternative (a_j)	$\text{Max}_j \, C_{ij} (\times \$10^5)$
a_1	18
a_2	16
a_3	20
a_4	17

If the minimax principle is adopted, a_2 is indicated because it results in minimum cost, assuming the worst possible conditions.

The mirror image of the minimax principle, the **maximin principle**, may be applied when the matrix contains profits or revenue measures. In this case the most pessimistic view suggests that the alternative to select is the one that *maximizes* the *minimum* profit or revenue associated with each alternative. The mathematical formulation of the maximin principle is

$$\underset{i}{\text{Max}} \left[\underset{j}{\text{Min}} \, R_{ij} \right] \tag{10.18b}$$

where R_{ij} is the revenue or profit resulting from the combination of a_i and s_j.

The Minimin (or Maximax) Principle

The **minimin principle** is based on the view that the best possible outcome will occur when a given alternative is selected. It is optimistic in the extreme. The minimum cost associated with each alternative is examined, and the alternative that *minimizes* the *minimum* cost is selected. The mathematical formulation is

$$\text{Min}_i \left[\text{Min}_j C_{ij} \right] \qquad (10.19a)$$

From the example,

Alternative (a_i)	$\text{Max}_j C_{ij} (\times \$10^5)$
a_1	10
a_2	16
a_3	8
a_4	9

Alternative a_3 minimizes the minimum cost.

As a corollary to the minimin principle, the **maximax principle** is appropriate when the decision maker is extremely optimistic and the matrix contains measures of profit or revenue. The maximum profit (or revenue) associated with each alternative is examined, and the alternative that *maximizes* the *maximum* profit (or revenue) is selected. The mathematical formulation is

$$\text{Max}_i \left[\text{Max}_j R_{ij} \right] \qquad (10.19b)$$

The Hurwicz Principle

It may be argued that decision makers need not be either completely optimistic or pessimistic, in which case the **Hurwicz principle** permits selection of a position between the two extremes. When evaluating costs, C_{ij}, the *Hurwicz criterion* for alternative a_i is given by

$$\text{Min} \left\{ H(a_i) = \alpha \left[\text{Min}_j C_{ij} \right] + (1 - \alpha) \left[\text{Max}_j C_{ij} \right] \right\} \qquad (10.20a)$$

where α is the *index of optimism* such that $0 \leq \alpha \leq 1$. Extreme pessimism is defined by $\alpha = 0$; extreme optimism is defined by $\alpha = 1$. The value of α used in any particular analysis is selected by the decision maker based on subjective judgment. The alternative that minimizes the quantity $H(a_i)$ is the preferred alternative.

When evaluating profits or revenue, R_{ij}, the expression for the Hurwicz criterion is

$$\text{Max} \left\{ H(a_i) = \alpha \left[\text{Max}_j R_{ij} \right] + (1 - \alpha) \left[\text{Min}_j R_{ij} \right] \right\} \qquad (10.20b)$$

The values of $H(a_i)$ are plotted in **Figure 10.10** on page 376 for the sample problem. Worth measures are stated as costs, so we want to

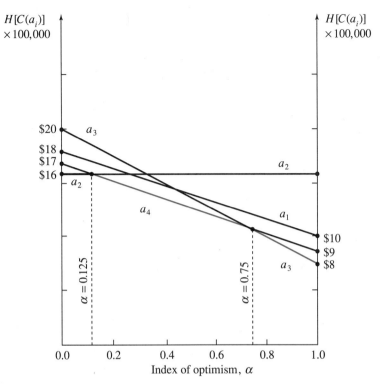

Figure 10.10 Sample Problem—Hurwicz Criterion as Function of Index of Optimism, α

minimize the Hurwicz criterion. We may determine, either graphically or algebraically, that a_2 will be chosen for $0 \leq \alpha \leq 0.125$, a_4 will be selected for $0.125 \leq \alpha \leq 0.75$, and a_3 is least costly for $0.75 \leq \alpha < 1.00$.

The Laplace Principle (Insufficient Reason)

The **Laplace principle**, sometimes known as the **principle of insufficient reason**, assumes that the probabilities of future events occurring are equal. That is, in the absence of any information to the contrary, it is assumed that all future outcomes are equally likely to occur. The expected cost (or profit/revenue) of each alternative is then computed, and the alternative that yields the minimum expected cost (or maximum expected profit/revenue) is selected. The mathematical expression for this principle is

$$\underset{i}{\text{Min}} \left\{ \left(\frac{1}{k}\right) \sum_{j=1}^{k} C_{ij} \right\} \qquad (10.21a)$$

when the figure of merit is expressed as a cost, or as

$$\underset{i}{\text{Max}} \left\{ \left(\frac{1}{k}\right) \sum_{j=1}^{k} R_{ij} \right\} \qquad (10.21b)$$

when the figure of merit is expressed as revenue or profit.

Returning to our example, the insufficient reason assumption yields $p_1 = p_2 = p_3 = p_4 = 0.25$. With these probabilities,

$$E[C(a_1)] = \$1,250,000 \qquad E[C(a_3)] = \$1,550,000$$
$$E[C(a_2)] = \$1,600,000 \qquad E[C(a_4)] = \$1,350,000$$

Alternative a_1 should therefore be selected because it results in the minimum expected annual cost.

The Savage Principle (Minimax Regret)

The **Savage principle**, or **principle of minimax regret**, is based on the assumption that the decision maker's primary interest is the *difference* between the actual outcome and the outcome that would have occurred had it been possible to predict the future accurately. Given these differences, or *regrets*, the decision maker then adopts a conservative position and selects the alternative that minimizes the maximum potential regret for each alternative.

A **regret matrix** is constructed, having for its cell values either

$$C_{ij} - \left[\underset{i}{\text{Min }} C_{ij} \right] \tag{10.22a}$$

for cost data, or

$$\left[\underset{i}{\text{Max }} R_{ij} \right] - R_{ij} \tag{10.22b}$$

for revenue or profit data. In either case, these cell values, or regrets, represent the differences between (1) the outcome if alternative a_i is selected and state of nature s_j subsequently occurs and (2) the outcome that would have been achieved had it been known in advance which state of nature would occur, so that the best alternative could have been selected. To illustrate, consider alternative a_1 and state of nature s_1: $C_{11} = \$1,800,000$. However, if we had known *a priori* that state s_1 would in fact occur, we would have selected a_4, incurring a cost of only $\$900,000$. The difference ($\$1,800,000 - \$900,000$) is a measure of "regret" about selecting a_1 when we could have selected a_4 (had we known the state of nature in advance). The complete regret matrix for the example is given in **Table 10.5** on page 378.

The alternative that *minimizes* the *maximum* regret is preferred. That is, for cost data,

$$\underset{i}{\text{Min }} \underset{j}{\text{Max }} \left\{ C_{ij} - \left[\underset{i}{\text{Min }} C_{ij} \right] \right\} \tag{10.23a}$$

or, when the cell values are based on revenue or profit data,

$$\underset{i}{\text{Min }} \underset{j}{\text{Max }} \left\{ \left[\underset{i}{\text{Max }} R_{ij} \right] - R_{ij} \right\} \tag{10.23b}$$

TABLE 10.5   Regret Matrix for Sample Problem
(Cell Values Are Multiples of $100,000)

		Possible Outcomes			
		s_1	s_2	s_3	s_4
Alternatives	a_1	$18-9=9$	$11-11=0$	$11-8=3$	$10-10=0$
	a_2	$16-9=7$	$16-11=5$	$16-8=8$	$16-10=6$
	a_3	$17-9=8$	$20-11=9$	$8-8=0$	$17-10=7$
	a_4	$9-9=0$	$12-11=1$	$17-8=9$	$16-10=6$

Equation (10.23a) is applicable for the sample problem, yielding

Alternative	Maximum Regret
a_1	$900,000
a_2	$800,000
a_3	$900,000
a_4	$900,000

Thus, according to this principle of choice, a_2 should be preferred.

   A major inconsistency in using this principle is that the solution can be altered by adding still another alternative that itself is not selected as the most desirable. To illustrate, suppose that we are considering another alternative, a_6, with the following annual costs ($\times$ $100,000):

	s_1	s_2	s_3	s_4
a_6	19	18	6	18

Including this new alternative results in a revised regret matrix:

	s_1	s_2	s_3	s_4
a_1	9	0	5	0
a_2	7	5	10	6
a_3	8	9	2	7
a_4	0	1	11	6
a_6	10	7	0	8

Now, minimizing the maximum regret:

a_1	⑨
a_2	10
a_3	⑨
a_4	11
a_6	10

In this instance, the addition of alternative a_6 has shifted the solution from a_2 to a_1 or a_3—an unreasonable result. Thus, by adding an alternative that is not selected under the Savage principle of minimax regret, the solution has been changed.

10.4.5   Summary of Results

There is no special reason why the principles of choice discussed in the preceding sections should yield the same solution. Indeed, each of the alternatives in this example problem was selected at least once.

Decisions Under Risk		Decisions Under Uncertainty	
Principle	Solution	Principle	Solution
Expectation	a_4	Minimax	a_2
Most probable future	a_1	Minimin	a_3
Aspiration level	a_1 or a_4	Hurwicz	
		$(0.125 < \alpha < 0.75)$	a_4
		Laplace	
		(insufficient reason)	a_1
		Savage	
		(minimax regret)	a_2

Is one principle more "correct" than any other? There is no simple answer to this question, since the choice of principle depends largely on the predisposition of the decision maker and the specific decision situation. Each principle has certain obvious advantages, and each is deficient in one or more desirable characteristics. Nevertheless, the principles in this section are useful because they shed some light on the subjective decision process and make the available information explicit to the decision process.

10.5   OTHER APPROACHES FOR DEALING WITH THE UNCERTAIN/RISKY FUTURE

As indicated at the beginning of this chapter, risk and uncertainty are inherent in the general problem of resource allocation, because all decisions depend on estimates about the noncertain future. Thus risk and uncertainty have occupied the attention of a great many theoreticians and practitioners. A substantial number of approaches have been proposed, several of which have been summarized above. Now, six additional approaches are briefly identified.

10.5.1   Decision Trees

Decision tree methodology is useful for evaluation of problems characterized by *sequential decisions*, each of which involves a variety of outcomes. The pictorial representation of this problem is suggestive of a "tree" lying on its side, with the branches of the tree representing succession of outcomes. The graphical portrayal of the problem structure is both its primary asset as well as its principal disadvantage. The ability to communicate complex dependencies is of great value, of course. However, the number of sequential decisions and outcomes (branches) is necessarily limited by the graphic medium (CRT screen, $8\frac{1}{2}'' \times 11''$ paper, etc.).

Each path through the tree represents a set of policy decisions. (In the case of a *stochastic* tree, that is, where certain intervening events over time are noncertain but for which the probabilities of occurrence can be estimated, each path reflects policy decisions coupled with the outcomes of the stochastic events.) Procedures exist for determining the least-cost or maximum-benefit path through the tree, thereby identifying the optimal set of sequential decisions.

10.5.2   Digital Computer (Monte Carlo) Simulation

The statistical procedures related to risk analysis suffer from at least one important drawback: The analytical techniques necessary to derive the mean, variance, and possibly the probability distribution of the figure of merit may be extremely difficult to implement. Indeed, the complexity of many real-world problems precludes the use of these computational techniques altogether; computations may be intractable, or the necessary underlying assumptions may not be met. Under these conditions, analysts may find **digital computer (Monte Carlo) simulation** especially useful. (Strictly speaking, *Monte Carlo simulation* and *digital computer simulation* are not synonymous. Monte Carlo simulation is a sampling technique used in the digital computer simulation of systems behavior. However, in recent years, practitioners have tended to blur this semantic distinction, using the terms interchangeably.)

The objective of digital computer simulation is to generate a probability distribution for the figure of merit, generally present worth or rate of return, given the probability distributions for the various components of the analysis. The decision maker can thus compare expected returns as well as the variability of returns for two or more alternatives. Moreover, probability statements can be made in this form: The probability is x that project y will result in a profit in excess of z.

The general procedure can be described in four steps.

Step 1. Determine the probability distribution(s) for the significant factors, as illustrated in **Figure 10.11.**

Step 2. Using Monte Carlo simulation, select random samples from these factors according to their relative probabilities of occurring in the future. (See **Figure 10.12**.) Note that the selection of one factor (price, for example) may determine the probability distribution of another factor (total amount demanded, for example).

Step 3. Determine the figure of merit (rate of return or present

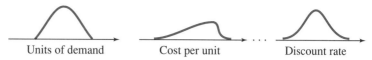

Figure 10.11 Probability Functions for Inputs

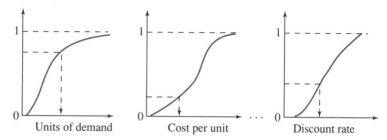

Figure 10.12 Cumulative Distribution Functions

worth, for example) for each combination of factors. One trial consists of one calculation of the figure of merit.

Step 4. Repeat the process, that is, conduct a series of trials, building a frequency histogram with the results, as in **Figure 10.13**. Continue until you are reasonably satisfied that the histogram yields a clear portrayal of the investment risk. (There is no universally accepted rule for determining the optimum number of trials. It is clearly less expensive to produce a smaller number of trials, yet a larger number of trials yields more information. Substantial literature is addressed to this interesting problem, but additional discussion is not warranted here.)

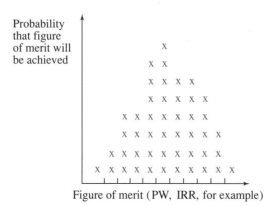

Figure 10.13 Frequency Histogram for Measures of Worth

10.5.3 Increasing the Minimum Attractive Rate of Return

Some analysts advocate adjusting the minimum attractive rate of return to compensate for risky investments, suggesting that, since the future is uncertain, stipulation of an MARR of, say, $i + \Delta i$, will ensure that i will be earned in the long run. Since some investments will not turn out as well as expected, they will be compensated for by the incremental "safety margin," Δ_i. This approach, however, fails to come to grips with the risk or uncertainty associated with estimates for specific alternatives, and thus an element Δ_i in the minimum attractive rate of return penalizes all alternatives equally.

10.5.4 Differentiating Rates of Return by Risk Class

Rather than building a "safety margin" into a single minimum attractive rate of return, some firms establish several risk classes with separate standards for each class. For example, a firm may require low-risk investments to yield at least 15%, medium-risk investments to yield at least 20%, and it may define a minimum attractive rate of return of 25% for high-risk proposals. The analyst then judges which class a specific proposal belongs in, and the relevant minimum attractive rate of return is used in the analysis. Although this approach is a step away from treating all alternatives equally, it fails to focus attention on the uncertainty associated with the individual proposals. No two proposals have precisely the same degree of risk, and grouping alternatives by class obscures distinctions within each class. Moreover, the attention of the decision maker should be directed to the causes of uncertainty, that is, to the individual estimates.

10.5.5 Decreasing the Expected Project Life

Still another procedure frequently employed to compensate for uncertainty is to decrease the expected project life. It is argued that estimates become less and less reliable as they occur farther and farther into the future; thus shortening project life is equivalent to ignoring those distant, less reliable estimates. Furthermore, distant consequences are more likely to be favorable than unfavorable; that is, distant estimated cash flows are generally positive (resulting from net revenues), and estimated cash flows near date zero are more likely to be negative (resulting from start-up costs). Reducing expected project life, however, has the effect of penalizing the proposal by precluding possible future benefits, thereby allowing for risk in much the same way that increasing the minimum attractive rate of return penalizes marginally attractive proposals. Again, this procedure is to be criticized on the basis that it obscures noncertain estimates.

10.5.6 Utility Models

In essence, **utility** is a single metric on the unit interval denoting the degree of desirability of an item or a quantity of items with respect to a completely defined collection of such items. Thus an item or group of items with the greatest desirability would have a utility of, say, 100, and at a least desirable item, a zero utility. All items and groups within the collection range between these extremes in an ordered fashion. Amounts of monetary receipts and disbursements would provide a utility function from zero to 100. A monetary gamble would be reviewed in this theory as a linear combination of the amount won and lost in the gamble, with the expected utility associated with winning. Once the decision maker's utility function is derived, the theory of utility describes how one should act in order to remain consistent with his or her goals. Accordingly, utility theory is a description of normative economic behavior based on several stated axioms.

A number of advocates of this theory have therefore recommended that utility functions be established and economic risk analysis conducted with respect to this theory. That is, projects with the greatest expected utility should be selected by rational economic decision makers. There are many compelling features to this approach. However, it also requires development of the utility function, which is not a simple task; the question of whose utility function should represent the firm; and other perplexing problems. Also, it has been shown that current methods of risky cash flow analysis do represent a reasonable and rational approximation of the utility theory approach. There are also challenges to the axioms of existing theories of utility. Because of these and other criticisms and limitations, the utility theory approach has not enjoyed popularity among practitioners.

SUMMARY

Limited capital is allocated to competing investment proposals on the basis of estimates of future consequences. Since the future can never be known with absolute certainty, it follows that procedures must be developed that analysts can use when dealing with risk and uncertainty in capital budgeting problems. Several of those procedures are discussed in this chapter.

Sensitivity (or breakeven) analysis is widely used because of its simplicity and ability to focus on particular estimates. It generally treats only one estimate at a time, however, and it is difficult to apply when determining possible effects of combinations of errors in the estimates.

Another approach is based on the notion that the underlying probability distributions of the variables of concern are known or can be estimated. Under certain conditions the means and variances of the measures of worth of competing investment alternatives can be determined and compared. Preferred alternatives are those that maximize return (as measured by the mean) and minimize risk (as measured by the variance). Moreover, if the probability density function of the worth measure can be assumed, statements can be made as to the likelihood that a figure of merit (e.g., present worth) will exceed some critical level.

Decision theory provides a number of principles that may be used to select from among mutually exclusive alternatives when various future outcomes are possible. They may be grouped into two classes: those that are applicable when nothing at all is known about the relative probabilities of future events (decisions under uncertainty), and those that are applicable when these probabilities can be defined with reasonable precision (decisions under risk). There are significant differences between the principles of choice employed, and in some instances principles are mutually contradictory. In any event, they are helpful in formalizing the decision process and thereby exposing the relative strengths and weaknesses of the various viewpoints.

Decision trees are sometimes useful in the evaluation of problems characterized by sequential decisions, each of which involves a variety of outcomes. This graphical display is necessarily limited, however, by the inherent confines of the graphic medium.

Digital computer (Monte Carlo) simulation is a technique that is becoming more widely used, especially in view of the recent dramatic increase in computing power and the corresponding decrease in the cost of simulation. However, this approach requires the analyst to provide estimates of the probability distributions of all the relevant parameters. This requirement may be difficult to fulfill in many applications.

Some firms adjust for risk and uncertainty by requiring prospective investments to promise somewhat higher returns than would normally be expected if future consequences were known with perfect certainty. In some cases the minimum attractive rate of return is increased above the risk-free rate; in others expected project life is truncated. These approaches are criticized primarily on the grounds that they penalize all alternatives equally and that they tend to obscure the degree of uncertainty associated with specific estimates.

There are no simple, widely accepted techniques for effectively dealing with risk and uncertainty in capital allocation decisions. Although some promising advances have been made in recent years in the application of probability theory and decision theory, practical application remains as much an art as a science. (This observation is arguable, of course; some observers hold the view that there has been notable progress over the past 20 years or so. Nevertheless, there is very little evidence that these advances have in fact been implemented by practitioners.)

PROBLEMS

Sensitivity Analysis

10.1 A phased program of plant expansion is being compared with a program wherein full expansion will be undertaken immediately with a total initial investment of $1,400,000. The phased program requires $800,000 now, $600,000 in 5 years, and another $600,000 in 10 years. If the full program is selected, estimated annual disbursements will be an additional $40,000 during the first 5 years and an additional $20,000 during the second 5 years. There are no other differences between the alternatives.

a. If the pretax minimum attractive rate of return is 10%, which alternative should be selected?

b. At what range of values for the minimum attractive rate of return is the phased program economically superior?

[*Answers*: a. phased program; b. greater than 4.9%.]

10.2 The research director and the controller of the Truline Trucking Company are discussing the merits of a proposed labor-saving device designed to eliminate a certain freight-handling operation. The proposal has a first cost of $25,000, an estimated life of 10 years, and zero net salvage value. The manual operation that it will replace currently requires 500 labor hours per year at a cost of $10 per hour. There are no other differences between alternatives. The minimum attractive rate of return is 20% before taxes.

a. The controller believes that the investment should not be made. Is he correct? Why or why not?

b. The research director claims that labor costs may be expected to increase by about 10% per year. That is, the cost per labor hour will be $10 the first year, $11 the second, $12.10 the third, and so on. Under these conditions, she says, the new device should be purchased. Is she correct? Why or why not?

10.3 An industrial sales firm is considering the purchase of a personal computer for $10,000. If this computer is purchased, clerical costs will be reduced by $1,000 per year. The computer will be used for 10 years, at the end of which time the net salvage value will be zero. The firm requires a minimum attractive rate of return of 15% before taxes.

a. Assuming no other relevant differences between alternatives, should the computer be purchased?

b. Test the sensitivity of the solution in part a to the assumption concerning annual labor costs by assuming that these costs will increase by $50 per year, that is, $1,000 the first year, $1,050 the second, $1,100 the third, and so on.

c. How much must the cost of labor increase per year for the proposed investment to be economically sound?

[*Answers*: a. No; b. PW $= -\$4,132$; c. $G > \$293$.]

10.4 A company has fixed costs of $80,000 with variable costs equal to 60% of net sales. The company is planning to increase its present capacity of $400,000 net sales by 30%, with a 20% increase in fixed costs. The tax rate on profits is 40%.

a. What would be net profit after taxes if the *present plant* were operated at *full capacity*?

b. Assume that the present plant is enlarged as described above. What must the new sales revenue be if the company is to "break even," that is, with revenues equal to expenses?

10.5 Recall the "sensitivity analysis" problem discussed in the text:

Initial cost (P)	$100,000
Revenue per unit sold (r)	$12
Cost per unit sold (c)	$10
No. of units sold/year (Q)	7,000
Salvage value after 15 years (S)	$20,000
Study period (N)	15 years
Discount rate (i)	10%

Suppose that our decision maker is concerned about the possible sensitivity of the analysis to estimates of cost per unit sold (c). It is estimated that costs may range from a low of $9 to a high of $12 per unit, although the "point estimate" is $10 per unit.

a. Determine the minimum annual worth (AW).
b. Determine the maximum annual worth (AW).
c. Determine the cost per unit to assure breakeven.
d. Let d represents the percent deviation from the estimated value ($10.00) for cost per unit. That is, $c = \$10(1 + d)$. Determine the percent deviation at breakeven.
e. Draw a graph of annual worth as a function of cost per unit.
f. Draw a graph of annual worth as a function of percent deviation for cost per unit.

[*Answers*: a. $-\$12,500$; b. $8,500$; c. 10.21; d. 2.14%.]

10.6 Recall the "sensitivity analysis" problem discussed in the text:

Initial cost (P)	$100,000
Revenue per unit sold (r)	$12
Cost per unit sold (c)	$10
	[Range: $9–$12]
No. of units sold/year (Q)	7,000
Salvage value after 15 years (S)	$20,000
Study period (N)	15 years
Discount rate (i)	10%
	[Range: 8%–12%]

Suppose that the decision maker is concerned about possible sensitivity to two variables: cost per unit sold (c) as well as the discount rate (i).
a. Determine the minimum annual worth (AW).
b. Determine the maximum annual worth (AW).
c. Draw the graph to reflect the sensitivity of AW to both c and i. Label the ordinate and abscissa as well as all curves drawn on the graph. (*Hint:* Use the "family of curves" concept discussed in the text.) Show the region of feasibility.

10.7 Recall the "sensitivity analysis" problem discussed in the text:
a. Suppose that the plant capacity (Q_{max}) is 10,000 units. How many units must be produced at breakeven? Assume the certainty estimates for all parameters as given in the text.
b. If the variable cost per unit can be reduced

to $9, determine the percent of plant utilization at breakeven.

[*Answers*: a. 6,250 units; b. 41.7%.]

10.8 Recall the "sensitivity analysis" problem discussed in the text. Let p = percent deviation from the most likely value.
a. Draw the "sensitivity graph" in which the interest rate (i) is the variable in question. Sketch annual worth as a function of p for deviations over the range $\pm 20\%$. Base your sketch on only 5 points: $p = -20\%$, $-10\%, 0\%, 10\%, 20\%$.
b. Determine the annual worth when $p = 0\%$.
c. Determine the breakeven percent deviation.
d. Determine the maximum value of the interest rate such that this project yields a positive annual worth.

10.9 Recall the "sensitivity analysis" problem discussed in the text. This investment (initial cost = $100,000) will be depreciated under the alternate MACRS method as a 10-year property with a 10-year recovery period.
a. Determine the depreciation expense in the second year of ownership.
b. Determine the capital recovery cost.
c. Assuming that "fixed costs" consist entirely of depreciation expense for this equipment, determine the breakeven point in the second year.
d. Assuming that "fixed costs" consist entirely of capital recovery cost, determine the breakeven point in the second year.

[*Answers*: a. $10,000; b. $12,500; c. 5,000 units; d. 6,250 units.]

10.10 Our problem is to select the horsepower for a motor to drive a pump. Motors ranging in size from 10 to 40 HP will have an equivalent annual cost of $120 plus $0.50 per HP. The operating cost of such motors for 1 HP-hour will be $0.005 divided by the horsepower. If 90,000 HP-hours will be needed per year, determine the HP that should be specified for a minimum annual equivalent cost. *Hint:* Write out the total cost equation as a function of the number of horsepower. Then differentiate the function to find the optimal value of the variable (HP) such that total cost is minimized.

10.11 [This problem is based on G. T. Stevens, Jr., *Economic Analysis of Capital Investments* (Reston Publishing Co., 1983), pp. 201–202.] Weekly sales (q units) is related to the unit selling price (p per unit) by

$$p = 180 - 0.5q \qquad 100 \le q \le 250$$

The weekly total cost (C) is also related to sales by

$$C = -0.25q^2 + 120q + 200 \qquad 100 \le q \le 250$$

Thus the weekly profit (P) is given by

$$P = pq - C = \text{(price/unit)(units sold)}$$
$$- \text{(total cost)}$$

a. Determine the sales volume yielding maximum weekly profit.
b. Determine the breakeven sales volume.

[*Answers*: a. 120 units/week; b. 237 units/week.]

10.12 A certain production plan is under consideration for which the annual cost and revenue functions are as follows:

Annual costs of operation and maintenance

$$= C(x)$$
$$= \$45x + \$6,500$$

Annual revenues

$$= R(x) = \$20,000 - \left(\frac{\$1,000,000}{x}\right)$$

where $x =$ number of units sold annually.

a. Determine the net annual benefits [$R(200) - C(200)$] when $x = 200$.
b. Determine the breakeven point(s).
c. Determine the value(s) of x such that net annual benefits are maximum.

[*Answers*: a. −$500; b. 133 and 167; c. 149.]

10.13 The Sunrise Tool Company is negotiating a contract with one of its customers that will result in net operating revenues of $100,000 per year. (Here, "net operating revenues" is defined as revenue from sales, less manufacturing and distribution costs other than capital recovery.) Under the terms of the contract, the customer agrees to purchase parts from Sunrise for not less than 5 years nor more than 8 years. Sunrise's MARR is 12%.

In order to produce parts under the contract, Sunrise will have to purchase equipment at a cost of $500,000. Salvage value of this equipment at the end of N years $= \$500,000(0.8)^N$.

a. Find the annual worth (AW) of the contract if $N = 8$ years. *Hint:* AW = $100,000 − CR($N$).
b. Is the decision to accept the contract sensitive to the contract period (N)? Explain.

[*Answers*: a. $6,170; b. yes, because the breakeven point lies within the range.]

10.14 Consider three mutually exclusive alternatives, each with a 20-year life and zero salvage value. The pretax minimum attractive rate of return is 8%.

	Alt. X	Alt. Y	Alt. Z
Initial cost	$2,000	$4,000	$6,000
Uniform annual benefits	400	700	900

a. Determine the PW of alternative X.
b. Determine the PW of alternative Y.
c. Determine the PW of alternative Z.
d. Suppose that there is some question as to the reliability of the estimate of benefits for alternative X. Using sensitivity analysis, determine the minimum annual benefits for alternative X to justify the conclusion that X is preferred to Y.
e. Suppose that there is some question as to the reliability of the estimate of the minimum attractive rate of return. Determine the range of values for the MARR such that Y is preferred to X.

10.15 The Little Manufacturing Co. is considering the introduction of a new product that has the following characteristics:

- Selling price: $60/unit
- Cost of direct labor: $20/unit
- Cost of direct materials: $10/unit
- Indirect costs, variable with production output: $5/unit
- Initial cost of manufacturing equipment: $200,000
- Life of project: 5 years

- Salvage value of manufacturing equipment after 5 years: zero
- Fixed costs other than manufacturing equipment: $80,000/year
- Expected sales volume: 10,000 units/year.

The firm's pretax MARR is 20%; the after-tax MARR is 12%. The manufacturing equipment required for this new product line, if acquired, will be depreciated by the straight line method over 5 years to a zero salvage value. The firm's marginal tax rate is 40%.

a. Determine the pretax annual cost of capital recovery for the manufacturing equipment.
b. Determine the total pretax variable cost per unit.
c. Determine the total annual pretax fixed cost.
d. Assuming the above certainty estimates, determine the annual pretax profit.
e. Assuming the above certainty estimates, determine the pretax annual worth (AW) of this proposed project.
f. (Breakeven points should properly be based on AW, not accounting profit.) Determine the pretax breakeven volume of sales.
g. Assuming the above certainty estimates, determine the after-tax AW of this project.
h. Determine the after-tax breakeven selling price per unit.

[*Answers*: a. $66,880; b. $35; c. $146,880; d. $130,000; e. $103,120; f. 5,875 units; g. $62,520; h. $48,548.]

10.16 Southwest Manufacturing Company is considering a new product line. Manufacturing and marketing costs are estimated as follows:

	Fixed Costs per Year	Variable Costs per Unit
Capital recovery	$300,000	
Insurance and property taxes	15,000	
Marketing	75,000	
Maintenance/repairs	8,000	$ 0.40
Materials/supplies		5.60
Direct labor		3.50
Other variable costs (e.g., power)		0.50
	$398,000	$10.00

a. If the product is sold for a price of $18.00 per unit, how many units must be sold annually to break even?
b. If 60,000 units are sold each year, what must the price per unit be in order to break even?

10.17

- Selling price per unit: $12
- Variable cost per unit: $7
- Fixed cost per day: $400

a. Using the above data, determine the breakeven point (number of units sold per day).
b. What level of fixed costs would justify reducing the breakeven point to 40 units per day?
c. What level of variable cost per unit would justify reducing the breakeven point to 40 units per day?
d. What selling price per unit would be required to reduce the breakeven point to 40 units?
e. Suppose that the number of units produced and sold is assumed to be a "beta-distributed" random variable with pessimistic, most likely, and optimistic estimates of 40, 60, and 90 units per day, respectively. Determine the expected pretax profit per day. (See Problem 10.22.)

[*Answers*: a. **80 units**; b. **$200**; c. **$2.00**; d. **$17.00**; e. **−$91.67**.]

10.18 A manufacturing firm is considering the purchase of a certain machining center. The firm's pretax minimum attractive rate of return is 15%. Estimates for the other relevant parameters are as follows:

	Optimistic	Most Likely	Pessimistic
Initial investment (*P*)	$95,000	$100,000	$110,000
Service life (*N*)	10 years	8 years	5 years
Salvage value (*S*)	$30,000	$20,000	0
Net annual savings (*A*)	$30,000	$25,000	$20,000

a. Determine the annual worth of this proposed investment using the "most likely" estimates.

b. Management is concerned about the sensitivity of the decision to the estimate for net annual savings (A). At what value of A is this a breakeven project?

c. Holding all other estimates at their most likely values, let p_A represent the percent deviation in the estimate of the net annual savings such that the project is at breakeven. Determine p_A.

d. Is this project sensitive to the estimated net annual savings? Explain.

e. Management is interested in testing the sensitivity of the solution to all four parameters (P, N, S, A) considered simultaneously. To do so, find the estimates of AW under the optimistic and pessimistic scenarios. That is, determine the largest and smallest values of AW possible under these estimates.

f. Do your results in part e indicate that the decision to invest in the machining center is or is not sensitive to the estimates when considered simultaneously? Explain.

Risk Analysis

10.19 [This problem was adapted from G. A. Taylor, *Managerial and Engineering Economy*, 3rd ed. (New York: D. Van Nostrand, 1980), pp. 312–313.] Engineering economy can be used to evaluate measure designed to reduce the cost of risk, as illustrated by this problem.

A firm currently pays a $550-per-year fire insurance premium to provide $100,000 coverage. If a fire were to occur, however, the uninsured loss would be $50,000. That is, the total cost of fire damage would be $150,000, of which $100,000 would be reimbursed by the insurance company and $50,000 would be a cost to the firm.

The firm is considering the installation of a sprinkler system that would reduce the premium to $300 annually. The sprinkler system costs $4,500, has a 15-year life, and will have no salvage value at the end of 15 years. Coverage would remain at $100,000, but the expected uninsured loss is only $20,000 with the sprinkler system in place.

The key to this problem is the question of the probability that a loss will occur. Assume

that the insurance premiums directly reflect this probability: annual insurance premium = (probability of loss)(expected damage). For example, the implied probability of loss with no sprinkler system is $550/$100,000 = 0.0055. Assume that this same probability applies to uninsured damage. Let $i = 0.10$.

Complete the analysis. Determine the (equivalent uniform) annual cost both with and without the sprinkler.

[*Answers*: AC (with) = **$952**; AC (without) = **$825**.]

10.20 A firm is planning to keep a certain old machine for another 2 years. The maintenance department advises that the machine will probably break down and require repair some time during the 2-year period unless an immediate overhaul is performed. It estimates that, without an overhaul, the probability is 0.5 that the machine will break down the first year and 0.3 that the machine will survive the first year but break down during the second year.

Repairing the machine after a breakdown would (1) cost an estimated $1,000, (2) guarantee against further breakdowns in the next 2 years, and (3) add $500 to the salvage value. Overhauling the machine now would (1) cost $800, (2) guarantee against further breakdown in the next 2 years, and (3) add $500 to the salvage value. The minimum attractive rate of return is 10% before taxes.

Use "expected cost" calculations and the end-of-year convention to determine whether the machine should be repaired now. Assume that cash expenses during any year will be concentrated at the end of that year.

10.21 (This problem was prepared by R. J. Borntraeger, U.S. Civil Service Commission, for use in economic analysis training by the Financial Management and FPB Training Center, Bureau of Training, U.S. Civil Service Commission.) The Military Air Transport Service (MATS) operates a fleet of transport aircraft that is periodically overhauled after so many hours of operation at three different facilities: Upton Park, Flyby Service, and Mitchel Air Force Base.

Major John Hardesty, a recently assigned cost analyst with MATS, notes while reviewing the maintenance plans for the three facilities that each facility follows a different procedure in overhauling the transports' pressurization system. Upton Park ascertains only whether the system is functioning properly. Flyby Service overhauls the system on each transport, at an average unit cost of $24,000. Mitchel AFB replaces the entire system, at an average unit cost of $42,000. Further examination by Major Hardesty discloses that, in a majority of cases, there are no major problems with the pressurization system between overhauls. Sometimes, however, there is a normal component malfunction during line operations that costs an average of $67,000 in repair and lost aircraft time. Occasionally, a system failure is more severe, causing damage not only to the pressurization system but also the aircraft structure and to other equipment housed near the pressurization components. When these malfunctions occur, special crews must be dispatched to make the repairs, and more aircraft time is lost. Historically, costs in such instances have averaged about $330,000.

Major Hardesty questions the incremental costs of overhauling and replacing the systems and suspects that the present practice employed by Upton may be the cheapest alternative. He directs his assistant, Lieutenant B. G. Hotshot, to compile past service data on systems serviced by each facility. Lt. Hotshot's sample investigation provided the following information:

	Upton	Flyby	Mitchel
Aircraft overhauled	87	160	100
Subsequent failures	48	48	10
Subsequent damage to other components	6	8	3
No subsequent problems	33	104	87

Based on the information available, is Major Hardesty's position sound? (*Hint:* Assume that the historical data provide the best estimates of future performance. Determine the expected cost per aircraft for each alternative.)
[*Answers:* E[cost for Upton] = **$59,950**, E[cost for Flyby] = **$60,600**, E[cost for Mitchel] = **$58,600.**]

10.22 A heat exchanger is being installed as part of a plant modernization program. It costs $80,000,

will last for 4 years, and is expected to reduce the overall plant fuel cost by x dollars per year, where

x	Prob[x]
$18,000	0.2
19,000	0.4
20,000	0.3
21,000	0.1

Estimates of the salvage value range from an optimistic $25,000 to a pessimistic $5,000. The most likely estimate is $10,000.
a. Find the expected annual reduction in fuel costs.
b. Find the Beta estimate expected salvage value. (*Hint:* An estimate of mean (expected) value of a *Beta-distributed random variable* is

$$m = \left(\frac{a + 4b + c}{6}\right)$$

where a is the minimum anticipated value, b is the most likely value, c is the maximum anticipated value, and m is the estimate of the mean value.)
c. Using the expected values of parts a and b and assuming a 10% discount rate, find the expected present worth for this proposal.

10.23 Recall the "sensitivity analysis" problem discussed in the text. Now suppose that we are considering the sensitivity to the number of units sold (Q), the discount rate (i), and the terminal salvage value (S). Three estimates for each of these variables are as follows:

Variable	Optimistic	Most Likely	Pessimistic
Q	7,500	7,000	6,000
i	8%	10%	12%
S	$22,000	$20,000	$15,000

Recall that in the "most likely" case, the annual worth is $1,500.
a. Find the "optimistic" annual worth.
b. Find the "pessimistic" annual worth.
c. Assuming that annual worth is a Beta-

distributed random variable, determine the expected annual worth. (See Problem 10.22.)

[*Answers*: a. $4,130; b. $-$2,278; c. $1,309.]

10.24 A manufactuer is considering a new product:

- Initial cost: $100,000
- Cost per unit sold: $10
- Discount rate: 20% before taxes
- Service life: 15 years

Suppose that there are three estimates for number of units sold, revenue per unit sold, and salvage value:

	Optimistic	Most Likely	Pessimistic
Number of units sold per year (x_1)	7,500	7,000	6,000
Revenue per unit sold (x_2)	$13	$12	$10
Salvage value at end of 15 years (x_3)	$22,000	$20,000	$15,000

a. Use the Beta distribution to provide estimates of the mean (expected) values for x_1, x_2, and x_3. (See Problem 10.22.)
b. Using your estimates from part a, find the expected (equivalent uniform) annual benefits.

10.25 A certain retail firm estimates that the probability that a theft will occur during any 1-month period is approximately 0.02. The probability that more than one theft will occur during any 1-month period is zero. If a theft occurs, the firm will incur a cost of $10,000. Theft insurance may be purchased at a cost of $x at the start of each month. The firm's effective minimum attractive rate of return is 1% per month before income taxes.

a. Based on the above assumptions, how much can the firm afford to pay for theft insurance?
b. Suppose that a more rigorous analysis of theft likelihood reveals the following estimates of probabilities:

Optimistic	0.018
Most likely	0.020
Pessimistic	0.025

Using the Beta distribution assumption, how much can the firm afford to pay for theft insurance? (See Problem 10.22 for the mean value of a Beta-distributed random variable.)

[*Answers*: a. $199; b. $204.]

10.26 Plant capacity is 100 units/day for a certain product. Additional data are as follows:

- Selling price per unit: $12.00
- Variable cost per unit: $7.00
- Fixed cost per day: $400.00

Suppose that both the selling price per unit (p) and the number of units sold (q) are random variables. They are related, of course. The joint probabilities are as follows:

	q	
p	60	70
$11.5	0.1	0.4
$12.5	0.3	0.2

Determine the expected pretax profit per day.

10.27 (Adapted from *AIIE, Engineering Economy Division Newsletter*, Vol. 9, No. 1, September 1974.) Suppose that your company is considering constructing additional warehouse space for the storage of raw materials and finished goods. The present need is for an additional 60,000 ft^2 at a total cost of $1.5 million. However, eventually there will be a requirement for a warehouse twice as large. A 120,000-ft^2 building could be built now for $2.25 million. For each year of delay in constructing a second 60,000-ft^2 building, the cost is expected to increase by $150,000 per year over and above the present cost of $1.5 million.

Summarizing, a large warehouse addition (120,000-ft^2) could be built now at a cost of $2.25 million. Alternatively, we could build only a 60,000-ft^2 addition now at a cost of $1.50 million; then delay the construction of the second 60,000-ft^2 addition until N years in the future. Future construction costs will be $1,500,000 + $150,000N$.

The probabilities associated with the need for the additional space by end of year N are:

Years from Now (N):	1	2	3	4	5
Probability:	0	0.1	0.2	0.3	0.4

Assume that the warehouse may be built at the start of the year in which it is needed; that is, construction is instantaneous. Also assume that we must have full capacity ($120,000 \text{ ft}^2$) at the end of year 5.

If the cost of capital is 10%, determine the expected PW of delayed construction and compare this value with the cost of building the larger warehouse now.

[*Answer*: PW(build now) = **$2.25 million and PW(postpone) = $2.93 million.**]

10.28 A firm is planning to keep a certain old machine for another 2 years. The maintenance department advises that the machine will probably break down and require repair some time during the 2 years unless an immediate overhaul is made. They estimate that without an overhaul, the probability is 0.4 that the machine will break down during the first year and 0.2 that the machine will survive the first year but break down during the second year.

Repairing the machine after a breakdown would (1) cost an estimated $2,500, (2) guarantee against further breakdowns in the next 2 years, and (3) add $600 to the salvage value.

Overhauling the machine now would (1) cost $1,000, (2) guarantee against further breakdowns in the next 2 years, and (3) add $600 to the salvage value. The minimum attractive rate of return is 10% before taxes.

Use the "expected cost" principle and the end-of-year convention to determine whether the machine should be repaired now. Assume that cash expenses during any year will be concentrated at the end of that year.

a. If we do not overhaul the machine now,

what is the probability that it will survive for 2 years with no breakdown?
b. If we do not overhaul now, and the machine should break down in the first year, determine the PW of costs.
c. Determine the expected PW of costs if we overhaul the machine now.
d. Determine the expected PW of costs if we do not overhaul now.

10.29 The Universal Fruit Company is considering the purchase of special-purpose canning equipment to use in the coming season. Three alternative systems are available: X, Y, and Z. Economic analyses have been prepared under two scenarios related to growing conditions. The results of these analyses are shown below. Figures in the table are present worths in thousands of dollars. Probabilities of occurrence for the three possible growing conditions are given in parentheses.

Alternative	Growing Conditions		
	Poor (0.3)	Fair (0.5)	Good (0.2)
X	50	100	200
Y	−100	200	250
Z	−200	100	400

a. Determine the expected present worth of each of the three alternatives.
b. Suppose that the analyst is confident that the probability of "poor" growing conditions is indeed 0.3, but there is some uncertainty as to the probability of "good" growing conditions. Let $p = \text{Prob}(\text{"good"})$. Find the value of p such that the firm will be indifferent between alternatives X and Y.

[*Answers*: a. **$105,000 for X, $120,000 for Y, $70,000 for Z; b. 0.50.**]

10.30 A business firm is considering three mutually exclusive investment alternatives (A, B, and C). As indicated next, the present worths associated with these alternatives depend on future business conditions. The probability for each condition is as shown:

	Business Condition		
Alternative	Prob["poor"] = 0.3	Prob["fair"] = 0.5	Prob["good"] = 0.2
A	0	0	0
B	− $100,000	0	$500,000
C	− 10,000	$20,000	100,000

Which alternative(s) would be preferred if the decision maker chooses to:

a. Maximize expected present worth
b. Maximize the probability that the firm will not experience a loss
c. Select the optimal policy assuming that the most probable future will in fact occur

10.31 Recall the "sensitivity analysis" problem discussed in the text. The nominal (most likely) values are:

Initial cost (P)	$100,000
Revenue per unit sold (r)	$12
Cost per unit sold (c)	$10
No. of units sold/year (Q)	7,000
Salvage value after 15 years (S)	$20,000
Study period (N)	15 years
Discount rate (i)	10%

Suppose we have reason to believe that the number of units sold is a continuous random variable uniformly distributed over the range 6,000–7,500. That is,

$$\text{Prob}[Q] = \frac{1}{7,500 - 6,000} = \frac{1}{1,500}$$

$$\text{for } 6,000 < Q < 7,500$$

Note that

$$\text{Prob}[Q < 7,000] = (7,000 - 6,000)\left(\frac{1}{1,500}\right)$$

$$= 0.67$$

Holding all other parameters at their nominal values, what is the probability that the proposed project will at least break even (that is, AW > $0)?
[*Answer*: **0.83.**]

10.32 Certain materials handling equipment is expected to be used over a 24-month period.

Toward the end of the second year a breakdown (machine failure) is possible. If such a breakdown should occur, it will cost $1,000 to repair, but the repair should be adequate to ensure that no further breakdown will occur. Estimates of the likelihood of a breakdown are as follows:

Month (n)	Probability of Breakdown in Month n $P(n)$
1–18	0.00
19	0.02
20	0.04
21	0.06
22	0.08
23	0.10
24	0.12
Total	0.42

a. What is the probability that there will be no breakdown during the 24 months?
b. If the firm's MARR is 1% per month, determine the expected PW of the cost of a breakdown. (Here, PW is at the beginning of the first month.)
c. The firm can purchase a maintenance contract for $X payable at the start of the second year, that is, at the start of the 13th month. The contract will ensure that, should there be a breakdown, the cost of repair will be fully recoverable. What is the maximum amount the firm should pay for this contract?

10.33 A company is considering renting certain materials handling equipment for a 24-month period. The rental charge is $400 per month, payable at the start of each month. Estimates of monthly savings, end of month, are as follows:

> Minimum: $350
>
> Most likely: $420
>
> Maximum: $450

The company's pretax MARR is 20% per year.
a. Assuming the Beta estimate, what is the expected monthly savings? (See Problem 10.22.)
b. At what MARR per month (i_m) would the proposed rental be economically justified? That is, what is the breakeven point for i_m?
c. Is the proposed rental justified? Why?

[*Answers*: a. $413; b. 3.3% per month; c. yes, because IRR > MARR.]

10.34 a. Ms. Ho recognizes a potential market for the rental of certain computer-aided design (CAD) hardware. If she purchases this hardware, she believes that she can realize the equivalent of $20,000 per year over a 5-year rental period. This $20,000 figure is net of all related operating expenses and is assumed to be end-of-year. But now she has to buy the CAD equipment that she will subsequently rent out. Her MARR, her cost of capital, is 12%. The CAD hardware, she believes, will have a 20% salvage value at the end of 5 years. How much can she afford to pay for this equipment so as to realize a 12% annual return on her investment?
b. The annual rental, $20,000 in part a, is now assumed to be a random variable with estimates as follows: Prob[$18,000] = 0.3; Prob[$20,000] = 0.5; Prob[$22,000] = 0.2. Determine the expected annual rental.
c. Although the salvage value is most likely to be 20% of the initial cost, it could range from a low of 0% to a high of 30%. Using the "Beta assumption," determine the estimate for this random variable. (See Problem 10.22.)
d. Assume that the initial cost of the CAD equipment is $75,000. Using your estimates (expected values) determined in parts b and c, determine the expected annual worth of this proposed project.

10.35 The Hall Manufacturing Company (HMC) is buying some property in Denver for a new production facility. The size of the property is more than necessary for production operations, and HMC is considering selling off the excess property, about one acre, for $100,000. Alternatively, HMC could wait several years before selling the property. HMC's pretax MARR is 20%.
a. Assume that this parcel of land will increase in value at the rate of 10% per year. Determine its value at the end of 4 years.
b. Assume that the annual rate of appreciation in land value is a random variable with density function as follows:

Annual Increase, x:	0.09	0.10	0.11
Prob. of Annual Increase, $P(x)$:	0.3	0.5	0.2

Determine the expected value of the property at the end of 4 years.
c. Now let us further assume that HMC might keep the property for either 3, 4, or 5 years before disposing of it. The relevant probabilities are:

Years Kept, N:	3	4	5
Prob. of Keeping N years, $p(N)$:	0.5	0.3	0.2

Determine the expected value of the property if it is kept for N years. *Hint:* Both the sale price (x) as well as the duration (N) are random variables, so the joint probabilities are of interest.

[*Answers*: a. $146,410; b. $145,914; c. $142,227.]

10.36 Our sensitivity analysis in Problem 10.19 is somewhat naive in that it assumes that the actual values of the parameters (P, N, S, A) are mutually independent. But this is not likely to be the case, especially with respect to the estimates for service life (N) and salvage value (S). It is reasonable to expect that the salvage value decreases as the service life increases. Assume that additional analysis of the six possible combinations yields the following:

Prob. of Occurrence	Service Life (N) Years	Salvage Value (S)
0.1	5	$30,000
0.1	6	28,000
0.2	7	25,000
0.3	8	20,000
0.2	9	10,000
0.1	10	0

Assuming an initial cost (P) of $100,000 and MARR = 15%, determine the expected cost of capital recovery.

10.37 It is anticipated that a $50,000 investment in an inspection device will result in substantial labor savings. The physical life of the device is very long—at least 20 years—but the product life for which it will be used is relatively short. Estimates of the useful life of this investment, together with relative likelihoods of occurrence, are summarized below. Also shown are estimates of salvage values associated with these lives. The firm's pretax MARR is 20%.

Life N	Probability p(N)	Salvage Value at End of Life		
		Minimum	Most Likely	Maximum
4	0.4	$10,000	$10,500	$11,000
5	0.4	8,600	9,200	10,000
6	0.2	8,000	8,300	9,000

a. Using the Beta estimate, determine the expected salvage value if the device should be in use for 5 years. (See Problem 10.22.)
b. Using the Beta estimates for expected salvage values, determine the expected cost of capital recovery if the device should be kept in service between 4 and 6 years with probabilities as given.

[*Answers*: a. $9,233; b. $15,974.]

10.38 Consider a $50,000 investment in a certain capital asset for which:

If Kept:	3 years	4 years	5 years
The Salvage Value Will Be:	$30,000	$20,000	$15,000

The probabilities of keeping the asset 3, 4, or 5 years are 0.5, 0.3, and 0.2, respectively.

The discount rate will be 9% with probability 0.6 and 10% with probability 0.4. Determine the expected cost of capital recovery.

10.39 An investment of $60,000 (now) is expected to return $10,000 at the end of each year for 10 years. The investor's MARR is expected to be 10% throughout this 10-year period.
a. With the "certainty estimates" as above, determine the PW of the investment.
b. Determine the maximum MARR such that this investment is attractive.
c. Assume that the discount rate is a random variable with minimum, most likely, and maximum values of 8%, 10%, and 11%, respectively. Using the Beta distribution assumption for the expected discount rate, determine the expected PW. (See Problem 10.22.)

[*Answers*: a. $1,450; b. 10.6%; c. $1,900.]

10.40 Our firm is considering an investment in a certain research project that requires cash outlays of $10,000 per month, each month, over a 24-month period. (We will consider these to be continuous cash flows.) The firm's nominal MARR is 1% per month. (Assume continuous compounding and discounting at nominal rate 1% per month.) The object of this research project, a certain product, will have value F at the end of 24 months.
a. What is the value of F at breakeven?
b. We have attempted to assess the likelihood that the new product will be successful. Our estimates are:

Success	Future Payoff (F)	Prob.[success]
None	0	0.5
Moderate	$ 300,000	0.3
Good	1,000,000	0.2

1. What is the expected *future* payoff (F)?
2. What is the probability that this research effort will *not* prove profitable?

Decision Theory Applications

10.41 The following table gives the costs that would result from alternative testing sequences required in connection with a certain research program. All costs are in thousands of dollars.

		Possible Outcomes				
		s_1	s_2	s_3	s_4	s_5
	a_1	18	16	10	14	15
Mutually	a_2	14	15	15	14	15
Exclusive	a_3	5	16	12	10	15
Alternatives	a_4	14	22	10	12	15
	a_5	10	12	15	10	15

Suppose that information is available suggesting that failure patterns will occur with the probabilities $\text{Prob}[s_1] = 0.20$, $\text{Prob}[s_2] = 0.15$, $\text{Prob}[s_3] = 0.40$, $\text{Prob}[s_4] = 0.10$, and $\text{Prob}[s_5] = 0.15$. Apply criteria for decision making when probabilities are known or estimated and comment on the results. Use the various principles of choice discussed in the chapter: (a) expectation, (b) expectation-variance, and (c) most probable future.

Suppose that nothing at all is known or can be estimated about the relative likelihoods of the future states of nature. Which alternative(s) would be selected under each of the following principles of choice: (d) minimax, (e) minimin, (f) Hurwicz with $\alpha = 0.25$, (g) Savage, and (h) Laplace?
[*Answers*: a. a_3; b. a_5; c. a_1 or a_4; d. a_5; e. a_3; f. a_3; g. a_3; h. a_3.]

10.42 Managers of a certain company are attempting to determine which cleaning equipment they will purchase for an Illinois plant. The decision depends on wage rates used in the analysis, but these rates are uncertain because they are the subject of current negotiations with the union. The investment decision must be made immediately, however, The following table represents the annual benefits in millions of dollars (reduction in labor and other operating costs) that may result from four different alternatives coupled with the four possible outcomes of the current wage negotiations.

		Possible Outcomes			
		s_1	s_2	s_3	s_4
	a_1	1	2	3	4
Mutually	a_2	3	2	1	0
Exclusive	a_3	2	3	4	2
Alternatives	a_4	3	4	2	1

Suppose there is reason to believe that $\text{Prob}[s_1] = \frac{1}{8}$, $\text{Prob}[s_2] = \frac{1}{8}$, $\text{Prob}[s_3] = \frac{1}{4}$, $\text{Prob}[s_4] = \frac{1}{2}$. Determine which alternative is indicated by the following decision rules: (a) expectation, (b) most probable future, and (c) expectation-variance.

10.43 The following questions refer to the maximin principle.

a. Consider the following profit matrix:

	s_1	s_2
a_1	\$1	\$100
a_2	\$2	\$2

Alternative a_2 is chosen under the maximin principle of choice. What in general may be inferred from this problem?

b. Consider the following profit matrix (case I):

	s_1	s_2
a_1	\$1	\$4
a_2	\$3	\$2

Suppose that additional information has been developed indicating that, if s_1 occurs, profits will increase by \$3 over the original values. Moreover, profits will increase by \$3 regardless of which alternative is selected. The revised profit matrix (case II) is

	s_1	s_2
a_1	\$4	\$4
a_2	\$6	\$2

Comment on the maximin solutions for these two cases.
[*Answers*: a. a_2; b. a_2 under case I and a_1 under case II.]

10.44 A manufacturing firm is planning to construct a large industrial smoke stack that, because of certain design characteristics, may be either 80 feet or 150 feet high. The tall stack can be accommodated only by a large footing, whereas the short stack can be built with either a large or a small footing. The initial costs are

Large footing: $40,000

Small footing: 20,000

Tall stack: $140,000

Short stack: 40,000

At first glance it appears that the company should build the short stack with the small footing with a total first cost of $60,000. Unfortunately, life is not quite so simple. If a short stack is built, the company risks being cited by the local air pollution control board and will have to build a tall stack.

The tall stack and the footing are permanent, but the short stack will have to be replaced by a tall one in 10 years. Using a minimum attractive rate of return of 10% before taxes, analyze this problem so that management will have a reasonable basis for decision. Assume that cash expenses during any year will be concentrated at the end of that year. (*Hint:* Construct a matrix of expected costs for the various combinations of alternatives and scenarios.)

10.45 A plant manager is considering two mutually exclusive alternatives: expand the existing warehouse (a_1) or build an additional warehouse (a_2). Three possible future events are likely: There will be no change in output other than normal growth (s_1), a new product, A, will be manufactured at the plant (s_2), or a new product, B, will be manufactured at the plant (s_3).
a. The matrix of (equivalent uniform) annual costs to be considered by the plant manager is

	s_1	s_2	s_3
a_1	$140,000	$300,000	$300,000
a_2	$200,000	$250,000	$250,000

Using the Laplace principle of insufficient reason, decide which alternative should be selected.
b. Suppose that the decision is reviewed by the comptroller at corporate headquarters. Noting that costs depend only on whether or not a new product line is manufactured at the plant, he simplifies the cost matrix as follows:

	s_1	s_1
a_1	$140,000	$300,000
a_2	$200,000	$250,000

In the revised matrix, s_2 represents the outcome "A new product line will be manufactured at the plant." Applying the Laplace principle, decide which alternative should be selected.
c. Comment on the results for parts a and b.
[*Answers (partial)*: a. a_2; b. a_1.]

10.46 The university is considering the construction of a parking structure to accommodate 700 automobiles. Four designs are under consideration, each of which will provide a certain degree of protection against earthquake damage. The economic analysis has considered four possible scenarios: (1) no significant earthquakes over the 20-year life of the structure; (2) one minor earthquake; (3) two minor earthquakes; and (4) one major earthquake. The present worth of costs (in millions) of each design–scenario combination are summarized below.

	Scenario (State of Nature)			
Design Alternative	No Earthquake	One Minor	Two Minor	One Major
a_1	$3	$5	$6	$9
a_2	4	4	6	8
a_3	5	5	5	7
a_4	7	7	7	7

a. Using the Hurwicz criterion with $\alpha = 0.4$, determine the preferred alternative.

b. Using the Savage principle of minimax regret, determine the preferred alternative.

c. If Prob["no"]=0.5, Prob[one minor]=0.3, Prob[two minor]=0.1, and Prob[one major]=0.1, which design is preferred using the principle of minimum expected cost?

[*Answers:* a. a_3; b. a_2; c. a_1.]

10.47 Four mutually exclusive plans are being considered for automating a certain manufacturing operation. The present worths for these alternatives have been evaluated under three scenarios related to anticipated rates of inflation over the study period. The results are as follows:

Automation Plan	Low Inflation	Moderate Inflation	High Inflation
a_1	$200,000	$180,000	$150,000
a_2	165,000	170,000	175,000
a_3	200,000	185,000	160,000
a_4	150,000	175,000	170,000

a. Is/are any alternative(s) dominated by any other(s)?

b. Using the Hurwicz principle with $\alpha = 0.5$, which alternative is preferred?

c. Using the Laplace principle of insufficient reason, which alternative is preferred?

d. Using the Savage principle of minimax regret, which alternative is preferred?

[*Answers:* a. a_3 dominates a_1; b. a_3; c. a_3 d. a_3.]

10.48 Analysis has been completed for four mutually exclusive investment proposals under four alternative scenarios, or states of nature. The present worths, given a 15% discount rate and 10-year planning horizon, are displayed below.

Alternative	State of Nature s_1	s_2	s_3	s_4
a_1	$ 70,000	$70,000	$100,000	$ 80,000
a_2	100,000	60,000	80,000	100,000
a_3	60,000	70,000	50,000	80,000
a_4	80,000	40,000	110,000	70,000

a. Which alternative(s), if any, is/are dominated?

b. According to the maximin principle, which alternative is preferred?

c. According to the maximax principle, which alternative is preferred?

d. According to the Hurwicz principle with $\alpha = 0.6$, which alternative is preferred?

e. According to the Savage principle of minimax regret, which alternative is preferred?

10.49 Three alternative equipment replacement policies are under consideration: (a_1) conservative, (a_2) moderate, and (a_3) aggressive. Present worth analyses of these mutually exclusive alternatives have been evaluated under three scenarios related to possible changes in federal tax law: (s_1) status quo, (s_2) reduction in tax rate, and (s_3) restoration of the investment tax credit accompanied by increases in the tax rate. Present worths are summarized below:

	s_1	s_2	s_3
a_1	$610,000	$620,000	$580,000
a_2	$650,000	$705,000	$550,000
a_3	$700,000	$726,000	$490,000

Which alternative(s) is/are preferred under:

a. The Laplace principle of insufficient reason

b. The Hurwicz principle

c. The Savage principle of minimax regret

[*Answers:* a. a_3; b. a_2; c. a_2.]

10.50 Three alternative investments are under consideration (a_1, a_2, a_3). A present worth analysis has been completed under three different scenarios (s_1, s_2, s_3). The present worths are summarized below. Figures shown are in thousands of equivalent dollars.

	s_1	s_2	s_3
a_1	40	39	20
a_2	25	25	25
a_3	20	30	40

a. Under the maximin principle, which alternative is preferred?

b. Under the maximax principle, which alternative is preferred?

c. Determine the range for the index of optimism, alpha (α), such that alternative a_1 is preferred to a_2.

d. If the states of nature (s_j) are equally likely, which alternative(s) is/are preferred under the expectation principle?

e. Under the Savage principle of minimax regret, complete the regret matrix and indicate the preferred alternative(s).

APPENDIX 1: *Fundamental Concepts—Probability and Expectation*

A1.1 PROBABILITY

There are a variety of approaches to the subject of probability, ranging from the most simplistic to the most elegant and esoteric. This discussion is not meant to be exhaustive. Rather, it introduces some fundamental notions of probability so that you can implement risk analysis with reasonable effectiveness. A more exhaustive treatment can be found in probability textbooks.*

A1.1.1 Probability Defined

Consider an event, say, rolling a 7 with a pair of dice. The **probability** of the event, E, is the ratio of the number of ways that E can occur divided by the total number of possible outcomes. Or

$$P[E] = \frac{\text{number of ways } E \text{ can occur}}{\text{total number of possible outcomes}} \qquad (10.A.1)$$

Given a pair of "fair" dice, fair in the sense that each face has an equal chance of appearing, there are 36 possible outcomes:

1, 1	2, 1	3, 1	4, 1	5, 1	6, 1
1, 2	2, 2	3, 2	4, 2	5, 2	6, 2
1, 3	2, 3	3, 3	4, 3	5, 3	6, 3
1, 4	2, 4	3, 4	4, 4	5, 4	6, 4
1, 5	2, 5	3, 5	4, 5	5, 5	6, 5
1, 6	2, 6	3, 6	4, 6	5, 6	6, 6

---

*A number of excellent textbooks deal with probability. See William Mendenhall and Richard L. Scheaffer, *Mathematical Statistics with Applications* (North Scituate, MA: Duxbury Press, 1973). For an elementary reference, see Seymour Lipschutz, *Theory and Problems of Probability*, Schaum's Outline Series (New York: McGraw-Hill, 1968).

Inspecting these outcomes indicates that exactly six of them yield a total of 7 when the two numbers are added. Thus, the probability of rolling a 7, given a pair of fair dice, is exactly $\frac{6}{36} = \frac{1}{6}$. Note that this is *a relative frequency* definition.

The probabilities of many events can be determined once the underlying physical mechanism is clearly understood. For example, the probability that a 6 will show on the upturned face of a fair die is exactly $\frac{1}{6}$; the probability that heads will show, given the flip of a fair coin, is exactly $\frac{1}{2}$, or 0.5. (Probabilities are generally written as decimals rather than fractions.) The probability of being dealt a 10 or a face card from a complete shuffled deck of playing cards is $\frac{16}{52}$, or 0.31. Unfortunately, these simple illustrations, treasured for their illustrative value in the classroom, do not reflect the kinds of situations of greatest interest to economic analysts. More relevant questions might be: What is the probability that rainfall in the South Coast Basin will exceed 5 in. next winter? What is the probability that a specific motor will be used for 4,000–4,200 hr 3 years from now? What is the probability that current negotiations with the union will result in wage increases of 7–8%? Clearly, probability estimates concerning future events can be exceedingly difficult to forecast, especially when there is little history from which to extrapolate into the future. *The quality of risk analysis can be no better than the quality of the underlying probability estimates.* The following discussion must be understood in light of this caveat.

A1.1.2 Probability Axioms

There are three fundamental probability axioms that analysts will find useful:

Axiom 1. For any event E, the probability of that event must be nonnegative.

$$P[E] \geq 0 \qquad (10.A.2)$$

Axiom 2. The sum of the probabilities of the set of all possible outcomes (the sample space) is exactly unity:

$$P[S] = 1 \qquad (10.A.3)$$

where S is the sum of all possible outcomes in the finite sample space.

Axiom 3. Given two mutually exclusive events, E and F, from a finite sample space, the probability that either E or F will occur is the sum of the probabilities of the two events:

$$P[E \cup F] = P[E] + P[F] \qquad (10.A.4)$$

In addition, if two events, E and F, are independent, then the probability of *both* events E *and* F occurring is the product of their respective probabilities, or

$$P[EF] = P[E]P[F] \qquad (10.A.5)$$

These axioms can be illustrated by the rolling of a pair of dice. It is evident that

Axiom 1: $P[E_i] > 0$ for $E_i = 2, 3, \ldots, 12$

$= 0$ for all other values of E_i

Axiom 2: $P[S] = P[2] + P[3] + \cdots + P[12] = 1$

To illustrate Axiom 3, consider any subset of outcomes, say rolling a 7 or an 11. These two events are mutually exclusive for any single roll of the dice. Thus the probability of rolling either a 7 or an 11 on any one roll is

$$P[7 \text{ or } 11] = P[7] + P[11] = \frac{6}{36} + \frac{2}{36} = \frac{2}{9}$$

Similarly, the probability of rolling a 2, a 3, or a 12 is

$$P[2, 3, 12] = P[2] + P[3] + P[12]$$

$$= \frac{(1 + 2 + 1)}{36} = \frac{1}{9}$$

A variety of probability calculations can be made, of course, using these probability axioms.

A1.1.3 Probability Functions

A **random variable** is a function that assigns a value to each outcome in an exhaustive set of all possible outcomes. For example, in a roll of a pair of fair dice, the random variable describing the total number of spots appearing on the upturned faces of the dice can have the values $2, 3, \ldots, 12$. When a random variable is *discrete*, as in this example, a **probability mass function**, $p(x)$, is used to describe the probability that a random variable will be equal to a particular value. Graphically, the probability mass function appears as in **Figure 10.A.1.**

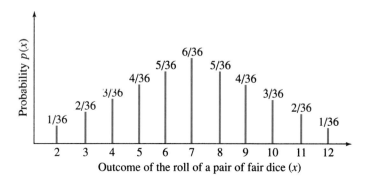

Figure 10.A.1

Certain random variables are *continuous* rather than discrete, as are temperature, weight, and distance. For continuous random variables, a **probability density function**, $f(x)$, is used to relate the probability of an event to a value (more properly, a *range* of values) for the random variable. Three well-known probability density functions are illustrated in **Figure 10.A.2**.

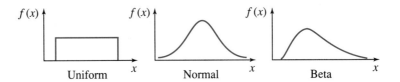

Figure 10.A.2

Consider a continuous random variable X, and let x represent a specific value of this random variable. The probability of x occurring between points a and b is illustrated in **Figure 10.A.3**. Strictly speaking, the probability of x occurring is zero, since X is a continuous random variable. For a short interval of length Δx with midpoint $x = c$, the corresponding probability is approximately equal to $p(c)(\Delta x)$, because this is the area of the rectangle of width Δx and height $p(c)$. If Δx approaches 0, then $P[X = c] = 0$.

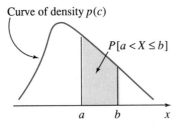

Figure 10.A.3

A1.2 EXPECTATION

The concept of **mathematical expectation** is useful in analyses involving random variables. The **expected value** of a *discrete* random variable X is given by

$$E[X] = \sum x \cdot p(x) \qquad (10.A.6)$$

where the probability mass function, $p(x)$, is defined for every number x

by $p(x) = P[X = x]$. The expected value of a *continuous* random variable X is given by

$$E[X] = \int_{-\infty}^{\infty} x \cdot f(x)\, dx \qquad (10.A.7)$$

where $f(x)$ is the associated probability density function. The expected value is frequently called the **mean**, or **mean value**, of the distribution and is denoted by the Greek letter μ.

To illustrate, let us compute the expected value of the discrete random variable X, where X represents the sum of the upturned faces of a pair of fair dice when rolled, or $x = 2, 3, \ldots, 12$. (It may be assumed that rolling the dice has the effect of randomizing the variable.) Then

$$E[X] = \sum x \cdot p(x)$$

$$= \frac{1}{36}(2) + \frac{2}{36}(3) + \cdots + \frac{1}{36}(12)$$

$$= \frac{252}{36} = 7$$

In this example, the mean value is the same as the most likely value, but this is not always the case.

Another example may be instructive. Suppose that John and Mary are gambling on the flip of a fair coin. They agree to wager \$1 on each flip: heads, John wins \$1 from Mary, and tails, Mary wins \$1 from John. (See **Table 10.A.1**.) John's expected winnings may be easily computed:

TABLE 10.A.1

Outcome	Probability	John Wins	Mary Wins
Heads	0.5	$+1$	-1
Tails	0.5	-1	$+1$

$$E[\text{winnings}] = \frac{1}{2}(\$1) + \frac{1}{2}(-\$1) = 0$$

Put somewhat differently, if John and Mary continue to play this game many times, then the *long-run* average winnings *per play* approaches zero. Note that on no single play is it possible to win exactly zero; John must either win \$1 or lose \$1 on each play. Nevertheless, the expected value is zero. This example clearly demonstrates that expectation is a mathematical concept and is not related directly to "anticipated" values.

As a final example, consider a certain random variable, namely annual operating savings, ranging from \$2,400 to \$3,300. The assumed probability distribution for this discrete random variable is as follows:

Annual Savings [A_i]	Probability $P[A_i]$
Less than $2,400	0
$2,400	0.04
2,500	0.08
2,600	0.10
2,700	0.12
2,800	0.14
2,900	0.16
3,000	0.16
3,100	0.10
3,200	0.08
3,300	0.02
More than $3,300	0
	1.00

The expected value of the annual operating savings is simply the sum of the products of the individual values and their respective probabilities:

$$E[\text{savings}] = \sum A_i P[A_i]$$

$$= \$2,400(0.04) + \$2,500(0.08) + \cdots$$

$$+ \$3,300(0.02)$$

$$= \$2,848$$

Again, note that $P[\$2,848] = 0$. Yet if we were to sample from this distribution a large number of times, the long-term average value per sample would approach $2,848.

APPENDIX: *Areas Under the Standard Normal Curve from 0 to z*

z	0	1	2	3	4	5	6	7	8	9
0.0	.0000	.0040	.0080	.0120	.0160	.0199	.0239	.0279	.0319	.0359
0.1	.0398	.0438	.0478	.0517	.0557	.0596	.0636	.0675	.0714	.0754
0.2	.0793	.0832	.0871	.0910	.0948	.0987	.1026	.1064	.1103	.1141
0.3	.1179	.1217	.1255	.1293	.1331	.1368	.1406	.1443	.1480	.1517
0.4	.1554	.1591	.1628	.1664	.1700	.1736	.1772	.1808	.1844	.1879
0.5	.1915	.1950	.1985	.2019	.2054	.2088	.2123	.2157	.2190	.2224
0.6	.2258	.2291	.2324	.2357	.2389	.2422	.2454	.2486	.2518	.2549
0.7	.2580	.2612	.2642	.2673	.2704	.2734	.2764	.2794	.2823	.2852
0.8	.2881	.2910	.2939	.2967	.2996	.3023	.3051	.3078	.3106	.3133
0.9	.3159	.3186	.3212	.3238	.3264	.3289	.3315	.3340	.3365	.3389
1.0	.3413	.3438	.3461	.3485	.3508	.3531	.3554	.3577	.3599	.3621
1.1	.3643	.3665	.3686	.3708	.3729	.3749	.3770	.3790	.3810	.3830
1.2	.3849	.3869	.3888	.3907	.3925	.3944	.3962	.3980	.3997	.4015
1.3	.4032	.4049	.4066	.4082	.4099	.4115	.4131	.4147	.4162	.4177
1.4	.4192	.4207	.4222	.4236	.4251	.4265	.4279	.4292	.4306	.4319
1.5	.4332	.4345	.4357	.4370	.4382	.4394	.4406	.4418	.4429	.4441
1.6	.4452	.4463	.4474	.4484	.4495	.4505	.4515	.4525	.4535	.4545
1.7	.4554	.4564	.4573	.4582	.4591	.4599	.4608	.4616	.4625	.4633
1.8	.4641	.4649	.4656	.4664	.4671	.4678	.4686	.4693	.4699	.4706
1.9	.4713	.4719	.4726	.4732	.4738	.4744	.4750	.4756	.4761	.4767
2.0	.4772	.4778	.4783	.4788	.4793	.4798	.4803	.4808	.4812	.4817
2.1	.4821	.4826	.4830	.4834	.4838	.4842	.4846	.4850	.4854	.4857
2.2	.4861	.4864	.4868	.4871	.4875	.4878	.4881	.4884	.4887	.4890
2.3	.4893	.4896	.4898	.4901	.4904	.4906	.4909	.4911	.4913	.4916
2.4	.4918	.4920	.4922	.4925	.4927	.4929	.4931	.4932	.4934	.4936
2.5	.4938	.4940	.4941	.4943	.4945	.4946	.4948	.4949	.4951	.4952
2.6	.4953	.4955	.4956	.4957	.4959	.4960	.4961	.4962	.4963	.4964
2.7	.4965	.4966	.4967	.4968	.4969	.4970	.4971	.4972	.4973	.4974
2.8	.4974	.4975	.4976	.4977	.4977	.4978	.4979	.4979	.4980	.4981
2.9	.4981	.4982	.4982	.4983	.4984	.4984	.4985	.4985	.4986	.4986
3.0	.4987	.4987	.4987	.4988	.4988	.4989	.4989	.4989	.4990	.4990
3.1	.4990	.4991	.4991	.4991	.4992	.4992	.4992	.4992	.4993	.4993
3.2	.4993	.4993	.4994	.4994	.4994	.4994	.4994	.4995	.4995	.4995
3.3	.4995	.4995	.4995	.4996	.4996	.4996	.4996	.4996	.4996	.4997
3.4	.4997	.4997	.4997	.4997	.4997	.4997	.4997	.4997	.4997	.4998
3.5	.4998	.4998	.4998	.4998	.4998	.4998	.4998	.4998	.4998	.4998
3.6	.4998	.4998	.4999	.4999	.4999	.4999	.4999	.4999	.4999	.4999
3.7	.4999	.4999	.4999	.4999	.4999	.4999	.4999	.4999	.4999	.4999
3.8	.4999	.4999	.4999	.4999	.4999	.4999	.4999	.4999	.4999	.4999
3.9	.5000	.5000	.5000	.5000	.5000	.5000	.5000	.5000	.5000	.5000

Source: Murray R. Spiegel, *Theory and Problems of Statistics*, Schaum's Outline Series (New York: Schaum Publishing Co., 1961).

11

Incorporating Price-Level Changes (Inflation) into the Analyses

When allocating limited capital resources among competing investment alternatives, the effects of price-level changes can be significant to the analysis. Cash flows, proxy measures of goods and services received and expended, are affected by both the *quantities* of goods and services as well as their *prices*. Thus, to the extent that changes in price levels affect cash flows, these changes must be incorporated into the analysis.

Prior to the mid-1970s, the traditional literature of economic analysis in the United States largely ignored price-level changes because they had not been historically significant. As measured by the Consumer Price Index (CPI), price levels in the United States were reasonably constant during the 20-year period prior to World War II, rose somewhat due to the World War II and Korean War years, and then were flat (about $1\frac{1}{2}\%$ per year) from 1953 to 1965. Only in recent years have price-level changes become so pronounced as to demand our attention. During the decade 1973–1982, for example, the general rate of **inflation** as measured by the CPI was about 8.8% per year.[1] (See **Figure 11.1**.) The average annual rate over the next decade, 1983–1992, declined to about 3.8%. As this is being written (mid-1993), the annualized rate of inflation is about 3% and is expected to remain at that level for some time to come. Our ability to see into the future notwithstanding, it is clear that the topic of inflation merits close attention.

---

[1]*Inflation* is a term commonly used to describe price-level changes. Although it generally implies price *increases*, we will use inflation synonymously with price-level changes to describe both increases and decreases in relative prices.

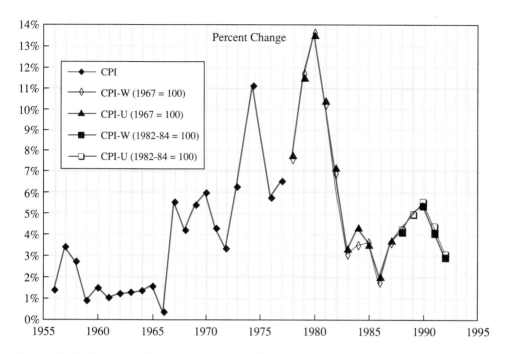

Figure 11.1   Consumer Price Indexes: U.S. City Average, All Item Indexes

A number of inflation-related issues are explored in this chapter. We will begin by presenting some terminology and concepts basic to our understanding of relative price changes. Next, we discuss briefly a variety of techniques for measuring price-level changes, emphasizing the most popular of these measures, the Consumer Price Index. Then the effect of inflation on two measures of effectiveness, present worth and internal rate of return, are discussed, and appropriate steps for incorporating price-level changes directly into the analysis are presented. The effect of inflation on after-tax economy studies is also discussed.

11.2   SIMPLE MEASUREMENT OF RELATIVE PRICE CHANGE

We need not dwell at length on the approximate meaning of "inflation." All adults—and many children, for that matter—are familiar with the phenomenon of changing prices of goods and services over time. Over the recent past, at least, we have experienced regular increases in the prices of certain commodities (e.g., shoes, gasoline, and tuition), whereas the relative prices of certain other items appear to have declined (e.g., video cameras, personal computers, and ballpoint pens).

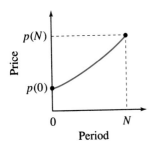

Figure 11.2

We can measure the periodic rate of change of relative prices of a single commodity rather simply. Let $p(0)$ represent the price of the commodity at a certain point in time, and let $p(N)$ represent the price of that same commodity exactly N periods later. (Of course, the commodity must be identically described, quantitatively and qualitatively, at both points in time, otherwise the price change is without significance.) See **Figure 11.2**. Analogous to compound interest calculations,

$$p(N) = p(0) \times (1 + g)^N \qquad (11.1)$$

where g is the average periodic rate of price change over the N periods. If $p(0)$ and $p(N)$ are known, g can be determined by solving

$$g = \left(\frac{p(N)}{p(0)} \right)^{1/N} - 1 \qquad (11.2)$$

EXAMPLE 11.1

The price of a certain commodity was $18.45/kg 5 years ago. Today its price is $23.20/kg. What was the average annual rate of price change? From Eq. (11.2):

$$g = \left(\frac{\$23.20}{\$18.45} \right)^{1/5} - 1$$

$$= 0.04688$$

EXAMPLE 11.2

Suppose that the rate of price change over the past 5 years is expected to continue over the next 2 years. What price can we expect for this commodity 2 years from now? From Eq. (11.1):

$$p(2) = \$23.20(1.04688)^2$$

$$= \$25.43/kg$$

As will be demonstrated, it is frequently more convenient to express prices relatively, that is, to convert the prices to a suitable *index*. (Normally, a *base year* is selected and the price in that base year is given an index number of 100.) Relative changes in prices are identical to relative changes in the corresponding index.

11.3 **MEASURES OF RELATIVE PRICE CHANGES**

11.3.1 The Consumer Price Index

A price index, in general, is simply a relative measure of prices, as measured over time, of specific goods and/or services. The **Consumer Price Index (CPI)**, in particular, is widely used to measure a set (*market basket*) of goods and services consumed by "typical" American consumers. It is a highly visible measure of general inflation, widely quoted in the print and broadcast media as well as the investment community. Its importance is further emphasized because the CPI is used to adjust income payments for a great many Americans, including (as of 1987):

- More than 3 million workers covered by collective bargaining agreements that provide for increases in wage rates based on increases in the CPI

- About 60 million persons receiving benefits through social programs funded by the federal government

- More than 38 million recipients of Social Security benefits

- More than 3 million retired military and federal civil service employees and their survivors

- About 24 million children through adjustments in the federally funded School Lunch Program

Moreover, beginning in 1985, the *CPI-U* (see below) has been used to adjust tax brackets of the federal income tax so as to prevent tax rate increases due to inflation.

The CPI was first published by the U.S. Bureau of Labor Statistics (BLS) in 1919 to help set new wage levels for workers in shipbuilding yards. There have been several subsequent major revisions. Among the major changes:

- 1940 — Reference base period updated from 1917–1919 to 1935–1939

- 1953 — Base period updated to 1947–1949

- 1964 — Consumption market basket changed to reflect a survey of consumer expenditures conducted in 1960–1961

- 1978 — Consumption market basket reflects consumer expenditures surveyed in 1972–1973
  - Original index revised. Now known as the **CPI-W**, this index includes only wage earners and clerical workers, about one-half of the urban population
  - New index published for urban consumers, the **CPI-U**. This index covers approximately 50% of the total noninstitutional civilian population.

- 1987—Consumption market basket updated to reflect consumer expenditures from the period 1982–1984. (See **Table 11.1.**) A new base period, corresponding to the base period for the expenditure patterns determined during the period 1982–1984, is established effective January 1988.

Prices are collected periodically (monthly or bimonthly) in most urban areas across the country. Individual CPIs are published for 27 urban areas. Regional indexes are available for urban areas of different population sizes.

Price changes for various items in the various locations are averaged together using weights that represent the importance of the consumption patterns of the appropriate population group. The CPI is a measure of the price change in a fixed market value of consumption goods and services of constant quality and quantity bought by either all urban consumers (CPI-U) or by urban wage earners and clerical workers (CPI-W). This **fixed-quantity price index** is

$$I_n = \frac{\sum_x p(x,\ n)q(x,\ b)}{\sum_x p(x,\ 0)q(x,\ 0)} \tag{11.3}$$

where

I_n = index in given period n

$p(x,\ n)$ = price of item x in given period n

$p(x,\ 0)$ = price of item x in reference period 0

$q(x,\ b)$ = quantity of the item x consumed in the base period

$q(x,\ 0)$ = quantity of the item x consumed in reference period 0

When the expenditure base period (b) and the reference period (0) coincide, this is the **Laspeyres price index** formula.[2] However, for the 1987 revision of the CPI they did not coincide and the formula was modified: Expenditure data from the 1982–1984 Consumer Expenditure Survey were updated for relative price changes to the end of 1986, when they were introduced into the CPI.

It should be emphasized that the CPI is a measure of *prices*, specifically, the prices of those goods and services included in the market baskets of urban consumers. *It does not measure the cost of living.* Income taxes, for example, are excluded. The CPI is an economic indicator, and as such it is useful to governmental officials in formulating fiscal and

---

[2]The Laspeyres index as well as other price indexes are discussed in a number of textbooks. See the Bibliography for Chapter 11 for references.

TABLE 11.1   Percent Distribution of the Consumer Price Index
by Major Expenditure Groups[a,b]

	Wage Earners and Clerical Workers (CPI-W)		All Urban Consumers (CPI-U)	
	1978 (1972–1973)	1987 (1982–1984)	1978 (1972–1973)	1987 (1982–1984)
Food and beverages	20.5	19.6	18.8	17.8
Housing	40.7	40.0	43.9	42.6
Apparel and upkeep	5.8	6.5	5.8	6.5
Transportation	20.2	20.9	18.0	18.7
Medical care	4.5	3.9	5.0	4.8
Entertainment	3.9	4.1	4.1	4.4
Other	4.4	5.0	4.4	5.1

[a]Expenditure survey periods are shown in parentheses.
[b]*Source:* U.S. Department of Labor, Bureau of Labor Statistics. *The Consumer Price Index: 1987 Revision,* Report 736 (Washington, D.C.: U.S. Government Printing Office, January 1987).

economic policies. It is used to adjust other economic statistics for price changes and to translate current-dollar amounts into "inflation-free" values. And, as noted previously, the CPI is widely used as a means of adjusting income payments. However, as a guide in economic decision making, the CPI must be used with great care. *The appropriate test of the usefulness of any price index should be the extent to which the index is representative of the expenditures (and receipts) expected to occur as the result of a specific investment decision.*

11.3.2   Other Standard Indexes of Relative Price Changes

The Consumer Price Index is only one of a large number of indexes that are used regularly to monitor and report relative price changes. The CPI is by far the most well known, yet it is not particularly useful for specific economic analyses. Analysts should be interested in relative price changes of goods and services that are germane to the particular investment alternatives under consideration. The appropriate indexes are those that are related, say, to construction materials, costs of certain labor skills, energy, and other cost and revenue factors.

Government agencies regularly produce a number of indexes and other price-level data that the analyst may find useful. Examples are the *Wholesale Price Index,* the *Implicit Price Index,* the *Producer Price Index,* the *Engineering News-Record Construction Cost Index,* and the *Industrial Production Indexes.* Industry trade associations are also excellent sources for price-level data, and other lending institutions often maintain useful databases for economic activity in their areas of operation. Still other indexes are available from private firms that specialize in econometric data, such as Data Resources, Inc., and Standard and Poors. Finally, if analysts can find no readily available indexes relevant to a specific investment problem or class of problems, they may develop their own through continual monitoring of actual prices over time. In fact, it is sound practice for an analyst to maintain a current file of appropriate indexes.

11.4 EFFECTS OF RELATIVE PRICE CHANGES ON BEFORE-TAX ECONOMY STUDIES

11.4.1 Measure of Worth When Inflation Is Negligible or Ignored

We begin our discussion with the simplest possible case: Inflation is *not* a consideration. (Here, and throughout the remainder of this chapter, *inflation* will be used to mean *relative price change*. Although the latter term is more precise, the former is equally significant.) Assuming end-of-period compounding/discounting, the present worth (PW) and future worth (FW) of a sequence of cash flows over a planning horizon of N periods are given by:

$$PW = \sum_{j=0}^{N} A_j(1 + i)^{-j} \tag{11.4}$$

and

$$FW = (PW)(1 + i)^N$$

$$= \sum_{j=0}^{N} A_j(1 + i)^{N-j} \tag{11.5}$$

where

A_j = cash flow at end of period j

i = interest rate used for compounding/discounting

The interest rate, i, when used to determine PW or FW, should reflect the rate of return available from alternative investment opportunities; that is, it is the decision maker's minimum attractive rate of return (MARR). These formulations are appropriate when the effects of inflation on the cash flows and the MARR are either assumed to be zero or are believed to be so small as to have no significant effect on the outcome.

11.4.2 Measure of Worth When Inflation Rate Is Nonzero

Effect on Cash Flows

Unfortunately, there is no universal agreement as to the terms used to describe "inflated" and "uninflated" cash flows. The following are among the most common terms used in the literature of economics, finance, and engineering economy:

"Uninflated"	"Inflated"
Real cash flow	Nominal cash flow
Real dollars	Actual dollars
Today's dollars	Current dollars
Constant dollars	Then-current dollars
Now dollars	Then dollars
Constant worth dollars	Actual cash flow

In the interest of consistency, we will use the term **actual dollars** to represent the cash flow at the time of the transaction; **constant dollars** will be used to represent the *equivalent* amount of dollars at some other point in time, normally, "now" or "today."

The relationship between actual dollars and constant dollars for a given transaction is given by

$$A_j^* = A_j(1 + g)^j \qquad (11.6)$$

where

A_j^* = cash flow at end of period j in *actual* dollars

A_j = equivalent cash flow at end of period j in constant dollars

g = rate of relative price change per period, assumed constant over the j periods

EXAMPLE 11.3

A certain operation is expected to require 350 gallons of lubricant 3 years from now. The current price is $12.00 per gallon, and it is expected to increase at the rate of 4% per year. What is the expected cost 3 years hence in actual dollars?

$$A_3^* = (350 \text{ gal})(\$12.00/\text{gal})(1.04)^3$$

$$= (\$4{,}200)(1.12486)$$

$$= \$4{,}724$$

Effect on the Compounding/Discounting Rate

Conceptually, the interest rate used for compounding/discounting in cash flow models is a direct measure of the percentage increase the borrower (the investment) pays to the lender (the investor) for the privilege of borrowing. In the literature of finance this is known as the **real rate of interest**. If X_0 is the amount borrowed and X_1 is the amount owed at the end of one period,

$$i = \text{real rate of interest}$$

$$= \frac{X_1 - X_0}{X_0}$$

$$= \frac{X_1}{X_0} - 1 \tag{11.7}$$

(The real rate is also known as the *unadjusted rate* or the *inflation-free rate*.)

In the presence of inflation, the **nominal rate of interest** is the percentage by which the borrower (the investment) repays the lender (the investor), making an adjustment for any change in the purchasing power of this money that results from inflation. Let f denote the rate of change in general purchasing power. If X_0 is the amount borrowed and $X_1 = X_0(1 + f)$ is the amount owed at the end of the period,

$$i^* = \text{nominal rate of interest}$$

$$= \frac{X_1^* - X_0}{X_0}$$

$$= \frac{X_1^*}{X_0} - 1 \tag{11.8}$$

where

$$X_1^* = X_1(1 + f) \tag{11.9}$$

Combining Eqs. (11.7)–(11.9) and simplifying,

$$i^* = \frac{X_1(1 + f)}{X_0} - 1$$

$$= (1 + i)(1 + f) - 1 \tag{11.10}$$

(The nominal rate of interest is also known as the *adjusted rate* or the *inflated rate*.)

It is a fundamental concept in engineering economy that the minimum rate by which the investor is to be repaid by the investment is the minimum attractive rate of return (MARR). Thus the nominal MARR (i^*) is related to the real MARR (i) by Eq. (11.10). In the subsequent development we will assume that the rates i and f remain constant over the study period.

EXAMPLE   11.4

Given an inflation-free (real) MARR of 10% and further assuming an inflationary effect of 5% per period, determine the nominal MARR.

$$i* = (1.10)(1.05) - 1$$

$$= 0.155$$

EXAMPLE   11.5

During the past 10 years our firm has experienced a 15% per year overall (nominal) rate of return. The general inflation rate during this same period has been 5% per year. What has been the real rate of return? Solving Eq. (11.10) for i:

$$i = \frac{1 + i*}{1 + f} - 1 \tag{11.11}$$

$$= \frac{1.15}{1.05} - 1$$

$$= 0.0952$$

Present Worth and Future Worth Models

Consider a set of actual cash flows, A_j^*, received at the end of each period j ($j = 1, 2, \ldots, N$). Reinvestment of these actual dollars at the nominal rate $i*$ over the remaining $N - j$ periods yields the **future worth** of the project:

$$FW = \sum_{j=0}^{N} A_j^*(1 + i*)^{N-j} \tag{11.12}$$

In terms of constant-dollar cash flows and the real interest rate, the future worth can be written as

$$FW = \sum_{j=0}^{N} A_j(1 + g)^j[(1 + i)(1 + f)]^{N-j} \tag{11.13}$$

$$= [(1 + i)(1 + f)]^N \sum_{j=0}^{N} A_j \left[\frac{(1 + i)(1 + f)}{1 + g} \right]^{-j} \tag{11.14}$$

The **present worth** is easily derived. In terms of actual cash flow and the nominal interest rate,

$$PW = \sum_{j=0}^{N} A_j^*(1 + i*)^{-j} \tag{11.15}$$

In terms of the constant-dollar cash flows and the real interest rate,

$$PW = \sum_{j=0}^{N} A_j \left[\frac{(1+i)(1+f)}{1+g} \right]^{-j} \tag{11.16}$$

In the event that the general rate of inflation affecting the MARR is the same as the rate of price change for the item(s) generating the cash flows, that is, if $f = g$, then the future worth becomes

$$FW = (1+i^*)^N \sum_{j=0}^{N} A_j (1+i)^{-j} \tag{11.17}$$

and the present worth becomes

$$PW = \sum_{j=0}^{N} A_j (1+i)^{-j} \tag{11.18}$$

This latter result is of special interest. We see here that, if $f = g$, the present worth model is identical to Eq. (11.4).

$$\left\{ \begin{array}{l} \text{PW of} \\ \text{actual cash flows } (A_j^*) \\ \text{discounted at} \\ \text{nominal rate } (i^*) \end{array} \right\} = \left\{ \begin{array}{l} \text{PW of} \\ \text{constant-dollar} \\ \text{cash flows } (A_j) \\ \text{discounted at real rate } (i) \end{array} \right\}$$

In other words, inflation can be ignored; we need only discount the A_j's at rate i. It should be emphasized, however, that this result does *not* hold for the future worth model. Moreover, with respect to the PW model, it is valid only in the very special case where the rates of price changes for *all* cash flow elements of the investment are equal to the general rate of inflation.

EXAMPLE 11.6 *Future Worth and Present Worth*

Consider an investment of $1,000 that is expected to return, in constant dollars, $580 at the end of each year for 2 years. This is proposal X, and, because of relative price changes, returns are expected to increase at the rate of 5% per year. If the real MARR is 10% per year and the general rate of inflation affecting the MARR is 5% per year, determine the future worth and present worth of the investment. The analysis is summarized in **Table 11.2**. In this case,

$$i^* = (1+i)(1+f) - 1$$
$$= (1.10)(1.05) - 1 = 0.155$$

From Eq. (11.12):

$$FW(X) = \sum_{j=0}^{2} A_{jX} (1.05)^j (1.155)^{2-j}$$
$$= \$8.82$$

TABLE 11.2   Determining the Future and Present Worths of Proposal X (Example 11.6)

End of Period j (1)	Constant Dollars A_{jx} (2)	Actual Dollars A_{jx}^{*} (3)	Future Value F_x (4)	Present Value P_x (5)
0	−$1,000	−$1,000.00	−$1,334.03	−$1,000.00
1	580	609.00	703.40	527.27
2	580	639.45	639.45	479.34
Total	$ 160	$ 248.45	$ 8.82	$ 6.61

Column 3: $A_{jx}^{*} = A_{jx}(1 + g_x)^j = $ (col. 2) $\times (1.05)^j$
Column 4: $F_x = A_{jx}^{*}(1 + i^{*})^{N-j} = $ (col. 3) $\times (1.155)^{2-j}$
Column 5: $P_x = A_{jx}^{*}(1 + i^{*})^{-j} = $ (col. 3) $\times (1.155)^{-j}$

Moreover,

$$\text{PW(X)} = [\text{FW(X)}](1 + i^{*})^{-N}$$

$$= \$8.82(1.155)^{-2} = \$6.61$$

If we had used only the constant-dollar cash flows (A_{jx}) and the real MARR, that is, if inflationary effects were simply disregarded, then

$$\text{FW(X)} = \sum_{j=0}^{2} A_j(1.10)^{2-j}$$

$$= -\$1,000(1.10)^2 + \$580(1.10) + \$580$$

$$= \$8.00$$

and

$$\text{PW(X)} = [\text{FW(X)}](1 + i)^{-N}$$

$$= \$8.00(1.10)^{-2} = \$6.61$$

Note that the present worth of the *actual* cash flows (A_j^{*}) discounted at the *nominal* rate (15.5%) is identical to the PW of the *constant-dollar* cash flows discounted at the *real* rate (10.0%).

$$\text{PW}(A_j \text{ @ } 10\%) = \text{PW}(A_j^{*} \text{ @ } 15.5\%) = \$6.61$$

However, the future worths are not equal:

$$\text{FW}(A_j \text{ @ } 10\%) = \$8.00 \quad \text{and} \quad \text{FW}(A_j^{*} \text{ @ } 15.5\%) = \$8.82$$

EXAMPLE 11.7 *Rate of Relative Price Change Not Equal to the General Inflation Rate*

Consider proposal Y, an investment alternative to proposal X discussed in the prior example. An investment of $1,000 is required now, which will result in returns in constant dollars of $560 at the end of each year for 2 years. Proposal Y is a capital-intensive project. Because of relative price changes, returns are expected to increase at the rate of 8% per year. The future worth and present worth calculations are summarized in **Table 11.3**.

TABLE 11.3   Determining the Future and Present Worths of Proposal Y (Example 11.7)

End of Period j (1)	Constant Dollars A_{jy} (2)	Actual Dollars A_{jy}^* (3)	Future Value F_y (4)	Present Value P_y (5)
0	−$1,000	−$1,000.00	−$1,334.03	−$1,000.00
1	560	604.80	698.54	523.64
2	560	653.18	653.18	489.63
Total	$ 120	$ 257.98	$ 17.69	$ 13.27

Column 3: $A_{jy}^* = A_{jy}(1 + g_y)^j = (\text{col. 2}) \times (1.08)^j$
Column 4: $F_y = A_{jy}^*(1 + i^*)^{N-j} = (\text{col. 3}) \times (1.155)^{2-j}$
Column 5: $P_y = A_{jy}^*(1 + i^*)^{-j} = (\text{col. 3}) \times (1.155)^{-j}$

Examination of the constant-dollar cash flow indicates that X is clearly superior to Y:

End of Period	Cash Flows	
	A_{jX}	A_{jY}
0	−$1,000	−$1,000
1	580	560
2	580	560

Yet when the effects of relative price changes (g_X, g_Y) and general inflation (f) are taken into account, the PW and FW models indicate that project Y is preferred:

	X	Y
PW	$6.61	$13.27
FW	$8.82	$17.69

The switch in ranking is due to the effects of the different rates of relative price changes for the two projects.

11.4.3 Differential Inflation

It may be useful to "standardize" relative price changes by specifying the difference between relative price changes for individual component elements of an investment alternative and the real MARR. This can be effected by noting that

$$FW = \sum_{j=0}^{N} A_j^*(1 + i^*)^{N-j}$$

$$= (1 + i^*)^N \sum_{j=0}^{N} A_j(1 + g)^j(1 + i^*)^{-j}$$

$$= (1 + i^*)^N \sum_{j=0}^{N} A_j(1 + d)^{-j} \qquad (11.19)$$

where

$$d = \left(\frac{1 + i^*}{1 + g}\right) - 1 \qquad (11.20)$$

The present worth formulation is given by

$$PW = \sum_{j=0}^{N} A_j(1 + d)^{-j} \qquad (11.21)$$

EXAMPLE 11.8

Recall Example 11.7, in which $g_Y = 8\%$ and $i^* = 15.5\%$. Thus

$$d = \left(\frac{1.155}{1.08}\right) - 1 = 0.06944$$

Given the constant-dollar cash flows for project Y, the present worth is determined from Eq. (11.21):

$$PW(Y) = -\$1,000 + \$560(1.06944)^{-1} + \$560(1.06944)^{-2}$$

$$= -\$1,000 + \$523.64 + \$489.63$$

$$= \$13.27$$

This is the same result as obtained previously (Table 11.3).

11.5 EFFECTS OF RELATIVE PRICE CHANGES ON AFTER-TAX ECONOMY STUDIES

Relative price-level changes have a peculiar effect on economy studies in which income taxes are considered. This may be illustrated through a numerical example. We assume here that the effective income tax rate (t_j)

is constant for all periods over the life of the asset ($j = 1, 2, \ldots, N$), and allowances for depreciation are computed using the straight line method. The inferences drawn from this example are not strictly dependent on either assumption, however.

Consider an investment opportunity, alternative Z, with the following characteristics: initial cost (P) of $100, life ($N$) of 5 years, and expected salvage value (S) of $20 at the end of its depreciable life. Thus the annual depreciation expense is

$$D_j = \frac{P - S}{N} = \frac{\$100 - \$20}{5} = \$16 \qquad \text{for } j = 1, 2, \ldots, 5$$

If it is purchased, this asset will result in savings (R_j) of $30 each year over the 5-year planning horizon. (For convenience, assume end-of-year cash flows.) The cash flows before taxes and the annual depreciation expenses are shown in columns 2 and 3 of **Table 11.4**.

11.5.1 Inflation Negligible (or Ignored)

The *internal rate of return before income taxes* (ρ_0) is that interest rate such that the net present value of all cash flows is equal to zero. That is,

$$\text{PW}(\rho_0) = 0 = \sum_{j=0}^{N} A_j (1 + \rho_0)^{-j} \qquad (11.22)$$

where

$$A_0 = -P$$
$$A_j = R_j \text{ for } j = 1, 2, \ldots, N-1$$
$$A_N = R_N + S$$

$\left.\begin{array}{l}\end{array}\right\}$ cash flows before income taxes

Solving Eq. (11.22), $\rho_0 = 19.04\%$.

TABLE 11.4   Cash Flow Table for Proposal Z in the Absence of Inflation

End of Year j (1)	Cash Flow Before Taxes $A_j = (P, R_j, S)$ (2)	Depreciation D_j (3)	Taxable Income $R_j - D_j$ (4)	Income Taxes $t_j(R_j - D_j)$ (5)	Cash Flow After Taxes $\hat{A}_j$ (6)	Notes
0	−$100	—	—	—	−$100	First cost
1	30	$16	$14	$5.6	$24.4⎫	
2	30	16	14	5.6	24.4⎪	Annual
3	30	16	14	5.6	24.4⎬	operations
4	30	16	14	5.6	24.4⎪	
5	30	16	14	5.6	24.4⎭	
5	20	$\langle 20 \rangle^a$	0	0	20	Salvage value

$^a\langle$Book value$\rangle$ at $j = 5$.

In the absence of inflation, the cash flows after income taxes (A_j) are shown in column 6 in Table 11.4. The effective income tax rate (t) is assumed to be 0.40. The after-tax cash flows are found from the following equation:

$$\hat{R}_j = R_j - t_j(R_j - D_j)$$
$$= R_j(1 - t_j) + t_j D_j \qquad \text{for } j = 1, 2, \dots, N \qquad (11.23)$$

The *after-tax internal rate of return* (ρ_1) is that interest rate such that

$$\text{PW}(\rho_1) = 0 = \sum_{j=0}^{N} \hat{A}_j(1 + \rho_1)^{-j} \qquad (11.24)$$

where

$$\hat{A}_0 = -P$$
$$\hat{A}_j = \hat{R}_j \qquad \text{for } j = 1, 2, \dots, N - 1$$
$$\hat{A}_N = \hat{R}_N + S$$

When R_j, t_j, and D_j are constant for $j = 1, 2, \dots, N$, Eq. (11.24) may be simplified to

$$\text{PW}(\rho_1) = 0 = [R(1-t) + tD](P/A, \rho_1, N) - P + S(P/F, \rho_1, N) \qquad (11.25)$$

Solving, $\rho_1 = 11.74\%$ for proposal Z.

Our notational scheme throughout the remainder of this section is as follows:

- The subscript 0 denotes pretax internal rate of return, and the subscript 1 denotes after-tax internal rate of return; that is, ρ_0 is the pretax IRR and ρ_1 is after-tax IRR.
- The circumflex, or hat ($\hat{A}$, for example), denotes after-tax cash flows.
- As before, the asterisk (*) denotes *actual* cash flows, that is, cash flows affected by relative price changes. (In the case of the interest rate, the asterisk denotes the *nominal* MARR.)

11.5.2   Inflation at Rate $g > 0$: Discounting Inflated Cash Flows

Now, assume that all future cash flows are affected by price-level changes at the rate (g) of 10% per annum. That is, future cash flows, as measured in actual dollars, are found from Eq. (11.6),

$$A_j^* = A_j(1 + g)^j$$

TABLE 11.5 Cash Flow Table for Proposal Z Assuming 10% Inflation per Year

End of Year j	Cash Flow Before Taxes $A_j^* = (P^*,\ R_j^*,\ S^*)$	Depreciation D_j	Taxable Income $A_j^* - D_j$	Income Taxes $t_j(R_j^* - D_j)$	Cash Flow After Taxes $\hat{A}_j^*$	Notes
(1)	(2)	(3)	(4)	(5)	(6)	Notes
0	$-\$100.00$	—	—	—	$-\$100$	First cost
1	33.00	$16	$17.00	$ 6.80	26.20	
2	36.30	16	20.30	8.12	28.18	Annual
3	39.93	16	23.93	9.57	30.36	operations
4	43.92	16	27.92	11.17	32.75	
5	48.32	16	32.32	12.93	35.39	
5	32.21	$\langle 20 \rangle^a$	12.21	4.88	27.33	Salvage value

$^a \langle \text{Book value} \rangle$ at $j = 5$.

The cash flow table is given in **Table 11.5**. The after-tax cash flows for the annual revenues are found from

$$\hat{R}_j^* = R_j^* - t_j(R_j^* - D_j)$$
$$= R_j(1+g)^j(1-t_j) + t_j D_j \qquad \text{for } j = 1, 2, \ldots, N \qquad (11.26)$$

If inflation persists at the rate of 10% per annum over 5 years, the expected salvage value will be $\$20(1.10)^5 = \$32.21$. The book value, however, will be only $20, resulting in a gain on disposal of $12.21. This gain will be taxed as ordinary income at the 40% rate, resulting in an increase in taxes of $4.88. Thus the after-tax cash flow will be $32.21 - \$4.88 = \$27.33$. In general,

$$\hat{S}^* = S^* - t_N(S^* - S)$$
$$= S[(1+g)^N(1-t_N) + t_N] \qquad (11.27)$$

where

$$S^* = S(1+g)^N$$

The "inflated" internal rate of return before income taxes is that value of α_0 such that

$$PW(\alpha_0) = 0 = \sum_{j=0}^{N} A_j^*(1+\alpha_0)^{-j} \qquad (11.28)$$

The actual dollar cash flows, before taxes, are shown in column 2 of Table 11.5. Solving Eq. (11.28), $\alpha_0 = 30.95\%$.

The "inflated" internal rate of return after income taxes is that value of α_1 such that

$$PW(\alpha_1) = 0 = \sum_{j=0}^{N} \hat{A}_j^*(1+\alpha_1)^{-j} \qquad (11.29)$$

Determination of the after-tax cash flows is summarized in Table 11.5.

Using these values in Eq. (11.29), we have

$$PW(\alpha_1) = 0 = -\$100 + \$26.20(1+\alpha_1)^{-1} + \cdots + (\$35.39 + \$27.33)(1+\alpha_1)^{-5}$$

Solving, $\alpha_1 = 19.99\%$.

The presence of inflation at the rate of 10% per annum has increased the pretax rate of return from 19% to 31%. This increase is due entirely to growth in the pretax cash flows.

11.5.3 Inflation at Rate $g > 0$: Discounting Deflated Cash Flows

The apparent inflated rates of return, either before or after income taxes, may be adjusted by converting cash flows from actual (inflated) dollars to constant dollars. For the pre-tax case, the internal rate of return after correcting for inflation is clearly the same as that which was obtained ignoring inflation:

$$\begin{Bmatrix} \text{Deflated pretax} \\ \text{cash flow} \end{Bmatrix} = \begin{Bmatrix} \text{actual inflated} \\ \text{cash flow} \end{Bmatrix} \times \begin{Bmatrix} \text{deflation} \\ \text{factor} \end{Bmatrix}$$

$$A_{jd}^* = A_j^*(1+g)^{-j}$$
$$= A_j \qquad (11.30)$$

Thus, the pretax internal rate of return, after applying this adjustment for inflation, is approximately 19%.

The after-tax internal rate of return, after adjusting for inflated cash flows, is determined similarly:

$$\hat{A}_{jd}^* = \hat{A}_j^*(1+g)^{-j} \qquad (11.31)$$

From the example, the deflated after-tax cash flows are

End of Year j	Actual Dollars After Taxes $\hat{A}_j^*$	Deflation Factor $(1+g)^{-j}$	Constant Dollars (Deflated) $\hat{A}_{jd}^*$
0	$-\$100.00$	1.000	$-\$100.00$
1	26.20	0.909	23.82
2	28.18	0.826	23.29
3	30.36	0.751	22.81
4	32.75	0.683	22.37
5	35.39	0.621	21.97
5	27.33	0.621	16.97

In general, the internal rate of return based on deflated after-tax dollars in that value of β_1 that satisfies the equation

$$PW(\beta_1) = 0 = \sum_{j=0}^{N} \hat{A}_{jd}^*(1+\beta_1)^{-j} \qquad (11.32)$$

Solving Eq. (11.32) for proposal Z, $\beta_1 = 9.1\%$.

11.5.4 The Effect of Indexing Depreciation Expenses

It has frequently been argued that current tax laws penalize investors during periods of inflation because allowable depreciation expenses are based on historical costs. If depreciation expenses could be indexed to inflation, that is, if the depreciation expenses used to determine taxable income could be allowed to increase with general price levels, income taxes would be reduced. The increased after-tax cash flows could then be used by investors to support the replacement of depreciated plant and equipment.

The effect of indexed inflation on the example problem is illustrated in **Table 11.6**. Again, we are assuming straight line depreciation with relative prices increasing at the rate of 10% per year. The pretax actual cash flows (A_j^*) are as before. However, the annual allowances for depreciation are *indexed* as follows:

$$D_j^* = D_j(1 + h)^j \qquad (11.33)$$

where h is an indexing rate permitted by the government agency having jurisdiction over tax policies. In this example, we assume that $h = g = 10\%$ per year, and thus

$$D_j^* = \$16(1.10)^j \qquad \text{for } j = 1, 2, \ldots, 5$$

These are the values shown in column 3 of Table 11.6.

The after-tax rate of return is that value of γ_1 such that

$$\text{PW}(\gamma_1) = 0 = \sum_{j=0}^{N} \hat{A}_{ji}^*(1 + \gamma_1)^{-j} \qquad (11.34)$$

TABLE 11.6   Cash Flow Table for Proposal Z Assuming 10% Inflation per Year and Depreciation Fully Indexed to Inflation at Rate $h = 10\%$

End of Year j (1)	($g = 10\%$) Cash Flow Before Taxes A_j^* (2)	($h = 10\%$) Indexed Depreciation D_j^* (3)	Taxable Income $A_j^* - D_j^*$ (4)	($t = 40\%$) Income Taxes $t_j(R_j^* - D_j^*)$ (5)	Cash Flow After Taxes $\hat{A}_{ji}^*$ (6)	Notes
0	−$100.00	—	—	—	−$100.00	First cost
1	33.00	$17.60	$15.40	$6.16	26.84	
2	36.30	19.36	16.94	6.78	29.52	Annual
3	39.93	21.30	18.63	7.45	32.48	operations
4	43.92	23.43	20.49	8.20	35.72	
5	48.32	25.77	22.55	9.02	39.30	
5	32.21	⟨32.21⟩[a]	—	—	32.21	Salvage value

[a] ⟨Book value⟩ at $N = 5$.

where $\hat{A}_{ji}^*$ is the after-tax cash flow, in actual dollars, resulting from depreciation indexed at rate h. The cash flows after taxes due to net operating revenues in years 1 through 5 are found from

$$\hat{A}_{ji}^* = A_j^* - t(A_j^* - D_j^*)$$
$$= A_j^*(1-t) + tD_j^*$$
$$= A_j(1+g)^j(1-t) + tD_j(1+h)^j \qquad (11.35)$$

When $g = h$,

$$\hat{A}_{ji}^* = [A_j(1-t) + tD_j](1+g)^j$$
$$= \hat{A}_j(1+g)^j \qquad (11.36)$$

Thus, if the index-adjusted after-tax cash flows are deflated by rate g,

$$\hat{A}_{jd}^* = \hat{A}_{ji}^*(1+g)^j$$

$$= \hat{A}_j \qquad (11.37)$$

Using the cash flows from column 6 of Table 11.6, solving Eq. (11.34) yields $\gamma_1 = 22.9\%$.

This solution for the after-tax IRR, using actual cash flows based on indexed depreciation, could also have been obtained by observing the after-tax IRR in the absence of inflation. From before,

$$PW(\rho_1) = 0 = \sum_{j=0}^{N} \hat{A}_j(1+\rho_1)^{-j} \qquad (11.24)$$

and

$$PW(\gamma_1) = 0 = \sum_{j=0}^{N} \hat{A}_{ji}^*(1+\gamma_1)^{-j} \qquad (11.34)$$

It can be shown that $PW(\rho_1) = PW(\gamma_1)$ when

$$(1+g)^j(1+\gamma_1)^{-j} = (1+\rho_1)^{-j}$$

Solving,

$$\gamma_1 = (1+g)(1+\rho_1) - 1 \qquad (11.38)$$

Given that g and ρ_1 are known, it is possible to solve for γ_1 directly. In the example,

$$\gamma_1 = (1.10)(1.1174) - 1$$
$$= 0.2291$$

This is the same solution as was obtained using Eq. (11.34).

11.6 INTERPRETATION OF INTERNAL RATES OF RETURN UNDER INFLATION

As demonstrated in the previous example, the internal rate of return as a function of constant dollars (A_j) is not the same as the IRR based on actual dollars (A_j^*). Project Z, above, yielded a pretax cash flow of 19% with constant dollars and 31% with actual dollars. What is the significance of these results? How can they be used to guide the decision maker? We now turn our attention to answering these questions.

The key issue here is the appropriate minimum attractive rate of return, that is, the rate of return available from the "do nothing" alternative. The IRR resulting from *constant* dollars (A_j) should be compared to the *real* MARR (i); the IRR resulting from *actual* dollars (A_j^*) should be compared to the *nominal* MARR (i^*). If the cash flows are such that the PW function is monotone decreasing, as in **Figure 11.3**, it follows that the proposed investment is justified if the IRR exceeds the MARR. Summarizing, the "accept" criterion is

$$IRR(A_j) > MARR(i)$$

or

$$IRR(A_j^*) > MARR(i^*)$$

To illustrate this view, consider proposal Z evaluated in the previous example. Recall that the pretax IRR based on constant dollars was $\rho_0 = \beta_0 = 19\%$; the pretax IRR based on actual dollars was $\alpha_0 = 31\%$. Now suppose that the pretax MARR, in real terms, is $i = 25\%$, and suppose further that inflation affects the MARR at rate $f = 10\%$. Thus the nominal MARR is

$$i^* = (1 + i)(1 + f) - 1$$
$$= (1.25)(1.10) - 1 = 0.375$$

Proposal Z is *not* preferred to the do-nothing alternative, because:

- Based on A_j's and i, $0.19 < 0.25$.
- Based on A_j^*'s and i^*, $0.31 < 0.375$.

These results are shown in Figure 11.3.

SUMMARY

Relative price changes (or "inflation," as this condition is more popularly known) can have a significant effect on cash flows and, by extension, on measures of project worth such as present worth and internal rate of

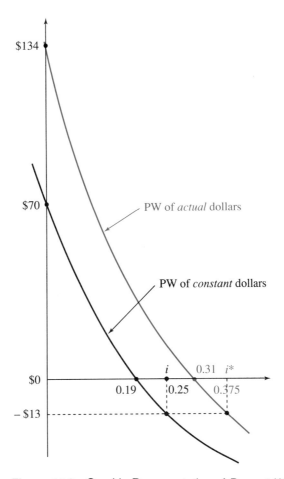

Figure 11.3 Graphic Representation of Present Worth Models for Example Problem (Project Z)

return. The inflationary measure that is best known to the general public is the Consumer Price Index. Although the CPI is of extraordinary importance because of its effect on the income of many Americans, it is of little relevance in most economy studies. Statistical measures of inflation should reflect the specific components, that is, labor, materials, and associated services related to the investment proposal being evaluated.

Proper consideration of inflation requires attention to two separate but related features. First, the then-current actual cash flows must be estimated. This will generally require adjusting constant-dollar values on the basis of expected relative price changes for the particular cash flow components of the project. Next, the present worth or future worth must be calculated using as discount rate the minimum attractive rate of return available to the investor at that time. If the internal rate of return is used as a worth measure, the IRR based on actual cash flows must be compared to the nominal MARR.

Under current tax law, the federal government does not permit taxpayers to index allowances for depreciation. Therefore, in an inflationary economic environment, taxable income will be overstated and income taxes will be greater than would otherwise be the case. The U.S. business community, of course, would like to see the indexing feature incorporated into the tax law.

| PROBLEMS

11.1 The nation of Otherland has experienced a high inflation rate (25% per year). Recently, the exchange rate was 11 pesos per U.S. dollar; that is, one dollar will buy 11 pesos in the foreign-exchange market.

Assume that Otherland will continue to experience an inflation rate of 25% per year and that the United States will experience an average rate of inflation of 10% per year over the next 5 years.

If the exchange rate varies with the rate of inflation, how many pesos will one U.S. dollar buy 5 years from now? (Note that this type of calculation enables the analyst to combine inflation rates for two or more countries.)

[*Answer:* $1 = 20.844 pesos.]

11.2 Assume that $1,000 is invested in an account earning 8% per year over a 5-year period. During this period, general inflation is expected to remain at 6% per year.

a. Determine the amount in the fund at the end of 5 years. That is, what is the future worth after 5 years in "then" (current) dollars?

b. Considering the eroding effect of inflation, what is the dollar value of the fund after 5 years in terms of today's buying power? In other words, what is the future worth after 5 years in "now" (present) dollars?

c. What is the interest rate at which "now" dollars will expand, with their same buying power, into equivalent "then" dollars? (In the literature of finance, this is known as the *real interest rate*.)

11.3 [This problem is adapted from L. T. Blank and A. J. Tarquin, *Engineering Economy*, 2nd ed. (New York: McGraw-Hill, 1983), p. 240.] An

alumnus has offered to make a donation to his alma mater. Three alternative plans for the donation have been offered: (1) donate $65,000 now, (2) donate $16,000 at the end of every year for 12 years, or (3) donate $50,000 three years from now with an additional $80,000 five years from now. Currently, the university can earn 6% per year on its "ready assets account." Inflation is expected to average 3% per year over the next 12 years.

a. Determine which plan yields maximum future worth at the end of 12 years, assuming that short-term investment of funds in the ready assets account will continue to earn at the rate of 6% per year.

b. Assume that the rate of return on funds in the ready assets account will increase with the rate of inflation; that is, today's rate, 6% per year, will increase at the rate of 3% per year. With this assumption, determine which plan yields maximum future worth at the end of 12 years.

[*Answers:* a. FW(2) = $269,919;
b. FW(1) = $325,736.]

11.4 Assume that the donation described in Problem 11.3a will be used to purchase microcomputers that currently cost $2,500 each. Assume that computers will be purchased as soon as the funds are available.

a. If this cost per unit remains constant over the next 12 years and if invested funds continue to earn at 6% per year, determine the number of microcomputers that can be purchased over the 12-year period under each of the three plans. Round your answers to the nearest integer.

b. If the cost of microcomputers were to increase at the rate of 3% per year and if invested funds continue to earn at 6% per year, determine how many microcomputers could be purchased over the 12-year period.

c. Solve part b assuming that the rate of return on funds in the ready assets account will also increase with the rate of inflation.

11.5 A small manufacturing firm is considering a proposal to fully automate an assembly operation that is currently done entirely by manual operation at a cost of $150,000 annually. The automatic equipment will cost $500,000, will have an estimated economic life of 10 years, and will have an expected net value of $20,000 on disposal. The cost of operation and maintenance for this equipment is estimated to be $50,000 per year. The firm's inflation-free pretax minimum attractive rate of return is 15% per year. The general rate of inflation is estimated to be 8% per year over the next 10 years, and the MARR is expected to reflect this inflationary effect. The cost of labor (for current assembly operations as well as for proposed operations and maintenance of the automatic equipment) is expected to increase at the rate of 10% per year over the study period. For simplicity, assume that all cash flows, other than the initial investment, are end-of-year.

After adjusting for inflation, determine the pretax (equivalent uniform) annual costs of the two alternatives.

[*Answers:* **$202,960 for manual and $203,670 for automated operation.**]

11.6 Consider a proposed investment with the following anticipated cash flows in constant dollars:

End of Year	Pretax Cash Flow
0	−$100,000
1	10,000
2	20,000
3	30,000
4	40,000
5	50,000

a. Determine the cash flows in actual dollars if the inflation rate is 7% per year.

b. Assuming that the firm's pretax MARR is 25%—including expectation of 7% inflation over the study period—use the rate of return method to determine whether this proposal is preferable to the do-nothing alternative.

11.7 One barrel of crude oil (of a certain grade) cost $30.50 in November 1990. In January 1993, 26 months later, the price was $26.50 per barrel.

a. What was the average rate of price change per month?

b. What was the equivalent average rate of price change per year?

[*Answers:* **a.** −0.539%; **b.** −6.28%.]

11.8 (In this problem, assume that all cash flows are end-of-year.) We are currently purchasing certain parts from a vendor at a cost of $900,000. Because of competition pressures, we believe that these costs will remain constant over the next 2 years, at the end of which time we will no longer be using these parts in our manufacturing operations.

Rather than purchase the parts, it is proposed that we manufacture them in-house using equipment that will cost $200,000 and that will have no salvage value after 2 years. The cost of labor will be $300,000 per year; the cost of materials will be $400,000 per year. All these estimates are based on current dollars.

Our MARR is now 20%, and it is expected to remain unchanged over the next 2 years. The cost of labor is expected to increase at the rate of 4% per year; the cost of materials is expected to increase at the rate of 6% per year.

a. What is the actual cost (then dollars) of labor at the end of year 1?

b. What is the actual cost (then dollars) of materials at the end of year 2?

c. Determine the present worth of this proposal.

d. Our current MARR (20%) reflects a 5% general rate of inflation. What "inflation-free" rate is implied?

11.9 A certain project requires an initial investment of $1,000 and annual returns of $300 at the end of each year for 5 years. The "inflation-free" (real) discount rate is 10% per year. These end-of-year returns are expected to increase at the rate of 6% per year. Assuming that the MARR will also be affected by the inflation rate of 6%, determine:

a. The present worth (PW)

b. The future worth (FW)

[*Answers:* a. $137; b. $296.]

11.10 A certain construction material cost $8.20/lb 30 months ago. The cost today is $7.90/lb.

a. Determine the average relative rate (compound) of price change per month over the past 30 months.

b. What is the equivalent effective annual rate of price change?

c. If prices continue to be affected over the next 30 months as they have during the past 30 months, what price can we expect 30 months from now?

d. Suppose that our discount rate (MARR) over the next 30 months is expected to be 1.5% per month. What is the equivalent present value (now) of 100 pounds of this construction material purchased 30 months from now?

11.11 An investment of $100 now promises to return $100 one period hence and another $100 two periods hence. (Estimates of cash flows are in terms of constant dollars; i.e., they are unadjusted for inflation.) The current MARR is 10% per period.

a. What is the FW of this investment?

b. What is the PW of this investment?

c. Assuming that inflation at the rate of 20% per period affects the future cash flows, but the MARR is not affected similarly by inflation, what are the PW and FW of this investment?

d. Assuming that the MARR as well as the cash flows are affected by inflation (20% per year), what are the PW and FW of this investment?

[*Answers:* a. FW = $89; b. PW = $73.55; c. FW = $155 and PW = $128.10; d. FW = $128.16 and PW = $73.55.]

11.12 In the absence of inflation, a $30,000 investment can be expected to earn $10,000 at the end of each year for 5 years. The "inflation-free" (real) MARR is 8%. If a 3% inflation rate

is assumed, and if the inflation rate is expected to affect cash flows as well as the MARR, then:

a. Ignoring inflation, determine the future worth (FW).

b. Determine the inflation-adjusted (nominal) MARR.

c. Including inflation, determine the future worth (FW).

11.13 [This problem is based on one given in Max Kurtz, *Handbook of Engineering Economics* (New York: McGraw-Hill, 1984), pp. 1–170.] A reservoir is to be constructed, and it is anticipated that the required capacity will increase until it attains its maximum value 35 years hence. Under Plan A, a large reservoir is to be built immediately to provide all future needs. The cost of construction is $18 million and annual maintenance is $62,000. Under Plan B, the reservoir is to be built in two stages, with capital expenditure of $13 million now and $11 million 20 years hence. Annual maintenance is $48,000 for the first 20 years and $62,000 thereafter. All data are based on current costs. The "inflation-free" cost of capital is 7% per year.

a. Based on the above estimates, determine the present worths of the costs of the two alternatives.

b. If the relevant estimates will be affected by inflation as follows:

  • Construction costs will increase by 8% per year.

  • Annual maintenance costs will increase by 8% per year.

  • Cost of capital will increase by 5% per year.

  (1) Determine the inflation-adjusted interest rate.

  (2) Using this rate as the reinvestment (discount) rate, determine the present worth of costs for Plan A.

[*Answers:* a. $18,083,000 for Plan A and $16,497,000 for Plan B; b. 12.35%, $19,153,000.]

11.14 A firm borrows $1,000 and agrees to pay 10% interest ($100) at the end of each year for 2 years. At the end of the second year the principal ($1,000) is repaid. This results in cash flows as follows:

End of Year:	0	1	2
Cash Flow:	$1,000	−$100	−$1,100

The annual rate of inflation during this 2-year period is 12%. The marginal income tax rate is 46%.

a. Determine the pretax cost of this loan, ignoring inflation.

b. Determine the after-tax cost of this loan, ignoring inflation.

c. Taking into consideration the effect of inflation, determine the after-tax cost of this loan. (*Note:* It is possible that the cost will be negative when inflation effects are considered!)

11.15 Consider the following certainty estimates for pre-tax cash flows:

Initial investment	$400,000
Labor savings at the end of months 1–72	$10,000
Equipment maintenance cost at end of years 1–5	$10,000
Salvage value at end of year 6	$20,000

The above estimates assume no inflation over the 6-year life of the project.

Now let us assume that, due to the effects of inflation, labor savings and equipment maintenance costs will increase at the rate of 0.5% per month, and the expected salvage value of the equipment will increase at the rate of 3.0% per year.

a. Determine the actual cash flow for salvage value, before come taxes.

b. If the firm's "inflation-free" (real) pretax MARR is 25% per year, and assuming that this will be affected by general inflation at the rate of 0.5% per month, determine the inflation-adjusted (nominal) reinvestment rate per year.

c. Find the inflation-adjusted present worth for this investment, before income tax.

[*Answers:* a. $23,881; b. 32.71%; c. −$29,402.]

11.16 The firm is considering an investment of $1,000 that promises to yield the following "inflation-free" (actual) returns over a 2-year study period:

j	A_j
0	−$1,000.00
1	588.00
2	617.40

Assuming that cash flows will "inflate" at the rate of 2.857% per year and the nominal reinvestment rate is 15.5% per year, find the future worth of the proposed investment.

11.17 I. M. Smart, an engineer for the Hardrock Construction Company, is reviewing a previous investment in a certain heavy-duty compressor. The compressor was purchased in October 1992 for $5,000. At that time the price index for compressors was 215.6; the base for this index was January 1989. Three months later, in January 1993, the price index was 220.2. The base for the index was revised in January 1993 to 100. In January 1994, twelve months later, the new index was at 114.6. In summary:

Time	Old Index	New Index
October 1992	215.6	—
January 1993	220.2	100.0
January 1994	—	114.6

a. What was the average inflation rate per month, stated as a percentage, over the 15-month period between October 1992 and January 1994?

b. Convert the answer from part a to an average inflation rate per year.

c. If the heavy-duty compressor cost $5,000 in October 1992, how much would it cost in January 1994? Assume that the price index for this specific compressor reflects the price index for compressors in general.

[*Answers:* a. 1.055%; b. 13.42%; c. $5,852.]

11.18 a. In January 19XX the price of a certain commodity was $85/kg. Exactly 5 years later the price was $140/kg. What was the average annual (compounded) rate of price change?

b. In January 19XX, the price index for a certain commodity was 115.6. If the average annual (compounded) rate of change was

-4.7% per year, what was the index exactly 5 years later?

c. A certain commodity costs $20/kg at the start of the planning horizon. It is estimated that 900 kg will be required 2 years after the start of the project. What will be the total cost in then-current dollars if the rate of price change is 7% per year?

11.19 Consider the geometric gradient series:

$$A_j = \$100{,}000(1.10)^{j-1} \qquad 1 \le j \le 5$$

a. With $i = 0.08$, find the equivalent present value of this sequence of cash flows.

b. Assume that these cash flows are affected by inflation at the rate of 5% per period. That is, $A_j^* = A_j(1.05)^j$. Assuming reinvestment at the inflation-affected rate, i^*, find the future worth of these cash flows A_j^* at the end of 5 periods.

[*Answers:* a. $480,400; b. $900,900.]

11.20 In an inflation-free environment, consider a $300 initial investment that is expected to result in returns of $100 one year hence and $200 two years hence. The firm's "inflation-free" MARR (i) is 5% per annum. It is expected that the general rate of inflation (f) will be 4% per year over these next 2 years and that the reinvestment rate will be affected accordingly. It is also expected that because of market conditions the revenues will increase at the rate of 6% per year.

a. What is the FW ignoring inflation, that is, using A_j's and i?

b. What is the FW if inflation is considered, that is, using A_j^*'s and i^*?

Summary of Notation

Included here are symbols and key abbreviations used throughout the book. Not included, however, is the notation used in Chapter 9 (Revenue Requirement Method). The notation specific to that chapter may be found in Appendix 9A.

A	Cash flow, or equivalent cash flow, occurring uniformly at the *end* of every period for specified number of periods.
A_j	*Before-tax* cash flow at *end* of period j in constant dollars.
A_s	Cash flow at the end of each subperiod.
$\overline{A}$	Amount of money (or equivalent value) flowing continuously and uniformly *during* each period for a specified number of periods.
$\overline{A}_j$	Cash flow occurring continuously and uniformly *during* the jth period.
$\hat{A}_j$	*After-tax* cash flow in *constant* dollars.
A_j^*	*Pre-tax* cash flow in *actual* dollars.
$\hat{A}_j^*$	*After-tax* cash flow in *actual* dollars.
A_{jd}^*	*Pre-tax* cash flow after adjustment for inflation. *Deflated* pre-tax cash flow.
$\hat{A}_{jd}^*$	*After-tax* cash flow after adjustment for inflation. *Deflated* after-tax cash flow.
A_{ji}^*	*After-tax* cash flow in *actual* dollars resulting from *depreciation indexed* at rate h.
a	Depreciation rate used with the declining balance method.
a_i	Alternative i.

AC	(Equivalent uniform) annual cost.
$AC(C, N)$	(Equivalent uniform) annual cost for challenger if kept in service N periods.
$AC(D, N)$	(Equivalent uniform) annual cost for defender if kept in service N periods.
$ACRS$	Accelerated Cost Recovery System.
$\overline{AP}$	Average annual accounting profit.
AW	(Equivalent uniform) annual worth.
B	(Incremental) benefits. Present worth of benefits. Unadjusted basis (when computing depreciation).
B_j	Book value after j years of depreciation. (Includes the depreciation expense for the jth year.)
B_o	Original book value of depreciable asset.
B_N	Book value of depreciable asset at end of period N.
$B:C$	Benefit-cost ratio.
C	(Incremental) costs. Present worth of costs.
c	Cost per unit (Chapter 10).
C_j	Negative cash flow (cost) at end of period j.
$\overline{C}_j$	Continuous negative cash flow (cost) during year j.
C_{ij}	Cost associated with the ith alternative and the jth state of nature.
CC	Capitalized cost: the equivalent present value of an infinite series of cash flows.
CDF	Cumulative distribution function.
CN^*	Challenger's economic life.
CR	Capital recovery.
$CR(C, j)$	Cost of capital recovery for challenger if kept j periods.
d	Imaginary uniform deposit into an imaginary sinking fund.
D_c	Cost depletion.
D_j	Depreciation expense for year j. (No subscript if constant for all j.)
D_p	Percentage depletion.
DDB	Double declining balance.
e	Base of the Napierian logarithm system, the "exponential," approximately equal to 2.71828.
$EUAC$	Equivalent uniform annual cost.

ERR	External rate of return.
$E[X]$	Expected value of random variable X.
F	Amount of cash flow at end of Nth period. Equivalent future value (measured at end of Nth period) of prior cash flows. Face (par) value of a bond.
$\bar{F}$	Amount of money (or equivalent value) flowing continuously and uniformly during the Nth period.
f	General rate of inflation: uniform rate of increase/decrease of prices.
FW	Future worth.
$f(x)$	Probability density function for continuous random variable.
G	Arithmetic gradient: *amount* of cash flow increase/decrease from period to period.
g	Geometric gradient: *rate* of cash flow increase/decrease from period to period. (In Chapter 11, the rate of relative price change per period.)
h	Inflation rate used to index depreciation.
$H(a_i)$	Hurwicz value for alternative a_i.
I	Interest paid on debt. Annual interest payment.
I_j	Loan interest accumulated in period j.
I_j^*	Inflated after-tax interest expense in year j.
$\bar{I}$	Average interest paid over N payments.
i	Effective interest rate per interest period.
i_a	Effective interest rate per year (per *annum*).
i_d	Pre-tax cost of loan (debt).
$\hat{i}_d$	After-tax cost of loan (debt).
i_m	Effective interest rate per subperiod.
i_s	Rate of interest "earned" by imaginary sinking fund.
i^*	Internal rate of return (IRR).
i_e^*	External rate of return (ERR).
$\hat{i}^*$	After-tax (internal) rate of return.
IRR	(Internal) rate of return. Sometimes written as RoR for "rate of return."
j	Index, generally used to denote interest period, for example, A_j.
K	Number of replacements over a given planning horizon.

k	Auxiliary interest rate used when computing the external rate of return. The minimum attractive rate of return. Index for ACRS or MACRS property class (Chapter 6). (In Chapter 10, the number of standard deviations the estimate is away from the mean.)
$k*$	Cost of capital, reflecting inflation.
k_e	Cost of capital—equity. The equity capitalization rate.
k_o	Weighted average cost of capital. (Usually written without subscript.)
L	Probability of loss. The probability that the measure of worth will not exceed some value h, where, usually, $h = 0$.
L_j	Lease payment at end of period j.
M	Number of compounding subperiods per period (each of which is assumed to be of equal length).
$MACRS$	Modified accelerated cost recovery system.
$MARR$	Minimum attractive rate of return.
N	Number of compounding periods (each of which is assumed to be of equal length): the length of the planning horizon (study period). Project life.
N_a	Actual service life.
N_d	Depreciable life.
N_k	Life of the kth replacement.
N_o	Current age of the defender.
N_s	Number of periods per superperiod.
N_{max}	Maximum physical life.
$N*$	Number of periods for "payback" of original investment (Chapter 5). Economic life (Chapter 8).
$n(k)$	Recovery period for property class k.
$NB:C$	Net benefit-cost ratio.
NPV	Net present value.
P	Initial investment. Equivalent present value of future cash flow(s). Loan principal.
P_j	Amount of loan principal unpaid at the start of period j.
p	Percent deviation from estimate.
p_c	Percent deviation for estimate of cost per unit.
p_N	Percent deviation for estimate of project life.
p_Q	Percent deviation for estimate of number of units sold.
p_r	Percent deviation for estimate of revenue per unit.

$p_j(k)$	Applicable depreciation percentage in year j for property class k.
$p(x)$	Probability mass function for discrete random variable.
$P(E)$	Discounted present value of earnings over the life of the project.
$P(I)$	Present value of initial investment.
$P(S)$	Present value of terminal salvage value.
$P(T)$	Present value of cash flows for income taxes.
$P[E]$	Probability of the event E. Sometimes written $\Pr[E]$ or $\text{Prob}[E]$.
$P[E \cup F]$	Probability of the union of events E and F.
$P[EF]$	Probability of the intersection of events E and F.
$P[S]$	Probability of the sum of all possible outcomes in the sample space.
PI	Profitability index.
PW	(Equivalent) present worth, or (net) present value, of present and future cash flows.
$PWOC$	Present worth of costs.
PWP	Premium worth percentage.
Q	Amount of loan. (In Chapter 10, quantity of units sold.)
r	Nominal interest rate per period; usually, the nominal interest rate per *year*. Rate of return under the "truth in lending formula" (Chapter 5); widely known as the *annual percentage rate*. (In Chapter 10, revenue per unit.)
r_b	Bond interest rate, nominal, per year.
R_j	Positive cash flow at end of period j.
R_{ij}	Revenue or profit associated with the ith alternative and the jth state of nature.
RND	Random normal deviate.
ROI	Return on investment.
RoR	Rate of return.
S	Net salvage value of capital investment.
s_j	Outcome, scenario, or state of nature, j.
S_a	Actual net salvage value.
S_d	Expected net salvage value for computing depreciation.
S_h	Semivariance.
S_N	Salvage value if kept in service N years.
SIR	Savings-investment ratio.

SYD	Sum of the years-digits.
TI	Taxable income.
t	Effective income tax rate. (In Chapter 8, the number of periods until the kth replacement.)
t_f	Effective *federal* income tax rate.
t_s	Effective *state* income tax rate.
U_e	Number of units of output expected over the depreciable life of the asset.
U_j	Number of units of actual output during period j.
Var[x]	Variance of x.
X	Random variable.
x	In Chapter 10, a number assumed by random variable X.
$\varnothing$	The do-nothing alternative.
α	Index of optimism.
Δ	Incremental.
α_0	*Before-tax* IRR based on *actual* dollars.
α_1	*After-tax* IRR based on *actual* dollars.
β_0	*Before-tax* IRR based on *deflated* dollars.
β_1	*After-tax* IRR based on *deflated* dollars.
γ_1	*After-tax* IRR based on deflated dollars and after adjusting for indexed depreciation.
μ	Mean, or expected value.
μ_j	Mean of distribution for cash flow at end of period j.
μ_p	Mean of present worth distribution.
ρ_0	*Before-tax* IRR based on *constant* dollars.
ρ_1	*After-tax* IRR based on *constant* dollars.
σ	Standard deviation. (σ^2 = variance.)

B

Compound Interest Tables

COMPOUND INTEREST FACTORS i = 0.01

	Single Payment			Uniform Series				Uniform Series		Gradient Series		
	Compound Amount	Present Worth		Compound Amount		Present Worth		Sinking Fund	Capital Recovery	Uniform Series	Present Worth	
N	F/P	P/F	P/F	F/A	F/A	P/A	P/A	A/F	A/P	A/G	P/G	N
1	1.010	0.9901	0.9950	1.000	1.005	0.990	0.995	1.0000	1.0100	0.000	0.000	1
2	1.020	0.9803	0.9852	2.010	2.020	1.970	1.980	0.4975	0.5075	0.498	0.980	2
3	1.030	0.9706	0.9754	3.030	3.045	2.941	2.956	0.3300	0.3400	0.993	2.921	3
4	1.041	0.9610	0.9658	4.060	4.081	3.902	3.921	0.2463	0.2563	1.488	5.804	4
5	1.051	0.9515	0.9562	5.101	5.126	4.853	4.878	0.1960	0.2060	1.980	9.610	5
6	1.062	0.9420	0.9467	6.152	6.183	5.795	5.824	0.1625	0.1725	2.471	14.321	6
7	1.072	0.9327	0.9374	7.214	7.250	6.728	6.762	0.1386	0.1486	2.960	19.917	7
8	1.083	0.9235	0.9281	8.286	8.327	7.652	7.690	0.1207	0.1307	3.448	26.381	8
9	1.094	0.9143	0.9189	9.369	9.415	8.566	8.609	0.1067	0.1167	3.934	33.696	9
10	1.105	0.9053	0.9098	10.462	10.514	9.471	9.519	0.0956	0.1056	4.418	41.843	10
11	1.116	0.8963	0.9008	11.567	11.625	10.368	10.419	0.0865	0.0965	4.901	50.807	11
12	1.127	0.8874	0.8919	12.683	12.746	11.255	11.311	0.0788	0.0888	5.381	60.569	12
13	1.138	0.8787	0.8830	13.809	13.878	12.134	12.194	0.0724	0.0824	5.861	71.113	13
14	1.149	0.8700	0.8743	14.947	15.022	13.004	13.069	0.0669	0.0769	6.338	82.422	14
15	1.161	0.8613	0.8656	16.097	16.177	13.865	13.934	0.0621	0.0721	6.814	94.481	15
16	1.173	0.8528	0.8571	17.258	17.344	14.718	14.791	0.0579	0.0679	7.289	107.273	16
17	1.184	0.8444	0.8486	18.430	18.522	15.562	15.640	0.0543	0.0643	7.761	120.783	17
18	1.196	0.8360	0.8402	19.615	19.713	16.398	16.480	0.0510	0.0610	8.232	134.996	18
19	1.208	0.8277	0.8319	20.811	20.915	17.226	17.312	0.0481	0.0581	8.702	149.895	19
20	1.220	0.8195	0.8236	22.019	22.129	18.046	18.136	0.0454	0.0554	9.169	165.466	20
21	1.232	0.8114	0.8155	23.239	23.355	18.857	18.951	0.0430	0.0530	9.635	181.695	21
22	1.245	0.8034	0.8074	24.472	24.594	19.660	19.759	0.0409	0.0509	10.100	198.566	22
23	1.257	0.7954	0.7994	25.716	25.845	20.456	20.558	0.0389	0.0489	10.563	216.066	23
24	1.270	0.7876	0.7915	26.973	27.108	21.243	21.349	0.0371	0.0471	11.024	234.180	24
25	1.282	0.7798	0.7837	28.243	28.384	22.023	22.133	0.0354	0.0454	11.483	252.894	25
26	1.295	0.7720	0.7759	29.526	29.673	22.795	22.909	0.0339	0.0439	11.941	272.196	26
27	1.308	0.7644	0.7682	30.821	30.975	23.560	23.677	0.0324	0.0424	12.397	292.070	27
28	1.321	0.7568	0.7606	32.129	32.289	24.316	24.438	0.0311	0.0411	12.852	312.505	28
29	1.335	0.7493	0.7531	33.450	33.617	25.066	25.191	0.0299	0.0399	13.304	333.486	29
30	1.348	0.7419	0.7456	34.785	34.959	25.808	25.937	0.0287	0.0387	13.756	355.002	30
31	1.361	0.7346	0.7382	36.133	36.313	26.542	26.675	0.0277	0.0377	14.205	377.039	31
32	1.375	0.7273	0.7309	37.494	37.681	27.270	27.406	0.0267	0.0367	14.653	399.586	32
33	1.389	0.7201	0.7237	38.869	39.063	27.990	28.129	0.0257	0.0357	15.099	422.629	33
34	1.403	0.7130	0.7165	40.258	40.459	28.703	28.846	0.0248	0.0348	15.544	446.157	34
35	1.417	0.7059	0.7094	41.660	41.868	29.409	29.555	0.0240	0.0340	15.987	470.158	35
40	1.489	0.6717	0.6750	48.886	49.130	32.835	32.999	0.0205	0.0305	18.178	596.856	40
45	1.565	0.6391	0.6422	56.481	56.763	36.095	36.275	0.0177	0.0277	20.327	733.704	45
50	1.645	0.6080	0.6111	64.463	64.785	39.196	39.392	0.0155	0.0255	22.436	879.418	50
55	1.729	0.5785	0.5814	72.852	73.216	42.147	42.358	0.0137	0.0237	24.505	1032.815	55
60	1.817	0.5504	0.5532	81.670	82.077	44.955	45.179	0.0122	0.0222	26.533	1192.806	60
65	1.909	0.5237	0.5263	90.937	91.391	47.627	47.864	0.0110	0.0210	28.522	1358.390	65
70	2.007	0.4983	0.5008	100.676	101.179	50.169	50.419	0.0099	0.0199	30.470	1528.647	70
75	2.109	0.4741	0.4765	110.913	111.466	52.587	52.850	0.0090	0.0190	32.379	1702.734	75
80	2.217	0.4511	0.4534	121.672	122.279	54.888	55.162	0.0082	0.0182	34.249	1879.877	80

COMPOUND INTEREST FACTORS $i = 0.02$

	Single Payment			Uniform Series				Uniform Series		Gradient Series		
	Compound Amount	Present Worth		Compound Amount		Present Worth		Sinking Fund	Capital Recovery	Uniform Series	Present Worth	
N	F/P	P/F	P/F	F/A	F/A	P/A	P/A	A/F	A/P	A/G	P/G	N
1	1.020	0.9804	0.9902	1.000	1.010	0.980	0.990	1.0000	1.0200	0.000	0.000	1
2	1.040	0.9612	0.9707	2.020	2.040	1.942	1.961	0.4950	0.5150	0.495	0.961	2
3	1.061	0.9423	0.9517	3.060	3.091	2.884	2.913	0.3268	0.3468	0.987	2.846	3
4	1.082	0.9238	0.9331	4.122	4.163	3.808	3.846	0.2426	0.2626	1.475	5.617	4
5	1.104	0.9057	0.9148	5.204	5.256	4.713	4.760	0.1922	0.2122	1.960	9.240	5
6	1.126	0.8880	0.8968	6.308	6.371	5.601	5.657	0.1585	0.1785	2.442	13.680	6
7	1.149	0.8706	0.8792	7.434	7.508	6.472	6.536	0.1345	0.1545	2.921	18.903	7
8	1.172	0.8535	0.8620	8.583	8.669	7.325	7.398	0.1165	0.1365	3.396	24.878	8
9	1.195	0.8368	0.8451	9.755	9.852	8.162	8.244	0.1025	0.1225	3.868	31.572	9
10	1.219	0.8203	0.8285	10.950	11.059	8.983	9.072	0.0913	0.1113	4.337	38.955	10
11	1.243	0.8043	0.8123	12.169	12.290	9.787	9.884	0.0822	0.1022	4.802	46.998	11
12	1.268	0.7885	0.7964	13.412	13.546	10.575	10.681	0.0746	0.0946	5.264	55.671	12
13	1.294	0.7730	0.7807	14.680	14.827	11.348	11.461	0.0681	0.0881	5.723	64.948	13
14	1.319	0.7579	0.7654	15.974	16.133	12.106	12.227	0.0626	0.0826	6.179	74.800	14
15	1.346	0.7430	0.7504	17.293	17.466	12.849	12.977	0.0578	0.0778	6.631	85.202	15
16	1.373	0.7284	0.7357	18.639	18.825	13.578	13.713	0.0537	0.0737	7.080	96.129	16
17	1.400	0.7142	0.7213	20.012	20.212	14.292	14.434	0.0500	0.0700	7.526	107.555	17
18	1.428	0.7002	0.7071	21.412	21.626	14.992	15.141	0.0467	0.0667	7.968	119.458	18
19	1.457	0.6864	0.6933	22.841	23.068	15.678	15.835	0.0438	0.0638	8.407	131.814	19
20	1.486	0.6730	0.6797	24.297	24.540	16.351	16.514	0.0412	0.0612	8.843	144.600	20
21	1.516	0.6598	0.6664	25.783	26.040	17.011	17.181	0.0388	0.0588	9.276	157.796	21
22	1.546	0.6468	0.6533	27.299	27.571	17.658	17.834	0.0366	0.0566	9.705	171.379	22
23	1.577	0.6342	0.6405	28.845	29.132	18.292	18.475	0.0347	0.0547	10.132	185.331	23
24	1.608	0.6217	0.6279	30.422	30.725	18.914	19.102	0.0329	0.0529	10.555	199.630	24
25	1.641	0.6095	0.6156	32.030	32.350	19.523	19.718	0.0312	0.0512	10.974	214.259	25
26	1.673	0.5976	0.6035	33.671	34.007	20.121	20.322	0.0297	0.0497	11.391	229.199	26
27	1.707	0.5859	0.5917	35.344	35.697	20.707	20.913	0.0283	0.0483	11.804	244.431	27
28	1.741	0.5744	0.5801	37.051	37.420	21.281	21.493	0.0270	0.0470	12.214	259.939	28
29	1.776	0.5631	0.5687	38.792	39.179	21.844	22.062	0.0258	0.0458	12.621	275.706	29
30	1.811	0.5521	0.5576	40.568	40.972	22.396	22.620	0.0246	0.0446	13.025	291.716	30
31	1.848	0.5412	0.5466	42.379	42.802	22.938	23.166	0.0236	0.0436	13.426	307.954	31
32	1.885	0.5306	0.5359	44.227	44.668	23.468	23.702	0.0226	0.0426	13.823	324.403	32
33	1.922	0.5202	0.5254	46.112	46.571	23.989	24.228	0.0217	0.0417	14.217	341.051	33
34	1.961	0.5100	0.5151	48.034	48.513	24.499	24.743	0.0208	0.0408	14.608	357.882	34
35	2.000	0.5000	0.5050	49.994	50.493	24.999	25.248	0.0200	0.0400	14.996	374.883	35
40	2.208	0.4529	0.4574	60.402	61.004	27.355	27.628	0.0166	0.0366	16.889	461.993	40
45	2.438	0.4102	0.4143	71.893	72.609	29.490	29.784	0.0139	0.0339	18.703	551.565	45
50	2.692	0.3715	0.3752	84.579	85.422	31.424	31.737	0.0118	0.0318	20.442	642.361	50
55	2.972	0.3365	0.3399	98.587	99.569	33.175	33.505	0.0101	0.0301	22.106	733.353	55
60	3.281	0.3048	0.3078	114.052	115.188	34.761	35.107	0.0088	0.0288	23.696	823.698	60
65	3.623	0.2761	0.2788	131.126	132.433	36.197	36.558	0.0076	0.0276	25.215	912.709	65
70	4.000	0.2500	0.2525	149.978	151.473	37.499	37.872	0.0067	0.0267	26.663	999.834	70
75	4.416	0.2265	0.2287	170.792	172.494	38.677	39.063	0.0059	0.0259	28.043	1084.639	75
80	4.875	0.2051	0.2072	193.772	195.703	39.745	40.141	0.0052	0.0252	29.357	1166.787	80

COMPOUND INTEREST FACTORS   *i* = 0.03

	Single Payment			Uniform Series				Uniform Series		Gradient Series		
	Compound Amount	Present Worth		Compound Amount		Present Worth		Sinking Fund	Capital Recovery	Uniform Series	Present Worth	
N	F/P	P/F	P/F	F/A	F/A	P/A	P/A	A/F	A/P	A/G	P/G	N
1	1.030	0.9709	0.9854	1.000	1.015	0.971	0.985	1.0000	1.0300	0.000	0.000	1
2	1.061	0.9426	0.9567	2.030	2.060	1.913	1.942	0.4926	0.5226	0.493	0.943	2
3	1.093	0.9151	0.9288	3.091	3.137	2.829	2.871	0.3235	0.3535	0.980	2.773	3
4	1.126	0.8885	0.9017	4.184	4.246	3.717	3.773	0.2390	0.2690	1.463	5.438	4
5	1.159	0.8626	0.8755	5.309	5.388	4.580	4.648	0.1884	0.2184	1.941	8.889	5
6	1.194	0.8375	0.8500	6.468	6.565	5.417	5.498	0.1546	0.1846	2.414	13.076	6
7	1.230	0.8131	0.8252	7.662	7.777	6.230	6.323	0.1305	0.1605	2.882	17.955	7
8	1.267	0.7894	0.8012	8.892	9.025	7.020	7.124	0.1125	0.1425	3.345	23.481	8
9	1.305	0.7664	0.7779	10.159	10.311	7.786	7.902	0.0984	0.1284	3.803	29.612	9
10	1.344	0.7441	0.7552	11.464	11.635	8.530	8.658	0.0872	0.1172	4.256	36.309	10
11	1.384	0.7224	0.7332	12.808	12.999	9.253	9.391	0.0781	0.1081	4.705	43.533	11
12	1.426	0.7014	0.7118	14.192	14.404	9.954	10.103	0.0705	0.1005	5.148	51.248	12
13	1.469	0.6810	0.6911	15.618	15.851	10.635	10.794	0.0640	0.0940	5.587	59.420	13
14	1.513	0.6611	0.6710	17.086	17.341	11.296	11.465	0.0585	0.0885	6.021	68.014	14
15	1.558	0.6419	0.6514	18.599	18.877	11.938	12.116	0.0538	0.0838	6.450	77.000	15
16	1.605	0.6232	0.6325	20.157	20.458	12.561	12.749	0.0496	0.0796	6.874	86.348	16
17	1.653	0.6050	0.6140	21.762	22.086	13.166	13.363	0.0460	0.0760	7.294	96.028	17
18	1.702	0.5874	0.5962	23.414	23.764	13.754	13.959	0.0427	0.0727	7.708	106.014	18
19	1.754	0.5703	0.5788	25.117	25.492	14.324	14.538	0.0398	0.0698	8.118	116.279	19
20	1.806	0.5537	0.5619	26.870	27.271	14.877	15.100	0.0372	0.0672	8.523	126.799	20
21	1.860	0.5375	0.5456	28.676	29.105	15.415	15.645	0.0349	0.0649	8.923	137.550	21
22	1.916	0.5219	0.5297	30.537	30.993	15.937	16.175	0.0327	0.0627	9.319	148.509	22
23	1.974	0.5067	0.5143	32.453	32.937	16.444	16.689	0.0308	0.0608	9.709	159.657	23
24	2.033	0.4919	0.4993	34.426	34.940	16.936	17.188	0.0290	0.0590	10.095	170.971	24
25	2.094	0.4776	0.4847	36.459	37.003	17.413	17.673	0.0274	0.0574	10.477	182.434	25
26	2.157	0.4637	0.4706	38.553	39.128	17.877	18.144	0.0259	0.0559	10.853	194.026	26
27	2.221	0.4502	0.4569	40.710	41.317	18.327	18.601	0.0246	0.0546	11.226	205.731	27
28	2.288	0.4371	0.4436	42.931	43.572	18.764	19.044	0.0233	0.0533	11.593	217.532	28
29	2.357	0.4243	0.4307	45.219	45.894	19.188	19.475	0.0221	0.0521	11.956	229.414	29
30	2.427	0.4120	0.4181	47.575	48.286	19.600	19.893	0.0210	0.0510	12.314	241.361	30
31	2.500	0.4000	0.4060	50.003	50.749	20.000	20.299	0.0200	0.0500	12.668	253.361	31
32	2.575	0.3883	0.3941	52.503	53.286	20.389	20.693	0.0190	0.0490	13.017	265.399	32
33	2.652	0.3770	0.3827	55.078	55.900	20.766	21.076	0.0182	0.0482	13.362	277.464	33
34	2.732	0.3660	0.3715	57.730	58.592	21.132	21.447	0.0173	0.0473	13.702	289.544	34
35	2.814	0.3554	0.3607	60.462	61.365	21.487	21.808	0.0165	0.0465	14.037	301.627	35
40	3.262	0.3066	0.3111	75.401	76.527	23.115	23.460	0.0133	0.0433	15.650	361.750	40
45	3.782	0.2644	0.2684	92.720	94.104	24.519	24.885	0.0108	0.0408	17.156	420.632	45
50	4.384	0.2281	0.2315	112.797	114.480	25.730	26.114	0.0089	0.0389	18.558	477.480	50
55	5.082	0.1968	0.1997	136.072	138.103	26.774	27.174	0.0073	0.0373	19.860	531.741	55
60	5.892	0.1697	0.1723	163.053	165.487	27.676	28.089	0.0061	0.0361	21.067	583.053	60
65	6.830	0.1464	0.1486	194.333	197.233	28.453	28.878	0.0051	0.0351	22.184	631.201	65
70	7.918	0.1263	0.1282	230.594	234.036	29.123	29.558	0.0043	0.0343	23.215	676.087	70
75	9.179	0.1089	0.1106	272.631	276.700	29.702	30.145	0.0037	0.0337	24.163	717.698	75
80	10.641	0.0940	0.0954	321.363	326.160	30.201	30.652	0.0031	0.0331	25.035	756.087	80

COMPOUND INTEREST FACTORS $i = 0.04$

	Single Payment			Uniform Series				Uniform Series		Gradient Series		
	Compound Amount	Present Worth		Compound Amount		Present Worth		Sinking Fund	Capital Recovery	Uniform Series	Present Worth	
N	F/P	P/F	P/F	F/A	F/A	P/A	P/A	A/F	A/P	A/G	P/G	N
1	1.040	0.9615	0.9806	1.000	1.020	0.962	0.981	1.0000	1.0400	0.000	0.000	1
2	1.082	0.9246	0.9429	2.040	2.081	1.886	1.924	0.4902	0.5302	0.490	0.925	2
3	1.125	0.8890	0.9067	3.122	3.184	2.775	2.830	0.3203	0.3603	0.974	2.703	3
4	1.170	0.8548	0.8718	4.246	4.331	3.630	3.702	0.2355	0.2755	1.451	5.267	4
5	1.217	0.8219	0.8383	5.416	5.524	4.452	4.540	0.1846	0.2246	1.922	8.555	5
6	1.265	0.7903	0.8060	6.633	6.765	5.242	5.346	0.1508	0.1908	2.386	12.506	6
7	1.316	0.7599	0.7750	7.898	8.055	6.002	6.121	0.1266	0.1666	2.843	17.066	7
8	1.369	0.7307	0.7452	9.214	9.397	6.733	6.867	0.1085	0.1485	3.294	22.181	8
9	1.423	0.7026	0.7165	10.583	10.793	7.435	7.583	0.0945	0.1345	3.739	27.801	9
10	1.480	0.6756	0.6890	12.006	12.245	8.111	8.272	0.0833	0.1233	4.177	33.881	10
11	1.539	0.6496	0.6625	13.486	13.754	8.760	8.935	0.0741	0.1141	4.609	40.377	11
12	1.601	0.6246	0.6370	15.026	15.324	9.385	9.572	0.0666	0.1066	5.034	47.248	12
13	1.665	0.6006	0.6125	16.627	16.957	9.986	10.184	0.0601	0.1001	5.453	54.455	13
14	1.732	0.5775	0.5889	18.292	18.655	10.563	10.773	0.0547	0.0947	5.866	61.962	14
15	1.801	0.5553	0.5663	20.024	20.421	11.118	11.339	0.0499	0.0899	6.272	69.735	15
16	1.873	0.5339	0.5445	21.825	22.258	11.652	11.884	0.0458	0.0858	6.672	77.744	16
17	1.948	0.5134	0.5236	23.698	24.168	12.166	12.407	0.0422	0.0822	7.066	85.958	17
18	2.026	0.4936	0.5034	25.645	26.155	12.659	12.911	0.0390	0.0790	7.453	94.350	18
19	2.107	0.4746	0.4841	27.671	28.221	13.134	13.395	0.0361	0.0761	7.834	102.893	19
20	2.191	0.4564	0.4655	29.778	30.370	13.590	13.860	0.0336	0.0736	8.209	111.565	20
21	2.279	0.4388	0.4476	31.969	32.604	14.029	14.308	0.0313	0.0713	8.578	120.341	21
22	2.370	0.4220	0.4303	34.248	34.928	14.451	14.738	0.0292	0.0692	8.941	129.202	22
23	2.465	0.4057	0.4138	36.618	37.345	14.857	15.152	0.0273	0.0673	9.297	138.128	23
24	2.563	0.3901	0.3979	39.083	39.859	15.247	15.550	0.0256	0.0656	9.648	147.101	24
25	2.666	0.3751	0.3826	41.646	42.473	15.622	15.932	0.0240	0.0640	9.993	156.104	25
26	2.772	0.3607	0.3679	44.312	45.192	15.983	16.300	0.0226	0.0626	10.331	165.121	26
27	2.883	0.3468	0.3537	47.084	48.020	16.330	16.654	0.0212	0.0612	10.664	174.138	27
28	2.999	0.3335	0.3401	49.968	50.960	16.663	16.994	0.0200	0.0600	10.991	183.142	28
29	3.119	0.3207	0.3270	52.966	54.019	16.984	17.321	0.0189	0.0589	11.312	192.121	29
30	3.243	0.3083	0.3144	56.085	57.199	17.292	17.636	0.0178	0.0578	11.627	201.062	30
31	3.373	0.2965	0.3024	59.328	60.507	17.588	17.938	0.0169	0.0569	11.937	209.956	31
32	3.508	0.2851	0.2907	62.701	63.947	17.874	18.229	0.0159	0.0559	12.241	218.792	32
33	3.648	0.2741	0.2795	66.210	67.525	18.148	18.508	0.0151	0.0551	12.540	227.563	33
34	3.794	0.2636	0.2688	69.858	71.246	18.411	18.777	0.0143	0.0543	12.832	236.261	34
35	3.946	0.2534	0.2585	73.652	75.116	18.665	19.035	0.0136	0.0536	13.120	244.877	35
40	4.801	0.2083	0.2124	95.026	96.914	19.793	20.186	0.0105	0.0505	14.477	286.530	40
45	5.841	0.1712	0.1746	121.029	123.434	20.720	21.132	0.0083	0.0483	15.705	325.403	45
50	7.107	0.1407	0.1435	152.667	155.700	21.482	21.909	0.0066	0.0466	16.812	361.164	50
55	8.646	0.1157	0.1180	191.159	194.957	22.109	22.548	0.0052	0.0452	17.807	393.689	55
60	10.520	0.0951	0.0969	237.991	242.719	22.623	23.073	0.0042	0.0442	18.697	422.997	60
65	12.799	0.0781	0.0797	294.968	300.829	23.047	23.505	0.0034	0.0434	19.491	449.201	65
70	15.572	0.0642	0.0655	364.290	371.529	23.395	23.859	0.0027	0.0427	20.196	472.479	70
75	18.945	0.0528	0.0538	448.631	457.545	23.680	24.151	0.0022	0.0422	20.821	493.041	75
80	23.050	0.0434	0.0442	551.245	562.198	23.915	24.391	0.0018	0.0418	21.372	511.116	80

COMPOUND INTEREST FACTORS $i = 0.05$

	Single Payment			Uniform Series				Uniform Series		Gradient Series		
	Compound Amount	Present Worth	Present Worth	Compound Amount	Compound Amount	Present Worth	Present Worth	Sinking Fund	Capital Recovery	Uniform Series	Present Worth	
N	F/P	P/F	P/F	F/A	F/A	P/A	P/A	A/F	A/P	A/G	P/G	N
1	1.050	0.9524	0.9760	1.000	1.025	0.952	0.976	1.0000	1.0500	0.000	0.000	1
2	1.103	0.9070	0.9295	2.050	2.101	1.859	1.906	0.4878	0.5378	0.488	0.907	2
3	1.158	0.8638	0.8853	3.153	3.231	2.723	2.791	0.3172	0.3672	0.967	2.635	3
4	1.216	0.8227	0.8431	4.310	4.417	3.546	3.634	0.2320	0.2820	1.439	5.103	4
5	1.276	0.7835	0.8030	5.526	5.663	4.329	4.437	0.1810	0.2310	1.903	8.237	5
6	1.340	0.7462	0.7647	6.802	6.971	5.076	5.202	0.1470	0.1970	2.358	11.968	6
7	1.407	0.7107	0.7283	8.142	8.344	5.786	5.930	0.1228	0.1728	2.805	16.232	7
8	1.477	0.6768	0.6936	9.549	9.786	6.463	6.623	0.1047	0.1547	3.245	20.970	8
9	1.551	0.6446	0.6606	11.027	11.300	7.108	7.284	0.0907	0.1407	3.676	26.127	9
10	1.629	0.6139	0.6291	12.578	12.890	7.722	7.913	0.0795	0.1295	4.099	31.652	10
11	1.710	0.5847	0.5992	14.207	14.559	8.306	8.512	0.0704	0.1204	4.514	37.499	11
12	1.796	0.5568	0.5706	15.917	16.312	8.863	9.083	0.0628	0.1128	4.922	43.624	12
13	1.886	0.5303	0.5435	17.713	18.152	9.394	9.627	0.0565	0.1065	5.322	49.988	13
14	1.980	0.5051	0.5176	19.599	20.085	9.899	10.144	0.0510	0.1010	5.713	56.554	14
15	2.079	0.4810	0.4929	21.579	22.114	10.380	10.637	0.0463	0.0963	6.097	63.288	15
16	2.183	0.4581	0.4695	23.657	24.244	10.838	11.107	0.0423	0.0923	6.474	70.160	16
17	2.292	0.4363	0.4471	25.840	26.481	11.274	11.554	0.0387	0.0887	6.842	77.140	17
18	2.407	0.4155	0.4258	28.132	28.830	11.690	11.979	0.0355	0.0855	7.203	84.204	18
19	2.527	0.3957	0.4055	30.539	31.296	12.085	12.385	0.0327	0.0827	7.557	91.328	19
20	2.653	0.3769	0.3862	33.066	33.886	12.462	12.771	0.0302	0.0802	7.903	98.488	20
21	2.786	0.3589	0.3678	35.719	36.605	12.821	13.139	0.0280	0.0780	8.242	105.667	21
22	2.925	0.3418	0.3503	38.505	39.460	13.163	13.489	0.0260	0.0760	8.573	112.846	22
23	3.072	0.3256	0.3336	41.430	42.458	13.489	13.823	0.0241	0.0741	8.897	120.009	23
24	3.225	0.3101	0.3178	44.502	45.606	13.799	14.141	0.0225	0.0725	9.214	127.140	24
25	3.386	0.2953	0.3026	47.727	48.911	14.094	14.443	0.0210	0.0710	9.524	134.228	25
26	3.556	0.2812	0.2882	51.113	52.381	14.375	14.732	0.0196	0.0696	9.827	141.259	26
27	3.733	0.2678	0.2745	54.669	56.025	14.643	15.006	0.0183	0.0683	10.122	148.223	27
28	3.920	0.2551	0.2614	58.403	59.851	14.898	15.268	0.0171	0.0671	10.411	155.110	28
29	4.116	0.2429	0.2490	62.323	63.868	15.141	15.517	0.0160	0.0660	10.694	161.913	29
30	4.322	0.2314	0.2371	66.439	68.086	15.372	15.754	0.0151	0.0651	10.969	168.623	30
31	4.538	0.2204	0.2258	70.761	72.515	15.593	15.979	0.0141	0.0641	11.238	175.233	31
32	4.765	0.2099	0.2151	75.299	77.166	15.803	16.195	0.0133	0.0633	11.501	181.739	32
33	5.003	0.1999	0.2048	80.064	82.049	16.003	16.399	0.0125	0.0625	11.757	188.135	33
34	5.253	0.1904	0.1951	85.067	87.176	16.193	16.594	0.0118	0.0618	12.006	194.417	34
35	5.516	0.1813	0.1858	90.320	92.560	16.374	16.780	0.0111	0.0611	12.250	200.581	35
40	7.040	0.1420	0.1456	120.800	123.795	17.159	17.585	0.0083	0.0583	13.377	229.545	40
45	8.985	0.1113	0.1141	159.700	163.660	17.774	18.215	0.0063	0.0563	14.364	255.315	45
50	11.467	0.0872	0.0894	209.348	214.539	18.256	18.709	0.0048	0.0548	15.223	277.915	50
55	14.636	0.0683	0.0700	272.713	279.475	18.633	19.096	0.0037	0.0537	15.966	297.510	55
60	18.679	0.0535	0.0549	353.584	362.351	18.929	19.399	0.0028	0.0528	16.606	314.343	60
65	23.840	0.0419	0.0430	456.798	468.125	19.161	19.636	0.0022	0.0522	17.154	328.691	65
70	30.426	0.0329	0.0337	588.529	603.122	19.343	19.822	0.0017	0.0517	17.621	340.841	70
75	38.833	0.0258	0.0264	756.654	775.416	19.485	19.968	0.0013	0.0513	18.018	351.072	75
80	49.561	0.0202	0.0207	971.229	995.312	19.596	20.082	0.0010	0.0510	18.353	359.646	80

COMPOUND INTEREST FACTORS $i = 0.06$

	Single Payment			Uniform Series				Uniform Series		Gradient Series		
	Compound Amount	Present Worth	Present Worth	Compound Amount	Compound Amount	Present Worth	Present Worth	Sinking Fund	Capital Recovery	Uniform Series	Present Worth	
N	F/P	P/F	P/F	F/A	F/A	P/A	P/A	A/F	A/P	A/G	P/G	N
1	1.060	0.9434	0.9714	1.000	1.030	0.943	0.971	1.0000	1.0600	0.000	0.000	1
2	1.124	0.8900	0.9164	2.060	2.121	1.833	1.888	0.4854	0.5454	0.485	0.890	2
3	1.191	0.8396	0.8646	3.184	3.278	2.673	2.752	0.3141	0.3741	0.961	2.569	3
4	1.262	0.7921	0.8156	4.375	4.505	3.465	3.568	0.2286	0.2886	1.427	4.946	4
5	1.338	0.7473	0.7695	5.637	5.805	4.212	4.338	0.1774	0.2374	1.884	7.935	5
6	1.419	0.7050	0.7259	6.975	7.183	4.917	5.063	0.1434	0.2034	2.330	11.459	6
7	1.504	0.6651	0.6848	8.394	8.643	5.582	5.748	0.1191	0.1791	2.768	15.450	7
8	1.594	0.6274	0.6461	9.897	10.192	6.210	6.394	0.1010	0.1610	3.195	19.842	8
9	1.689	0.5919	0.6095	11.491	11.833	6.802	7.004	0.0870	0.1470	3.613	24.577	9
10	1.791	0.5584	0.5750	13.181	13.572	7.360	7.579	0.0759	0.1359	4.022	29.602	10
11	1.898	0.5268	0.5424	14.972	15.416	7.887	8.121	0.0668	0.1268	4.421	34.870	11
12	2.012	0.4970	0.5117	16.870	17.371	8.384	8.633	0.0593	0.1193	4.811	40.337	12
13	2.133	0.4688	0.4828	18.882	19.443	8.853	9.116	0.0530	0.1130	5.192	45.963	13
14	2.261	0.4423	0.4554	21.015	21.639	9.295	9.571	0.0476	0.1076	5.564	51.713	14
15	2.397	0.4173	0.4297	23.276	23.967	9.712	10.001	0.0430	0.1030	5.926	57.555	15
16	2.540	0.3936	0.4053	25.673	26.435	10.106	10.406	0.0390	0.0990	6.279	63.459	16
17	2.693	0.3714	0.3824	28.213	29.051	10.477	10.789	0.0354	0.0954	6.624	69.401	17
18	2.854	0.3503	0.3608	30.906	31.824	10.828	11.149	0.0324	0.0924	6.960	75.357	18
19	3.026	0.3305	0.3403	33.760	34.763	11.158	11.490	0.0296	0.0896	7.287	81.306	19
20	3.207	0.3118	0.3211	36.786	37.878	11.470	11.811	0.0272	0.0872	7.605	87.230	20
21	3.400	0.2942	0.3029	39.993	41.181	11.764	12.114	0.0250	0.0850	7.915	93.114	21
22	3.604	0.2775	0.2857	43.392	44.681	12.042	12.399	0.0230	0.0830	8.217	98.941	22
23	3.820	0.2618	0.2696	46.996	48.392	12.303	12.669	0.0213	0.0813	8.510	104.701	23
24	4.049	0.2470	0.2543	50.816	52.325	12.550	12.923	0.0197	0.0797	8.795	110.381	24
25	4.292	0.2330	0.2399	54.865	56.494	12.783	13.163	0.0182	0.0782	9.072	115.973	25
26	4.549	0.2198	0.2263	59.156	60.914	13.003	13.389	0.0169	0.0769	9.341	121.468	26
27	4.822	0.2074	0.2135	63.706	65.598	13.211	13.603	0.0157	0.0757	9.603	126.860	27
28	5.112	0.1956	0.2014	68.528	70.564	13.406	13.804	0.0146	0.0746	9.857	132.142	28
29	5.418	0.1846	0.1900	73.640	75.828	13.591	13.994	0.0136	0.0736	10.103	137.310	29
30	5.743	0.1741	0.1793	79.058	81.407	13.765	14.174	0.0126	0.0726	10.342	142.359	30
31	6.088	0.1643	0.1691	84.802	87.321	13.929	14.343	0.0118	0.0718	10.574	147.286	31
32	6.453	0.1550	0.1596	90.890	93.590	14.084	14.502	0.0110	0.0710	10.799	152.090	32
33	6.841	0.1462	0.1505	97.343	100.235	14.230	14.653	0.0103	0.0703	11.017	156.768	33
34	7.251	0.1379	0.1420	104.184	107.279	14.368	14.795	0.0096	0.0696	11.228	161.319	34
35	7.686	0.1301	0.1340	111.435	114.745	14.498	14.929	0.0090	0.0690	11.432	165.743	35
40	10.286	0.0972	0.1001	154.762	159.360	15.046	15.493	0.0065	0.0665	12.359	185.957	40
45	13.765	0.0727	0.0748	212.744	219.064	15.456	15.915	0.0047	0.0647	13.141	203.110	45
50	18.420	0.0543	0.0559	290.336	298.961	15.762	16.230	0.0034	0.0634	13.796	217.457	50
55	24.650	0.0406	0.0418	394.172	405.882	15.991	16.466	0.0025	0.0625	14.341	229.322	55
60	32.988	0.0303	0.0312	533.128	548.967	16.161	16.642	0.0019	0.0619	14.791	239.043	60
65	44.145	0.0227	0.0233	719.083	740.446	16.289	16.773	0.0014	0.0614	15.160	246.945	65
70	59.076	0.0169	0.0174	967.932	996.688	16.385	16.871	0.0010	0.0610	15.461	253.327	70
75	79.057	0.0126	0.0130	1300.949	1339.598	16.456	16.945	0.0008	0.0608	15.706	258.453	75
80	105.796	0.0095	0.0097	1746.600	1798.489	16.509	17.000	0.0006	0.0606	15.903	262.549	80

COMPOUND INTEREST FACTORS $\qquad i = \quad 0.07$

	Single Payment			Uniform Series				Uniform Series		Gradient Series		
	Compound Amount	Present Worth		Compound Amount		Present Worth		Sinking Fund	Capital Recovery	Uniform Series	Present Worth	
N	F/P	P/F	P/F	F/A	F/A	P/A	P/A	A/F	A/P	A/G	P/G	N
1	1.070	0.9346	0.9669	1.000	1.035	0.935	0.967	1.0000	1.0700	0.000	0.000	1
2	1.145	0.8734	0.9037	2.070	2.142	1.808	1.871	0.4831	0.5531	0.483	0.873	2
3	1.225	0.8163	0.8445	3.215	3.326	2.624	2.715	0.3111	0.3811	0.955	2.506	3
4	1.311	0.7629	0.7893	4.440	4.594	3.387	3.504	0.2252	0.2952	1.416	4.795	4
5	1.403	0.7130	0.7377	5.751	5.950	4.100	4.242	0.1739	0.2439	1.865	7.647	5
6	1.501	0.6663	0.6894	7.153	7.401	4.767	4.931	0.1398	0.2098	2.303	10.978	6
7	1.606	0.6227	0.6443	8.654	8.953	5.389	5.576	0.1156	0.1856	2.730	14.715	7
8	1.718	0.5820	0.6021	10.260	10.615	5.971	6.178	0.0975	0.1675	3.147	18.789	8
9	1.838	0.5439	0.5628	11.978	12.392	6.515	6.741	0.0835	0.1535	3.552	23.140	9
10	1.967	0.5083	0.5259	13.816	14.295	7.024	7.267	0.0724	0.1424	3.946	27.716	10
11	2.105	0.4751	0.4915	15.784	16.330	7.499	7.758	0.0634	0.1334	4.330	32.466	11
12	2.252	0.4440	0.4594	17.888	18.507	7.943	8.218	0.0559	0.1259	4.703	37.351	12
13	2.410	0.4150	0.4293	20.141	20.838	8.358	8.647	0.0497	0.1197	5.065	42.330	13
14	2.579	0.3878	0.4012	22.550	23.331	8.745	9.048	0.0443	0.1143	5.417	47.372	14
15	2.759	0.3624	0.3750	25.129	25.999	9.108	9.423	0.0398	0.1098	5.758	52.446	15
16	2.952	0.3387	0.3505	27.888	28.853	9.447	9.774	0.0359	0.1059	6.090	57.527	16
17	3.159	0.3166	0.3275	30.840	31.907	9.763	10.101	0.0324	0.1024	6.411	62.592	17
18	3.380	0.2959	0.3061	33.999	35.176	10.059	10.407	0.0294	0.0994	6.722	67.622	18
19	3.617	0.2765	0.2861	37.379	38.672	10.336	10.693	0.0268	0.0968	7.024	72.599	19
20	3.870	0.2584	0.2674	40.995	42.414	10.594	10.961	0.0244	0.0944	7.316	77.509	20
21	4.141	0.2415	0.2499	44.865	46.418	10.836	11.210	0.0223	0.0923	7.599	82.339	21
22	4.430	0.2257	0.2335	49.006	50.702	11.061	11.444	0.0204	0.0904	7.872	87.079	22
23	4.741	0.2109	0.2182	53.436	55.285	11.272	11.662	0.0187	0.0887	8.137	91.720	23
24	5.072	0.1971	0.2040	58.177	60.190	11.469	11.866	0.0172	0.0872	8.392	96.255	24
25	5.427	0.1842	0.1906	63.249	65.438	11.654	12.057	0.0158	0.0858	8.639	100.676	25
26	5.807	0.1722	0.1782	68.676	71.053	11.826	12.235	0.0146	0.0846	8.877	104.981	26
27	6.214	0.1609	0.1665	74.484	77.061	11.987	12.402	0.0134	0.0834	9.107	109.166	27
28	6.649	0.1504	0.1556	80.698	83.490	12.137	12.557	0.0124	0.0824	9.329	113.226	28
29	7.114	0.1406	0.1454	87.347	90.369	12.278	12.703	0.0114	0.0814	9.543	117.162	29
30	7.612	0.1314	0.1359	94.461	97.730	12.409	12.838	0.0106	0.0806	9.749	120.972	30
31	8.145	0.1228	0.1270	102.073	105.605	12.532	12.965	0.0098	0.0798	9.947	124.655	31
32	8.715	0.1147	0.1187	110.218	114.032	12.647	13.084	0.0091	0.0791	10.138	128.212	32
33	9.325	0.1072	0.1109	118.933	123.049	12.754	13.195	0.0084	0.0784	10.322	131.643	33
34	9.978	0.1002	0.1037	128.259	132.697	12.854	13.299	0.0078	0.0778	10.499	134.951	34
35	10.677	0.0937	0.0969	138.237	143.021	12.948	13.396	0.0072	0.0772	10.669	138.135	35
40	14.974	0.0668	0.0691	199.635	206.544	13.332	13.793	0.0050	0.0750	11.423	152.293	40
45	21.002	0.0476	0.0493	285.749	295.638	13.606	14.076	0.0035	0.0735	12.036	163.756	45
50	29.457	0.0339	0.0351	406.529	420.597	13.801	14.278	0.0025	0.0725	12.529	172.905	50
55	41.315	0.0242	0.0250	575.929	595.859	13.940	14.422	0.0017	0.0717	12.921	180.124	55
60	57.946	0.0173	0.0179	813.520	841.673	14.039	14.525	0.0012	0.0712	13.232	185.768	60
65	81.273	0.0123	0.0127	1146.755	1186.439	14.110	14.598	0.0009	0.0709	13.476	190.145	65
70	113.989	0.0088	0.0091	1614.134	1669.992	14.160	14.650	0.0006	0.0706	13.666	193.519	70
75	159.876	0.0063	0.0065	2269.657	2348.200	14.196	14.688	0.0004	0.0704	13.814	196.104	75
80	224.234	0.0045	0.0046	3189.063	3299.421	14.222	14.714	0.0003	0.0703	13.927	198.075	80

COMPOUND INTEREST FACTORS   i =   0.08

	Single Payment			Uniform Series				Uniform Series		Gradient Series		
	Compound Amount	Present Worth		Compound Amount		Present Worth		Sinking Fund	Capital Recovery	Uniform Series	Present Worth	
N	F/P	P/F	P/F	F/A	F/A	P/A	P/A	A/F	A/P	A/G	P/G	N
1	1.080	0.9259	0.9625	1.000	1.039	0.926	0.962	1.0000	1.0800	0.000	0.000	1
2	1.166	0.8573	0.8912	2.080	2.162	1.783	1.854	0.4808	0.5608	0.481	0.857	2
3	1.260	0.7938	0.8252	3.246	3.375	2.577	2.679	0.3080	0.3880	0.949	2.445	3
4	1.360	0.7350	0.7641	4.506	4.684	3.312	3.443	0.2219	0.3019	1.404	4.650	4
5	1.469	0.6806	0.7075	5.867	6.098	3.993	4.150	0.1705	0.2505	1.846	7.372	5
6	1.587	0.6302	0.6551	7.336	7.626	4.623	4.805	0.1363	0.2163	2.276	10.523	6
7	1.714	0.5835	0.6065	8.923	9.275	5.206	5.412	0.1121	0.1921	2.694	14.024	7
8	1.851	0.5403	0.5616	10.637	11.057	5.747	5.974	0.0940	0.1740	3.099	17.806	8
9	1.999	0.5002	0.5200	12.488	12.981	6.247	6.494	0.0801	0.1601	3.491	21.808	9
10	2.159	0.4632	0.4815	14.487	15.059	6.710	6.975	0.0690	0.1490	3.871	25.977	10
11	2.332	0.4289	0.4458	16.645	17.303	7.139	7.421	0.0601	0.1401	4.240	30.266	11
12	2.518	0.3971	0.4128	18.977	19.726	7.536	7.834	0.0527	0.1327	4.596	34.634	12
13	2.720	0.3677	0.3822	21.495	22.344	7.904	8.216	0.0465	0.1265	4.940	39.046	13
14	2.937	0.3405	0.3539	24.215	25.171	8.244	8.570	0.0413	0.1213	5.273	43.472	14
15	3.172	0.3152	0.3277	27.152	28.224	8.559	8.897	0.0368	0.1168	5.594	47.886	15
16	3.426	0.2919	0.3034	30.324	31.522	8.851	9.201	0.0330	0.1130	5.905	52.264	16
17	3.700	0.2703	0.2809	33.750	35.083	9.122	9.482	0.0296	0.1096	6.204	56.588	17
18	3.996	0.2502	0.2601	37.450	38.929	9.372	9.742	0.0267	0.1067	6.492	60.843	18
19	4.316	0.2317	0.2409	41.446	43.083	9.604	9.983	0.0241	0.1041	6.770	65.013	19
20	4.661	0.2145	0.2230	45.762	47.569	9.818	10.206	0.0219	0.1019	7.037	69.090	20
21	5.034	0.1987	0.2065	50.423	52.414	10.017	10.412	0.0198	0.0998	7.294	73.063	21
22	5.437	0.1839	0.1912	55.457	57.647	10.201	10.604	0.0180	0.0980	7.541	76.926	22
23	5.871	0.1703	0.1770	60.893	63.298	10.371	10.781	0.0164	0.0964	7.779	80.673	23
24	6.341	0.1577	0.1639	66.765	69.401	10.529	10.945	0.0150	0.0950	8.007	84.300	24
25	6.848	0.1460	0.1518	73.106	75.993	10.675	11.096	0.0137	0.0937	8.225	87.804	25
26	7.396	0.1352	0.1405	79.954	83.112	10.810	11.237	0.0125	0.0925	8.435	91.184	26
27	7.988	0.1252	0.1301	87.351	90.800	10.935	11.367	0.0114	0.0914	8.636	94.439	27
28	8.627	0.1159	0.1205	95.339	99.103	11.051	11.487	0.0105	0.0905	8.829	97.569	28
29	9.317	0.1073	0.1116	103.966	108.071	11.158	11.599	0.0096	0.0896	9.013	100.574	29
30	10.063	0.0994	0.1033	113.283	117.756	11.258	11.702	0.0088	0.0888	9.190	103.456	30
31	10.868	0.0920	0.0956	123.346	128.216	11.350	11.798	0.0081	0.0881	9.358	106.216	31
32	11.737	0.0852	0.0886	134.214	139.513	11.435	11.887	0.0075	0.0875	9.520	108.857	32
33	12.676	0.0789	0.0820	145.951	151.714	11.514	11.969	0.0069	0.0869	9.674	111.382	33
34	13.690	0.0730	0.0759	158.627	164.890	11.587	12.044	0.0063	0.0863	9.821	113.792	34
35	14.785	0.0676	0.0703	172.317	179.121	11.655	12.115	0.0058	0.0858	9.961	116.092	35
40	21.725	0.0460	0.0478	259.057	269.286	11.925	12.395	0.0039	0.0839	10.570	126.042	40
45	31.920	0.0313	0.0326	386.506	401.768	12.108	12.587	0.0026	0.0826	11.045	133.733	45
50	46.902	0.0213	0.0222	573.770	596.427	12.233	12.717	0.0017	0.0817	11.411	139.593	50
55	68.914	0.0145	0.0151	848.923	882.445	12.319	12.805	0.0012	0.0812	11.690	144.006	55
60	101.257	0.0099	0.0103	1253.213	1302.699	12.377	12.865	0.0008	0.0808	11.902	147.300	60
65	148.780	0.0067	0.0070	1847.248	1920.190	12.416	12.906	0.0005	0.0805	12.060	149.739	65
70	218.606	0.0046	0.0048	2720.080	2827.488	12.443	12.934	0.0004	0.0804	12.178	151.533	70
75	321.205	0.0031	0.0032	4002.557	4160.605	12.461	12.953	0.0002	0.0802	12.266	152.845	75
80	471.955	0.0021	0.0022	5886.935	6119.393	12.474	12.966	0.0002	0.0802	12.330	153.800	80

COMPOUND INTEREST FACTORS i = 0.09

	Single Payment			Uniform Series				Uniform Series		Gradient Series		
	Compound Amount	Present Worth		Compound Amount		Present Worth		Sinking Fund	Capital Recovery	Uniform Series	Present Worth	
N	F/P	P/F	P/F	F/A	F/A	P/A	P/A	A/F	A/P	A/G	P/G	N
1	1.090	0.9174	0.9581	1.000	1.044	0.917	0.958	1.0000	1.0900	0.000	0.000	1
2	1.188	0.8417	0.8790	2.090	2.183	1.759	1.837	0.4785	0.5685	0.478	0.842	2
3	1.295	0.7722	0.8064	3.278	3.423	2.531	2.644	0.3051	0.3951	0.943	2.386	3
4	1.412	0.7084	0.7398	4.573	4.776	3.240	3.383	0.2187	0.3087	1.393	4.511	4
5	1.539	0.6499	0.6788	5.985	6.250	3.890	4.062	0.1671	0.2571	1.828	7.111	5
6	1.677	0.5963	0.6227	7.523	7.857	4.486	4.685	0.1329	0.2229	2.250	10.092	6
7	1.828	0.5470	0.5713	9.200	9.609	5.033	5.256	0.1087	0.1987	2.657	13.375	7
8	1.993	0.5019	0.5241	11.028	11.518	5.535	5.780	0.0907	0.1807	3.051	16.888	8
9	2.172	0.4604	0.4808	13.021	13.599	5.995	6.261	0.0768	0.1668	3.431	20.571	9
10	2.367	0.4224	0.4411	15.193	15.867	6.418	6.702	0.0658	0.1558	3.798	24.373	10
11	2.580	0.3875	0.4047	17.560	18.339	6.805	7.107	0.0569	0.1469	4.151	28.248	11
12	2.813	0.3555	0.3713	20.141	21.034	7.161	7.478	0.0497	0.1397	4.491	32.159	12
13	3.066	0.3262	0.3406	22.953	23.971	7.487	7.819	0.0436	0.1336	4.818	36.073	13
14	3.342	0.2992	0.3125	26.019	27.173	7.786	8.131	0.0384	0.1284	5.133	39.963	14
15	3.642	0.2745	0.2867	29.361	30.663	8.061	8.418	0.0341	0.1241	5.435	43.807	15
16	3.970	0.2519	0.2630	33.003	34.467	8.313	8.681	0.0303	0.1203	5.724	47.585	16
17	4.328	0.2311	0.2413	36.974	38.614	8.544	8.923	0.0270	0.1170	6.002	51.282	17
18	4.717	0.2120	0.2214	41.301	43.133	8.756	9.144	0.0242	0.1142	6.269	54.886	18
19	5.142	0.1945	0.2031	46.018	48.060	8.950	9.347	0.0217	0.1117	6.524	58.387	19
20	5.604	0.1784	0.1863	51.160	53.429	9.129	9.533	0.0195	0.1095	6.767	61.777	20
21	6.109	0.1637	0.1710	56.765	59.282	9.292	9.704	0.0176	0.1076	7.001	65.051	21
22	6.659	0.1502	0.1568	62.873	65.662	9.442	9.861	0.0159	0.1059	7.223	68.205	22
23	7.258	0.1378	0.1439	69.532	72.616	9.580	10.005	0.0144	0.1044	7.436	71.236	23
24	7.911	0.1264	0.1320	76.790	80.196	9.707	10.137	0.0130	0.1030	7.638	74.143	24
25	8.623	0.1160	0.1211	84.701	88.458	9.823	10.258	0.0118	0.1018	7.832	76.926	25
26	9.399	0.1064	0.1111	93.324	97.463	9.929	10.369	0.0107	0.1007	8.016	79.586	26
27	10.245	0.0976	0.1019	102.723	107.279	10.027	10.471	0.0097	0.0997	8.191	82.124	27
28	11.167	0.0895	0.0935	112.968	117.979	10.116	10.565	0.0089	0.0989	8.357	84.542	28
29	12.172	0.0822	0.0858	124.135	129.641	10.198	10.651	0.0081	0.0981	8.515	86.842	29
30	13.268	0.0754	0.0787	136.308	142.353	10.274	10.729	0.0073	0.0973	8.666	89.028	30
31	14.462	0.0691	0.0722	149.575	156.209	10.343	10.802	0.0067	0.0967	8.808	91.102	31
32	15.763	0.0634	0.0663	164.037	171.313	10.406	10.868	0.0061	0.0961	8.944	93.069	32
33	17.182	0.0582	0.0608	179.800	187.775	10.464	10.929	0.0056	0.0956	9.072	94.931	33
34	18.728	0.0534	0.0558	196.982	205.719	10.518	10.984	0.0051	0.0951	9.193	96.693	34
35	20.414	0.0490	0.0512	215.711	225.278	10.567	11.035	0.0046	0.0946	9.308	98.359	35
40	31.409	0.0318	0.0332	337.882	352.869	10.757	11.234	0.0030	0.0930	9.796	105.376	40
45	48.327	0.0207	0.0216	525.859	549.183	10.881	11.364	0.0019	0.0919	10.160	110.556	45
50	74.358	0.0134	0.0140	815.084	851.236	10.962	11.448	0.0012	0.0912	10.430	114.325	50
55	114.408	0.0087	0.0091	1260.092	1315.982	11.014	11.503	0.0008	0.0908	10.626	117.036	55
60	176.031	0.0057	0.0059	1944.792	2031.051	11.048	11.538	0.0005	0.0905	10.768	118.968	60
65	270.846	0.0037	0.0039	2998.288	3131.274	11.070	11.561	0.0003	0.0903	10.870	120.334	65
70	416.730	0.0024	0.0025	4619.223	4824.103	11.084	11.576	0.0002	0.0902	10.943	121.294	70
75	641.191	0.0016	0.0016	7113.232	7428.731	11.094	11.586	0.0001	0.0901	10.994	121.965	75
80	986.552	0.0010	0.0011	10950.574	11436.273	11.100	11.592	0.0001	0.0901	11.030	122.431	80

COMPOUND INTEREST FACTORS   i = 0.10

	Single Payment			Uniform Series				Uniform Series		Gradient Series		
	Compound Amount	Present Worth		Compound Amount		Present Worth		Sinking Fund	Capital Recovery	Uniform Series	Present Worth	
N	F/P	P/F	P/F	F/A	F/A	P/A	P/A	A/F	A/P	A/G	P/G	N
1	1.100	0.9091	0.9538	1.000	1.049	0.909	0.954	1.0000	1.1000	0.000	0.000	1
2	1.210	0.8264	0.8671	2.100	2.203	1.736	1.821	0.4762	0.5762	0.476	0.826	2
3	1.331	0.7513	0.7883	3.310	3.473	2.487	2.609	0.3021	0.4021	0.937	2.329	3
4	1.464	0.6830	0.7166	4.641	4.869	3.170	3.326	0.2155	0.3155	1.381	4.378	4
5	1.611	0.6209	0.6515	6.105	6.406	3.791	3.977	0.1638	0.2638	1.810	6.862	5
6	1.772	0.5645	0.5922	7.716	8.095	4.355	4.570	0.1296	0.2296	2.224	9.684	6
7	1.949	0.5132	0.5384	9.487	9.954	4.868	5.108	0.1054	0.2054	2.622	12.763	7
8	2.144	0.4665	0.4895	11.436	11.999	5.335	5.597	0.0874	0.1874	3.004	16.029	8
9	2.358	0.4241	0.4450	13.579	14.248	5.759	6.042	0.0736	0.1736	3.372	19.421	9
10	2.594	0.3855	0.4045	15.937	16.722	6.145	6.447	0.0627	0.1627	3.725	22.891	10
11	2.853	0.3505	0.3677	18.531	19.443	6.495	6.815	0.0540	0.1540	4.064	26.396	11
12	3.138	0.3186	0.3343	21.384	22.437	6.814	7.149	0.0468	0.1468	4.388	29.901	12
13	3.452	0.2897	0.3039	24.523	25.729	7.103	7.453	0.0408	0.1408	4.699	33.377	13
14	3.797	0.2633	0.2763	27.975	29.352	7.367	7.729	0.0357	0.1357	4.996	36.800	14
15	4.177	0.2394	0.2512	31.772	33.336	7.606	7.980	0.0315	0.1315	5.279	40.152	15
16	4.595	0.2176	0.2283	35.950	37.719	7.824	8.209	0.0278	0.1278	5.549	43.416	16
17	5.054	0.1978	0.2076	40.545	42.540	8.022	8.416	0.0247	0.1247	5.807	46.582	17
18	5.560	0.1799	0.1887	45.599	47.843	8.201	8.605	0.0219	0.1219	6.053	49.640	18
19	6.116	0.1635	0.1716	51.159	53.676	8.365	8.777	0.0195	0.1195	6.286	52.583	19
20	6.727	0.1486	0.1560	57.275	60.093	8.514	8.932	0.0175	0.1175	6.508	55.407	20
21	7.400	0.1351	0.1418	64.002	67.152	8.649	9.074	0.0156	0.1156	6.719	58.110	21
22	8.140	0.1228	0.1289	71.403	74.916	8.772	9.203	0.0140	0.1140	6.919	60.689	22
23	8.954	0.1117	0.1172	79.543	83.457	8.883	9.320	0.0126	0.1126	7.108	63.146	23
24	9.850	0.1015	0.1065	88.497	92.852	8.985	9.427	0.0113	0.1113	7.288	65.481	24
25	10.835	0.0923	0.0968	98.347	103.186	9.077	9.524	0.0102	0.1102	7.458	67.696	25
26	11.918	0.0839	0.0880	109.182	114.554	9.161	9.612	0.0092	0.1092	7.619	69.794	26
27	13.110	0.0763	0.0800	121.100	127.059	9.237	9.692	0.0083	0.1083	7.770	71.777	27
28	14.421	0.0693	0.0728	134.210	140.814	9.307	9.765	0.0075	0.1075	7.914	73.650	28
29	15.863	0.0630	0.0661	148.631	155.944	9.370	9.831	0.0067	0.1067	8.049	75.415	29
30	17.449	0.0573	0.0601	164.494	172.588	9.427	9.891	0.0061	0.1061	8.176	77.077	30
31	19.194	0.0521	0.0547	181.943	190.896	9.479	9.945	0.0055	0.1055	8.296	78.640	31
32	21.114	0.0474	0.0497	201.138	211.035	9.526	9.995	0.0050	0.1050	8.409	80.108	32
33	23.225	0.0431	0.0452	222.252	233.188	9.569	10.040	0.0045	0.1045	8.515	81.486	33
34	25.548	0.0391	0.0411	245.477	257.556	9.609	10.081	0.0041	0.1041	8.615	82.777	34
35	28.102	0.0356	0.0373	271.024	284.360	9.644	10.119	0.0037	0.1037	8.709	83.987	35
40	45.259	0.0221	0.0232	442.593	464.371	9.779	10.260	0.0023	0.1023	9.096	88.953	40
45	72.890	0.0137	0.0144	718.905	754.279	9.863	10.348	0.0014	0.1014	9.374	92.454	45
50	117.391	0.0085	0.0089	1163.909	1221.180	9.915	10.403	0.0009	0.1009	9.570	94.889	50
55	189.059	0.0053	0.0055	1880.591	1973.128	9.947	10.437	0.0005	0.1005	9.708	96.562	55
60	304.482	0.0033	0.0034	3034.816	3184.147	9.967	10.458	0.0003	0.1003	9.802	97.701	60
65	490.371	0.0020	0.0021	4893.707	5134.506	9.980	10.471	0.0002	0.1002	9.867	98.471	65
70	789.747	0.0013	0.0013	7887.470	8275.579	9.987	10.479	0.0001	0.1001	9.911	98.987	70
75	1271.895	0.0008	0.0008	12708.954	13334.309	9.992	10.484	0.0001	0.1001	9.941	99.332	75
80	2048.400	0.0005	0.0005	20474.002	21481.443	9.995	10.487	0.0000	0.1000	9.961	99.561	80

COMPOUND INTEREST FACTORS $i = 0.12$

	Single Payment			Uniform Series				Uniform Series		Gradient Series		
	Compound Amount	Present Worth		Compound Amount		Present Worth		Sinking Fund	Capital Recovery	Uniform Series	Present Worth	
N	F/P	P/F	P/F	F/A	F/A	P/A	P/A	A/F	A/P	A/G	P/G	N
1	1.120	0.8929	0.9454	1.000	1.059	0.893	0.945	1.0000	1.1200	0.000	0.000	1
2	1.254	0.7972	0.8441	2.120	2.245	1.690	1.790	0.4717	0.5917	0.472	0.797	2
3	1.405	0.7118	0.7537	3.374	3.573	2.402	2.543	0.2963	0.4163	0.925	2.221	3
4	1.574	0.6355	0.6729	4.779	5.061	3.037	3.216	0.2092	0.3292	1.359	4.127	4
5	1.762	0.5674	0.6008	6.353	6.727	3.605	3.817	0.1574	0.2774	1.775	6.397	5
6	1.974	0.5066	0.5365	8.115	8.593	4.111	4.353	0.1232	0.2432	2.172	8.930	6
7	2.211	0.4523	0.4790	10.089	10.683	4.564	4.832	0.0991	0.2191	2.551	11.644	7
8	2.476	0.4039	0.4277	12.300	13.024	4.968	5.260	0.0813	0.2013	2.913	14.471	8
9	2.773	0.3606	0.3818	14.776	15.645	5.328	5.642	0.0677	0.1877	3.257	17.356	9
10	3.106	0.3220	0.3409	17.549	18.582	5.650	5.983	0.0570	0.1770	3.585	20.254	10
11	3.479	0.2875	0.3044	20.655	21.870	5.938	6.287	0.0484	0.1684	3.895	23.129	11
12	3.896	0.2567	0.2718	24.133	25.554	6.194	6.559	0.0414	0.1614	4.190	25.952	12
13	4.363	0.2292	0.2427	28.029	29.679	6.424	6.802	0.0357	0.1557	4.468	28.702	13
14	4.887	0.2046	0.2167	32.393	34.299	6.628	7.018	0.0309	0.1509	4.732	31.362	14
15	5.474	0.1827	0.1935	37.280	39.474	6.811	7.212	0.0268	0.1468	4.980	33.920	15
16	6.130	0.1631	0.1727	42.753	45.270	6.974	7.385	0.0234	0.1434	5.215	36.367	16
17	6.866	0.1456	0.1542	48.884	51.761	7.120	7.539	0.0205	0.1405	5.435	38.697	17
18	7.690	0.1300	0.1377	55.750	59.032	7.250	7.676	0.0179	0.1379	5.643	40.908	18
19	8.613	0.1161	0.1229	63.440	67.174	7.366	7.799	0.0158	0.1358	5.838	42.998	19
20	9.646	0.1037	0.1098	72.052	76.294	7.469	7.909	0.0139	0.1339	6.020	44.968	20
21	10.804	0.0926	0.0980	81.699	86.508	7.562	8.007	0.0122	0.1322	6.191	46.819	21
22	12.100	0.0826	0.0875	92.503	97.948	7.645	8.095	0.0108	0.1308	6.351	48.554	22
23	13.552	0.0738	0.0781	104.603	110.761	7.718	8.173	0.0096	0.1296	6.501	50.178	23
24	15.179	0.0659	0.0698	118.155	125.111	7.784	8.243	0.0085	0.1285	6.641	51.693	24
25	17.000	0.0588	0.0623	133.334	141.183	7.843	8.305	0.0075	0.1275	6.771	53.105	25
26	19.040	0.0525	0.0556	150.334	159.184	7.896	8.360	0.0067	0.1267	6.892	54.418	26
27	21.325	0.0469	0.0497	169.374	179.345	7.943	8.410	0.0059	0.1259	7.005	55.637	27
28	23.884	0.0419	0.0443	190.699	201.925	7.984	8.454	0.0052	0.1252	7.110	56.767	28
29	26.750	0.0374	0.0396	214.583	227.215	8.022	8.494	0.0047	0.1247	7.207	57.814	29
30	29.960	0.0334	0.0353	241.333	255.539	8.055	8.529	0.0041	0.1241	7.297	58.782	30
31	33.555	0.0298	0.0316	271.293	287.263	8.085	8.561	0.0037	0.1237	7.381	59.676	31
32	37.582	0.0266	0.0282	304.848	322.793	8.112	8.589	0.0033	0.1233	7.459	60.501	32
33	42.092	0.0238	0.0252	342.429	362.587	8.135	8.614	0.0029	0.1229	7.530	61.261	33
34	47.143	0.0212	0.0225	384.521	407.157	8.157	8.637	0.0026	0.1226	7.596	61.961	34
35	52.800	0.0189	0.0201	431.663	457.074	8.176	8.657	0.0023	0.1223	7.658	62.605	35
40	93.051	0.0107	0.0114	767.091	812.248	8.244	8.729	0.0013	0.1213	7.899	65.116	40
45	163.988	0.0061	0.0065	1358.230	1438.185	8.283	8.770	0.0007	0.1207	8.057	66.734	45
50	289.002	0.0035	0.0037	2400.018	2541.300	8.304	8.793	0.0004	0.1204	8.160	67.762	50
55	509.321	0.0020	0.0021	4236.005	4485.366	8.317	8.807	0.0002	0.1202	8.225	68.408	55
60	897.597	0.0011	0.0012	7471.641	7911.474	8.324	8.814	0.0001	0.1201	8.266	68.810	60
65	1581.872	0.0006	0.0007	13173.937	13949.447	8.328	8.818	0.0001	0.1201	8.292	69.058	65
70	2787.800	0.0004	0.0004	23223.332	24590.419	8.330	8.821	0.0000	0.1200	8.308	69.210	70
75	4913.056	0.0002	0.0002	40933.799	43343.447	8.332	8.822	0.0000	0.1200	8.318	69.303	75
80	8658.483	0.0001	0.0001	72145.693	76392.690	8.332	8.823	0.0000	0.1200	8.324	69.359	80

COMPOUND INTEREST FACTORS   i = 0.15

	Single	Payment		Uniform	Series			Uniform	Series	Gradient	Series	
	Compound Amount	Present	Worth	Compound Amount			Present Worth	Sinking Fund	Capital Recovery	Uniform Series	Present Worth	
N	F/P	P/F	P/F	F/A	F/A	P/A	P/A	A/F	A/P	A/G	P/G	N
1	1.150	0.8696	0.9333	1.000	1.073	0.870	0.933	1.0000	1.1500	0.000	0.000	1
2	1.323	0.7561	0.8115	2.150	2.307	1.626	1.745	0.4651	0.6151	0.465	0.756	2
3	1.521	0.6575	0.7057	3.473	3.727	2.283	2.450	0.2880	0.4380	0.907	2.071	3
4	1.749	0.5718	0.6136	4.993	5.359	2.855	3.064	0.2003	0.3503	1.326	3.786	4
5	2.011	0.4972	0.5336	6.742	7.236	3.352	3.598	0.1483	0.2983	1.723	5.775	5
6	2.313	0.4323	0.4640	8.754	9.395	3.784	4.062	0.1142	0.2642	2.097	7.937	6
7	2.660	0.3759	0.4035	11.067	11.877	4.160	4.465	0.0904	0.2404	2.450	10.192	7
8	3.059	0.3269	0.3508	13.727	14.732	4.487	4.816	0.0729	0.2229	2.781	12.481	8
9	3.518	0.2843	0.3051	16.786	18.015	4.772	5.121	0.0596	0.2096	3.092	14.755	9
10	4.046	0.2472	0.2653	20.304	21.791	5.019	5.386	0.0493	0.1993	3.383	16.979	10
11	4.652	0.2149	0.2307	24.349	26.133	5.234	5.617	0.0411	0.1911	3.655	19.129	11
12	5.350	0.1869	0.2006	29.002	31.126	5.421	5.818	0.0345	0.1845	3.908	21.185	12
13	6.153	0.1625	0.1744	34.352	36.868	5.583	5.992	0.0291	0.1791	4.144	23.135	13
14	7.076	0.1413	0.1517	40.505	43.472	5.724	6.144	0.0247	0.1747	4.362	24.972	14
15	8.137	0.1229	0.1319	47.580	51.066	5.847	6.276	0.0210	0.1710	4.565	26.693	15
16	9.358	0.1069	0.1147	55.717	59.799	5.954	6.390	0.0179	0.1679	4.752	28.296	16
17	10.761	0.0929	0.0997	65.075	69.842	6.047	6.490	0.0154	0.1654	4.925	29.783	17
18	12.375	0.0808	0.0867	75.836	81.392	6.128	6.577	0.0132	0.1632	5.084	31.156	18
19	14.232	0.0703	0.0754	88.212	94.674	6.198	6.652	0.0113	0.1613	5.231	32.421	19
20	16.367	0.0611	0.0656	102.444	109.948	6.259	6.718	0.0098	0.1598	5.365	33.582	20
21	18.822	0.0531	0.0570	118.810	127.513	6.312	6.775	0.0084	0.1584	5.488	34.645	21
22	21.645	0.0462	0.0496	137.632	147.714	6.359	6.824	0.0073	0.1573	5.601	35.615	22
23	24.891	0.0402	0.0431	159.276	170.944	6.399	6.868	0.0063	0.1563	5.704	36.499	23
24	28.625	0.0349	0.0375	184.168	197.659	6.434	6.905	0.0054	0.1554	5.798	37.302	24
25	32.919	0.0304	0.0326	212.793	228.381	6.464	6.938	0.0047	0.1547	5.883	38.031	25
26	37.857	0.0264	0.0284	245.712	263.711	6.491	6.966	0.0041	0.1541	5.961	38.692	26
27	43.535	0.0230	0.0247	283.569	304.341	6.514	6.991	0.0035	0.1535	6.032	39.289	27
28	50.066	0.0200	0.0214	327.104	351.066	6.534	7.012	0.0031	0.1531	6.096	39.828	28
29	57.575	0.0174	0.0186	377.170	404.799	6.551	7.031	0.0027	0.1527	6.154	40.315	29
30	66.212	0.0151	0.0162	434.745	466.592	6.566	7.047	0.0023	0.1523	6.207	40.753	30
31	76.144	0.0131	0.0141	500.957	537.654	6.579	7.061	0.0020	0.1520	6.254	41.147	31
32	87.565	0.0114	0.0123	577.100	619.375	6.591	7.073	0.0017	0.1517	6.297	41.501	32
33	100.700	0.0099	0.0107	664.666	713.355	6.600	7.084	0.0015	0.1515	6.336	41.818	33
34	115.805	0.0086	0.0093	765.365	821.431	6.609	7.093	0.0013	0.1513	6.371	42.103	34
35	133.176	0.0075	0.0081	881.170	945.719	6.617	7.101	0.0011	0.1511	6.402	42.359	35
40	267.864	0.0037	0.0040	1779.090	1909.415	6.642	7.128	0.0006	0.1506	6.517	43.283	40
45	538.769	0.0019	0.0020	3585.128	3847.752	6.654	7.142	0.0003	0.1503	6.583	43.805	45
50	1083.657	0.0009	0.0010	7217.716	7746.440	6.661	7.148	0.0001	0.1501	6.620	44.096	50
55	2179.622	0.0005	0.0005	14524.148	15588.093	6.664	7.152	0.0001	0.1501	6.641	44.256	55
60	4383.999	0.0002	0.0002	29219.992	31360.460	6.665	7.153	0.0000	0.1500	6.653	44.343	60
65	8817.787	0.0001	0.0001	58778.583	63084.322	6.666	7.154	0.0000	0.1500	6.659	44.390	65

COMPOUND INTEREST FACTORS $i = 0.20$

	Single	Payment		Uniform	Series			Uniform	Series	Gradient	Series	
	Compound Amount	Present Worth		Compound Amount		Present Worth		Sinking Fund	Capital Recovery	Uniform Series	Present Worth	
N	F/P	P/F	P/F	F/A	F/A	P/A	P/A	A/F	A/P	A/G	P/G	N
1	1.200	0.8333	0.9141	1.000	1.097	0.833	0.914	1.0000	1.2000	0.000	0.000	1
2	1.440	0.6944	0.7618	2.200	2.413	1.528	1.676	0.4545	0.6545	0.455	0.694	2
3	1.728	0.5787	0.6348	3.640	3.993	2.106	2.311	0.2747	0.4747	0.879	1.852	3
4	2.074	0.4823	0.5290	5.368	5.888	2.589	2.840	0.1863	0.3863	1.274	3.299	4
5	2.488	0.4019	0.4408	7.442	8.163	2.991	3.281	0.1344	0.3344	1.641	4.906	5
6	2.986	0.3349	0.3674	9.930	10.893	3.326	3.648	0.1007	0.3007	1.979	6.581	6
7	3.583	0.2791	0.3061	12.916	14.168	3.605	3.954	0.0774	0.2774	2.290	8.255	7
8	4.300	0.2326	0.2551	16.499	18.099	3.837	4.209	0.0606	0.2606	2.576	9.883	8
9	5.160	0.1938	0.2126	20.799	22.816	4.031	4.422	0.0481	0.2481	2.836	11.434	9
10	6.192	0.1615	0.1772	25.959	28.476	4.192	4.599	0.0385	0.2385	3.074	12.887	10
11	7.430	0.1346	0.1476	32.150	35.268	4.327	4.747	0.0311	0.2311	3.289	14.233	11
12	8.916	0.1122	0.1230	39.581	43.418	4.439	4.870	0.0253	0.2253	3.484	15.467	12
13	10.699	0.0935	0.1025	48.497	53.199	4.533	4.972	0.0206	0.2206	3.660	16.588	13
14	12.839	0.0779	0.0854	59.196	64.936	4.611	5.058	0.0169	0.2169	3.817	17.601	14
15	15.407	0.0649	0.0712	72.035	79.020	4.675	5.129	0.0139	0.2139	3.959	18.509	15
16	18.488	0.0541	0.0593	87.442	95.921	4.730	5.188	0.0114	0.2114	4.085	19.321	16
17	22.186	0.0451	0.0494	105.931	116.202	4.775	5.238	0.0094	0.2094	4.198	20.042	17
18	26.623	0.0376	0.0412	128.117	140.539	4.812	5.279	0.0078	0.2078	4.298	20.680	18
19	31.948	0.0313	0.0343	154.740	169.744	4.843	5.313	0.0065	0.2065	4.386	21.244	19
20	38.338	0.0261	0.0286	186.688	204.790	4.870	5.342	0.0054	0.2054	4.464	21.739	20
21	46.005	0.0217	0.0238	225.026	246.845	4.891	5.366	0.0044	0.2044	4.533	22.174	21
22	55.206	0.0181	0.0199	271.031	297.311	4.909	5.385	0.0037	0.2037	4.594	22.555	22
23	66.247	0.0151	0.0166	326.237	357.870	4.925	5.402	0.0031	0.2031	4.647	22.887	23
24	79.497	0.0126	0.0138	392.484	430.541	4.937	5.416	0.0025	0.2025	4.694	23.176	24
25	95.396	0.0105	0.0115	471.981	517.746	4.948	5.427	0.0021	0.2021	4.735	23.428	25
26	114.475	0.0087	0.0096	567.377	622.392	4.956	5.437	0.0018	0.2018	4.771	23.646	26
27	137.371	0.0073	0.0080	681.853	747.967	4.964	5.445	0.0015	0.2015	4.802	23.835	27
28	164.845	0.0061	0.0067	819.223	898.658	4.970	5.452	0.0012	0.2012	4.829	23.999	28
29	197.814	0.0051	0.0055	984.068	1079.486	4.975	5.457	0.0010	0.2010	4.853	24.141	29
30	237.376	0.0042	0.0046	1181.882	1296.480	4.979	5.462	0.0008	0.2008	4.873	24.263	30
31	284.852	0.0035	0.0039	1419.258	1556.873	4.982	5.466	0.0007	0.2007	4.891	24.368	31
32	341.822	0.0029	0.0032	1704.109	1869.345	4.985	5.469	0.0006	0.2006	4.906	24.459	32
33	410.186	0.0024	0.0027	2045.931	2244.311	4.988	5.471	0.0005	0.2005	4.919	24.537	33
34	492.224	0.0020	0.0022	2456.118	2694.270	4.990	5.474	0.0004	0.2004	4.931	24.604	34
35	590.668	0.0017	0.0019	2948.341	3234.221	4.992	5.476	0.0003	0.2003	4.941	24.661	35
40	1469.772	0.0007	0.0007	7343.858	8055.940	4.997	5.481	0.0001	0.2001	4.973	24.847	40
45	3657.262	0.0003	0.0003	18281.310	20053.920	4.999	5.483	0.0001	0.2001	4.988	24.932	45
50	9100.438	0.0001	0.0001	45497.191	49908.734	4.999	5.484	0.0000	0.2000	4.995	24.970	50

COMPOUND INTEREST FACTORS   *i* = 0.25

	Single Payment			Uniform Series				Uniform Series		Gradient Series		
	Compound Amount	Present Worth		Compound Amount		Present Worth		Sinking Fund	Capital Recovery	Uniform Series	Present Worth	
N	F/P	P/F	P/F	F/A	F/A	P/A	P/A	A/F	A/P	A/G	P/G	N
1	1.250	0.8000	0.8963	1.000	1.120	0.800	0.896	1.0000	1.2500	0.000	0.000	1
2	1.563	0.6400	0.7170	2.250	2.521	1.440	1.613	0.4444	0.6944	0.444	0.640	2
3	1.953	0.5120	0.5736	3.813	4.271	1.952	2.187	0.2623	0.5123	0.852	1.664	3
4	2.441	0.4096	0.4589	5.766	6.460	2.362	2.646	0.1734	0.4234	1.225	2.893	4
5	3.052	0.3277	0.3671	8.207	9.195	2.689	3.013	0.1218	0.3718	1.563	4.204	5
6	3.815	0.2621	0.2937	11.259	12.614	2.951	3.307	0.0888	0.3388	1.868	5.514	6
7	4.768	0.2097	0.2350	15.073	16.888	3.161	3.542	0.0663	0.3163	2.142	6.773	7
8	5.960	0.1678	0.1880	19.842	22.230	3.329	3.730	0.0504	0.3004	2.387	7.947	8
9	7.451	0.1342	0.1504	25.802	28.908	3.463	3.880	0.0388	0.2888	2.605	9.021	9
10	9.313	0.1074	0.1203	33.253	37.255	3.571	4.000	0.0301	0.2801	2.797	9.987	10
11	11.642	0.0859	0.0962	42.566	47.689	3.656	4.096	0.0235	0.2735	2.966	10.846	11
12	14.552	0.0687	0.0770	54.208	60.732	3.725	4.173	0.0184	0.2684	3.115	11.602	12
13	18.190	0.0550	0.0616	68.760	77.035	3.780	4.235	0.0145	0.2645	3.244	12.262	13
14	22.737	0.0440	0.0493	86.949	97.414	3.824	4.284	0.0115	0.2615	3.356	12.833	14
15	28.422	0.0352	0.0394	109.687	122.888	3.859	4.324	0.0091	0.2591	3.453	13.326	15
16	35.527	0.0281	0.0315	138.109	154.731	3.887	4.355	0.0072	0.2572	3.537	13.748	16
17	44.409	0.0225	0.0252	173.636	194.534	3.910	4.381	0.0058	0.2558	3.608	14.108	17
18	55.511	0.0180	0.0202	218.045	244.287	3.928	4.401	0.0046	0.2546	3.670	14.415	18
19	69.389	0.0144	0.0161	273.556	306.480	3.942	4.417	0.0037	0.2537	3.722	14.674	19
20	86.736	0.0115	0.0129	342.945	384.220	3.954	4.430	0.0029	0.2529	3.767	14.893	20
21	108.420	0.0092	0.0103	429.681	481.395	3.963	4.440	0.0023	0.2523	3.805	15.078	21
22	135.525	0.0074	0.0083	538.101	602.864	3.970	4.448	0.0019	0.2519	3.836	15.233	22
23	169.407	0.0059	0.0066	673.626	754.701	3.976	4.455	0.0015	0.2515	3.863	15.362	23
24	211.758	0.0047	0.0053	843.033	944.496	3.981	4.460	0.0012	0.2512	3.886	15.471	24
25	264.698	0.0038	0.0042	1054.791	1181.741	3.985	4.464	0.0009	0.2509	3.905	15.562	25
26	330.872	0.0030	0.0034	1319.489	1478.296	3.988	4.468	0.0008	0.2508	3.921	15.637	26
27	413.590	0.0024	0.0027	1650.361	1848.990	3.990	4.471	0.0006	0.2506	3.935	15.700	27
28	516.988	0.0019	0.0022	2063.952	2312.358	3.992	4.473	0.0005	0.2505	3.946	15.752	28
29	646.235	0.0015	0.0017	2580.939	2891.568	3.994	4.474	0.0004	0.2504	3.955	15.796	29
30	807.794	0.0012	0.0014	3227.174	3615.581	3.995	4.476	0.0003	0.2503	3.963	15.832	30
31	1009.742	0.0010	0.0011	4034.968	4520.597	3.996	4.477	0.0002	0.2502	3.969	15.861	31
32	1262.177	0.0008	0.0009	5044.710	5651.866	3.997	4.478	0.0002	0.2502	3.975	15.886	32
33	1577.722	0.0006	0.0007	6306.887	7065.953	3.997	4.479	0.0002	0.2502	3.979	15.906	33
34	1972.152	0.0005	0.0006	7884.609	8833.561	3.998	4.479	0.0001	0.2501	3.983	15.923	34
35	2465.190	0.0004	0.0005	9856.761	11043.072	3.998	4.480	0.0001	0.2501	3.986	15.937	35
40	7523.164	0.0001	0.0001	30088.655	33709.976	3.999	4.481	0.0000	0.2500	3.995	15.977	40

COMPOUND INTEREST FACTORS $i = 0.30$

	Single	Payment		Uniform	Series			Uniform	Series	Gradient	Series	
	Compound Amount	Present	Worth	Compound	Amount	Present	Worth	Sinking Fund	Capital Recovery	Uniform Series	Present Worth	
N	F/P	P/F	P/F	F/A	F/A	P/A	P/A	A/F	A/P	A/G	P/G	N
1	1.300	0.7692	0.8796	1.000	1.143	0.769	0.880	1.0000	1.3000	0.000	0.000	1
2	1.690	0.5917	0.6766	2.300	2.630	1.361	1.556	0.4348	0.7348	0.435	0.592	2
3	2.197	0.4552	0.5205	3.990	4.562	1.816	2.077	0.2506	0.5506	0.827	1.502	3
4	2.856	0.3501	0.4004	6.187	7.075	2.166	2.477	0.1616	0.4616	1.178	2.552	4
5	3.713	0.2693	0.3080	9.043	10.340	2.436	2.785	0.1106	0.4106	1.490	3.630	5
6	4.827	0.2072	0.2369	12.756	14.586	2.643	3.022	0.0784	0.3784	1.765	4.666	6
7	6.275	0.1594	0.1822	17.583	20.105	2.802	3.204	0.0569	0.3569	2.006	5.622	7
8	8.157	0.1226	0.1402	23.858	27.280	2.925	3.344	0.0419	0.3419	2.216	6.480	8
9	10.604	0.0943	0.1078	32.015	36.607	3.019	3.452	0.0312	0.3312	2.396	7.234	9
10	13.786	0.0725	0.0829	42.619	48.733	3.092	3.535	0.0235	0.3235	2.551	7.887	10
11	17.922	0.0558	0.0638	56.405	64.497	3.147	3.599	0.0177	0.3177	2.683	8.445	11
12	23.298	0.0429	0.0491	74.327	84.989	3.190	3.648	0.0135	0.3135	2.795	8.917	12
13	30.288	0.0330	0.0378	97.625	111.629	3.223	3.686	0.0102	0.3102	2.889	9.314	13
14	39.374	0.0254	0.0290	127.913	146.261	3.249	3.715	0.0078	0.3078	2.969	9.644	14
15	51.186	0.0195	0.0223	167.286	191.283	3.268	3.737	0.0060	0.3060	3.034	9.917	15
16	66.542	0.0150	0.0172	218.472	249.812	3.283	3.754	0.0046	0.3046	3.089	10.143	16
17	86.504	0.0116	0.0132	285.014	325.899	3.295	3.767	0.0035	0.3035	3.135	10.328	17
18	112.455	0.0089	0.0102	371.518	424.812	3.304	3.778	0.0027	0.3027	3.172	10.479	18
19	146.192	0.0068	0.0078	483.973	553.399	3.311	3.785	0.0021	0.3021	3.202	10.602	19
20	190.050	0.0053	0.0060	630.165	720.562	3.316	3.791	0.0016	0.3016	3.228	10.702	20
21	247.065	0.0040	0.0046	820.215	937.874	3.320	3.796	0.0012	0.3012	3.248	10.783	21
22	321.184	0.0031	0.0036	1067.280	1220.379	3.323	3.800	0.0009	0.3009	3.265	10.848	22
23	417.539	0.0024	0.0027	1388.464	1587.636	3.325	3.802	0.0007	0.3007	3.278	10.901	23
24	542.801	0.0018	0.0021	1806.003	2065.071	3.327	3.804	0.0006	0.3006	3.289	10.943	24
25	705.641	0.0014	0.0016	2348.803	2685.735	3.329	3.806	0.0004	0.3004	3.298	10.977	25
26	917.333	0.0011	0.0012	3054.444	3492.600	3.330	3.807	0.0003	0.3003	3.305	11.005	26
27	1192.533	0.0008	0.0010	3971.778	4541.523	3.331	3.808	0.0003	0.3003	3.311	11.026	27
28	1550.293	0.0006	0.0007	5164.311	5905.123	3.331	3.809	0.0002	0.3002	3.315	11.044	28
29	2015.381	0.0005	0.0006	6714.604	7677.803	3.332	3.810	0.0001	0.3001	3.319	11.058	29
30	2619.996	0.0004	0.0004	8729.985	9982.288	3.332	3.810	0.0001	0.3001	3.322	11.069	30
31	3405.994	0.0003	0.0003	11349.981	12978.118	3.332	3.810	0.0001	0.3001	3.324	11.078	31
32	4427.793	0.0002	0.0003	14755.975	16872.697	3.333	3.811	0.0001	0.3001	3.326	11.085	32
33	5756.130	0.0002	0.0002	19183.768	21935.649	3.333	3.811	0.0001	0.3001	3.328	11.090	33
34	7482.970	0.0001	0.0002	24939.899	28517.487	3.333	3.811	0.0000	0.3000	3.329	11.094	34
35	9727.860	0.0001	0.0001	32422.868	37073.877	3.333	3.811	0.0000	0.3000	3.330	11.098	35

COMPOUND INTEREST FACTORS $i = 0.35$

	Single	Payment		Uniform	Series			Uniform	Series	Gradient	Series	
	Compound Amount	Present	Worth	Compound	Amount	Present	Worth	Sinking Fund	Capital Recovery	Uniform Series	Present Worth	
N	F/P	P/F	P/F	F/A	F/A	P/A	P/A	A/F	A/P	A/G	P/G	N
1	1.350	0.7407	0.8639	1.000	1.166	0.741	0.864	1.0000	1.3500	0.000	0.000	1
2	1.823	0.5487	0.6399	2.350	2.741	1.289	1.504	0.4255	0.7755	0.426	0.549	2
3	2.460	0.4064	0.4740	4.173	4.866	1.696	1.978	0.2397	0.5897	0.803	1.362	3
4	3.322	0.3011	0.3511	6.633	7.736	1.997	2.329	0.1508	0.5008	1.134	2.265	4
5	4.484	0.2230	0.2601	9.954	11.609	2.220	2.589	0.1005	0.4505	1.422	3.157	5
6	6.053	0.1652	0.1927	14.438	16.839	2.385	2.782	0.0693	0.4193	1.670	3.983	6
7	8.172	0.1224	0.1427	20.492	23.899	2.508	2.924	0.0488	0.3988	1.881	4.717	7
8	11.032	0.0906	0.1057	28.664	33.430	2.598	3.030	0.0349	0.3849	2.060	5.352	8
9	14.894	0.0671	0.0783	39.696	46.296	2.665	3.108	0.0252	0.3752	2.209	5.889	9
10	20.107	0.0497	0.0580	54.590	63.666	2.715	3.166	0.0183	0.3683	2.334	6.336	10
11	27.144	0.0368	0.0430	74.697	87.116	2.752	3.209	0.0134	0.3634	2.436	6.705	11
12	36.644	0.0273	0.0318	101.841	118.773	2.779	3.241	0.0098	0.3598	2.520	7.005	12
13	49.470	0.0202	0.0236	138.485	161.509	2.799	3.265	0.0072	0.3572	2.589	7.247	13
14	66.784	0.0150	0.0175	187.954	219.204	2.814	3.282	0.0053	0.3553	2.644	7.442	14
15	90.158	0.0111	0.0129	254.738	297.091	2.825	3.295	0.0039	0.3539	2.689	7.597	15
16	121.714	0.0082	0.0096	344.897	402.240	2.834	3.305	0.0029	0.3529	2.725	7.721	16
17	164.314	0.0061	0.0071	466.611	544.190	2.840	3.312	0.0021	0.3521	2.753	7.818	17
18	221.824	0.0045	0.0053	630.925	735.822	2.844	3.317	0.0016	0.3516	2.776	7.895	18
19	299.462	0.0033	0.0039	852.748	994.526	2.848	3.321	0.0012	0.3512	2.793	7.955	19
20	404.274	0.0025	0.0029	1152.210	1343.777	2.850	3.324	0.0009	0.3509	2.808	8.002	20
21	545.769	0.0018	0.0021	1556.484	1815.265	2.852	3.326	0.0006	0.3506	2.819	8.038	21
22	736.789	0.0014	0.0016	2102.253	2451.774	2.853	3.328	0.0005	0.3505	2.827	8.067	22
23	994.665	0.0010	0.0012	2839.042	3311.061	2.854	3.329	0.0004	0.3504	2.834	8.089	23
24	1342.797	0.0007	0.0009	3833.706	4471.099	2.855	3.330	0.0003	0.3503	2.839	8.106	24
25	1812.776	0.0006	0.0006	5176.504	6037.150	2.856	3.330	0.0002	0.3502	2.843	8.119	25
26	2447.248	0.0004	0.0005	6989.280	8151.318	2.856	3.331	0.0001	0.3501	2.847	8.130	26
27	3303.785	0.0003	0.0004	9436.528	11005.446	2.856	3.331	0.0001	0.3501	2.849	8.137	27
28	4460.109	0.0002	0.0003	12740.313	14858.518	2.857	3.331	0.0001	0.3501	2.851	8.143	28
29	6021.148	0.0002	0.0002	17200.422	20060.165	2.857	3.332	0.0001	0.3501	2.852	8.148	29
30	8128.550	0.0001	0.0001	23221.570	27082.390	2.857	3.332	0.0000	0.3500	2.853	8.152	30

COMPOUND INTEREST FACTORS $i = 0.40$

| | Single Payment | | | Uniform Series | | | | Uniform Series | | Gradient Series | | |
| | Compound Amount | Present Worth | | Compound Amount | | Present Worth | | Sinking Fund | Capital Recovery | Uniform Series | Present Worth | |
N	F/P	P/F	P/F	F/A	F/A	P/A	P/A	A/F	A/P	A/G	P/G	N
1	1.400	0.7143	0.8491	1.000	1.189	0.714	0.849	1.0000	1.4000	0.000	0.000	1
2	1.960	0.5102	0.6065	2.400	2.853	1.224	1.456	0.4167	0.8167	0.417	0.510	2
3	2.744	0.3644	0.4332	4.360	5.183	1.589	1.889	0.2294	0.6294	0.780	1.239	3
4	3.842	0.2603	0.3095	7.104	8.445	1.849	2.198	0.1408	0.5408	1.092	2.020	4
5	5.378	0.1859	0.2210	10.946	13.012	2.035	2.419	0.0914	0.4914	1.358	2.764	5
6	7.530	0.1328	0.1579	16.324	19.406	2.168	2.577	0.0613	0.4613	1.581	3.428	6
7	10.541	0.0949	0.1128	23.853	28.357	2.263	2.690	0.0419	0.4419	1.766	3.997	7
8	14.758	0.0678	0.0806	34.395	40.889	2.331	2.771	0.0291	0.4291	1.919	4.471	8
9	20.661	0.0484	0.0575	49.153	58.433	2.379	2.828	0.0203	0.4203	2.042	4.858	9
10	28.925	0.0346	0.0411	69.814	82.995	2.414	2.869	0.0143	0.4143	2.142	5.170	10
11	40.496	0.0247	0.0294	98.739	117.382	2.438	2.899	0.0101	0.4101	2.221	5.417	11
12	56.694	0.0176	0.0210	139.235	165.523	2.456	2.920	0.0072	0.4072	2.285	5.611	12
13	79.371	0.0126	0.0150	195.929	232.921	2.469	2.935	0.0051	0.4051	2.334	5.762	13
14	111.120	0.0090	0.0107	275.300	327.278	2.478	2.945	0.0036	0.4036	2.373	5.879	14
15	155.568	0.0064	0.0076	386.420	459.378	2.484	2.953	0.0026	0.4026	2.403	5.969	15
16	217.795	0.0046	0.0055	541.988	644.319	2.489	2.958	0.0018	0.4018	2.426	6.038	16
17	304.913	0.0033	0.0039	759.784	903.235	2.492	2.962	0.0013	0.4013	2.444	6.090	17
18	426.879	0.0023	0.0028	1064.697	1265.718	2.494	2.965	0.0009	0.4009	2.458	6.130	18
19	597.630	0.0017	0.0020	1491.576	1773.194	2.496	2.967	0.0007	0.4007	2.468	6.160	19
20	836.683	0.0012	0.0014	2089.206	2483.660	2.497	2.968	0.0005	0.4005	2.476	6.183	20
21	1171.356	0.0009	0.0010	2925.889	3478.312	2.498	2.969	0.0003	0.4003	2.482	6.200	21
22	1639.898	0.0006	0.0007	4097.245	4870.826	2.498	2.970	0.0002	0.4002	2.487	6.213	22
23	2295.857	0.0004	0.0005	5737.142	6820.346	2.499	2.971	0.0002	0.4002	2.490	6.222	23
24	3214.200	0.0003	0.0004	8032.999	9549.673	2.499	2.971	0.0001	0.4001	2.493	6.229	24
25	4499.880	0.0002	0.0003	11247.199	13370.730	2.499	2.971	0.0001	0.4001	2.494	6.235	25
26	6299.831	0.0002	0.0002	15747.079	18720.211	2.500	2.972	0.0001	0.4001	2.496	6.239	26
27	8819.764	0.0001	0.0001	22046.910	26209.485	2.500	2.972	0.0000	0.4000	2.497	6.242	27

COMPOUND INTEREST FACTORS $i = 0.45$

	Single Payment			Uniform Series				Uniform Series		Gradient Series		
	Compound Amount	Present Worth		Compound Amount		Present Worth		Sinking Fund	Capital Recovery	Uniform Series	Present Worth	
N	F/P	P/F	P/F	F/A	F/A	P/A	P/A	A/F	A/P	A/G	P/G	N
1	1.450	0.6897	0.8352	1.000	1.211	0.690	0.835	1.0000	1.4500	0.000	0.000	1
2	2.103	0.4756	0.5760	2.450	2.967	1.165	1.411	0.4082	0.8582	0.408	0.476	2
3	3.049	0.3280	0.3973	4.553	5.514	1.493	1.809	0.2197	0.6697	0.758	1.132	3
4	4.421	0.2262	0.2740	7.601	9.206	1.720	2.083	0.1316	0.5816	1.053	1.810	4
5	6.410	0.1560	0.1889	12.022	14.559	1.876	2.271	0.0832	0.5332	1.298	2.434	5
6	9.294	0.1076	0.1303	18.431	22.322	1.983	2.402	0.0543	0.5043	1.499	2.972	6
7	13.476	0.0742	0.0899	27.725	33.578	2.057	2.492	0.0361	0.4861	1.661	3.418	7
8	19.541	0.0512	0.0620	41.202	49.900	2.109	2.554	0.0243	0.4743	1.791	3.776	8
9	28.334	0.0353	0.0427	60.743	73.566	2.144	2.596	0.0165	0.4665	1.893	4.058	9
10	41.085	0.0243	0.0295	89.077	107.881	2.168	2.626	0.0112	0.4612	1.973	4.277	10
11	59.573	0.0168	0.0203	130.162	157.639	2.185	2.646	0.0077	0.4577	2.034	4.445	11
12	86.381	0.0116	0.0140	189.735	229.787	2.196	2.660	0.0053	0.4553	2.082	4.572	12
13	125.252	0.0080	0.0097	276.115	334.403	2.204	2.670	0.0036	0.4536	2.118	4.668	13
14	181.615	0.0055	0.0067	401.367	486.095	2.210	2.677	0.0025	0.4525	2.145	4.740	14
15	263.342	0.0038	0.0046	582.982	706.049	2.214	2.681	0.0017	0.4517	2.165	4.793	15
16	381.846	0.0026	0.0032	846.324	1024.982	2.216	2.684	0.0012	0.4512	2.180	4.832	16
17	553.676	0.0018	0.0022	1228.170	1487.434	2.218	2.686	0.0008	0.4508	2.191	4.861	17
18	802.831	0.0012	0.0015	1781.846	2157.991	2.219	2.688	0.0006	0.4506	2.200	4.882	18
19	1164.105	0.0009	0.0010	2584.677	3130.298	2.220	2.689	0.0004	0.4504	2.206	4.898	19
20	1687.952	0.0006	0.0007	3748.782	4540.143	2.221	2.690	0.0003	0.4503	2.210	4.909	20
21	2447.530	0.0004	0.0005	5436.734	6584.419	2.221	2.690	0.0002	0.4502	2.214	4.917	21
22	3548.919	0.0003	0.0003	7884.264	9548.619	2.222	2.691	0.0001	0.4501	2.216	4.923	22
23	5145.932	0.0002	0.0002	11433.182	13846.708	2.222	2.691	0.0001	0.4501	2.218	4.927	23
24	7461.602	0.0001	0.0002	16579.115	20078.938	2.222	2.691	0.0001	0.4501	2.219	4.930	24

COMPOUND INTEREST FACTORS   *i* =   0.50

	Single Payment			Uniform Series				Uniform Series		Gradient Series		
	Compound Amount	Present Worth		Compound Amount		Present Worth		Sinking Fund	Capital Recovery	Uniform Series	Present Worth	
N	F/P	P/F	P/F	F/A	F/A	P/A	P/A	A/F	A/P	A/G	P/G	N
1	1.500	0.6667	0.8221	1.000	1.233	0.667	0.822	1.0000	1.5000	0.000	0.000	1
2	2.250	0.4444	0.5481	2.500	3.083	1.111	1.370	0.4000	0.9000	0.400	0.444	2
3	3.375	0.2963	0.3654	4.750	5.857	1.407	1.736	0.2105	0.7105	0.737	1.037	3
4	5.063	0.1975	0.2436	8.125	10.019	1.605	1.979	0.1231	0.6231	1.015	1.630	4
5	7.594	0.1317	0.1624	13.188	16.262	1.737	2.142	0.0758	0.5758	1.242	2.156	5
6	11.391	0.0878	0.1083	20.781	25.626	1.824	2.250	0.0481	0.5481	1.423	2.595	6
7	17.086	0.0585	0.0722	32.172	39.673	1.883	2.322	0.0311	0.5311	1.565	2.947	7
8	25.629	0.0390	0.0481	49.258	60.742	1.922	2.370	0.0203	0.5203	1.675	3.220	8
9	38.443	0.0260	0.0321	74.887	92.347	1.948	2.402	0.0134	0.5134	1.760	3.428	9
10	57.665	0.0173	0.0214	113.330	139.753	1.965	2.424	0.0088	0.5088	1.824	3.584	10
11	86.498	0.0116	0.0143	170.995	210.863	1.977	2.438	0.0058	0.5058	1.871	3.699	11
12	129.746	0.0077	0.0095	257.493	317.528	1.985	2.447	0.0039	0.5039	1.907	3.784	12
13	194.620	0.0051	0.0063	387.239	477.524	1.990	2.454	0.0026	0.5026	1.933	3.846	13
14	291.929	0.0034	0.0042	581.859	717.520	1.993	2.458	0.0017	0.5017	1.952	3.890	14
15	437.894	0.0023	0.0028	873.788	1077.513	1.995	2.461	0.0011	0.5011	1.966	3.922	15
16	656.841	0.0015	0.0019	1311.682	1617.503	1.997	2.463	0.0008	0.5008	1.976	3.945	16
17	985.261	0.0010	0.0013	1968.523	2427.487	1.998	2.464	0.0005	0.5005	1.983	3.961	17
18	1477.892	0.0007	0.0008	2953.784	3642.464	1.999	2.465	0.0003	0.5003	1.988	3.973	18
19	2216.838	0.0005	0.0006	4431.676	5464.928	1.999	2.465	0.0002	0.5002	1.991	3.981	19
20	3325.257	0.0003	0.0004	6648.513	8198.626	1.999	2.466	0.0002	0.5002	1.994	3.987	20
21	4987.885	0.0002	0.0002	9973.770	12299.172	2.000	2.466	0.0001	0.5001	1.996	3.991	21
22	7481.828	0.0001	0.0002	14961.655	18449.991	2.000	2.466	0.0001	0.5001	1.997	3.994	22

Bibliography

GENERAL REFERENCES

The American Society for Engineering Education, and Institute of Industrial Engineers. *The Engineering Economist*. Norcross, GA: Institute of Industrial Engineers. Published quarterly by the Engineering Economy Division of the American Society for Engineering Education and the Institute of Industrial Engineers.

American Telephone and Telegraph, Engineering Department. *Engineering Economy*. 3rd ed. New York: American Telephone and Telegraph, 1980.

Au, Tung, and Thomas P. Au. *Engineering Economics for Capital Investment Analysis*. 2nd ed. Englewood Cliffs, NJ: Prentice-Hall, 1992.

Bierman, H., and S. Smidt. *The Capital Budgeting Decision*. 6th ed. New York: Macmillan, 1984.

Brown, R. J., and R. R. Yanuck. *Introduction to Life Cycle Costing*. Englewood Cliffs, NJ: Prentice-Hall, 1985.

Bussey, Lynn E., and T. E. Eschenbach. *The Economic Analysis of Industrial Projects*. Englewood Cliffs, NJ: Prentice-Hall, 1992.

Canada, John R., and John A. White, Jr. *Capital Investment Decision Analysis for Management and Engineering*. Englewood Cliffs, NJ: Prentice-Hall, 1980.

Canada, J. R., and W. T. Sullivan. *Economic and Multiattribute Evaluation of Advanced Manufacturing Systems*. Englewood Cliffs, NJ: Prentice-Hall, 1989.

Clark, F. D., and A. B. Lorenzoni. *Applied Cost Engineering*. 2nd ed. New York: Marcel Dekker, 1985.

Collier, C. A., and W. B. Ledbetter. *Engineering Cost Analysis*. 2nd ed. New York: Harper & Row, 1987.

De La Mare, R. F. *Manufacturing Systems Economics*. East Sussex, England: Holt-Rinehart and Winston, 1982.

English, J. Morley. *Project Evaluation*. New York: Macmillan, 1984.

Fabrycky, W. J., and B. S. Blanchard. *Life-Cycle Cost and Economic Analysis*. Englewood Cliffs, NJ: Prentice-Hall, 1991.

Fabrycky, Wolter, J., and Gerald J. Thuesen. *Economic Decision Analysis*. 2nd ed. Englewood Cliffs, NJ: Prentice-Hall, 1980.

Fish, J. C. L. *Engineering Economics*. 2nd ed. New York: McGraw-Hill, 1923. (This book is of historical importance; it was the first general engineering economy textbook.)

Fleischer, Gerald A. "Engineering Economy," in *Maynard's Industrial Engineering Handbook*. 4th ed. William K. Hodson, Editor-in-Chief. New York: McGraw-Hill, 1992, pp. 9.3–9.33.

Grant, Eugene L., W. Grant Ireson, and Richard S. Leavenworth. *Principles of Engineering Economy*. 8th ed. New York: John Wiley & Sons, 1990.

Halpin, D. W. *Financial and Cost Concepts for Construction Management.* New York: John Wiley & Sons, 1985.

Jelen, F. C., and J. H. Black. *Cost and Optimization Engineering.* 2nd ed. New York: McGraw-Hill, 1983.

Kleinfeld, Ira H. *Engineering and Managerial Economics.* New York: Holt, Rinehart and Winston, 1986.

Newnan, Donald G. *Engineering Economic Analysis.* 3d ed. San Jose, CA: Engineering Press, 1988.

Ostwald, P. F. *Cost Estimating.* 3rd ed. Englewood Cliffs, NJ: Prentice-Hall, 1992.

Park, Chan S. *Contemporary Engineering Economics.* New York: Addison-Wesley, 1993.

Park, Chan S., and Gunter P. Sharp-Bette. *Advanced Engineering Economics.* New York: John Wiley & Sons, 1990.

Peters, M. S., and K. D. Timmerhaus. *Plant Design and Economics for Chemical Engineers.* 4th ed. New York: McGraw-Hill, 1980.

Riggs, James L., and Tom M. West. *Engineering Economy.* 3rd ed. New York: McGraw-Hill, 1987.

Sepulveda, J. A., W. E. Souder, and B. S. Gottfried. *Theory and Problems of Engineering Economics.* Schaum's Outline Series. New York: John Wiley and Sons, 1984.

Smith, Gerald W. *Engineering Economy: The Analysis of Capital Expenditures.* 4th ed. Ames, IA: Iowa State University Press, 1987.

Sprague, J. C., and J. D. Whittaker. *Economic Analysis for Engineers and Managers.* Englewood-Cliffs, NJ: Prentice-Hall, 1986.

Steiner, H. M. *Public and Private Investments—Socioeconomic Analysis.* New York: John Wiley & Sons, 1980.

Stermole, F. J., and J. M. Stermole. *Economic Evaluation and Investment Decision Methods.* 6th ed. Golden, CO: Investment Evaluation Corp., 1987.

Stevens, G. T. *The Economic Analysis of Capital Investments for Engineers and Managers.* Reston, VA: Reston Publishing Co., 1983.

Stewart, R. D. *Cost Estimating.* New York: John Wiley & Sons, 1982.

Stewart, R. D., and R. M. Wyskida. *Cost Estimator's Reference Manual.* New York: John Wiley & Sons, 1987.

Tarquin, Anthony J., and Leland T. Blank. *Engineering Economy: A Behavioral Approach.* 2nd ed. New York: McGraw-Hill, 1983.

Taylor, George. *Managerial and Engineering Economy.* 3rd ed. New York: D. Van Nostrand, 1980.

Thuesen, G. J., and W. J. Fabrycky. *Engineering Economy.* 8th ed. Englewood Cliffs, NJ: Prentice-Hall, 1993.

Wellington, Arthur M. *The Economic Theory of Railway Location.* 2nd ed. New York: John Wiley & Sons, 1887. (This book is of historical importance; it was the first to address the issue of economic evaluation of capital investments due to engineering design decisions. Wellington is widely considered to be "the father of engineering economy.")

White, John A., Marvin H. Agee, and Kenneth E. Case. *Principles of Engineering Economic Analysis.* 3rd ed. New York: John Wiley & Sons, 1989.

∣ JOURNALS

Accounting Review
The Appraisal Journal
Chemical Engineering
Decision Science
The Engineering Economist
Financial Management
Harvard Business Review
IEEE Transactions on Engineering Mgmt
IIE Transactions
Industrial Engineering
Journal of Accountancy

Journal of Business
Journal of Finance
Journal of Financial & Quantitative Analysis
Journal of Industrial Economics
Journal of Manufacturing Systems
Journal of Taxation
Management Science
Plant Engineering
Power Engineering
Process Engineering
Public Utilities Fortnightly

Chapter 1   INTRODUCTION TO ENGINEERING ECONOMY: AN OVERVIEW

Cook, Thomas J., and Ronald J. Rizzuto. "Capital Budgeting Practices for R&D: A Survey and Analysis of Business Week's R&D Scoreboard," *The Engineering Economist*, Vol. 34, No. 4, Summer 1989, pp. 291–304. (Results of a survey of U.S. corporations regarding the extent of utilization of standard capital budgeting measures of investment worth to evaluate R&D projects.)

Cooper, Robin, and Robert S. Kaplan. "Measure Costs Right: Make the Right Decisions," *Harvard Business Review*, September–October 1988, pp. 96–103. (An early proposal for Activity Based Costing, which incorporates appropriate overhead costs into product costs.)

Dewhurst, P., and G. Boothroyd. "Early Cost Estimating in Product Design," *Journal of Manufacturing Systems*, Vol. 7, No. 3, 1988, pp. 183–192. (Describes product costing procedures which are intended to form the basis for a design analysis method for product design for efficient manufacture.)

Khan, Aman. "Capital Budgeting Practices in Large U.S. Cities," *The Engineering Economist*, Vol. 32, No. 2, Fall 1987, pp. 1–12. (Results from 107 U.S. cities with population ⩾ 50,000.)

Mills, Roger W. "Capital Budgeting — The State of the Art," *Long Range Planning,* Vol. 21, No. 4, August 1988, pp. 76–81. (Reviews 57 studies dating from 1976 to 1987 on capital budgeting practices of the largest companies in the United Kingdom.)

Mukherjee, Tarun K., and Glen V. Henderson. "The Capital Budgeting Process: Theory and Practice," *Interfaces*, Vol. 17, March–April 1987, pp. 78–90. (Contrasts capital budgeting practices with relevant theory.)

Park, Chan S., and Young K. Son. "An Economic Evaluation Model for Advanced Manufacturing Systems," *The Engineering Economist*, Vol. 34, No. 1, Fall 1988, pp. 1–26. (A multistage investment decision model that considers explicitly the nonconventional costs of quality, inventory, part waiting, idle equipment, and so forth.)

Pike, Richard H. "Do Sophisticated Capital Budgeting Approaches Improve Investment Decision-Making Effectiveness?" *The Engineering Economist*, Vol. 32, No. 2, Winter 1989, pp. 149–162. (Examines changes in selected capital budgeting practices over eleven years in 100 large U.S. corporations.)

Troxler, Joel W. "A Comprehensive Methodology for Manufacturing System Evaluation and Comparison," *Journal of Manufacturing Systems*, Vol. 8, No. 3, 1989, pp. 175–184.

Chapter 3   EQUIVALENT MEASURES OF WORTH

Beaves, Robert G. "Net Present Value and Rate of Return: Implicit and Explicit Reinvestment Assumptions," *The Engineering Economist*, Vol. 34, No. 4, Summer 1988, pp. 275–302.

Bernhard, Richard H. "A Comprehensive Comparison and Critique of Discounting Indices Proposed for Capital Investment Evaluation," *The Engineering Economist*, Vol. 16, No. 3, Spring 1971, pp. 157–186.

Bernhard, Richard H. "'Modified' Rates of Return for Investment Project Evaluation — A Comparison and Critique," *The Engineering Economist*, Vol. 24, No. 3, Spring 1979, pp. 161–167.

Charnes, A., W. W. Cooper, and M. H. Miller. "Application of Linear Programming to Financial Budgeting and the Costing of Funds," *Journal of Business*, Vol. 32, No. 1, January 1959, pp. 20–46.

Fleischer, Gerald A. "Two Major Issues Associated with the Rate of Return Method for Capital Allocation: The 'Ranking Error' and 'Preliminary Selection,'" *The Journal of Industrial Engineering*, Vol. 17, No. 4, April 1966, pp. 202–208.

Fleischer, Gerald A. "Discounting an Intraperiod Cash Flow," *The Engineering Economist*, Vol. 32, No. 1, Fall 1986, pp. 56–58.

Fleischer, Gerald A., and Lawrence C. Leung. "On

Future Worth and Its Relation to Present Worth as an Investment Criterion," *The Engineering Economist*, Vol. 35, No. 4, Summer 1990, pp. 323–332.

Freeland, J., and M. J. Rosenblatt. "An Analysis of Linear Programming Formulation for the Capital Rationing Problem," *The Engineering Economist*, Vol. 24, No. 4, Fall 1978, pp. 49–61.

Gordon, Lawrence A., and Michelle M. Hamer. "Rate of Return and Cash Flow Profiles: An Extension," *Accounting Review*, Vol. 63, No. 3, July 1988, pp. 514–521.

Kaplan, S. "A Note on a Method for Precisely Determining the Uniqueness or Nonuniqueness of the Internal Rate of Return for a Proposed Investment," *The Journal of Industrial Engineering*, Jan.–Feb. 1965, pp. 70–71.

Kim, Suk, and Henry Guithnes. *Capital Expenditure Analysis.* Washington DC: University Press, 1981.

Landau, H. J. "On Comparison of Cash Flow Streams," *Management Science*, Vol. 26, No. 12, December 1980, pp. 1218–1226.

Lorie, J. H., and L. J. Savage. "Three Problems in Rationing Capital," *Journal of Business*, Vol. 28, No. 4, October 1955, pp. 229–239.

Mahoney, J. F. "The Rule of 78s," *The Engineering Economist*, Vol. 35, No. 3, Spring 1990, pp. 231–238.

McMath, H. Kent. "Correction Constants for Present Values of Sub-Annual Cash Flows," *Decision Sciences*, Vol. 21, No. 4, Fall 1990, pp. 842–852.

Norstrom, C. J. "A Sufficient Condition for a Unique Nonnegative Internal Rate of Return," *Journal of Financial and Quantitative Analysis*, June 1972, pp. 1835–1839.

Swalm, Ralph O., and Jose L. Lopez-Leautaud. *Engineering Economic Analysis: A Future Wealth Approach.* New York: Wiley, 1984.

Unger, V. E., Jr. "Capital Budgeting and Mixed Zero-One Integer Programming," *AIIE Transactions*, Vol. 2, No. 1, March 1970, pp. 28–36.

Vernor, James D. "A Comparative Lease Analysis Using a Discounted Cash Flow Approach," *The Appraisal Journal*, Vol. 56, No. 3, July 1988, pp. 391–398.

Weingartner, H. Martin. *Mathematical Programming and the Analysis of Capital Budgeting Problems.* Englewood Cliffs, NJ: Prentice-Hall, 1963. (Also, Chicago, Markham Publishing, 1967.)

————."Capital Budgeting of Interrelated Projects: Survey and Synthesis," *Management Science*, Vol. 12, No. 7, March 1966, pp. 485–516.

Chapter 4   ECONOMIC ANALYSIS IN THE PUBLIC SECTOR

American Association of State Highway and Transportation Officials. *Manual on Road User Benefit Analysis of Highway and Bus-Transit Improvements.* Washington DC: American Association of State Highway and Transportation Officials, 1978.

American Association of State Highway Officials. *Road User Benefit Analysis for Highway Improvements.* Washington DC: American Association of State Highway Officials, 1960. (The "Red Book.")

Brown, Richard S., et al. *Economic Analysis Handbook.* Alexandria, VA: Naval Facilities Engineering Command, 1975, pp. 23–28.

Kendall, M. G., ed. *Cost-Benefit Analysis.* New York: American Elsevier, 1971.

Maas, Arthur. "Benefit-Cost Analysis: Its Relevance to Public Investment Decisions." In A. V. Kneese and S. C. Smith, *Water Research.* Baltimore, MD: The Johns Hopkins Press, 1966, pp. 311–328.

Oglesby, Clark H. *Highway Engineering*, 4th ed. New York: Wiley, 1973.

Prest, A. R., and R. Turvey. "Cost-Benefit Analysis: A Survey." In *Surveys of Economic Theory: Vol. 3, Resource Allocation,* prepared for the American Economic Association and the Royal Economic Society. New York: St. Martin's Press, 1966.

United Nations. *Guidelines for Project Evaluation*, Sales No. E.72-II.B.11. New York: United Nations, 1972.

U.S. Congress. Joint Economic Committee, Subcommittee on Priorities in Government,

Benefit-Cost Analyses of Federal Programs, A Compendium of Papers, 92nd Cong., 2nd Session, 1973.

U.S. Government. *The Analysis and Evaluation of Public Expenditures: The PPB System.* Washington DC: Government Printing Office, 1969.

U.S. Government. Interagency Committee on Water Resources, *Proposed Practices for Economic Analysis of River Basin Projects.* Washington DC: Government Printing Office, 1958. (The "Green Book.")

Winfrey, Robley. *Economic Analysis for Highways.* Scranton, PA: International Textbook, 1969.

Chapter 5   PAYBACK, RETURN ON INVESTMENT, AND OTHER MEASURES OF DESIRABILITY

Baldwin, R. H. "How to Assess Investment Proposals,"*Harvard Business Review*, Vol. 37, No. 3, May/June 1959, pp. 98–104.

Darvish, Tikva, and Shlomo Eckstein. "A Model for Simultaneous Sensitivity Analysis of Projects," *Applied Economics*, Vol. 20, No. 1, January 1988, pp. 113–124.

Ijiri, Yuji. "Approximations to Interest Formulas," *Journal of Business*, Vol. 45, No. 3, July 1972, pp. 398–402.

Longmore, Dean R. "The Persistence of the Payback Method: A Time-Adjusted Decision Rule Perspective," *The Engineering Economist*, Vol. 34,

No. 3, Spring 1989, pp. 185–194.

Ward, Thomas J. "Estimating Profitability Using Net Return Rate," *Chemical Engineering*, Vol. 96, No. 3, March 1989, pp. 151–155.

Weingartner, H. Martin. "The Excess Present Value Index — A Theoretical Basis and Critique," *The Journal of Accounting Research*, Vol. 1, No. 2, 1963, pp. 213–224.

Wortham, A. W., and R. J. McNichols. "Return Analysis on Equipment Payout," *International Journal of Production Research*, Vol. 7, No. 3, 1969, pp. 183–187.

Chapters 6 and 7   DEPRECIATION, TAXATION, AND AFTER-TAX ECONOMY STUDIES

Auerbach, Alan J., and James R. Hines, Jr. "Investment Tax Incentives and Frequent Tax Reforms," *American Economic Review*, Vol. 78, No. 2, May 1988, pp. 211–216.

Berg, Menachem, and Giora Moore. "The Choice of Depreciation Method Under Uncertainty," *Decision Sciences*, Vol. 20, No. 4, Fall 1989, p. 12.

Dammon, Robert M. "The Effect of Taxes and Depreciation on Corporate Investment and Financial Leverage," *Journal of Finance*, Vol. 43, No. 2, June 1988, p. 357.

DeMoville, Wig. "Impact of the Tax Reform Act of 1986 on Capital Budgeting," *Taxes*, Vol. 65, No. 7,

Nov. 1987, pp. 758–763.

Farid, Foad. "Composite Corporate Income Tax Rates," *The Engineering Economist*, Vol. 33, No. 2, Winter 1988, pp. 122–129.

Fleischer, G. A., A. K. Mason, and L. C. Leung. "Optimal Depreciation Under the Tax Reform Act of 1986," *IIE Transactions*, Vol. 22, No. 4, December 1990, pp. 330–339.

Frair, L. C., and M. D. Devine. "The Consideration of Depletion in After-Tax Optimization Models," *The Engineering Economist*, Vol. 24, No 1, Fall 1978, pp. 13–27.

Grant, E. L., and Paul T. Norton. *Depreciation.* New York: The Ronald Press, 1955, Revised printing.

Grasso, Albert L. "Depreciation Deductions for Short Tax Years," *Taxation for Accountants*, Vol. 43, November 1989, pp. 312–315.

Kalfayan, Garo. "How to Maximize First Year Depreciation Using the Section 179 Expense Election," *The Practical Accountant*, Vol. 23, May 1990, pp. 48–52.

Kaufman, Daniel J., and Lawrence J. Gitman. "The Tax Reform Act of 1986 and Corporate Investment Decisions," *The Engineering Economist*, Vol. 33, No. 2, Winter 1988, pp. 95–108.

Lasser, J. K. *Your Income Tax.* New York: Simon and Schuster. (See latest edition.)

Leung, Lawrence C., and Gerald A. Fleischer. "Depreciation and Tax Policies in Certain East Asian Countries," *Engineering Costs and Production Economics*, Vol. 16, No. 2, April 1989, p. 125.

Nadeau, Serge. "A Model to Measure the Effects of Taxes on the Real and Financial Decisions of the Firms," *National Tax Journal.* Vol. 41, No. 4, December 1988, p. 9.

Pechman, Joseph A. *Comparative Tax Systems: Europe, Canada and Japan.* The Brookings Institution, 1987.

Plostock, Mark A. "Automating Depreciation Calculations," *The CPA Journal*, Vol. 61, November 1991, pp. 18–20.

Prentice-Hall. *Cumulative Changes in Tax Regulations Under the 1986 Code.* Paramus, NJ: Prentice-Hall, 1987.

Prentice-Hall. *Complete Internal Revenue Code.* Paramus, NJ: Prentice-Hall, 1990.

Ronen, Joshua. "Depreciation Policies in Regulated Companies: Which Policies Are the Most Efficient?" *Management Science*, Vol. 35, No. 5, May 1989, pp. 515–526.

Schoomer, B. Alva. "Optimal Depreciation Strategies for Income Tax Purposes," *Management Science*, Vol. 1, No. 12, August 1966, pp. 552–580.

U.S. Congress. Joint Committee on Taxation. *General Explanation of the Tax Reform Act of 1986.* Washington, DC: U.S. Government Printing Office, 1987.

U.S. Department of the Treasury. *Your Federal Income Tax*, IRS Publication 17. Washington, DC: Government Printing Office. (Revised annually.)

————. *Tax Guide for Small Business*, IRS Publication 334. Washington, DC: Government Printing Office. (Revised annually.)

————. *Depreciation*, IRS Publication 534. Washington, DC: Government Printing Office. (See latest edition.)

————. *Investment Credit*, IRS Publication 572. Washington, DC: Government Printing Office. (See latest edition.)

Webster, Joseph A. "A Simplified Method for Deciding on Whether or Not to Make Fourth Quarter Asset Purchases," *Taxation for Accountants*, Vol. 43, October 1989, pp. 218–222.

| Chapter 8 RETIREMENT AND REPLACEMENT

Alchian, A. A. *Economic Replacement Policy.* Santa Monica, CA: The Rand Corporation, RM-2153, 1958.

Bean, James C., Jack L. Lohmann, and Robert L. Smith. "A Dynamic Infinite Horizon Replacement Economy Decision Model," *The Engineering Economist*, Vol. 30, No. 2, Winter 1985, pp. 99–120.

Bernhard, Richard H. "Improving the Economic Logic Underlying Replacement Age Decisions for Multiple Garbage Trucks: Case Study," *The Engineering Economist*, Vol. 35, No. 2, Winter 1990, pp. 129–147.

Dean, Burton V. "Replacement Theory," *Progress in Operation Research*, Vol. 1, ORSA Publication No. 5 (edited by R. L. Ackoff), 1963, Chapter 9.

Dreyfus, Stuart E. "A Generalized Equipment Replacement Study," *Journal of the Society for Industrial and Applied Mathematics*, Vol. 8, No. 3, September 1960, pp. 425–435.

Feldstein, Martin S., and Michael Rothschild. "Towards an Economic Theory of Replacement Investment," *Econometica*, Vol. 42, No. 3, May 1974, pp. 393–424.

Grinyer, Peter H. "The Effects of Technological Change on the Economic Life of Capital Equipment," *AIIE Transactions*, Vol. 5, No. 3, September 1973, pp. 203–213.

Hackamack, L. C. *Making Equipment Decisions.* American Management Association, 1969.

Hackamack, Lawrence C. "Making Realistic-Machine Replacement Decisions," *Manufacturing Engineering and Management*, Vol. 73, No. 5, November 1974, pp. 36–38.

Lake, D. H., and A. P. Muhlemann. "An Equipment Replacement Problem," *Journal of the Operational Research Society*, Vol. 30, No. 5, 1979, pp. 405–411.

Leung, Lawrence C., and J. M. A. Tanchoco. "Multiple Machine Replacement Within an Integrated System Framework," *The Engineering Economist*, Vol. 32, No. 2, Winter 1987, pp. 89–114.

Matsuo, Haroshi. "A Modified Approach to the Replacement of an Existing Asset," *The Engineering Economist*, Vol. 33, No. 2, Winter 1988, pp. 109–120.

Morris, William T. "Replacement Policy," Chapter 9 in *Engineering Economic Analysis*. Reston Publishing Co., 1976.

Oakford, R. V., J. R. Lohmann, and A. Salazar. "A Dynamic Replacement Economy Decision Model," *IIE Transactions*, Vol. 16, 1984, pp. 65–72.

Terborgh, George. *Dynamic Equipment Policy.* Washington, DC: Machinery and Allied Products Institute, 1949.

————. *Realistic Depreciation Policy.* Washington, DC: Machinery and Allied Products Institute, 1954.

————. *Business Investment Policy.* Washington, DC: Machinery and Allied Products Institute, 1958.

————. *Business Investment Management.* Washington, DC: Machinery and Allied Products Institute, 1967.

Chapter 9   THE REVENUE RETIREMENT METHOD

Brigham, Eugene F., Louis C. Gapenski, and Dana A. Aberwald. "Capital Structure, Cost of Capital, and Revenue Requirements," *Public Utilities Fortnightly*, January 1987, pp. 15–24.

Bronner, Kevin M. "Costs and Benefits with Public and Investor-Owned Electric Systems," *Public Utilities Fortnightly*, Vol. 125, No. 7, March 29, 1990, pp. 25–27.

Commonwealth Edison Company. *Engineering Economics.* Chicago, IL: Commonwealth Edison Company, 1975.

Gallinger, George W., and Glenn V. Henderson, Jr. "Public Utility Cost Models: An Examination of Assumptions," *The Engineering Economist*, Vol. 34, No. 3, Spring 1989, pp. 177–184.

Mayer, R. R. "Finding Your Minimum Revenue Requirements," *Industrial Engineering*, Vol. 9, No. 4, April 1977, pp. 16–22.

Stevens, G. T., Jr. "Revenue Requirement Analysis with Working Capital Changes: A Tutorial," *The Engineering Economist*, Vol. 32, No. 1, Fall 1986, pp. 1–16.

Stoll, H. G. *Least Cost Electric Utility Planning.* New York: Wiley, 1987.

Ward, T. L., and W. G. Sullivan. "Equivalence of the Present Worth and Revenue Requirement Methods of Capital Investment Analysis," *AIIE Transactions*, Vol. 13, No. 1, March 1981, pp. 29–40.

Chapter 10   RISK AND UNCERTAINTY

Bennion, Edward G. "Capital Budgeting and Game Theory," *Harvard Business Review*, Vol. 34, No. 6, November/December 1956, pp. 115–123.

Buck, James R. *Economic Risk Decisions in Engineering and Management*, Iowa State University Press, Ames, 1989.

Estes, J. H., W. C. Moor, and D. A. Rollier. "Stochastic Cash Flow Evaluation Under Conditions of Uncertain Timing," *Engineering Costs and Production Economics*, Vol. 18, 1989, pp. 65–70.

Farrar, Donald. *The Investment Decision Under Uncertainty*. Englewood Cliffs, NJ: Prentice-Hall, 1962.

Fishburn, Peter C. *Decision and Value Theory*. New York: John Wiley & Sons, 1964.

Fleischer, G. A., Ed. *Risk and Uncertainty: Non-Deterministic Decision Making in Engineering Economy*, Monograph Series No. 2, American Institute of Industrial Engineers, Norcross, GA, 1975.

Fleischer, Gerald A. "Economic Risk Analysis," Ch. 52 in *Handbook of Industrial Engineering*, Gavriel Salvendy, editor, Wiley, 1992, pp. 1343–1375. Published in cooperation with the Institute of Industrial Engineers, Norcross, GA.

Hertz, David B. "Risk Analysis in Capital Investment," *Harvard Business Review*, Vol. 42, No. 1, January/February 1964, pp. 89–106.

————. "Investment Policies That Pay Off," *Harvard Business Review*, Vol. 42, No. 1, January/February 1968, pp. 96–108.

Hertz, David B., and H. Thomas. *Risk Analysis and Its Application*. New York: Wiley, 1983.

Hillier, F. S. "The Derivation of Probabilistic Information for Evaluation of Risky Investments," *Management Science*, Vol. 9, No. 3, April 1963, pp. 443–457.

————. Supplement to "The Derivation of Probabilistic Information for the Evaluation of Risky Investments," *Management Science*, Vol. 11, No. 3, January 1965, pp. 485–487.

————. *The Evaluation of Risky Interrelated Investments*. Amsterdam: North Holland Publishing, 1969.

Howard, R. A. "Decision Analysis: Practice and Promise," *Management Science*, Vol. 34, No. 6, June 1988, pp. 679–695.

Lin, Edward Y. H., and James R. Buck. "Measuring Conditional Partial Expected Loss Under Uniformly Distributed Uncertain Timing," *The Engineering Economist*, Vol. 34, No. 1, Fall 1988, pp. 61–78.

Luce, R. Duncan, and Howard Raiffa. *Games and Decisions: Introduction and Critical Survey*. New York: John Wiley & Sons, 1957.

Magee, J. F. "Decision Trees for Decision Making," *Harvard Business Review*, Vol. 42, No. 4, July/August 1964, pp. 126–138.

Oakford, R. V., A. Salazar, and H. A. DiGuilio. "The Long-Term Effectiveness of Expected Net Present Value Maximization in an Environment of Incomplete and Uncertain Information," *AIIE Transactions*, Vol. 13, No. 3, September 1981, pp. 265–276.

Park, C. S., and G. P. Sharp-Bette, *Advanced Engineering Economics*. New York: Wiley, 1990, pp. 353–576.

Perrakis, S., and C. Henin, "The Evaluation of Risky Investments with Random Timing of Cash Returns," *Management Science*, Vol. 21, No. 1, 1974, pp. 79–86.

Rose, L. M. *Engineering Investment Decisions: Planning Under Uncertainty*. New York: Elsevier, 1976.

Smith, L. D. "Quantifying Risk for Establishing Rates of Return in Regulated Industries," *The Engineering Economist*, Vol. 28, No. 4, Summer 1983.

Sullivan, W. G., and R. G. Orr. "Monte Carlo Simulation Analyzes Alternatives in Uncertain Economy," *Industrial Engineering*, Vol. 14, No. 11, November 1982, pp. 43–49.

Teichroew, D., Robichek, and Montalbano. "Mathematical Analysis of Rates of Return Under Uncertainty," *Management Science*, Vol. 11, No. 3, January 1965, pp. 395–404.

Van Horne, J. "Capital Budgeting Decisions Involving Combinations of Risky Investments," *Management Science*, Vol. 13, No. 10, October 1966, pp. 84–92.

von Neumann, J., and O. Morgenstern. *Theory of Games and Economic Behavior*. Princeton, NJ: Princeton University Press, 1947.

Young, D., and L. E. Contreras, "Expected Present Worths of Cash Flows Under Uncertain Timing," *The Engineering Economist*, Vol. 20, No. 4, Summer 1975, pp. 257–268.

Zinn, C. D., W. G. Lesso, and R. Motazed. "A Probabilistic Approach to Risk Analysis in Capital Investment Projects," *The Engineering Economist*, Vol. 22, No. 4, Summer 1977, pp. 239–260.

Chapter 11   PRICE LEVEL CHANGES (INFLATION)

Allen, R. G. D. *Index Number in Theory and Practice.* Chicago: Aldine, 1975.

Davidson, Sidney, Clyde Stickney, and Roman Weil. *Inflation Accounting.* New York: McGraw-Hill, 1976.

Dunn, Richard L. "Plant Engineering Cost Index," *Plant Engineering*, Vol. 42, No. 1, January 21, 1988, pp. 62–65.

Duvall, R. M., and James Bulloch. "Adjusting Rate of Return and Present Value for Price Level Changes," *The Accounting Review*, Vol. 40, No. 3, July 1965, pp. 569–573.

Engineering News Record. *Construction Costs.* New York: McGraw-Hill. (Published weekly.) For subscriptions, write to Fulfillment Manager, *Engineering News Record*, P.O. Box 430, Highstown, NJ 08520.

Ferris, Stephen P., and Anil K. Makhija. "Inflation Effect on Corporate Capital Investment," *Journal of Business Research*, Vol. 16, No. 3, May 1988, pp. 251–260.

Foster, Earl M. "Impact of Inflation on Capital Budgeting Decisions," *Quarterly Review of Economics and Business*, Vol. 10, No. 3, Fall 1970, pp. 19–24.

Freidenfelds, John, and Michael Kennedy. "Price Inflation and Long-Term Present Worth Studies," *The Engineering Economist*, Vol. 24, No. 3, Spring 1979, pp. 143–160.

Hartley, Ronald V. "Inflation, Maintenance of Capital, and the IRR Model: An Appraisal and Alternative," *Decision Sciences*, Vol. 19, No. 1, Winter 1988, pp. 93–103.

Industrial Engineering, Vol. 12, No. 3, March 1980. The entire issue is devoted to "The Industrial Engineer and Inflation." Of particular interest are:
a. Estes, C. B., W. C. Turner, and K. E. Case. "Inflation — Its Role in Engineering-Economic Analysis," pp. 18–22.
b. Bontadelli, J. A., and W. G. Sullivan. "How an IE Can Account for Inflation in Decision-Making," pp. 24–33.
c. Lohmann, Jack R. "The Consumer Price Index: Concepts and Content Over the Years," pp. 6–17.
d. Ward, Tom L., "Leasing During Inflation: A Two-Edged Sword," pp. 34–37.

Jones, Byron W. *Inflation in Engineering Economic Analysis*, USA and Canada: Wiley-Interscience, 1982.

Jordan, Joe D. "Aspects of Engineering Economy Studies Under Inflation," Ph.D. Dissertation, Dept of Industrial Engineering, Mississippi State University, May 1981.

Miller, Elwood L. "What's Wrong with Price Level Accounting?" *Harvard Business Review*, November/December 1978, pp. 111–118.

Oakford, R. V., and A. Salazar. "The Arithmetic of Inflation Corrections in Evaluating 'Real' Present Worths," *The Engineering Economist*, Vol. 27, No. 2, Winter 1982, pp. 127–146.

Oskounejad, M. Mehdi. "An Investigation of the Effects of Depreciation Methods on Pricing Policy During Inflation," Ph.D. Dissertation, Dept of Industrial Engineering, Mississippi State University, May 1979.

Remer, D. S., and S. A. Ganiy. "The Role of Interest and Inflation Rates in Present-Worth Analysis in the United States," *The Engineering Economist*, Vol. 28, No. 3, Spring 1983, pp. 173–190.

Ritter, Lawrence S., and William Silber. *Principles of Money, Banking, and Financial Markets*, Basic Books, Inc., 1989.

Spiegel, M. R. *Statistics*, Schaum's Outline Series, New York: Schaum Publishing, 1961. (See Chapter 17, "Index Numbers.")

Stuart, David O., and Robert D. Meckna. "Accounting for Inflation in Present Worth Studies," *Power Engineering*, February 1975, pp. 45–47.

U.S. Department of Commerce, Bureau of Economic Analysis. *Business Conditions Digest.* Washington, DC: U.S. Department of Commerce, Bureau of Economic Analysis. (Published monthly.)

U.S. Department of Labor, Bureau of Labor Statistics. *Handbook of Labor Statistics.* Washington, DC: Government Printing Office. (Published periodically; obtain latest issue.)

U.S. Department of Labor, Bureau of Labor Statistics. *The Consumer Price Index: Concepts and Contents Over the Years*, Report 517. Washington, DC: Government Printing Office, May 1978 (Revised).

Watson, F. A., and F. A. Holland. "Profitability Assessment of Projects Under Inflation," *Engineering and Process Economics*, Vol. 2, No. 3, 1976, pp. 207–221.

Index

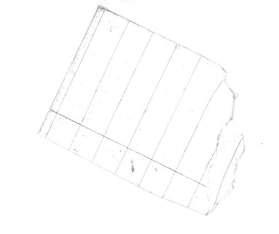

Discrete Cash Flows

End-of-period Compounding Effective Interest Rate (i)		Continuous Compounding Nominal Interest Rate (r)

1*

$F = P(F/P, i, N)$

$\quad = P(1 + i)^N$

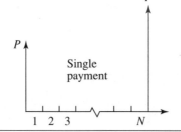

Single payment

2

$F = P(F/P, r, N)$

$\quad = Pe^{rN}$

3*

$P = F(P/F, i, N)$

$\quad = F(1 + i)^{-N}$

4

$P = F(P/F, r, N)$

$\quad = Fe^{-rN}$

5*

$F = A(F/A, i, N)$

$\quad = A\left[\dfrac{(1 + i)^N - 1}{i}\right]$

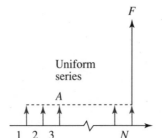

Uniform series

6

$F = A(F/A, r, N)$

$\quad = A\left[\dfrac{e^{rN} - 1}{e^r - 1}\right]$

7*

$A = F(A/F, i, N)$

$\quad = F\left[\dfrac{i}{(1 + i)^N - 1}\right]$

8

$A = F(A/F, r, N)$

$\quad = F\left[\dfrac{e^r - 1}{e^{rN} - 1}\right]$

9*

$P = A(P/A, i, N)$

$\quad = A\left[\dfrac{(1 + i)^N - 1}{i(1 + i)^N}\right]$

Uniform series

10

$P = A(P/A, r, N)$

$\quad = A\left[\dfrac{e^{rN} - 1}{e^{rN}(e^r - 1)}\right]$

11*

$A = P(A/P, i, N)$

$\quad = P\left[\dfrac{i(1 + i)^N}{(1 + i)^N - 1}\right]$

12

$A = P(A/P, r, N)$

$\quad = P\left[\dfrac{e^{rN}(e^r - 1)}{e^{rN} - 1}\right]$

13*

$P = G(P/G, i, N)$

$\quad = G\left[\dfrac{(1 + i)^N - iN - 1}{i^2(1 + i)^N}\right]$

(Arithmetic) gradient series

$(N - 1)G$

$(N - 2)G$

$2G$

G

0

14

$P = G(P/G, r, N)$

$\quad = G\left[\dfrac{e^{rN} - Ne^r + N - 1}{e^{rN}(e^r - 1)^2}\right]$

15*

$A = G(A/G, i, N)$

$\quad = G\left[\dfrac{(1 + i)^N - iN - 1}{i(1 + i)^N - i}\right]$

16

$A = G(A/G, i, N)$

$\quad = G\left[\dfrac{e^{rN} - Ne^r + N - 1}{(e^r - 1)(e^{rN} - 1)}\right]$

---

*Tabulated in Appendix B at effective interest rate i. Under conditions of continuous compounding, nominal and effective interest rates are related by $i = e^r - 1$, or $r = \ln(1 + i)$.